T0137307

Lecture Notes in Electrical Engineering

Volume 659

The book series *Lecture Notes in Electrical Engineering* (LNEE) publishes the latest developments in Electrical Engineering—quickly, informally and in high quality. While original research reported in proceedings and monographs has traditionally formed the core of LNEE, we also encourage authors to submit books devoted to supporting student education and professional training in the various fields and applications areas of electrical engineering. The series cover classical and emerging topics concerning:

- Communication Engineering, Information Theory and Networks
- Electronics Engineering and Microelectronics
- Signal, Image and Speech Processing
- Wireless and Mobile Communication
- Circuits and Systems
- Energy Systems, Power Electronics and Electrical Machines
- Electro-optical Engineering
- Instrumentation Engineering
- Avionics Engineering
- Control Systems
- Internet-of-Things and Cybersecurity
- Biomedical Devices, MEMS and NEMS

For general information about this book series, comments or suggestions, please contact leontina.dicecco@springer.com.

To submit a proposal or request further information, please contact the Publishing Editor in your country:

China

Jasmine Dou, Associate Editor (jasmine.dou@springer.com)

India, Japan, Rest of Asia

Swati Meherishi, Executive Editor (Swati.Meherishi@springer.com)

Southeast Asia, Australia, New Zealand

Ramesh Nath Premnath, Editor (ramesh.premnath@springernature.com)

USA, Canada:

Michael Luby, Senior Editor (michael.luby@springer.com)

All other Countries:

Leontina Di Cecco, Senior Editor (leontina.dicecco@springer.com)

** Indexing: The books of this series are submitted to ISI Proceedings, EI-Compendex, SCOPUS, MetaPress, Web of Science and Springerlink **

More information about this series at http://www.springer.com/series/7818

Nilesh Goel · Shazia Hasan · V. Kalaichelvi
Editors

Modelling, Simulation and Intelligent Computing

Proceedings of MoSICom 2020

 Springer

Editors
Nilesh Goel
Department of Electrical
and Electronics Engineering
Birla Institute of Technology
and Science Pilani, Dubai Campus
Dubai, United Arab Emirates

Shazia Hasan
Department of Electrical
and Electronics Engineering
Birla Institute of Technology
and Science Pilani, Dubai Campus
Dubai, United Arab Emirates

V. Kalaichelvi
Department of Electrical
and Electronics Engineering
Birla Institute of Technology
and Science Pilani, Dubai Campus
Dubai, United Arab Emirates

ISSN 1876-1100   ISSN 1876-1119   (electronic)
Lecture Notes in Electrical Engineering
ISBN 978-981-15-4777-5   ISBN 978-981-15-4775-1   (eBook)
https://doi.org/10.1007/978-981-15-4775-1

This Springer imprint is published by the registered company Springer Nature Singapore Pte Ltd.
The registered company address is: 152 Beach Road, #21-01/04 Gateway East, Singapore 189721, Singapore

Preface

Electrical and Electronics Engineering Department of BITS Pilani, Dubai Campus, organized an International Conference on Modelling, Simulation and Intelligent Computing from 29 January to 31 January 2020. MoSICom 2020 aims to bring together students, researchers, academia and industry practitioners to demonstrate their research work. It also provides a platform for researchers, professionals and educators to confer the most recent innovations, trends, researches and practical challenges encountered and the solutions proposed in modelling, simulation and intelligent computing. This conference was able to attract participants from academia and researchers of reputed universities from UAE and abroad.

One plenary and four outstanding keynote speeches were presented at MoSICom 2020 followed by three parallel sessions running simultaneously, with total of 24 sessions, during the three days of the conference. The first plenary talk was delivered by Dr. Madan M. Gupta, Emeritus Professor of engineering at the University of Saskatchewan and Director of the Intelligent Systems Research Laboratory. Dr. Madan M. Gupta has presented about the design of robust neuro-controller for complex dynamic systems.

The first keynote was presented by Dr. Hussam Amrouch, Research Group Leader at the Chair for Embedded Systems (CES), Karlsruhe Institute of Technology (KIT), Germany. He talked about negative capacitance transistor to rescue technology scaling from physics to the system level. Dr. Sukumar Mishra, Professor at the Indian Institute of Technology, Delhi, has presented a keynote on shift towards net-zero demand distribution system using distributed PV and EV charging station. Dr. Lotfi Romdhane, Professor and Associate Dean of Graduate Affairs and Research, American University of Sharjah, has presented about robotics being applied to several fields ranging from construction to health care and several ongoing applications of robotic projects applied to rehabilitation of the human body. The fourth keynote was given by Dr. Navneet Gupta on in-memory computing, emerging memory technologies and architectures.

This book is a compilation of current developments in the variety of fields related to IoT and applications; robotics and automation; switching circuits and power converters; big data analytics; antennas and coding techniques;

meta-heuristic algorithms; computer networks; image processing and applications; advanced CMOS and beyond; neural networks and deep learning; renewable energy sources and technology; artificial intelligence and applications; microgrids and distributed generation; biomedical and health informatics; block chain and applications; cloud computing; circuits and systems; device modelling and characterization; and emerging semiconductor technologies.

In addition to this, intelligent computational techniques are becoming important for interdisciplinary engineering applications. This book offers latest research findings in the field of engineering applications. It will act as definitive reference for researchers, professors and practitioners interested in exploring the advanced techniques in the field of modelling, simulation and computing.

Dubai, United Arab Emirates   Nilesh Goel
                                                                                      Shazia Hasan
                                                                                     V. Kalaichelvi

Contents

About the Editors

Nilesh Goel was born in India, in 1983. He received his bachelor's degree in Electronics and Communication Engineering from Motilal Nehru NIT Allahabad, India, in 2007. He received his Ph.D. degree in Electrical Engineering from the Indian Institute of Technology at Bombay (IIT Bombay), India, in 2015. In 2007, he joined ST Microelectronics as Design Engineer, where he worked in design and development of highspeed digital protocol (HDMI). After his Ph.D. from IIT Bombay, he joined SanDisk as Staff Device Engineer in 2015. At SanDisk, his primary job is to characterization and do failure analysis for NAND flash memory and qualify NAND flash for various reliability issues like program disturb, read disturb and data retention for SLC, MLC and TLC (2D and 3D NAND). He is currently serving as an Assistant Professor at BITS Pilani Dubai Campus from July 2017.

Dr. Goel has over 20 publications in peer-reviewed journals and highly reputed international conferences. He had also authored/co-authored 5 book chapters in Springer series. He is Reviewer of several international journals. His current research interest is semiconductor device reliability with focus on reliability aware circuit design.

Shazia Hasan was born in India, in 1980. She received her bachelor's degree in Electronics and Telecommunication from UCE, Burla, India, in 2002 and Ph.D. in Engineering from Biju Patnaik University of Technology, India, in 2012. She served as a Lecturer in Silicon Institute of Technology Bhubaneswar, India, from 2003 to 2010. Then, she served at ITER, Bhubaneswar as an Assistant Professor from 2010 to 2015. Currently, she is an Assistant Professor at BITS Pilani Dubai Campus from 2015. Dr. Shazia Hasan won 'Young Scientist Award' from VIFRA in 2015. She is the author of several papers in peer-reviewed international/national journals and international/national conferences. She also served as a Reviewer of international journals. She organized several conference/seminars/workshops in the past. Her research interest includes digital signal processing, biomedical signal processing, and statistical filter design. She successfully completed the industrial consultancy project from 'Star Cement' as Principal Investigator.

V. Kalaichelvi is an Associate Professor in the Department of Electrical and Electronics Engineering at BITS Pilani Dubai Campus. She received her Ph.D. degree in Instrumentation Engineering from Annamalai University in the year 2007. She has 16 years of teaching experience at Annamalai University, India. She is working with BITS, Pilani Dubai Campus since 2008. She has her research work published in refereed international journals and many international conferences. She has reviewed many papers in International Journals and Conferences. She has successfully completed two industrial consultancy projects from 'Star Cement Co., L.L.C' as Co-Principal Investigator. Her research area of expertise includes process control, control systems, neural networks, fuzzy logic, genetic algorithm, control system aspects applied to robotics, mechatronics. She has acted as Reviewer in the International Journal of Intelligent and Fuzzy Systems, Journal of Energy and Power Engineering, USA. She is currently an Editorial Board Member in the Journal of Instrumentation Technology and Innovations. She has also reviewed many papers in international conferences and journals. She is also a Faculty In-charge for IFOR team (Intelligent Flying Object for Reconnaissance), and it is a robotics team which is involved in doing R&D in aerial robotics. She is guiding students in the area of intelligent control systems applied to robotics field.

Modified Design Approach Using Cuckoo Search Optimization Algorithm for Mitigation of Harmonics and Improvement of Efficiency in Back Light LED TV Power Supply

C. Komathi[1](✉) ⓘ and M. G. Umamaheswari[2] ⓘ

[1] Sri Sairam Engineering College, Chennai, India
komathicll@gmail.com
[2] Rajalakshmi Engineering College, Chennai, India

Abstract. High power factor solutions are needed in Light Emitting Diode (LED) Television (TV) which otherwise results in the following problems: (i) Power is recycled from the backlight LED to the power source. (ii) Harmonics from backlight LEDs degrade the line which, in turn, affects the performance of other devices on the line. (iii) Additional losses are generated in the load which reduces efficiency. This paper proposes Cuckoo Search (CS) optimization algorithm based Interleaved DC-DC Single Ended Primary Inductance Converter (SEPIC) Converter for reducing the harmonics and thereby improving the efficiency of the power supply used for driving backlight LED. Cuckoo search (CS) algorithm using different fitness functions like Integral Square Error (ISE), Integral Absolute Error (IAE), Integral Time Square Error (ITSE) and Integral Time Absolute Error (ITAE) are employed for finding the optimal controller parameters. Various characteristics like rise time, peak time, peak overshoot, settling time, input power factor, % THD and % efficiency are used to analyse the performance of the proposed scheme using MATLAB/Simulink software tool.

Keywords: Power factor correction · Cuckoo search algorithm · Interleaved DC-DC SEPIC converter

1 Introduction

LED TVs are being preferred by the customers in recent years since it offers the following advantages: (i) Picture illumination clarity (ii) Low power consumption (iii) Less Distortion, etc. [1–3]. Various control strategies are employed in LED backlighting to provide and ensure the aforementioned advantages. PFC [3] is one of the techniques adopted to provide power factor close to unity which, in turn, reduces % THD and improves efficiency thereby assuring minimum power loss. In the present scenario, PFC control schemes for backlight LED TV have been implemented with Boost- and Flyback-based topologies. Boost topology has the drawback of operating only in step-up mode and flyback topology needs an isolation transformer for its operation.

© Springer Nature Singapore Pte Ltd. 2020
N. Goel et al. (eds.), *Modelling, Simulation and Intelligent Computing*,
Lecture Notes in Electrical Engineering 659,
https://doi.org/10.1007/978-981-15-4775-1_1

This work uses an interleaved DC-DC SEPIC converter as single-stage active PFC circuit since it reduces % THD by effective cancellation of harmonic currents at the input side [4, 5]. Cascade control strategy is implemented for the proposed converter applied for PFC in backlight LED power supply which consists of an outer Proportional Integral (PI) controller for regulating the output voltage and an inner PI controller for shaping the input current. Various conventional methods such as Zeigler Nichols (ZN) can be employed for designing PI controller which doesn't provide a satisfactory response in the presence of disturbance at load or line side [6].

Various optimization algorithms like Genetic Algorithm (GA), Particle Swarm Optimization (PSO), Ant Colony Optimization (ACO), etc. can be employed for finding the optimal parameters of PI controller. CS algorithm is chosen in this work for the following advantages: (i) Easy to implement as it involves a smaller number of parameters (ii) Faster convergence rate as there is an interaction among the solutions [7–10]. CS algorithm uses the parasitic nature of cuckoo birds which lay their eggs in the nest of other birds to find the best nest for reproduction and find the global optimal solution of the given problem.

This work aims to employ CS algorithm using various error criteria as fitness function for determining the optimal PI controller parameters of Interleaved DC-DC SEPIC single-stage active PFC circuit employed in backlight LED TV to improve the performance. The performance of the proposed scheme for backlight LED TV is analysed in terms of dynamic performance characteristics like rise time, peak time, peak overshoot, settling time, input power factor, % THD and % efficiency.

2 Modelling and Design of the Proposed Converter

Figure 1 shows the circuit of an Interleaved DC-DC SEPIC converter [9, 10].

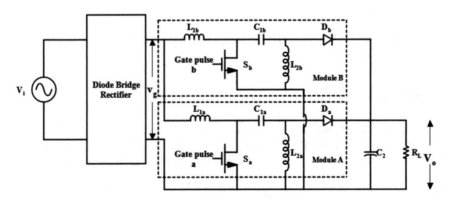

Fig. 1 Proposed converter

2.1   Derivation of State Space Model

There are four modes of operation in the proposed converter (Mode 1: S_a-ON, Mode II, D_a-ON, Mode III-S_b-ON, Mode IV-D_b-ON). The seven-state variables of the proposed converter are $i_{L_{1a}}, i_{L_{1b}}, i_{L_{2a}}, i_{L_{2b}}, v_{C_{1a}}, v_{C_{1b}}, v_{C_2}$. From the state equations, the equivalent state matrix A for the converter is given by Eq. (1).

$$A = \begin{bmatrix} 0 & 0 & 0 & 0 & -\frac{1-d}{L_{1a}} - \frac{1}{L_{1a}+L_{2a}} & 0 & 0 \\ 0 & 0 & 0 & 0 & \frac{d}{L_{2a}} & 0 & -\frac{(1-d)}{L_{2a}} \\ 0 & 0 & 0 & 0 & 0 & -\frac{1}{L_{1b}+L_{1b}} - \frac{d}{L_{2b}} & -\frac{(1-d)}{L_{1b}} \\ 0 & 0 & 0 & 0 & 0 & \frac{d}{L_{2b}} & -\frac{(1-d)}{L_{2b}} \\ \frac{(2-d)}{C_{1a}} & \frac{-d}{C_{1a}} & 0 & 0 & 0 & 0 & 0 \\ 0 & 0 & \frac{2-d}{C_{1b}} & \frac{-d}{C_{1b}} & 0 & 0 & 0 \\ -\frac{(1-d)}{C_2} & -\frac{(1-d)}{C_2} & -\frac{(1-d)}{C_2} & -\frac{(1-d)}{C_2} & 0 & 0 & -\frac{2}{C_2 R_L} \end{bmatrix} \tag{1}$$

The equivalent input matrix B for the converter is given by Eq. (2).

$$B = \begin{bmatrix} \frac{1}{L_{1a}} + \frac{1}{L_{1a}+L_{2a}} \\ 0 \\ \frac{1}{L_{1b}} + \frac{1}{L_{1b}+L_{2b}} \\ 0 \\ 0 \\ 0 \\ 0 \end{bmatrix} \tag{2}$$

The equivalent output matrix C for the converter is given by Eq. (3).

$$C = \begin{bmatrix} 0 & 0 & 0 & 0 & 0 & 0 & 2 \end{bmatrix} \tag{3}$$

The seventh-order transfer function $G_7(s)$ of the proposed converter is given by Eq. (4).

$$G_7(s) = \frac{2.6e^5 s^5 - 2.1e^{-7} s^4 + 6.5e^{13} s^3 + 18.1 s^2 + 4.1e^{21} s - 8.7e^8}{s^7 + 6.5s^6 + 2.5e^8 s^5 + 1.6e^9 s^4 + 1.4e^{16} s^3 + 9.8e^{16} s^2 + 9.2e^{20} s - 9.8e^7} \tag{4}$$

The seventh-order system is reduced to third-order system using Hankel Matrix method [9].

The reduced third-order transfer function $G_3(s)$ of the proposed converter after model order reduction is given by Eq. (5) [9].

$$G_3(s) = \frac{-0.00214s^2 + 21000s + 5.871}{s^3 + 6.57s^2 + 52500s + 1.468} \tag{5}$$

3   Implementation of Cuckoo Search Optimization Algorithm for the Proposed Scheme

Figure 2 shows the block diagram of the proposed scheme. The optimal parameters of outer PI controller are determined using CS for effective output voltage regulation. The current error signal determined using multiplier approach is fed as input to the inner CS-based PI controller which provides the gating pulses for the switches S_a and S_b to obtain UPF.

3.1   Design of CS-Based PI Controller for the Proposed Scheme

CS algorithm is used for determining the optimal PI controller parameters to improve the dynamic performance of the proposed converter [6, 7]. To design a PI controller using CS algorithm, the open-loop transfer function of the reduced-order model $G_3(s)$ in Eq. (5) is considered. The performance of different error criteria such as ISE, IAE, ITSE, ITAE used as a fitness function of CS algorithm are compared to find the best optimal PI controller parameters as shown in Fig. 3.

CS algorithm assumes every egg in a nest as a solution and then finds the optimal value till the best fitness value is obtained after various iterations. Each cuckoo lays one

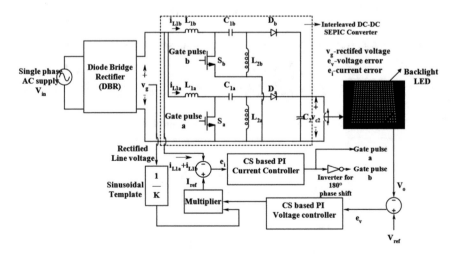

Fig. 2 Block diagram of the proposed scheme

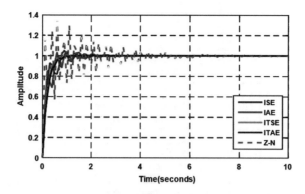

Fig. 3 Dynamic response of CS-based PI controller for various error criteria

egg at a time and dumps it in the nest of a host bird chosen randomly. The various steps used in CS-based PI controller are illustrated in Fig. 4.

Initialize the parameters as specified in Table 1. The proposed converter is simulated for the initial parameters found using the conventional ZN method. The error is calculated as the difference between the measured voltage and the reference voltage. The fitness value is calculated for the initial population of n host. Using Levy fight, the solutions (good nests) are selected for the next iteration based on fitness value. The solutions with the worst fitness values (best nests) are abandoned. The process is repeated till the solution with the best fitness value is obtained.

4 Simulation Results

Subsequent section discusses the simulation results of CS-based PI controller for different error criteria as fitness function of the proposed converter.

4.1 Performance Analysis of CS-Based PI Controller for the Proposed Scheme

Table 2 shows the optimal parameters obtained using CS-based PI controller for the proposed converter using different error criteria as fitness function.

It also lists the various dynamic response specifications like rise time (T_r), peak time (T_p), % Peak overshoot (M_p), settling time (T_s) of the closed-loop Reduced-order Interleaved DC-DC SEPIC converter for the optimal parameters obtained for each criterion.

From Table 2 and Fig. 3, it is seen that CS algorithm with ITAE as fitness function yields the best optimal parameters with reduced overshoot of 0.57% and settling time of 0.9 s.

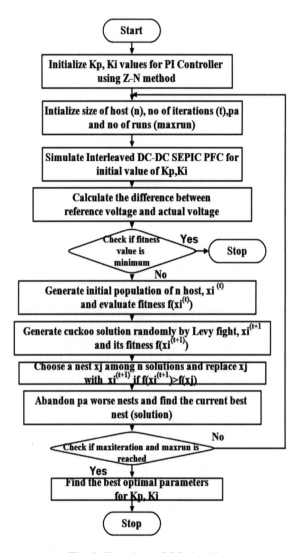

Fig. 4 Flowchart of CS algorithm

Table 1 Parameters of CS algorithm

Parameter	Value
Nest size, n	25
No of iterations (t)	10
No of variables (m)	2
No of runs (maxrun)	5
Probability, pa	0.25
Range of K_p	0–1
Range of K_i	0–3

Table 2 Dynamic response of CS-based PI controller for the proposed converter

Performance criterion	K_p	K_i	T_r (sec)	T_p (sec)	M_p (%)	T_s (sec)
Z-N method	0.3	3	0.04	0.342	38.6	5.93
CS-ISE	0.077	1.304	0.271	0.85	5.29	1.92
CS-IAE	0.091	1.283	0.265	0.857	5.1	1.86
CS-ITSE	0.06	1.373	0.383	1.17	1.22	1.3
CS-ITAE	0.08	1.071	0.438	0.7	0.57	0.9

4.2 Simulation Results and Discussion of the Proposed Scheme

The closed-loop circuit of the proposed scheme is constructed using Matlab/Simulink and the simulation results are shown in Figs. 5, 6. Table 3 summarizes the various performance parameters of the proposed system for the optimal PI controller parameters obtained for each criterion.

From Fig. 5a it is evident that the source voltage is in phase with the source current maintaining power factor close to unity. It is inferred from Fig. 5b that CS-ITAE method gives the minimum THD of 1.15%. Figures 5c, d show that CS-ITAE method yields the best optimum performance of output voltage and current in terms of settling time as 0.4 s, overshoot as 0.5%. Thus PI controller tuned by CS-ITAE method provides the best optimal parameters of the proposed converter with minimum settling time, % peak overshoot, % THD and maximum efficiency.

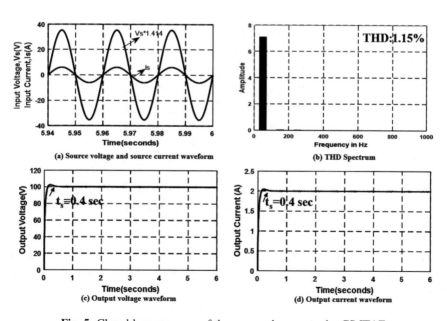

Fig. 5 Closed-loop response of the proposed converter by CS-ITAE

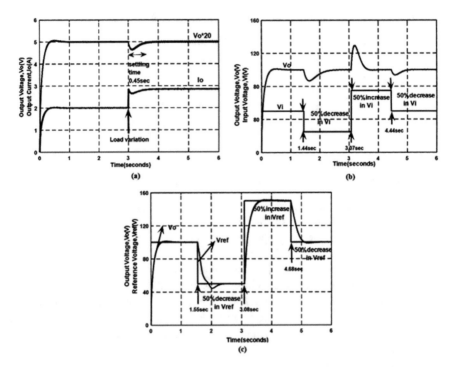

Fig. 6 Closed loop response of the proposed converter using CS-ITAE for load, line and reference voltage variations

Table 3 Closed-loop steady-state performance analysis of the proposed scheme

Performance criterion	K_p	K_i	THD (%)	Efficiency (%)
Z-N	0.3	3	3.7	90.48
CS-ISE	0.077	1.304	2.16	93.51
CS-IAE	0.091	1.283	1.76	93.52
CS-ITSE	0.06	1.373	1.52	93.54
CS-ITAE	0.08	1.071	1.15	94

Interleaved DC-DC SEPIC PFC converter tuned by CS-ITAE method is tested for variation in load, line and reference voltage and the results are shown in Figs. 6a–c. Results reveal that the proposed system is robust enough to track the variations effectively.

Table 4 shows the performance measures of the proposed converter compared with the existing topologies such as Buck converter, Boost converter, Flyback converter and SEPIC converter.

Table 4 Comparison of performance measures with existing topologies

Parameter	Buck	Boost	Flyback	SEPIC	Proposed
THD (%)	33.08	7.4	4.9	5.09	1.15
PF	0.954	0.995	0.998	0.996	0.999
Efficiency (%)	89.5	90.1	91.8	92.18	94

5   Conclusion

In this paper, the optimal PI controller parameters of the closed-loop Interleaved DC-DC SEPIC converter were obtained by CS algorithm. The performance analysis was done using different error criteria like ISE, IAE, ITSE, ITAE as fitness function and ITAE provides the optimal parameters for enhancing single-stage active PFC in backlight LED TV. Simulation results reveal that CS-based outer and inner PI voltage and current controller using ITAE as fitness function provides optimal performance with less overshoot, less settling time, minimum % THD which meets IEC61000-3-2 standard, maximum % efficiency and power factor very close to unity as compared to the existing PFC controllers available for driving backlight LED TV.

References

1. Choi WY, Kwon JM, Kwon BH (2007) Efficient LED back-light power supply for liquid-crystal-display. IET Electr Power Appl 1(2):133–142
2. Lee J-J, Kwon B-H (2007) High-performance light emitting diode backlight driving system for large-screen liquid crystal display. IET Electr Power Appl 1(6):946–955
3. ST, '185 W power supply with PFC and standby supply for LED TV using the L6564, L6599A, and VIPER27LN', pp 1–44 (2012)
4. Ye Z et al (2009) Single-stage offline SEPIC converter with power factor correction to drive high brightness LEDs. In: Twenty-Fourth annual IEEE applied power electronics conference and exposition, APEC 2009, pp 546–553 (2009)
5. Vishwakarma CB (2014) Modified Hankel Matrix approach for model order reduction in time domain. Int J Math Comput Phys Electr Comput Eng 8(2):404–410
6. Umamaheswari and Uma (2013) Analysis and design of reduced order linear quadratic regulator control for three phase power factor correction using Cuk rectifiers. Electr Power Sys Res 96:1–8
7. Barbosa RS (2014) Optimization of control systems by cuckoo search. In: 11th Portuguese conference on automatic control, pp 113–122 (2014)
8. Civicioglu P, Besdok E (2013) A conceptual comparison of the Cuckoo-search, particle swarm optimization, differential evolution and artificial bee colony algorithms. Artif Intell Rev 39(4):315–346

9. Komathi C, Umamaheswari MG (2019) Design of grey wolf optimizer algorithm based fractional order PI controller for power factor correction in SMPS applications. In: IEEE transactions on power electronics, (early access), digital object identifier https://doi.org/10.1109/tpel.2019.2920971)

10. Komathi C, Umamaheswari MG (2019) Analysis and design of genetic algorithm based cascade control strategy for improving the dynamic performance of interleaved DC-DC SEPIC PFC converter. Neural Comput Appl (online). https://doi.org/10.1007/s00521-018-3944-9

Start-Up Community Using Blockchain

S. Subhiksha$^{(\boxtimes)}$ (ID), Sonia Prakash (ID), S. Samundeswari (ID),
and A. Sangeerani Devi

Department of Computer Science and Engineering, Sri Sairam Engineering
College, Chennai, India
{Subhikshasudhakar03,Soniaprakash04}@gmail.com,
{Samundeswari.cse,Sangeerani.cse}@sairam.edu.in

Abstract. In India, 90% start-ups fail within the initial five years due to lack of huge capital and innovations. So, our idea is to develop a blockchain-based social network for start-ups, which will allow the collaboration between different start-ups and entrenched organizations who can provide some social and financial support to the start-ups. A token will be required to access this network, the user will earn the token through the Proof-of-Value protocol. In the proof-of-Value approach, the user needs to attach proof of their skills the subject matter expertise will verify it while stopping the spread of fake data. Thus, this will persuade the other member within the community and also the venture capitalist that the users are trustworthy and resourceful. In addition, the token will allow the users to get access to posting, commenting, and voting. The blockchain technology will ensure that the verification process to endorse skills is decentralized. The venture capitalist or a stranger will be able to trust the user more with this all-encompassing platform. And no second-guessing is required. Or delve into your social media profile to discover what sort of individual you are outside of work.

Keywords: Blockchain · Start-Ups · Social network · Proof-of-Value · Venture · Decentralized

1 Introduction

A blockchain is an increasing list of information, referred to as blocks, using cryptography it can be linked. Blockchain network employs decentralized trust network and secure smart contracts. Blockchain technology has received reputation mainly in industries worried about cyber security and payments, because of its potential to execute smart contracts and secure, fast transactions [1]. The key trait of blockchain technology are as follows:

- Decentralization.
- Persistency.
- Anonymity.
- Auditability.

The start-up failure rate is very high. A Fortune article approximately calculated that 90% of start-ups eventually failed. The main reasons for failure are:

© Springer Nature Singapore Pte Ltd. 2020
N. Goel et al. (eds.), *Modelling, Simulation and Intelligent Computing,*
Lecture Notes in Electrical Engineering 659,
https://doi.org/10.1007/978-981-15-4775-1_2

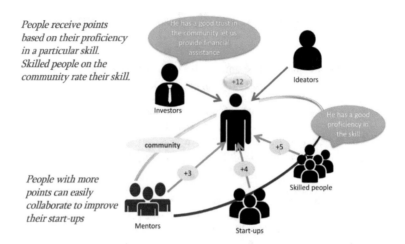

People receive points based on their proficiency in a particular skill. Skilled people on the community rate their skill.

People with more points can easily collaborate to improve their start-ups

Fig. 1 Start-up community

- Lack of consumer interest within the product or administration (42%).
- Funding issues (29%).
- Personnel or staffing issues (23%).

This paper aims to build a common social network platform through which different entrepreneurs across the world can meet and collaborate with each other [2]. By collaborating, people will tend to get more ideas and methods to improve their commercial enterprise. Each person will be given points based on their skills. Based on these points, entrepreneurs can choose which person is suitable to collaborate with.

We make use of blockchain technology to assure that all the works of entrepreneurs are original, unique and not copied. With these traits, Blockchain can greatly improve the efficiency of our platform (Fig. 1).

2   Proposed Solution

Our proposed solution is to create a decentralized blockchain-based social network platform. This platform is built on the Ethereum blockchain [3]. The two building blocks of this platform are as follows:

- **Blockchain**: Using blockchain technology, the network is decentralized (i.e. a group of nodes maintain the network and not by a single authority).
- **Token**: A token will be required to access this network. Token has two features: First: Users are offered a variety of token rewards which they earn by using their brain and skills. Second: A voting system using the token they earned.

It also provides users with the capacity to communicate with other start-up companies as well as the other well-known companies, create communities and clans and share and comment on the post. And also allow financial transactions between the companies for financial support.

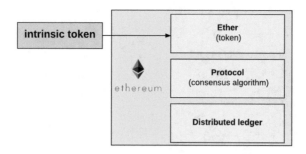

Fig. 2 Intrinsic token

2.1 Tokens

Tokens are used in blockchain projects to represent an asset digitally. It is also used as a payment method in blockchain applications. It provides the right for the user to participate in the community. Tokens can be shared with anyone. Tokens also represent assets and loyalty points. A token is created using smart contracts and programs which do not involve any third party for verification (Fig. 2).

We build a token using Ethereum. Once we make an Ethereum account we get a wallet known as MyEtherWallet. Using this wallet and smart contract we build the token.

Smart Contract: A blockchain program for verifying and authenticating a contract. In blockchain, Smart Contracts are utilized to validate a contract between any two parties [4]. Smart contracts without the third-parties will allow the occurrence of transactions.

MyEtherWallet (MEW) is a free, client-side interface. It enables interaction between user and Ethereum blockchain [5]. The token is a standard ERC20, it will have the principle functions and can be utilized as a general base for our application.

2.2 Token Rewards

The users who pose skills and post content are increasing value to the network by making and producing material or content that will attract new users to the platform and also the existing users engaged [6]. These help in issuing the token to a large set of users and grow the network effect. The subject matter experts play a crucial role in issuing the token to the users who are increasing the value to the network by taking time to evaluate and vote on the content of their skills. The blockchain rewards token for both for their activities respective to the value collected through the Proof of Stake voting system (Fig. 3).

2.3 Voting with Delegated Proof of Stake (Dpos)

Delegated Proof of Stake (DPoS): A consensus algorithm to check the accuracy and truth of the data by the voting method. This voting is combined with the social platform of reputation to reach consensus [7].

Fig. 3 Token rights

Every user who has tokens can comment, post or vote (if they are experts in the following content). Thus, every token holder can practice a degree of influence about what occurs on the network [8] (Fig. 4).

3   Battling Fake Data

Fake data is an extremely complex and multi-faceted hassle, and there's no simple solution. A prime part of this platform is to battle deluding content by encouraging well-researched and valuable material or skills. Some of the several strategies are:

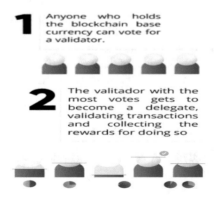

Fig. 4 Delegated proof of stake

3.1   Proof-of-Value

The protocol of Proof-of-Value is our technique of differentiating and rewarding valuable skills while stopping the spread of fake data and content.

Users for each community has a reputation rating that they participate in within the platform. The reputation rating changes when the contribution of the users is evaluated which reflects their expertise and reliability. Users to gain a reputation within the community will have to spend time as well as effort contributing to a community. Within a community, users with higher notoriety will have more effect and will almost certainly advance valuable content.

3.2   Identity Verification

Verification of identity is another important problem in diminishing the spread of deception. This platform will store the identities of the users on the blockchain for identification purposes. Well-known users will also have to verify their identity, making it extensively extra hard for trolls to dupe users into thinking that they are someone they're not.

3.3   Machine Learning

The spread of fake data has ended up so sophisticated that it can be very hard to figure fake from actual. We will use one algorithm called stance detection [9].

This will analyze the skills or posts and flag those that vary with most of the people of concern professionals at the platform. Taking under consideration the reputation of the user who created the content, the algorithm will label it with the probability of whether the data is real or fake, leaving users to trust it at their own choice. Users may also be able to report posts they suppose are suspicious or fake.

4   Result

A decentralized simple application (DApp) built on Ethereum blockchain. We use Solidity, a language used to compose smart contracts running on the Ethereum network [10] and Truffle for compilation and migration, and a system to be used for our front-end [11].

Vue.js as front-end library. It is a JavaScript open-source software to create user interfaces and single-page applications. Using uPort, we enable the users to enter the Ethereum blockchain with a globally unique identifier, giving users control over their identities, private keys, user accounts, and private information [5].

For storage to be decentralized, we use IPFS (InterPlanetary File System)-a peer-to-peer hypermedia protocol where you can store anything that http can: files, images, html, etc. Finally, uPort, Vue.js and IPFS are integrated using smart contracts.

This DApp stores the user name and their status securely on the blockchain. The key features:

- Allows reading and posting blogs.
- Ranking or voting members of the network based on their skill.
- Allows collaboration of different people and organizations (Figs. 5, 6).

Fig. 5 DApp profile

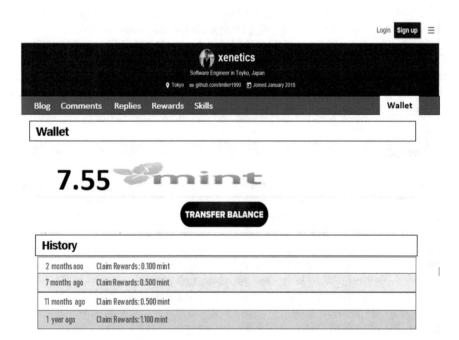

Fig. 6 Wallet balance and history of claim rewards

This application is built using smart contracts. Smart contracts are intended to give the user authority over their own information and not even the smart contract creator can control any data.

The wallet displays the current wallet balance of ether tokens and shows the latest transactions of the user. Users can send a request to another party for a specific number of tokens.

5   Conclusion

Blockchain has established itself in terms of transparency, immutability, security. The blockchain technology permits verification while not having to be depending on third-parties. In blockchain the data structure is append-only. It uses protected cryptography to secure the data ledgers.

Our motive is to bring a common community for entrepreneurs across the globe. With the use of our platform, we unite everyone in such a way that everyone is trustworthy and no fake data can enter. Proof-of-value and proof of identity [12] helps to verify a user and their content whereas we use machine learning techniques to verify the shared posts whether they are true. We use tokens to help user access the platform and perform voting. Token is also used to share data and help people in the community with their suggestions. This platform may help many entrepreneurs to improve their company and achieve great heights in their lives.

References

1. Monrat AA, Schelén O, Andersson K (2019) A survey of blockchain from the perspectives of applications, challenges, and opportunities. IEEE'19
2. Chakravorty A, Rong C (2017) Ushare: user controlled social media based on blockchain. Int Conf Ubiquitous Inf Manage Commun IMCOM
3. Wood DD Ethereum: a secure decentralised generalised transaction ledger
4. Westerkamp M (2019) Verifiable smart contract portability. In: IEEE international conference on blockchain and cryptocurrency (ICBC)
5. Hong Z, Xie S, Chen X An overview of blockchain technology: architecture, consensus, and future trends
6. Manjunatha MS, Usha S, Chaya BC, Manu R, Kavya S (2019) Blockchain based loyalty platform. Int J Recent Technol Eng IJRTE
7. Saleh FA (2018) Blockchain without waste: proof-of-stake. Soc Sci Res Netw SSRN
8. Nguyen CT, Hoang DT, Nguyen DN (2019) Dusit migration, and a system to be used for our front-end. In: Niyato D, Nguyen HT, Dutkiewicz E (eds) Proof-of-stake consensus mechanisms for future blockchain networks: fundamentals, applications and opportunities. IEEE'19
9. Zhang H, Wei X, Yu J Tian C, Li W Trusted intelligent salary paying method and system based on blockchain

10. Molina-Jimenez C, Sfyrakis I, Solaiman E, Ng I, Wong MW, Chun A, Crowcroft J (2018) Implementation of smart contracts using hybrid architectures with on- and off-blockchain components. In: IEEE 8th international symposium on cloud and service computing (SC2)
11. Bucher M Physical commodity-based currency system
12. Chen Y, Li Q, Wang H Towards trusted social networks with blockchain technology

Comparative Study on Load Frequency Control of a Single Microgrid Coupled with Thermal Power Plant Using Fuzzy and PID Controllers

Ranjit Singh[1(✉)] and Shazia Hasan[2]

[1] Dr. M.G.R Educational and Research Institute, Chennai, India
rsa.ranjit@gmail.com
[2] Birla Institute of Technology and Science, Pilani, Dubai Campus, Dubai, UAE
shazia.hasan@dubai.bits-pilani.ac.in

Abstract. In this paper, thermal power plant is connected to a microgrid and frequency deviation is observed and the results of the fuzzy and PID controllers are compared with fuzzy and the best controller giving best results is considered. Also, the different parameters are taken care for maintaining stability of the system like area control error (ACE) of individual areas, load demand, etc. This paper focuses on reducing error in quick time that developed due to mismatch between generation and demand. Because if error is not minimized in quick time, then stability of the system is affected and also the power flow through the tie-lines is not uniform. Software used is MATLAB 2014b. Results are shown and compared using necessary graphs. The Simulink model is prepared by selecting the transfer function blocks from the library and assigning the values to them. PID controller is used to minimize the error in quick time. Also, PID controller is replaced with fuzzy PID controller.

Keywords: Microgrid · Load frequency control · Load demand · Area control error (ACE) · Fuzzy logic control

1 Introduction

Microgrid is a localized group of electricity sources and loads which normally operates connected to and synchronous with the traditional wide area synchronous grid but can also disconnect to "island mode"—and function autonomously as physical or economic conditions dictate [1, 2]. Also, it is a discrete energy system comprising of distributed energy sources (including demand management, storage, and generation) and loads capable of operating in parallel with, or independently from, the main power grid.

Figure 1 shows the structure of a microgrid. Microgrids provide efficient, low-cost, clean energy, enhance local resiliency, and improve the operation and stability of the regional electric grid.

© Springer Nature Singapore Pte Ltd. 2020
N. Goel et al. (eds.), *Modelling, Simulation and Intelligent Computing*,
Lecture Notes in Electrical Engineering 659,
https://doi.org/10.1007/978-981-15-4775-1_3

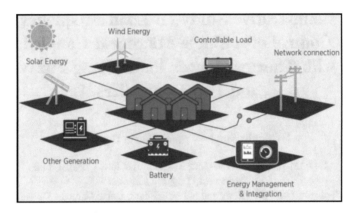

Fig. 1 Structure of a microgrid

The following are the important components of microgrid:

(i)   Generation: For a microgrid to provide energy supply to its connected loads without help from the utility, there must be a source of generation within the microgrid [3, 4]. This could be solar PV, wind, combustion turbines, reciprocating engines, cogeneration, or any other form of generation.

(ii) Energy storage: Most microgrids will have an element of energy storage that allows the microgrid to absorb and store energy that is produced when supply exceeds demand, and to return that energy (net of storage inefficiencies and line losses) when the demand exceeds supply (e.g., during evening hours when solar production is not available) [5–7]. Energy storage can also be used to provide arbitrage opportunities where wholesale power markets exist or when time-based rate schedules (e.g., time of use, real-time pricing, critical peak pricing, etc.) are available [8, 9].

(iii) Load Control: More sophisticated microgrids will incorporate the ability to control end-uses in a manner that allows the generation and storage resources to be optimized [10–12].

(iv) Utility interconnection: A key design feature of a microgrid includes the interface with the utility's power grid. During interconnected operation, the microgrid-utility interconnection must be designed for safe and reliable parallel operation of the microgrid and the power system [13–17].

(v)   Microgrid control system: A microgrid control system ties all of the components together and maintains the real-time balance of generation and load. In a very simple microgrid, a control system is typically a governor control on a diesel generator [18–20].

During interconnected operation, the control system must be able to manage the utility interface and communicate with the utility's (or independent system operator's) system operations center (including demand-response management systems) in near real time [21].

2 Load Frequency Control

Power sharing between two areas occurs through these tie-lines. Load frequency control, as the name signifies, regulates the power flow between different areas while holding the frequency constant.

To ensure the reliability of electric power, it is useful to have frequency and voltage constant. This can be achieved by load frequency controller and automatic voltage regulator [22, 23].

3 Fuzzy Logic

Fuzzy logic is originated as logic of ambiguous or inexact concepts.

The method of fuzzification has found increasing applications in power systems. The applications of fuzzy sets signify a major enhancement of power system analysis by avoiding heuristic assumptions in practical cases. This is because fuzzy sets could be deployed properly to represent power system uncertainties [24, 25].

The concept of fuzzy logic (FL) was developed by Zadeh in 1965 to address uncertainty and imprecision which widely exists in engineering problems. His process approach emphasized modeling uncertainties that arise commonly in human thought processes [26, 27].

Advantages of FLC:

- Flexible and intuitive knowledge base design.
- The consistency and completeness of the results can be checked in knowledge base.
- FL conceptual model can be used in many other paradigms.
- FLC can incorporate a conventional design and can be fine tune to system non-linearities; in other words, FLC gives consistent result even with nonlinearities.

Figure 2 shows the rule base design fuzzy rules which are conditional statement that specifies the relationship among fuzzy variables. With the knowledge of pervious system behavior, fuzzy rules are developed, for example, if the frequency deviation is more, then more controller gain is needed.

		ACE						
		NB	NM	NS	ZO	PS	PM	PB
	NB	PB	PB	PB	PB	PM	PM	PS
	NM	PB	PM	PM	PM	PS	PS	PS
$\dot{ACE}$	NS	PM	PM	PS	PS	PS	PS	ZO
	ZO	NS	NS	NS	ZO	PS	PS	PS
	PS	ZO	NS	NS	NS	NS	NM	NM
	PM	NS	NS	NM	NM	NM	NB	NB
	PB	NS	NM	NB	NB	NB	NB	NB

Fig. 2 Rule base design fuzzy rules

In the fuzzy rule format:
IF ACE is NB and ACE x is NS then output is PM.
IF ACE is PB and ACE x is PS then output is NM.

4   Controllers

A PID controller is a combination of proportional, integral, and derivative controllers. They take up the error as input and produces the controlled output [28].

A PID controller is widely used in industrial control systems and a variety of other applications requiring continuously modulated control. They are used in most automatic process control applications in industry. PID controllers can be used to regulate flow, temperature, pressure, level, and many other industrial process variables [29].

Figure 3 shows the area control error (ACE) which is the difference between scheduled and actual electrical generation within a control area on the power grid, taking frequency bias into account, is fed to the PID controller, and by adjusting the gains of the controller, the error is brought to zero in quick time. Trial and error method is employed till the best results are achieved.

4.1   Design of PID Controllers

Proportional control in engineering and process control is a type of linear feedback control system in which a correction is applied to the controlled variable which is proportional to the difference between the desired value (set point, SP) and the measured value (process value, PV) [30].

Integral Control. The integral controller produces an output, which is integral of the error signal. Therefore, the transfer function of the integral controller is KIs.

Derivative control. The derivative control is used to reduce the magnitude of the overshoot produced by the integral component and to improve process stability [31].

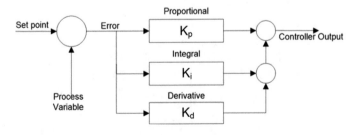

Fig. 3 PID controller

5 Modeling of the System

The whole system modeling is shown below with necessary equations:

5.1 Wind Turbine Generators

In "wind turbine generator," the wind pushes directly against the blades of the turbine that converts the linear motion of the wind into the rotary motion which is necessary to spin the generators rotor and the harder the wind pushes; the more electrical energy can be generated.

The wind power formula is given below.

$$P_{WT} = \frac{1}{2} \rho A_R C_P V_W^3 \tag{1}$$

where

P is the air density (kg/m^3); A_R is the swept area of blade (m^2); C_P is the power coefficient, a function of tip speed ratio λ and blade pitch angle β; and ρ is wind speed. The transfer function of the WTG is given by a simple first-order lag, neglecting all nonlinearities, as given below.

$$G_{WTH}(s) = \frac{\Delta P_{WTG}}{\Delta P_{WT}} = \frac{K_{wtg}}{1 + sT_{wtg}} \tag{2}$$

5.2 Alkaline Electrolysis

The alkaline electrolysis uses some of the power generated in the system for producing hydrogen, which is used up by fuel cells for the generation of power.

$$G_{AE}(s) = \frac{\Delta P_{AE}}{u_2} = \frac{K_{ae}}{1 + sT_{ae}} \tag{3}$$

5.3 Fuel Cells Power Generation

The fuel cells are electrochemical devices that convert chemical energy of fuel into electrical energy. The transfer function model can be written as below.

$$G_{FC}(s) = \frac{\Delta P_{FC}}{u_2} = \frac{K_{fc}}{1 + sT_{fc}} \tag{4}$$

5.4   Diesel Generator (DG)

The transfer function for diesel generator is given below.

$$G_{DG}(s) = \frac{\Delta P_{DEG}}{u_2} = \frac{K_{dg}}{1 + sT_{dg}} \tag{5}$$

5.5   Battery Energy Storage Systems

The transfer function of battery energy storage systems is given below.

$$G_{BESS}(s) = \frac{\Delta P_{BESS}}{u_2} = \frac{K_{bess}}{1 + sT_{bess}} \tag{6}$$

5.6   Power Deviation and System Frequency Variation

The power balance equation is expressed as below.

$$\Delta P_e = \Delta P_{MG} + \Delta P_{TH} - \Delta P_L \tag{7}$$

The output power of microgrid is given as.

$$\Delta P_{MG} = \Delta P_{WTH} + \Delta P_{FC} - \Delta P_{AE} + \Delta P_{DEG} \pm \Delta P_{BESS} \tag{8}$$

The frequency deviation is given as follows.

$$\Delta f = \frac{\Delta P_e}{K_{sys}} \tag{9}$$

The transfer function for the system frequency variation is stated as follows.

$$G_{sys} = \frac{\Delta f}{\Delta P_e} = \frac{1}{K_{sys}(1 + sT_{sys})} = \frac{K_P}{1 + sT_P} \tag{10}$$

6   MATLAB Simulink Model

The system frequency deviations are considered the error inputs to the controllers in the Simulink diagram. There are triangular membership functions which are used with five fuzzy linguistic variables such as negative big (NB), negative medium (NM), negative small (NS), zero (ZO), positive small (PS), positive medium (PM) and positive big (PB) for both the inputs and the output. The objective function K for controller parameters optimization of the interconnected power system is depicted as below.

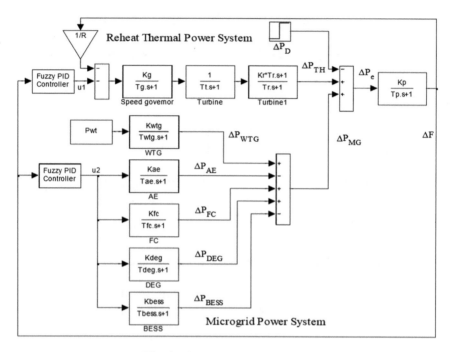

Fig. 4 MATLAB Simulink Model

$$K = \text{ITAE} = \int_0^{t_{\text{sim}}} |\Delta F|\, t.\text{d}t \tag{11}$$

Figure 4 shows the MATLAB Simulink model in which a microgrid is interconnected with a thermal power plant.

7 Necessary Graphs

Figure 5 shows the graph of deviation in frequency using PID controller and fuzzy PID controller. It is clear from the graph that fuzzy PID is best in minimizing the frequency deviation to zero which is good for maintaining stability.

8 Results and Conclusion

Figure 5 shows the graph of frequency deviation in both the areas, i.e., in microgrid and thermal power plant. The comparison of area control error (ACE) is done and seen that with PID controller the settling time is beyond 40 s which is harmful for the system stability, whereas with fuzzy PID controller, the settling time is 15 s which is quite good.

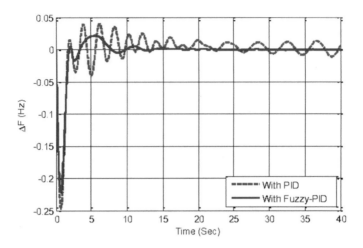

Fig. 5 Comparison of fuzzy and PID controller for frequency deviation

Thus, it was observed that by the application of fuzzy PID, the result was more accurate and quick response was observed than in case of simple PID controller. The system frequency error was quickly reduced to zero and stability of the entire system was brought to its normal.

Acknowledgements. I would like to express my sincere gratitude to my institution (Dr. M.G.R Educational and Research Institute, Chennai) for giving me permission to carry out this research work. I am also very thankful to my research guide Dr. L. Ramesh for his valuable suggestions on the topic.

Appendix

Thermal power system parameter.

$$K_G = 1.0$$
$$T_g = 0.09\,s$$
$$T_t = 0.35\,s$$
$$K_r = 0.4$$
$$T_r = 10s$$
$$K_p = 120\,Hz/p.u.\ MW$$
$$T_p = 20\,s$$
$$R = 2.5\,Hz/p.u.\ MW$$

The following are the system parameters used in power generation of microgrid.

$$K_{\text{wtg}} = 1$$
$$T_{\text{wtg}} = 1.5\,\text{s}$$
$$K_{\text{ae}} = 1.0$$
$$T_{\text{ae}} = 0.08\,\text{s}$$
$$K_{\text{fc}} = 0.01$$
$$T_{\text{fc}} = 4\,\text{s}$$
$$K_{\text{dg}} = 0.004$$
$$T_{\text{dg}} = 2.5\,\text{s}$$
$$K_{\text{bess}} = -0.004$$
$$T_{\text{bess}} = -0.15\,\text{s}$$
$$\Delta P_{\text{WT}} = 0.55\,\text{p.u.}$$

References

1. Shariatzadeh F, Kumar N, Srivastava A (2016) Optimal control algorithms for reconfiguration of shipboard microgrid distribution system using intelligent techniques. IEEE Trans. Indus Appl. https://doi.org/10.1109/tia.2016.2601558
2. Salam AA, Mohamed A, Hannan MA (2008) Technical challenges on microgrids. ARPN J Eng Appl Sci 3(6)
3. Singh VP, Mohanty SR, Kishor N, Ray PK (2013) Robust H-infinity load frequency control in hybrid distributed generation system. Int J Electr Power Energy Syst 46:294–305
4. Bevrani H, Habibi F, Babahajyani P, Watanabe M, Mitani Y (2012) Intelligent frequency control in an acmicrogrid: online PSO-based fuzzy tuning approach. IEEE Trans Smart Grid 3(4):1935–1944
5. Lal DK, Barisal AK, Tripathy M (2016) Grey wolf optimizer algorithm based fuzzy PID controller for AGC of multi-area power system with TCPS. Procedia Comput Sci 92:99–105
6. John T, Lam SP (2017) Voltage and frequency control during microgrid islanding in a multi-area multi-microgrid system. IET Gener Transm Distrib IEEE Trans 11(6):420
7. Golsorkhi MS, Hill DJ, Karshenas HR (2018) Distributed voltage control and power management of networked microgrids. IEEE J Emerg Sel Topics Power Electron 6(4)
8. Ferraro P, Crisostomi E, Shorten R, Milano F (2018) Stochastic frequency control of grid-connected microgrids. IEEE Trans Power Syst 33(5)
9. Department of Energy Office of Electricity Delivery and Energy Reliability. (2012). Summary Report: 2012 DOEMicrogridWorkshop
10. McDermott TE, Dugan RC (2002) Distributed generation impact on reliability and power quality. In: Proceedings of IEEE conference on rural electric power, pp D3–D37
11. Chiredeja P, Ramakumar R (2004) An approach to quantify the technical benefits of distributed generation. IEEE Trans Energy Convers 19(4):746–773

12. Brow RE, Pan J, Feng X, Koutlev K (2001) Sitting distributed generation to defer T&D expansion. In: Proceedings of 2001 IEEE PES transmission and distribution conference on expo, vol 2, pp 622–627
13. Lasseter R, Akhil A, Marnay C, Stephens J, Dagle J, Guttromson R, Meliopoulous AS, Yinger R, Eto J (2002) White paper on integration of distributed energy resources. The CERTS MICROGRID concept consultant report California Energy Commission, U.S. Department of Energy, Berkeley, CA, LBNL-0829, Apr 2002
14. Hatziargyiou N, Asano H, Iravani R, Marnay C (2007) Microgrids. IEEE Power Energy Mag 5(4):78–94
15. Kelly J (2010) Microgrids: a critical component of U.S. Energy Policy. http://www. slideshare.net/galvinpower/john-kellymicrogrid-briefing-on-capitol-hill-5202010, 20 May 2010
16. Lasseter RH (2002) Microgrid. In: Proceedings of IEEE power engineering society winter meeting, vol 1. New York, pp 305–308
17. Katiraei F, Iravani MR, Lehn PW (2005) Micro-grid autonomous operation during and subsequent to islanding process. IEEE Trans Power Del 20(1)
18. Piagi P, Lasseter RH (2006) Autonomous control of microgrids. In: IEEE PES meeting, Montreal, June 2006
19. Chakraborty S, Simões MG (2009) Experimental evaluation of active filtering in a single-phase high-frequency AC microgrid. IEEE Trans Energy Convers 24(3):673–682
20. Chakraborty S, Weiss MD, Simões MG (2007) Distributed intelligent energy management system for a single phase high frequency AC Microgrid. IEEE Trans Ind Electron 54(1):97–109
21. Kakigano H, Miura Y, Ise T, Momose T, Hayakawa H (2008) Fundamental characteristics of DC microgrid for residential houses with cogeneration system in each house. In: Proceedings of IEEE PESGM, 08GM0500
22. Kwasinski A, Onwuchekwa CN (2011) Dynamic behavior and stabilization of DC microgrids with instantaneous constant-power loads. IEEE Trans Power Electron 26(3):822–834
23. Kwasinski A (2011) Quantitative evaluation of DC microgrids availability: effect of system architecture and converter topology design choices. IEEE Trans Power Electron 26(3):835–851
24. Balog RS, Krein PT (2011) Bus selection in multibus DC microgrid. IEEE Trans Power Electron 26(3):860–867
25. Kakigano H, Miura Y, Ise T (2010) Loss evaluation of DC distribution for residential houses compared with AC system. In: The 2010 international power electronics conference-ECCE Asia-(IPECSapporo), 22A1-3, pp 480–486
26. Kakigano H, Miura Y, Ise T (2010) Low-voltage bipolar-type DC microgrid for super high quality distribution. IEEE Trans Power Electron 25(12):3066–3075
27. Kakigano H, Miura Y, Ise T (2009) Configuration and control of a DC microgrid for residential houses. In: Transmission and distribution conference and exposition. Asia and Pacific, pp 1–4, Oct (2009)
28. Kakigano H, Miura Y, Ise T (2013) Distribution voltage control for a DC microgrid using fuzzy control and a gain—scheduling technique. IEEE Trans Power Electron 28(5):2246–2258
29. Lopes JAP, Vasiljevska J, Ferreira R, Moreira C, Madureira A (2009) Advanced architectures and control concepts for more microgrids

30. Madureira AG, Pereira JC, Gil NJ, Lopes JAP, Korres GN, Hatziargyriou ND (2011) Advanced control and management functionalities for multi-microgrids. Eur Trans Electr Power 21:1159–1177
31. Palizban O, Kauhaniemi K, Guerrero JM Microgrids inactive network management—Part I: hierarchical control, energy storage

Star Schema-Based Data Warehouse Model for Education System Using Mondrian and Pentaho

Sweta Suman[(⊠)] [iD], Pallavi Khajuria [iD], and Siddhaling Urolagin [iD]

Birla Institute of Technology and Science Pilani, Dubai Campus, Dubai, UAE
{h20180906, h20180911, siddhaling}@dubai.bits-pilani.
ac.in

Abstract. Multiple strategic challenges are being faced by educational institutions across the globe which is of interest to both researchers and decision-makers. These challenges can be successfully addressed by analyzing the vast amount of data stored in multiple, unorganized, and unstructured operational databases in the educational institutes. Practitioners, researchers, and students would need data warehousing techniques to be able to utilize the knowledge stored in different archives. Data warehousing techniques include assimilating disparate sources of data, analysis of the requirements, designing the data, development, implementation, and deployment of the data. In this paper, a data warehouse (DW) for solving operational challenges of the center of higher education has been developed, which encompasses system design, ETL data processing, and online analytical processing analysis. The designing of this model is done using Mondrian and Pentaho business intelligence tool.

Keywords: Data warehouse · ETL · Online analytical processing (OLAP) · Mondrian · Pentaho

1 Introduction

Centers of higher education face multiple strategic challenges. Some of these challenges include (i) attract the right candidate, (ii) enhance student's academic performance, (iii) improve the financial health of center, (iv) compete for public grant, (v) attract right faculty, (vi) promote cutting-edge research, (vii) improve ranking among other colleges, (viii) improve operational efficiency of college. These challenges can be addressed by analyzing the large volume of data that institutions regularly collect. Some of the typical data collected are about students, faculty and department, administration details, revenue contributions, and enrollments. Most of these data are stored in operational databases, which is not useful for decision-making since the stand-alone database cannot provide such information immediately and efficiently. In a traditional information system, paper-based reports are developed based on which decisions are made. Running analytics on these reports is slow and causes delays in retrieving information. To overcome all the challenges of traditional systems, we use data warehousing. Data warehousing is based on the idea of online analytical

© Springer Nature Singapore Pte Ltd. 2020
N. Goel et al. (eds.), *Modelling, Simulation and Intelligent Computing*,
Lecture Notes in Electrical Engineering 659,
https://doi.org/10.1007/978-981-15-4775-1_4

processing (OLAP). OLAP is a tool that enables users to analyze data in different dimensions and graphical formats enabling faster decisions.

The content of the paper is organized into the following sections; Sect. 2 describes the literature survey briefing the related work done by various researchers in areas of education using data warehouse and related methods. Section 3 describes the architecture of the education system, and the various analytical tools integrated into a data warehouse. Section 4 describes our methods used and the design of a data warehouse for education. Section 5 describes the implementation details and results obtained by our method using multidimensional expressions (MDX), and finally, conclusions are provided in Sect. 6.

2 Literature Survey

There are different fundamentals for IT professionals to implement data warehouse and its components are provided by authors [1]. In [2], research has primarily focused on developing the dimension model and design of the student progression system using OLAP operations. Through [3], the solution to a decision-making support system has been discussed by authors with the help of an education-based case study. In [4], the author states the working aspects of extract transform and load (ETL) tools and various comparisons of ETL modeling. Through [5], the design considerations and data warehouse implementation in academics have been described using different educational data mining (EDM) techniques which are useful to design a DW for predictive analysis. In [6], the architecture and end-to-end process of data warehousing have been outlined by authors and explain about OLAP technology with the emphasis on data maintenance applications. In [7], the study focuses on the use of ETL tools to extract the data from different data sources in OLAP followed by the transformation and loading of data [8]. Pentaho is modern data integration and business analytics platform that captures data using a consistent format that is accessible to end users.

3 Data Warehouse for Higher Education

A "data warehouse" is an organization-wide snapshot of data, typically used for business decision-making and forecasting. The process of data warehousing is carried out in three main steps, also known as extraction, transformation, and loading (ETL). Relevant data from various sources are transformed into useful information by ETL tools and then stored in a data warehouse. As shown in Fig. 1, the data is first extracted from external or internal sources. The data that is extracted is usually raw and is transformed into an appropriate format for analysis. After extraction and transformation, the data is loaded in smaller chunks in a data warehouse. In data warehouse architecture, meta-data plays an important role as it specifies the source, usage, values, and features of data warehouse data. It is used for building, maintaining, and managing a data warehouse. A data mart is an access layer that is used to get data out to the users

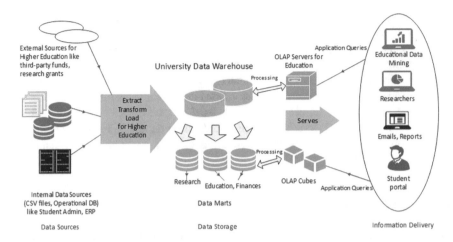

Fig. 1 Data warehouse architecture for higher education

through a variety of front-end tools: query tools, report writers, analysis tools, and data mining tools. From warehouse and marts, data is transferred to OLAP servers. OLAP servers provide the multidimensional analysis with powerful calculations and quick access. It provides the ability to perform complex query calculations and comparisons and present results in several ways like charts and graphs. The captured data through data analytic tools can help see trends, capture patterns, and also predict future outcomes.

4   Design of Data Warehouse for Higher Education

In a data warehouse, the user requirements are shown through Table 1 which contains business dimensions and fact measures of the education system. A representative set of dimensions that must be confirmed by the education institution includes student, course, calendar date, faculty, administration, and department. The set of fact measures includes total number of departments, total students enrolled in an individual course, a total sum of all contributions received for revenue or funds. In institutions, analysis can be done by first getting the descriptive analysis which means running analysis on current and previous year's performance sheet of every student. Based on such analysis, different professional programs could be provided to individuals for leadership qualities, technical improvements, professional communication, etcetera.

There are two modeling techniques named "Star Schema" and "Snowflake Schema" to represent multidimensional data [1]. Star schema is adopted here mainly because of its clarity, convenience, and rapid indexing ability that make the star schema more precise and understandable. Star schema can be defined as a specific type of

Table 1 Dimensions of the education system

Dimensions	Primary Key	Attributes
Student dimension	*StudentID*	*StudentID*, student name, nationality, gender, student contact number, address, email id, guardian name, guardian contact number
Course dimension	*CourseID*	*CourseID*, course no, course name, credit hrs, course time, professor assigned, course days, lab credits, lecture credits
Faculty dimension	*FacultyID*	*FacultyID*, faculty name, faculty type, faculty rank, department, faculty specialization, contact details, faculty address details, room no., email address, start date of service, end date of service
Department dimension	*DeptID*	*DeptID*, department name, department head details, department laboratory details, no. of laboratories, contact details
Account dimension	*AccountID*	*AccountID*, staff name, shift time, shift days, staff salary, staff type, no. of workshops, no. of lab equipment, laboratory number
Time dimension	*TimeID*	*TimeID*, assigned date, day of the week, month, quarter, academic year, semester, location
Fact measures	*FactID*	*FactID*, *StudentID*, *CourseID*, *FacultyID*, *DeptID*, *AccountID*, *TimeID*, no. of enrollments, no. of courses, no. of departments, no. of faculty members, no. of courses, grade GPA, donation funds amount, research grants, no. of staffs, tuition fees amount, hostel fees amount, alumni contributions, no of published papers, no of rejected papers

database design which includes a specific set of denormalized tables. Figure 2 shows a star schema model representation. This model consists of a central table (Fact table) and dimension tables that are directly linked to it. The star schema of the education system has six dimension tables and one fact table. As shown in Table 1, the dimension table contains the primary key of that table, e.g., in students dimension table student key is the primary key of that table. The attributes of the dimension table contain details about that dimension, e.g., in student dimension table student name, gender, nationality are the attributes. The fact table contains the primary key of all dimension tables as the foreign key. The star schema in this paper has been developed for operational analysis such as the breakdown of fees, faculty–student ratio, and acceptance rate of research papers published every year.

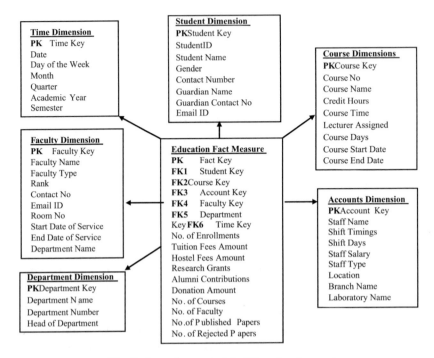

Fig. 2 Star schema model of the education system

5   Experimental Results

In this research work, a data warehouse is developed for the education system after completing the design part. We have used the following tools: Pentaho data integration for the ETL, Data Cleaner Tool (for data profiling) and an OLAP cube, MySQL workbench, and Mondrian. The ETL processes are implemented using the Pentaho data integration tool that uses spoon GUI to help the user define the ETL process to perform extraction and loading of the data source and transformed it into the target database [8]. The data has been generated to populate a data warehouse with values of StudentID, GPA, no. of courses, no. of faculty, no. of published papers, no. of rejected papers, no. of departments, etcetera with the help of Gaussian distribution. The care has been taken to generate data such that attribute values are most realistic, for example, StudentID is generated with alphanumerical combinations which are randomly produced. Tuition fees, hostel fees, research grants, donation funds, calendar date, academic year were produced with suitable conditions. All together a data warehouse has 5000 sample records. These samples were taken for each dimension and fact measures for analysis and OLAP operations. In our case study, the data is taken from spreadsheets (excel sheet), performed ETL, and loaded the filtered data (CSV Files) into a MySQL database. Database schema named "star schema" is created in Mondrian workbench to integrate and store all the students, department details after establishing the connection to the database. Figure 2 shows the star schema model that has a single fact table (education fact table) in units and six dimension tables (student, course, faculty, department, accounts, and time).

Mondrian workbench provides a user-friendly interface by allowing the user to decide the dimension and target of analysis, creating a cube, finally exhibiting data in different ways such as tables and figures. The dashboard can be presented with various aggregate calculations like total sum, count, average using Pentaho business intelligence (BI) server. End users can visit Mondrian workbench to carry out the analysis by writing MDX queries within an OLAP cube by slice, dice, and drill operations. Table 2 shows various fact measures that contribute to the statistics of total revenue, i.e., hostel fees and tuition fees, institute revenue (research grant and alumni details), the contribution of research paper per department, research grants received per department, student to faculty ratio for an individual year. In our analysis, every fact measure we create in Mondrian is backed by a predefined aggregate function, i.e., SUM, MIN, MAX, AVERAGE, or COUNT, that determines the fact measure's operation. Table 2a–f shows the MDX queries which are generated from the Mondrian tool. Based on the analysis, different types of aggregation can be selected for comparisons. Mondrian workbench executes these queries on star schema to generate the precomputed summary of data. Considering query execution from Table 2a, d, the result appears in single aggregated value, i.e., the sum of hostel fees of students in all courses for the year (2013–2019), total no. of enrollments for the given six years. These query reports and summaries would help the decision-makers (directors, department heads, and administrators) to understand the needs of their students and make more informed business decisions. The results of these queries are analyzed and presented in graphical reports below.

Table 2 MDX queries extracted from Mondrian workbench

(a) Select {[Measures].[hostelfees],[Measures]. [tuitionfees]} ON columns, {([student. studentid].[Allstudent.studentids], [time.timeid].[All time.timeids])} ON rows from starschema	(b) Select {[Measures].[hostelfees],[Measures]. [tuitionfees]} ON columns,{([course.courseid].[All course.courseids],[time.timeid].[All time.timeids])} ON rows from starschema
(c) Select {[Measures].[hostelfees],[Measures]. [tuitionfees]} ON columns, NONEMPTY ({[time].[academic year].[academic year], [student.studentid].[All student.studentids]}) ON rows from [starschema];	(d) Select {[Measures].[no_of_faculty],[Measures]. [no_of_enrollments]} ON columns, {([faculty.facultyid].[All faculty.facultyids], [student.studentid].[All student.studentids], [time.timeid].[All time.timeids])} ON rows from starschema
(e) Select {[Measures].[research_grants], [Measures]. [no_of_research_papers_published], [Measures].[no_of_departments]} ON columns,{([department.deptid].[All department.deptids],[time.timeid].[All time.timeids])} ON rows from starschema	(f) Select {[Measures].[alumnicontribution], [Measures].[donations_fund],[Measures]. [research_grants]} ON columns,{([time.timeid].[All time.timeids], [student.studentid].[All student.studentids])} ON rows from starschema

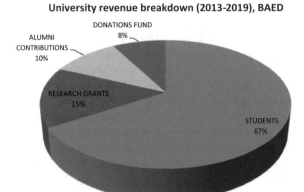

Fig. 3 Breakdown of revenue in billion AED

5.1   Aggregate Reports

Four aggregate reports have been generated. Figure 3 shows the breakdown of university total revenue from various sources in the period 2013–2019. The resultant graph is generated using the query Table 2a, f by selecting SUM and MAX aggregate functions. The analysis shows that 67% of university revenue came from students through tuition and hostel fees. The rest of the contributions are through research grants, alumni contributions, and donations fund.

Figure 4 breaks down the revenue from students to understand the contribution of tuition and hostel fees. The results have been generated from the above queries from Table 2b, c. In query Table 2b, we have used Mondrian cube wizard to pull out the details of fact measures (tuition fees, hostel fees) and using aggregate functions we calculated the total of tuition and hostel fees contributed for total six years of a period (2013–2019). In query Table 2c, the result set for each academic year (2013–2019) is calculated to drill down at every level. The NONEMPTY function used in the MDX query removes the null values from the result set. Similarly, many other options like "where" clause can be used to find the data for a particular year.

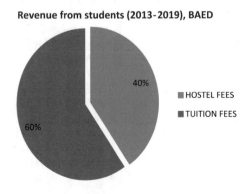

Fig. 4 Graph showing revenue details

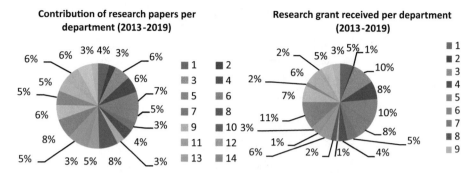

Fig. 5 Aggregate of research papers

Fig. 6 Aggregate of research grants

Figure 5 represents the number of research papers published by each department—an indication of research focus and department's contribution to building university brand, and Fig. 6 shows department-wise research grants received. The trend is in line with the contribution in research papers for the studied university. These results have been generated using query Table 2e, f where each measure is retrieved for each department. The total no. of the department is 20. Figure 5 represents various contributions of a research paper in each department where it is high as 8% and as low as 3%. While Fig. 6 shows various research grants received per department which is high as 11% and as low as 1%.

5.2 Trend Reports

Two trend reports have been generated. They are presented below:

Figure 7 shows that the revenues from research grants and alumni contributions have decreased over the years and should be the area of focus for the university. Figure 8 shows the number of students per faculty for the year 2013–2019. For the university under analysis, it is at a healthy ratio of 17:1. The results have been computed using the aggregation function (COUNT) to find the total using the query in Table 2d.

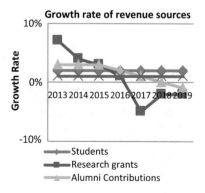

Fig. 7 Bar chart showing aggregate research grants per department

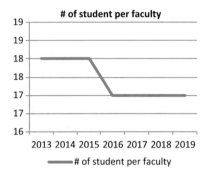

Fig. 8 Graph showing the ratio of student to faculty for the period 2013–2019

6 Conclusion

This paper proposes a thorough statistical analysis with the help of various OLAP techniques and methods that generates "smart, tailor-made" business reports. The universities today operate in a very competitive environment; hence, it has to be reliant on the quality and robust information to respond to business requirements. Directors, departments can have a multidimensional view of data implemented on millions of records. This schema model makes the data search and analysis process very effective and efficient while allowing the administrator to add both current and university historical data. The presented multidimensional analysis performed using Pentaho and Mondrian can spot historical trends and isolate problems that in turn help universities improve their performance over time. Further, the right set of data and queries can help universities identify the right candidates, right faculty, promote research work, improve sources of revenue, and enhance student's performance by personalized feedback. In this paper, we developed the overall report for students, faculty, total revenue, alumni contributions, research funds, and department details for the year 2013–2019. Similarly, we can obtain various other reports for specific timelines without changing the structure of the system. This paper will serve as a practical guide for professionals and researchers to implement a data warehouse and to experiment with Pentaho data integration and Mondrian methods in educational institutions.

References

1. Ponniah P (2009) Data warehousing fundaments for IT professionals. India
2. Singh RP, Singh K (2016) Design and research of data analysis system for student education improvement (case study: Student Progression System in University). In: 2016 International conference on micro-electronics and telecommunication engineering (ICMETE). IEEE Press, Ghaziabad, pp 508–512. https://doi.org/10.1109/icmete.2016.80
3. Dong P, Dong J, Huang T (2006) Application of data warehouse technique in educational decision support system. In: 2006 IEEE international conference on service operations and logistics, and informatics. IEEE, Shanghai, pp 818–822 (2006). https://doi.org/10.1109/soli.2006.328961

4. Sharma S, Jain R (2014) Modeling ETL Process for data warehouse: an exploratory study. In: 2014 fourth international conference on advanced computing and communication technologies. IEEE Press, Rohtak, pp 271–276 (2014). https://doi.org/10.1109/acct.2014.100
5. Moscoso-Zea O, Sampedro A, Luján-Mora S (2016) Datawarehouse design for educational data mining. In: 2016 15th international conference on information technology based higher education and training (ITHET), IEEE, Istanbul, pp 1–6 (2016). https://doi.org/10.1109/ithet.2016.7760754
6. Tripathy A, Das K (2011) A descriptive approach towards data warehouse and OLAP technology: an overview. In: Das VV, Stephen J, Chaba Y (eds) Computer networks and information technologies. CNC 2011. Communications in computer and information science, vol 142. Springer, Berlin, Heidelberg (2011)
7. Hanlin Q, Xianzhen J, Xianrong Z (2012) Research on extract, transform and load (ETL) in land and resources star schema data warehouse. In: 2012 Fifth international symposium on computational intelligence and design, Hangzhou. IEEE, Hangzhou, China, pp 120–123 (2012). https://doi.org/10.1109/iscid.2012.38
8. Pentaho-Data Integration Homepage. www.pentaho.com. Last accessed 15 Nov 2019

First Principle Calculation Based Investigation on the Two-Dimensional Sandwiched Tri-Layer van der Waals Heterostructures of MoSe$_2$ and SnS$_2$

Debapriya Som[1], Ankita Paul[1], Tanu[1], Arnab Mukhopadhyay[2], Neha Thakur[3], and Sayan Kanungo[1($\boxtimes$)]

[1] Birla Institute of Technology and Science, Pilani, Hyderabad Campus, Telengana, India
Ksayan_29@yahoo.co.in
[2] C.V. Raman College of Engineering, Bhubaneswar, Odisha, India
[3] Malla Reddy College of Engineering and Technology, Secunderabad, Telengana, India

Abstract. In this work, for the first time, the tri-layer sandwiched Van der Waals heterostructures of SnS$_2$ and MoSe$_2$ are investigated using first principle calculations. In such heterostructures, a monolayer of SnS$_2$ (MoSe$_2$) is sandwiched between two MoSe$_2$ (SnS$_2$) layers. Subsequently, such heterostructures are considered in two different stacking orders, i.e. ABA and AAA corresponding to the natural stacking of bulk MoSe$_2$ and SnS$_2$, respectively. The structural and electronic properties of such tri-layer heterostructures are extensively analyzed in comparison with the homogeneous SnS$_2$ and MoSe$_2$ tri-layers. In this context, emphasis has been given on the bond length and bond angles of metal and chalcogen atoms at the sandwiched layers of heterostructures. Finally, the influences of the homogenous and heterogeneous sandwiched layers on the energy band structures have been analyzed in detail from the orbital projections of electronic states at the conduction band and valence band.

Keywords: 2D material · Van der waals heterostructures · MoSe$_2$ · SnS$_2$ · DFT

1 Introduction

After the successful synthesize of Graphene in 2004, an ever-growing research activity has been observed in thin two-dimensional (2D) semiconducting materials having one or few atomic layer thicknesses. These 2D materials exhibit noteworthy physical and chemical properties that are distinctly different from their 3D counterparts. In this context, the Transition Metal Dichalcogenides (TMDs) have emerged as one of the most promising candidates in the family of semiconducting 2D materials. The TMDs are represented as MX$_2$, where a transition metal M (Mo, W) is covalently sandwiched between two chalcogen atoms X (S, Se, Te) in a hexagonal honeycomb structure forming a single layer. In bulk TMDs, the individual layers are held together by the

© Springer Nature Singapore Pte Ltd. 2020
N. Goel et al. (eds.), *Modelling, Simulation and Intelligent Computing*,
Lecture Notes in Electrical Engineering 659,
https://doi.org/10.1007/978-981-15-4775-1_5

relatively weaker Van der Waals force that allows the exfoliations of few or more layers from bulk TMDs. The TMDs are considered particularly suitable for electronic applications owing to their high surface to volume ratio, pristine surface quality, and the presence of intrinsic semiconducting energy band-gaps (1–2 eV).

Furthermore, the energy band-gaps and other electronic properties of the TMD can be tuned by varying the layer thickness, which offers an additional flexibility for electronic, optoelectronic, sensing and energy harvesting applications [1, 2].

In this line, the Van der Waals hetero-structure of two or more TMD materials, i.e. vertically aligned layers of two or more TMDs, has recently emerged as a potential aspirant in the rapidly growing family of 2D materials. Apart from first principle simulation based theoretical investigations, the sophisticated fabrication techniques like Van der Waals epitaxy leads to the experimental realizations of such hetero-structure TMDs [3]. Subsequently, it has been found that some of these heterostructures can offer distinct electronic properties that are not observed in their constituent TMDs [3–6]. Also, the electronic properties (energy band-gaps, density of states) of such heterostructures can be tuned based on their composition, stacking orientations and number of layers [5, 6]. This presents an exciting opportunity for exploiting such tunable and distinctive electronic properties of different HSTMD materials for electronic applications. This further inspires the investigations on the possibility of Van der Waals heterostructures of TMDs and other 2D materials having similar structural specifications. In this context, the semiconducting Metal (M) Dichalcogenides (X = S, Se) like Tin Disulfide (SnS_2) show considerable structural similarity with TMDs and yet offer distinct electronic properties in their monolayer and multilayer configurations compared to that of TMDs [7–9]. Very recently, few reports presented Van der Waals hetero-structure of SnS_2 with TMDs. It should be noted that these heterostructures are mostly investigated in their bilayer configurations which indicates a dramatic change in the electronic properties of such hetero-structure through the Van der Waals interactions of individual layers [10–13]. However, till date, only a few reports are available on the tri-layer Van der Waals heterostructures of SnS_2 and TMDs [13].

Also, to the best of the author's knowledge, till date, there are no reports available on the tri-layer Van der Waals heterostructures of SnS_2 and $MoSe_2$, which has relatively closer in-plane lattice constants compared to the other stable TMD materials. Subsequently, in this work, the structural and electronic properties of $MoSe_2/SnS_2/MoSe_2$ and $SnS_2/MoSe_2/SnS_2$ heterostructures are thoroughly investigated by emphasizing the influences of sandwiched hetero-layers.

2   Computational Method

In this work, the first principle calculations are performed using the commercially available Atomixtix Tool Kit (ATK) and Virtual Nano Lab (VNL) simulation package from Synopsys QuantumWise [14]. To avoid the artificial interactions from the periodic images in the out-of-plane directions, a vacuum of 20 and 40 Å are considered in these directions for the monolayers and tri-layers, respectively. The individual tri-layers are relaxed using geometry optimization with LBFGS methods within the force and pressure tolerance of 0.01 eV/Å and 0.0001 eV/Å^3, respectively. For the density functional theory (DFT) calculations, a linear combination of atomic orbitals (LCAO) basis set with Purdew Bruke, and Ernzerhof (PBE) Genaralized Gradient Approximation (GGA) exchange correlation functional is considered [9, 15, 16]. Further, the empirical correction based on Grimme DFT-D2 methods is incorporated to account for the weak Van der Waals forces for tri-layer structures [9]. The density mesh cut-off is considered as 125 Hartree, and the Brillouin zone is sampled using a Monkhorst-Pack grid of $10 \times 10 \times 1$. Further, in this work, no spin-orbit interaction is incorporated as it shows no major modulations in electronic band structures of $MoSe_2$ and SnS_2 [13]. The lattice parameters calculated for monolayers of $MoSe_2$ and SnS_2 are 3.341 Å and 3.697 Å, respectively, which closely resemble with the previously reported values of 3.33 Å ($MoSe_2$) [16] and 3.70 Å (SnS_2) [9]. Subsequently, the calculated energy band-gaps of $MoSe_2$ and SnS_2 are 1.443 eV and 1.576 eV, respectively, which are also in good agreement with reported energy band-gaps of 1.45 eV ($MoSe_2$) [16] and 1.57 eV (SnS_2) [9]. In this context, it should be noted that the GGA-PBE-D2 method usually underestimates the experimentally observed energy band-gaps in layered materials. However, such a method is often considered for DFT calculations owing to the reasonable trade-off between the computational cost and accuracy [9, 17].

3   Results and Discussions

3.1   Structural Properties

In this section, the equilibrium configurations of the tri-layer sandwiched heterostructures of $MoSe_2/SnS_2$, i.e. $MoSe_2/SnS_2/MoSe_2$ and $SnS_2/MoSe_2/SnS_2$, are analyzed in comparison with that of homogeneous tri-layer $MoSe_2$ and SnS_2.

The unit cells of tri-layer $MoSe_2$ and SnS_2 are illustrated in Fig. 1a, b, respectively. The semiconducting $MoSe_2$ is considered in its 2H phase with its characteristics stacking (ABA) between individual layers, as shown in Fig. 1a. On the other hand, Fig. 1b shows a different stacking (AAA) in SnS_2 which corresponds to its semiconducting 1T phase. Subsequently, in this work, the $MoSe_2/SnS_2/MoSe_2$ and $SnS_2/MoSe_2/SnS_2$ materials are considered in SnS_2- (AAA) and $MoSe_2$- (ABA) like stacking which are illustrated in Fig. 1c, d respectively. In this context, it should be noted that all the heterostructures considered in this work are found to be energetically stable. Furthermore, it can be observed from Table 1 that the heterostructures have intermediate values of lattice constant compared to that of tri-layer $MoSe_2$ and SnS_2 and the stacking orientations show marginal impact on the lattice constants of heterostructures.

The present work also takes into account the study of the bond length (M–X) and the bond angle (M–X–M) of the middle layers for all the tri-layer structures. The change in the electronic environment resulted in change of both the bond lengths and bond angles of the middle layers which are indicated in Table 1. It can be observed that the bond length of middle layer SnS_2 is reducing in heterogeneous environment compared to that of homogeneous environment. However, the bond angle of such layer is increasing in heterogeneous environment. It is interesting to note that such trends are exactly opposite for middle layer $MoSe_2$ in homogeneous and heterogeneous environments. In this context, it can be further observed that the stacking orders having marginal influence over such bond lengths and bond angles.

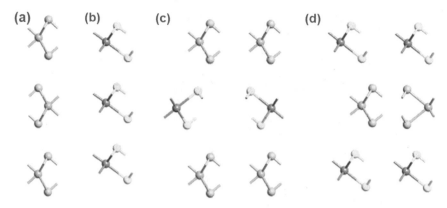

Fig. 1 The atomic structures (unit cell) of: **a** 3L-MoSe$_2$, **b** 3L-SnS$_2$, **c** 3L-MoSe$_2$/SnS$_2$/MoSe$_2$ in AAA/ABA stacking, **d** 3L SnS$_2$/MoSe$_2$/SnS$_2$ in AAA/ABA stacking

Table 1 The calculated structural properties of different tri-layers

Tri-layer structures	Lattice constant, a (Å)	Middle layer M–X bond length (Å)	Middle layer M–X–M bond angle (°)
MoSe$_2$ (ABA)	3.345	2.57	82.61
SnS$_2$ (AAA)	3.705	2.61	89.36
MoSe$_2$/SnS$_2$/ MoSe$_2$ (ABA)	3.468	2.55	93.82
MoSe$_2$/SnS$_2$/ MoSe$_2$ (AAA)	3.467	2.55	93.84
SnS$_2$/MoSe$_2$/ SnS$_2$ (ABA)	3.594	2.64	76.18
SnS$_2$/MoSe$_2$/ SnS$_2$ (AAA)	3.591	2.64	76.26

3.2 Electronic Properties

In this section, the band structures (E–K profiles) of the tri-layer heterostructures are investigated in comparison with their homogeneous counterparts and are illustrated in Fig. 2. Subsequently, the electronic properties of interest like energy band-gaps (E_g), electron effective mass (m_e) and hole effective mass (m_h) are calculated which are tabulated in Table 2. In this context, the key observations are analyzed from the atomic orbital projections of electronic states which are indicated in Fig. 3. Figure 2 indicates that similar to tri-layer MoSe$_2$ and SnS$_2$, all the hetero-structure tri-layers are indirect band-gap semiconductors. However, the heterostructures exhibit smaller direct and indirect band-gaps compared to their homogeneous tri-layer counterparts. Further, similar to tri-layer MoSe$_2$, the valence band maximum is observed at the Gamma (G) points for all the heterostructures. However, the introduction of a sandwiched SnS$_2$

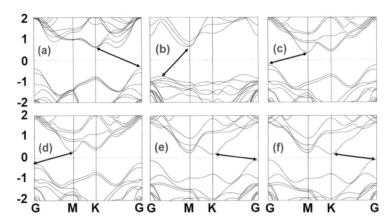

Fig. 2 The calculated band structures of tri-layer: **a** MoSe$_2$; **b** SnS$_2$; MoSe$_2$/SnS$_2$/MoSe$_2$ in **c** AAA, **d** ABA-stacking; SnS$_2$/MoSe$_2$/SnS$_2$ in **e** AAA and **f** ABA-stacking

layer shifts the conduction band minimum from K-point (tri-layer MoSe$_2$) to M-point (MoSe$_2$/SnS$_2$/MoSe$_2$). Whereas, the sandwiched hetero-layer of MoSe$_2$ shifts the conduction band minimum from M-point (tri-layer SnS$_2$) to K-point (SnS$_2$/MoSe$_2$/SnS$_2$).

On the other hand, Table 2 indicates that a reduction in conduction band effective masses can be observed in the heterostructures compared to that of homogeneous tri-layers. However, no such consistent trend can be observed for valence band effective mass. Interestingly, the changes in such electronic properties remain marginal between different stacking in both the heterostructures.

In order to analyze these observed trends in the band structures, the total and projected (orbital) density of states for all the tri-layer structures are considered. For heterostructures, only ABA stacking is considered as Fig. 2 shows marginal variations of energy band structures between ABA and AAA stacking. Also, the projections of s, p, and d orbitals of individual atoms are considered, as the contributions from f orbitals remain insignificant in all cases. It has been observed that the electronic states in the conduction band and valence band edges are primarily populated by the p-Se, and d-Mo for tri-layer MoSe$_2$. Whereas for tri-layer SnS$_2$, the s-Sn and p-S primarily

Table 2 The calculated electronic properties of different tri-layers

Tri-layer structures	Indirect E_g (eV)	Direct E_g (eV)	m_e (m_0)	m_h (m_0)
MoSe$_2$ (ABA)	0.981	1.381	0.577	0.885
SnS$_2$ (AAA)	1.332	1.678	0.549	1.790
MoSe$_2$/SnS$_2$/MoSe$_2$ (ABA)	0.593	1.071	0.527	1.501
MoSe$_2$/SnS$_2$/MoSe$_2$ (AAA)	0.573	1.060	0.542	1.463
SnS$_2$/MoSe$_2$/SnS$_2$ (ABA)	0.286	0.934	0.515	2.561
SnS$_2$/MoSe$_2$/SnS$_2$ (AAA)	0.297	0.942	0.510	3.056

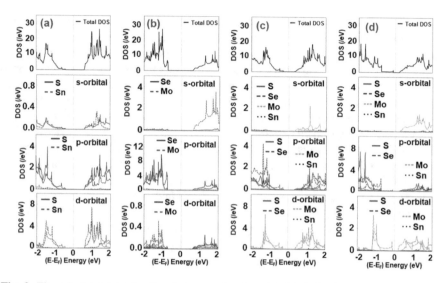

Fig. 3 The calculated total density of states and projected (orbital) density of states for tri-layer: **a** MoSe$_2$ (ABA-stacking), **b** SnS$_2$ (AAA-stacking), **c** MoSe$_2$/SnS$_2$/MoSe$_2$ (ABA-stacking), **d** SnS$_2$/MoSe$_2$/SnS$_2$ (ABA-stacking)

contributed to the states in conduction band edge and p-S contributed to that of the valence band. These observations are consistent with previously reported orbital composition analysis of the MoSe$_2$ and SnS$_2$ [9, 13, 15, 18].

Figure 3 indicates that the electronic states in the conduction band edge of the MoSe$_2$/SnS$_2$/MoSe$_2$ are predominantly populated by s-Sn, p-S and d-Mo, whereas the valence band is populated by p-Se and d-Mo. In the SnS$_2$/MoSe$_2$/SnS$_2$, the conduction band edge states are primarily contributed by s-Sn, p-S and d-Mo and the valence band edge is contributed by p-Se and d-Mo. In general, it can be observed that the predominant orbital contributions in heterostructures are nearer to Fermi-level compared to that of homogenous tri-layers. This leads to the observed reductions in the indirect band-gap of the heterostructures. Further, the difference in orbital contributions at the band edges between hetero-structure and homogenous tri-layers results into the observed difference in total DOS between such tri-layers.

4   Conclusions

An extensive study on the structural and electronic properties of sandwiched Van der Waals heterostructures of SnS$_2$ and MoSe$_2$ has been performed in different stacking configurations. In this context, emphasis has been given to the comparative analysis of such parameters with their homogenous counterparts. The results indicate replacement of the sandwiched MoSe$_2$ by SnS$_2$ layer in tri-layer MoSe$_2$ leads to an increase of M–X–M bond angle of the middle layer, whereas an opposite trend can be observed for tri-layer SnS$_2$. The study further indicates that unlike homogeneous tri-layers, the geometric configurations of heterostructures lead to a non-uniformity in inter-planer distances between different layers. On the other hand, all the hetero-structure tri-layers show a dramatic reduction in energy band-gaps compared to the homogeneous tri-layers. Typically, a very low energy band-gap can be observed for $SnS_2/MoSe_2/SnS_2$ hetero-structure. Interestingly, with the introduction of a specific sandwiched hetero-layer, a characteristic shift in the conduction band minima has been observed from the homogeneous tri-layer to the respective hetero-structure. The orbital projection analysis indicates that for any studied hetero-structure, the p orbitals of chalcogen atoms (S/Se) either contributes to the valence or conduction band edges, but not at both the edges as can be observed in homogeneous tri-layers. However, such orbital contributions are always relatively nearer to the Fermi-level in the heterostructures. Finally, it has been observed that the stacking orientation has a marginal effect on the electronic and structural properties of the studied heterostructures. In essence, the work introduces novel 2D electronic materials having low semiconducting band-gaps and suitable electron effective masses with tunable electronic properties that can have the potential for different electronic and optoelectronic applications.

Acknowledgements. The calculations were carried out using computational resources through the Central Computational Lab of BITS-Pilani, Hyderabad campus. The work is supported by an OPERA grant from BITS-Pilani.

References

1. Choi W, Choudhary N, Han GH, Park J, Akinwande D, Lee YH (2017) Recent development of two-dimensional transition metal dichalcogenides and their applications. Mater Today 20 (3):116–130
2. Li H, Wu J, Yin Z, Zhang H (2014) Preparation and applications of mechanically exfoliated single-layer and multilayer MoS_2 and WSe_2 Nanosheets. Acc Chem Res 47:1067–1075
3. Terrones H, López-Urías F, Terrones M (2013) Novel hetero-layered materials with tunable direct band gaps by sandwiching different metal disulfides and diselenides. Sci Rep 1549
4. Li M-Y, Chen C-H, Shi Y, Li L-J (2016) Heterostructures based on two-dimensional layered materials and their potential applications. Mater Today 19(6):322–335
5. Gong Y, Lin J, Wang X, Shi G, Lei S, Lin Z, Zou X, Ye G, Vajtai R, Yakobson BI, Terrones H, Terrones M, Tay BK, Lou J, Pantelides ST, Liu Z, Zhou W, Ajayan PM (2014) Vertical and in-plane heterostructures from WS_2/MoS_2 monolayers. Nat Mater 13:1135–1142
6. Wang F, Wang J, Guo S, Zhang J, Hu Z, Chu J (2017) Tuning coupling behavior of stacked heterostructures based on MoS_2, WS_2, and WSe_2. Sci Rep 17(7):44712
7. Huang Y, Sutter E, Sadowski JT, Cotlet M, Monti OL, Racke DA, Neupane MR, Wickramaratne D, Lake RK, Parkinson BA (2014) Tin disulfide, an emerging layered metal dichalcogenide semiconductor: materials properties and device characteristics. ACS Nano 8 (10):10743–10755
8. Xia C, Zhao X, Peng Y, Zhang H, Wei S, Jia Y (2015) First-principles study of group V and VII impurities in SnS_2. Superlattice Microst 85:664–671
9. Gonzalez JM, Oleynik II (2016) Layer-dependent properties of SnS_2 and $SnSe_2$ two-dimensional materials. Phys Rev B 94(12):125443

10. Zhou X, Zhou N, Li C, Song H, Zhang Q, Hu X, Gan L, Li H, Lü J, Luo J, Xiong J, Zhai T (2017) Vertical heterostructures based on SnS_2/MoS_2 for high performance photodetectors. 2D Mater 4(2):025048

11. Wang J, Jia R, Huang Q, Pan C, Zhu J, Wang H, Chen C (2018) Vertical WS_2/SnS_2 van der Waals heterostructure for tunneling transistors. Sci Rep 8(1): 17755

12. Li B, Huang L, Zhong M, Li Y, Wang Y, Li J, Wei Z (2016) Direct vapor phase growth and optoelectronic application of large band offset SnS_2/MoS_2 vertical bilayer heterostructures with high lattice mismatch. Adv Elec Mat 2(11):1600298

13. Mabiala-Poatya HB, Douma DH, Malonda-Boungou BR, Mapasha RE, Passi-Mabiala BM (2018) First-principles studies of SnS_2, MoS_2 and WS_2 stacked van der Waals hetero-multilayers. Comput Condensed Matter 16: e00303

14. Quantum ATK (2012) Version 2018.06, Synopsys Inc., Mountain View, CA, USA, June 2012 (Online). Available http://www.synopsys.com

15. Shafqata A, Iqbal T, Majid A (2017) A DFT study of intrinsic point defects in monolayer $MoSe_2$. AIP Adv 7:105306

16. Zhang C, Gong C, Nie Y, Min K-A, Liang C, Oh YJ, Zhang H, Wang W, Hong S, Colombo L, Wallace RM, Cho K (2017) Systematic study of electronic structure and band alignment of monolayer transition metal dichalcogenides in Van der Waals heterostructures. 2D Mater 4(1): 015026

17. Surrente A, Dumcenco D, Yang Z, Kuc A, Jing Y, Heine T, Kung Y-C, Maude DK, Kis A, Plochocka P (2017) Defect healing and charge transfer-mediated valley polarization in $MoS_2/MoSe_2/MoS_2$ Trilayer van der Waals Heterostructures. Nano Lett 17(7): 4130–4136

18. Abdul Wasey AHM, Chakrabarty S, Das GP (2014) Substrate induced modulation of electronic, magnetic and chemical properties of $MoSe_2$ monolayer. AIP Adv 4: 047107

Plasmonic Sensor Based on Graphene, Black and Blue Phosphorous in the Visible Spectrum

Jitendra Bahadur Maurya[1] ⓘ, Alka Verma[2], and Y. K. Prajapati[3(✉)]

[1] ECED, Bundelkhand Institute of Engineering and Technology,
Jhansi, U.P 284128, India
[2] ED, Institute of Engineering and Rural Technology, Pryagraj, U.P, India
[3] ECED, Motilal Nehru National Institute of Technology Allahabad, Pryagraj,
U.P, India
`yogendrapra@mnnit.ac.in`

Abstract. Numerical and theoretical analysis of plasmonic sensors based on Graphene, Black Phosphorous, and Blue Phosphorous are accomplished in visible spectrum from 480 to 650 nm. Attenuated total reflection intensity is obtained using a transfer matrix method. Angular interrogation is used for the surface plasmon resonance curve. Performance defining parameters viz shift in resonace angle, minimum reflection intensity, and beam width are obtained in visible spectrum for sensors having monolayer Graphene, five layers Black Phosphorous, and five layers Blue Phosphorous. It is found that the sensor is best suited in a far visible region. Further, addition of nanolayers increases minimum reflection intensity and beamwidth. Black Phosphorous has the highest shift in resonance angle followed by Blue Phosphorous and Graphene but at the penalty of higher minimum reflection intensity and beamwidth.

Keywords: Surface plasmon resonance sensor · Graphene · Black phosphorous · Blue phosphorous · Sensitivity

1 Introduction

Plasmonic thin film supported by a dielectric substrate is essential in basic plasmonic sensor chip [1]. Highest resolution and quality factor can be achieved by silver out of the plasmonic metals gold, silver, copper, and aluminium [2]. The plasmonic sensor detects the change in the refractive index (RI) of the sensing medium (SM) placed just above the sensor chip. Plasmonic sensors have several advantages e.g. fast, repeatable, reusable, highly reliable, high sensitivity, real-time monitoring, label-free detection. In addition, plasmonic sensors need small amount of sample. Hence, plasmonic sensor is widely used in food safety, in molecular interaction study of different biomolecules, e.g. proteins, nucleic acids, peptides, receptors, antibodies and lipids, in viral binding for surface functionalization, in antiviral drug discovery tools, etc. [3, 4]. Also, plasmonic sensor has vast applications in mines to sense leakages of toxic and hazardous gas [5, 6]. Plasmonic sensor can also be used to detect DNA hybridization [7].

Life sciences are being potentially facilitated by plasmonic sensor because of its capability to detect the presence of single biomolecule. The single-stranded

N. Goel et al. (eds.), *Modelling, Simulation and Intelligent Computing*,
Lecture Notes in Electrical Engineering 659,
https://doi.org/10.1007/978-981-15-4775-1_6

Deoxyribonucleic acid (ssDNA) virus-based human pathogens such as Human Boca virus (HBoV) and Parvoviridae B19 must be detected in the early stage of infections before the severity of infections in order to protect the patient from death. Human Boca virus is responsible for lower respiratory tract infections [8, 9] whereas Parvoviridae B19 is responsible for infectious disease in the pediatric population which is also known as the fifth disease or slapped cheek syndrome [10, 11]. The DNA hybridization detection capability of surface plasmon resonance (SPR) sensor makes it suitable to detect these ssDNA based human pathogens [7, 12]. In the DNA hybridization detection, the complementary ssDNA is immobilized on the top surface of sensor, and then target ssDNA is allowed to make pair via hybridization with complementary ssDNA, which results in double-stranded DNA (dsDNA), which eventually increases the refractive index near the top surface of SPR sensor. This local increase in refractive index can be detected by SPR sensor.

Plasmonic metals need functionalization for efficient attachment of antibody or antigen e.g. ssDNA on their surfaces. In addition, these plasmonic metals are corroded and oxidized in the environment, which affects the performance of the sensor [13]. Hence, these metals must be protected by appropriate material that has high affinity towards these metals as well as towards antigen or antibody and also shows chemical inert nature. These requirements can be fulfilled by Graphene, Black Phosphorous (BlackP), and Blue Phosphorous (BlueP) because of their chemical inert nature and natural binding property with ssDNA. The Graphene binds the ssDNA with the π-stacking bonding property with carbon rings of nucleobases present in ssDNA [14]. The BlackP can bind the ssDNA after the functionalization with cationic polymer poly-L-lysine (PLL) [15]. Here, PLL works as linker between Phosphorene and ssDNA. The BlueP can adsorb oxygen gas (O_2) on its surface [16]. Above mentioned utilization of Graphene, BlackP and BlueP for gas and biochemical sensing motivated us to investigate the influence of their optical properties on sensor performance.

2   Theory

2.1   Sensor Structure and Design Parameter

The diagram of nanomaterial-based surface plasmon resonance sensor structure is presented in Fig. 1. Here, the silver is considered as plasmonic metal layer because of its highest performance among gold, silver, copper, aluminium [2]. If the sensing medium is directly on the silver layer, sensor shown in Fig. 1 will become the conventional SPR sensor. At a time, anyone of the Graphene, BlackP, and BlueP can be placed above the silver layer of the conventional sensor.

The thickness of silver layer, Graphene, BlackP, and BlueP are 50 nm, 0.34 nm, 5 nm, and 5 nm, respectively. The refractive index of BK-7 prism can be given as [17];

$$n_{prism} = \left(1 + \frac{1.03961212\lambda^2}{\lambda^2 - 0.00600069867} + \frac{0.231792344\lambda^2}{\lambda^2 - 0.0200179144} + \frac{1.01046945\lambda^2}{\lambda^2 - 103.560653}\right)^{1/2}$$

(1)

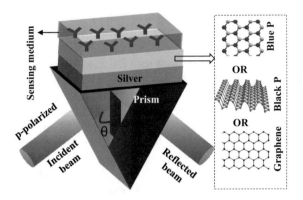

Fig. 1 Prism coupled Kretscmanmann configured plasmonic sensor

where λ represents operating wavelength in μm.

The refractive index of silver is calculated through the Drude–Lorentz model [18, 19];

$$n_{Ag} = \left(1 - \frac{\lambda^2 \lambda_c}{\lambda_p^2(\lambda_c + \lambda)}\right)^{1/2} \tag{2}$$

where λ_c and λ_p are collision and plasma wavelengths and given as 1.4541×10^{-7} and 1.7614×10^{-5} m.

The refractive index of Graphene can be given as [20];

$$n_{Gr} = 3 + i\frac{C}{3}\lambda \tag{3}$$

where C is 5.446 μm^{-1}.

The refractive indices of BlackP and BlueP at different wavelengths in the visible region are taken from Fig. 4f of [21] and from Fig. 6a of [22] respectively.

2.2   Mathematical Modeling of Reflectance Curve

The matrix method for N-layer model is applied for the calculation of reflection intensity of the reflected light [23]. This method is efficient and does not consider any approximation. The thicknesses of the layers, d_k, are considered along the z-axis. The dielectric constant and RI of the kth layer are considered as ε_k and n_k, respectively. By applying the boundary condition, the tangential fields at $Z = Z_1 = 0$ are presented in terms of the tangential field at $Z = Z_{N-1}$ as follows;

$$\begin{bmatrix} U_1 \\ V_1 \end{bmatrix} = \mathrm{H} \begin{bmatrix} U_{N-1} \\ V_{N-1} \end{bmatrix} \tag{4}$$

where U_1 and V_1 represent the tangential components of electric and magnetic fields respectively at the boundary of the first layer and U_{N-1} and V_{N-1} are the corresponding fields for the boundary at Nth layer. The H_{ij} presents the characteristics matrix of the combined structure of the sensor, and for p- polarized light it can be given as;

$$H_{ij} = \left(\prod_{k=2}^{N-1} H_k \right)_{ij} = \begin{bmatrix} H_{11} & H_{12} \\ H_{21} & H_{22} \end{bmatrix} \tag{5}$$

with

$$H_k = \begin{bmatrix} \cos \beta_k & (-i \sin \beta_k)/q_k \\ -i q_k \sin \beta_k & \cos \beta_k \end{bmatrix} \tag{6}$$

where

$$q_k = \left(\frac{\mu_k}{\varepsilon_k} \right)^{1/2} \cos \theta_k = \frac{\left(\varepsilon_k - n_1^2 \sin^2 \theta_1 \right)^{1/2}}{\varepsilon_k} \tag{7}$$

and

$$\beta_k = \frac{2\pi}{\lambda} n_k \cos \theta_k (z_k - z_{k-1}) = \frac{2\pi d_k}{\lambda} \left(\varepsilon_k - n_1^2 \sin^2 \theta_1 \right)^{1/2} \tag{8}$$

After some straightforward mathematical steps, one can obtain the reflection coefficient for p-polarized light which is given below;

$$r_p = \frac{(H_{11} + H_{12} q_N) q_1 - (H_{21} + H_{22} q_N)}{(H_{11} + H_{12} q_N) q_1 + (H_{21} + H_{22} q_N)} \tag{9}$$

The reflection intensity R_p of the defined multilayer configuration is given as;

$$R_p = |r_p|^2 \tag{10}$$

2.3   Principle of Operation

A p-polarized light wave of wavelength λ, generated from laser, is focused on the lateral plane of the prism. This prism couples the light to the deposited silver thin film. With the effect of attenuated total reflection (ATR), the evanescent field of coupled light excites the plasmons at silver surface. These excited plasmons propagate along the interface of metal-dielectric at the resonance angle and termed as surface plasmon wave (SPW). The wave vector of this SPW can be varied with the incidence angle (θ) of light at the prism. The resonance condition is achieved after the incidence angle greater than critical angle $\{\theta c = \sin^{-1}(n_{SM}/n_{prism})\}$ at which wave vector of SPW is exactly matched with the incident light. The incidence angle at which resonance condition is achieved known as resonance angle (θ_{res}), i.e. $\theta_{res} > \theta_c$. As the RI of sensing medium is changed, the resonance condition is disturbed, hence again one has to change the incidence angle to achieve the resonance condition. Thus, the change in RI of sensing medium due to adsorption of biomolecules can be measured by measuring the shift in resonance angle ($\Delta\theta_{res}$). The intensity of ATR light ($R = R_p$) corresponding to different incident angles can be measured on the opposite face of the prism by the photo-detector followed by Lock-in-amplifier, and the same can be plotted as incidence angle versus reflection intensity. This plot is known as the reflectance curve or SPR curve which has a dip at resonance angle. The incidence angle corresponding to the minimum reflection intensity is known as the resonance angle (θ_{res}).

2.4 Results and Discussions

The beam width of SPR curve is determined by the method adopted by Maurya and Prajapati [24]. In Fig. 2, reflection intensity is varied in accordance with the incidence angle for conventional, monolayer Graphene, five layers BlackP, and five layers BlueP at 630 nm wavelength. Since the adsorption of biomolecules accounts for very less change in the refractive index of sensing medium, Δn_{SM} is considered very less. As the n_{SM} is changed from 1.33 to 1.335, right shift of resonance angle for all of the nanomaterials can be easily observed from Fig. 2, which signifies the adsorption of biomolecules on the surface of nanomaterials. Since the Δn_{SM} is constant for all the nanomaterials, amount of right shift, i.e., $\Delta\theta$ will define the sensitivity ($S = \Delta\theta/\Delta n_{SM}$) order of different nanomaterial. However, the sensitivity will also depend on the physisorption property of different nanomaterials with respect to a particular biomolecule, which is out of scope for numerical analysis. The $\Delta\theta$ for conventional, Graphene, BlackP and BlueP are; 0.58, 0.59, 1.00, and 0.62 respectively. Hence, sensitivity order is BlackP > BlueP > Graphene ≈ Conventional.

In Fig. 3, shift in resonance angle, minimum reflection intensity, and BW are varied in accordance with the wavelength for conventional, monolayer Graphene, 5 layers Black Phosphorous, and five layers Blue Phosphorous at 1.335 refractive index of the sensing medium. Damping of surface plasmons in the layer above the plasmonic metal, i.e., nanolayers are directly proportional to the wavelength-dependent extinction coefficient of these nanolayers [25]. Rmin and BW are directly proportional to these surface plasmon damping [25] whose final effect is shown in Fig. 3. It can be observed from Fig. 3 that $\Delta\theta$, Rmin, and BW for conventional, Graphene, and BlueP are high near-ultraviolet (UV) region and decreases as shift towards infrared (IR) region. But, for a good sensor $\Delta\theta$ should be as high as possible whereas Rmin and BW should be as low as possible. Although Rmin and BW of conventional and Graphene are decreasing from UV to IR, they have acceptable very low value throughout the range. But, $\Delta\theta$ is also very low at UV and further decreases towards IR. Hence, conventional and Graphene-based sensors are not suitable for high sensitivity. Further, although the $\Delta\theta$ for BlueP is far high near UV region at the same time its Rmin and BW is also very high. Hence, BlueP shows contradictory nature between $\Delta\theta$ and (Rmin and BW). On comparing the BlackP with conventional and Graphene, it can be found that BlackP has far high $\Delta\theta$ at an acceptable very low value of Rmin and BW from mid of visible region (VR) to near IR. Hence, BlackP based sensor is far better than Graphene-based sensor and which is better than a conventional sensor for mid of VR to near IR.

In contrast to conventional, Graphene, and BlueP, $\Delta\theta$ for BlackP is very low near UV which increases rapidly from UV to IR and becomes maximum at 630 nm in the VR. Simultaneously, the Rmin and BW are very high near UV which decreases rapidly

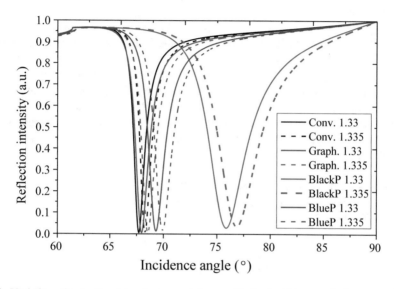

Fig. 2 Variation of reflection intensity in accordance with the incidence angle for conventional, monolayer Graphene, 5 layers Black Phosphorous, and five layers Blue Phosphorous at 630 nm wavelength

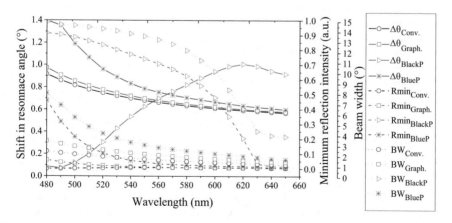

Fig. 3 Variation of shift in resonance angle, minimum reflection intensity, and BW in accordance with the wavelength for conventional, monolayer Graphene, 5 layers Black Phosphorous, and five layers Blue Phosphorous at 1.335 refractive index of the sensing medium

from UV to IR and becomes a minimum at 630 nm in the VR. Hence, BlackP does not show any contradiction between $\Delta\theta$ and (Rmin and BW) at 630 nm. Hence, all the three needs for a good sensor, i.e. high $\Delta\theta$ and low (Rmin and BW) are strongly fulfilled by BlackP at 630 nm, which is near to 632.8 nm generated by He–Ne laser.

3 Conclusion

Conventional, Graphene, BlackP and BlueP based plasmonic sensor is numerically analyzed. It is found that conventional, Graphene, and BlueP show contradiction nature between $\Delta\theta$ and (Rmin and BW), whereas BlackP does not show. Further, conventional, Graphene, BlueP can be used for mid-VR to near IR and has the sensitivity order BlueP < Graphene < conventional. In contrast, Black is the best at 630 nm in the VR. Hence, Graphene can be replaced by Black Phosphorous and Blue Phosphorous for better sensitivity provided that their molecular binding ability favours desired molecule.

Acknowledgements. One of the authors Y. K. Prajapati is thankful to the Ministry of Electronics & Information Technology (Meity), India for the fellowship (YFRF) under the Vishvesvaraya scheme.

References

1. Kretschmann E, Raether, H (1968) Radiative decay of non radiative surface plasmons excited by light. Zeitschrift für Naturforschung A 23(12):2135–2136
2. Maurya JB, Prajapati YK (2016) A comparative study of different metal and prism in the surface plasmon resonance biosensor having MoS_2-graphene. Opt Quant Electron 48(5):280–291
3. Victoria S (2012) Application of surface plasmon resonance (SPR) for the detection of single viruses and single biological nano-objects. J Bacteriol Parasitol 3(7):1–3
4. Homola J (2008) Surface plasmon resonance sensors for detection of chemical and biological species. Chem Rev 108(2):462–493
5. Liedberg B, Nylander C, Lunström I (1983) Surface plasmon resonance for gas detection and biosensing. Sensors and actuators 4:299–304
6. Maurya JB, Prajapati YK, Raikwar S, Saini JP (2018) A silicon-black phosphorous based surface plasmon resonance sensor for the detection of NO_2 gas. Optik 160:428–433
7. Zhang Q, Zou XN, Chu LQ (2018) Surface plasmon resonance studies of the hybridization behavior of DNA-modified gold nanoparticles with surface-attached DNA probes. Plasmonics 13(3):903–913
8. Ricour C, Goubau P (2008) Human bocavirus, a newly discovered parvovirus of the respiratory tract. Acta Clin Belg 63(5):329–334
9. Allander T, Jartti T, Gupta S, Niesters HG, Lehtinen P, üsterback R, Vuorinen T, Waris M, Bjerkner A, Tiveljung-Lindell A, van den Hoogen BG (2007) Human bocavirus and acute wheezing in children. Clin Infectious Diseases 44(7):904–910
10. Lunardi C, Tinazzi E, Bason C, Dolcino M, Corrocher R, Puccetti A (2008) Human parvovirus B19 infection and autoimmunity. Autoimmun Rev 8(2):116–120

11. Wawina TB, Tshiani OM, Ahuka SM, Pukuta ES, Aloni MN, Kasanga CJ, Muyembe JJT (2017) Detection of human parvovirus B19 in serum samples from children under 5 years of age with rash–fever illnesses in the Democratic Republic of the Congo. Int J Infectious Diseases 65:4–7

12. Endo T, Kerman K, Nagatani N, Takamura Y, Tamiya E (2005) Label-free detection of peptide nucleic acid-dna hybridization using localized surface plasmon resonance based optical biosensor. Anal Chem 77(21):6976–6984

13. Kravets VG, Jalil R, Kim YJ, Ansell D, Aznakayeva DE, Thackray B, Britnell L, Belle BD, Withers F, Radko IP, Han Z (2014) Graphene-protected copper and silver plasmonics. Sci Rep 4:5517

14. McGaughey GB, Gagné M, Rappé AK (1998) π-Stacking interactions alive and well in proteins. J Biol Chem 273(25):15458–15463

15. Kumar V, Brent JR, Shorie M, Kaur H, Chadha G, Thomas AG, Lewis EA, Rooney AP, Nguyen L, Zhong XL, Burke MG (2016) Nanostructured aptamer-functionalized black phosphorus sensing platform for label-free detection of myoglobin, a cardiovascular disease biomarker. ACS Appl Mater Interfaces 8(35):22860–22868

16. Liu N, Zhou S (2017) Gas adsorption on monolayer blue phosphorus: implications for environmental stability and gas sensors. Nanotechnology 28(17):75708

17. SCHOTT Zemax catalog. http://www.schott.com. Accessed 20 Jan 2017b

18. Raether H (1988) Surface Plasmons on Smooth and Rough Surfaces and on Gratings. Springer-Verlag, Berlin

19. Yang HU, D'Archangel J, Sundheimer ML, Tucker E, Boreman GD, Raschke MB (2015) Optical dielectric function of silver. Phys Rev B 91(23):235137

20. Bruna M, Borini S (2009) Optical constants of graphene layers in the visible range. Appl Phys Lett 94:031901

21. Mao N, Tang J, Xie L, Wu J, Han B, Lin J, Deng S, Ji W, Xu H, Liu K, Tong L (2015) Optical anisotropy of black phosphorus in the visible regime. J Am Chem Soc 138(1):300–305

22. Peng Q, Wang Z, Sa B, Wu B, Sun Z (2016) Electronic structures and enhanced optical properties of blue phosphorene/transition metal dichalcogenides van der Waals heterostructures. Sci Rep 6:31994

23. Maurya JB, Prajapati YK, Singh V, Saini JP, Tripathi R (2016) Improved performance of the surface plasmon resonance biosensor based on graphene or MoS_2 using silicon. Opt Commun 359:426–434

24. Maurya JB, Prajapati YK (2017) A Novel Method to Calculate Beam Width of SPR Reflectance Curve: A Comparative Analysis. IEEE Sens Lett 1(4):1–4

25. Pockrand I (1978) Surface plasma oscillations at silver surfaces with thin transparent and absorbing coatings. Surf Sci 72:577–588

Measurement, Modeling and Simulation of Photovoltaic Degradation Rates

Mohamed Shaik Honnurvali[1](✉) ⓘ, Naren Gupta[2], Keng Goh[2] ⓘ,
and Tariq umar[1] ⓘ

[1] Asharqiyah University, Ibra 400, Sultanate of Oman
honnur.vali420@gmail.com
[2] Edinburgh Napier University, Edinburgh EH105DT, UK

Abstract. Photovoltaic degradation rates play a vital role in visualizing and analyzing the performance of the PV modules over the long run. A site survey is conducted to calculate PV degradation rates. The results have shown that for the first three years since the initial installation, the degradation rates have remained in line with the manufacturer values (i.e., less than 0.6%), while the next two years the degradation rates have almost increased by 40%. This is due to discoloration of the encapsulant causing the reduction of the short circuit current (I_{sc}). Mathematically, modeling such visual loss factors has not considered so far. The visual loss factor equation is developed and incorporated in the output current equation of the PV module. Further, the I-V curves are simulated and compared with the measured I-V curves. The results have shown an acceptable error percentage of around 0.3%.

Keywords: PV (Photovoltaics) · STC (Standard test Conditions) · Short circuit current *(I_{sc})*

1 Introduction

The key initiatives adopted by most of the countries globally, to limit the carbon emissions and further rise in the global temperatures, have created an impact in revising the energy generation methods [1]. Different renewable energy technologies are being integrated into the current energy generation system and also in planning future energy needs [2–4. The progress on the renewable energy still suffers from a number of challenges in many countries including the Arabian Gulf region [5]. One of the most prominent and widely deployed renewable technologies currently is solar photovoltaics systems [6]. The solar photovoltaic systems have attracted the largest investments share in renewable energies around 161 billion USD dollars [7] which accounted for 58% of the new renewable energy investments. These investments are majorly witnessed in developing countries rather than in developed countries. As a result, production capabilities have increased far more, thereby reducing the cost of PV systems. According to the market predictions, the solar photovoltaic system installations will be tripled in the next four to five years. An emphasis on the solar market globally in terms of cost, employment, environmental preserving factors is very well explained in [7, 8]. It is also anticipated that by 2050 most of the future smart cities globally would be able

© Springer Nature Singapore Pte Ltd. 2020
N. Goel et al. (eds.), *Modelling, Simulation and Intelligent Computing,*
Lecture Notes in Electrical Engineering 659,
https://doi.org/10.1007/978-981-15-4775-1_7

to meet their energy demands by 100% renewable energy generation methods [9]. Considering the bright future and huge investments made in photovoltaics, it is vital to understand the performance of the photovoltaic systems over a period of time. Earlier research studies conducted in the different parts of the globe [10–12] have witnessed photovoltaic degradation due to the influence of various factors such as local environment conditions and other socioeconomic factors. A recent research study by Mohamed shaik et al. [13, 14] conducted a survey of around 130 PV modules to estimate the degradation rates in Oman. The observed degradation rate overall is 1.96% which is almost the double the degration rates seen in European countries. Further, their study has observed that the modules installed in hot and humid climatic zones have seen with higher degradation rate than that of other climatic zones. On the other hand, by PV technology type, thin film modules have seen with the lowest degradation rates. In this paper, a mathematical function estimating the short circuit current (I_{sc}) reduction due to visual effects is developed and used to calculate the final output current of the PV module. The reduction in short circuit current further decreases the maximum output power (P_{max}). Section two explains the experimental setup and the methodology to measure the I-V and P-V curves from the site. Section three presents the mathematical model to calculate the visual loss factor, modified output current equation for the PV module and the simulation performed in MATLAB to estimate the degradation rates. Section four discusses and compares the results observed in detail. Section five presents the conclusion of the paper.

2 Experimental Setup and Estimation of PV *(P_{max})* Degradation Rates

Sultanate of Oman, one of the GCC countries with abundant irradiance, long day duration with clear skies are all favorable factors to deploy solar technology to generate energy. Despite the hot and robust climatic conditions [15, 16], the performance and reliability of the photovoltaic modules have not been investigated thoroughly in Oman. On the other hand, several studies have been conducted in this regard in different parts of the world. According to the research study report by [17, 18], different PV technologies in different parts of the world have observed degradation. To understand the performance and reliability of the photovoltaic modules in the sultanate of Oman, a survey is conducted every year in the month of July since 2014 to record the I-V curves. The measurements are taken at two different irradiances from mid-noon around 12 PM and the other at post noon around 2:00–2:30 PM.

2.1 Site Description

The site is located near the Muscat city (google coordinates: 23°35'00.6"N 58°21'37.9"E). The purpose of this solution is to provide energy to the base transceiver station (BTS) supporting the local telecom network (Omantel). This solution is an off-grid solution which has a system size of around 1.4 KW. The energy is stored using a 24 V battery connected with the MPPT charge controller and inverter. Six SUNTECH 240W_p PV Panels are mounted to the aluminum stand to supply the required energy. The

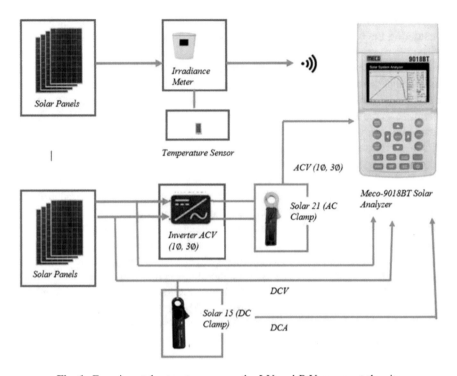

Fig. 1 Experimental setup to measure the I-V and P-V curves at the site

experimental setup to measure the I-V and P-V curves is shown in Fig. 1. To measure the I-V and P-V curves of a PV module, "MECO 9018 BT" portable solar analyzer is used. Besides that, an irradiance meter, temperature sensor, current and voltage sensors are used to record the corresponding values during the measurement. The environmental conditions in the Sultanate of oman usually remain hot and dusty (due to desert climate) for the most part of the year. Before performing measurements, the PV modules are cleaned with water and sodium detergent to remove the dust accumulation. The study of dust effects on the maximum output power is not considered in our study.

The maintenance of the PV modules is highly recommended by the client for the smooth and efficient performance of the PV modules. However, a recent study [19] conducted in six cities of northern Oman has noted the PV output power drop is around 60% if not cleaned for a month. The I-V and P-V curves are recorded for each PV module at two different irradiances. The irradiance measurements recorded at noontime stood around 800 w/m^2 with a tolerance level of ±50 w/m^2. Similarly, in the post noon, the recording stood at around 600 w/m^2 with the same tolerance level. "Photovoltaic degradation" is a common phenomenon that is observed in PV modules. These degradations mainly occur due to several factors that are discussed in the literature [15, 20]. The PV degradation rates are highly influenced by the local environmental factors and the environmental conditions vary from region to region all across the globe. PV degradation rate is calculated by the Eq. (1) as given below:

$$P_{max}Degradationrate = \frac{P_{(max,intialvalue)} - P_{(max,finalvalue)}}{P_{(max,initialvalue)} \times Ageof themodule} \%/year \qquad (1)$$

The average degradation rate for the six PV modules under the irradiance 800 w/m^2 is observed to be at 1.09% since the initial installation. while with irradiance 600 w/m^2, the degradation rates seen are 1.10%. Both values at different irradiances show good accuracy which confirms the correctness in the estimation of the degradation rates. The observed degradation rates seem to be higher than the degradation rates seen in other countries [21, 22]. The reasons are mainly due to discoloration of the encapsulant which causes the reduction in the short circuit current (I_{sc}), thus decreasing the maximum output power (P_{max}). The hot and humid climatic conditions play a very prominent role in discoloration of encapsulant and delamination of the front panel due to high salt sodium formations.

3   Modeling and Simulation of PV Degradation Rates Using MATLAB

Simulation tools have become an integral part of the design process to simulate the system performance that will correlate to the actual practical implementation in real time. However, there are challenges to the extent of incorporating all the real lifetime contributing factors into the simulation model. One such challenge is to integrate the effects of the reduction of I_{sc} due to visual defects seen in the real world as discussed in Sect. 2.1. The visual defects observed under the influence of the local climatic conditions may vary from region to region. For instance, in our study, PV modules are seen with discoloration of encapsulation and some modules with delamination of the front panel which has contributed to the reduction of short circuit current I_{sc}. In order to model the visual defects seen in our study, the following Eq. (2) is developed using exponential curve fitting methods:

$$V_{LF} = e^{-0.035 \times (ILF)} \qquad (2)$$

where V_{LF} is the visual defects loss factor, and ILF is a short circuit current (I_{sc}) loss factor and is given by ($\frac{I_{scmeasuredatirradiance}}{I_{sc}atthatirradianceunderSTCconditions}$).

Earlier, many studies have been reported [23, 24] considering MATLAB and Simulink for modeling the PV cell, panel and also found to be a good match between the simulated and the experimental results. However, those studies did not include the physical factors that commonly play a vital role in the degradation of the PV module. The physical factors are region-dependent and may vary from place to place. Therefore, one can develop similar mathematical loss factor models according to the factors which influence the performance of the PV module. A single diode model is considered in this study. Each solar cell in a PV module is modeled by the five parameters single diode model. The complete methodology to simulate a PV module using one diode model is very well explained in the paper [25]. Using equations (1–9 from [25]), a Simulink model to estimate the total output current of the PV module which includes the visual

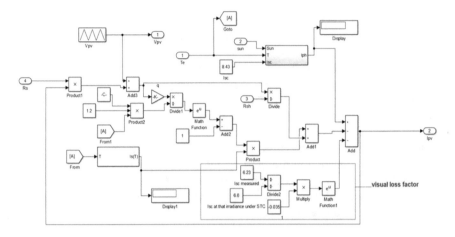

Fig. 2 Simulation of output current *I* using Simulink with *VLF (subsystem)*

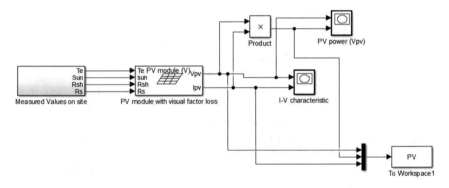

Fig. 3 Simulation of I-V and P-V curves of a PV module *(Full System)*

losses observed on site is shown in Fig. 2. Further, Fig. 3. shows complete PV module simulation setup to estimate the output power of the PV module and I-V curves at different irradiances.

4   Results and Discussion

The I-V curves for a sample 5 PV module measured during the initial installation year (i.e., 2014) at two different irradiances along with the simulated I-V curves are presented in Fig. 4. The maximum power output P_{max} measured during the initial installation at 800 w/m^2 irradiance is 178w, while the simulated I-V curve has shown a close match with an error 0.28%. At 600 w/m^2 irradiance, the error value is 0.56% which is also at a considerable level. Similarly, the I-V curves for the same PV module

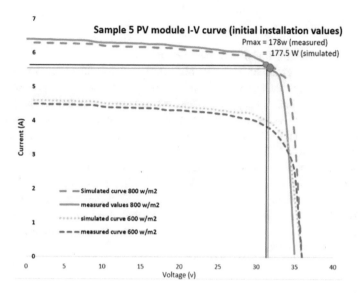

Fig. 4 Sample 5 PV module I-V curves along with the simulated curves during the initial year of installation

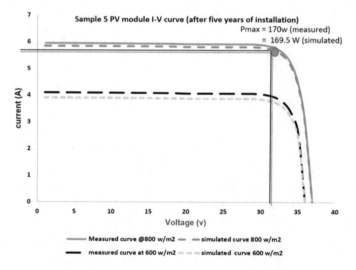

Fig. 5 Sample 5 PV module I-V curves along with the simulated curves after five years of field exposure

measured after five years of field exposure are presented in Fig. 5. The measured maximum output power is 170 W which has degraded at a rate of 0.89%/year which is higher than the manufacturer given rate (0.5% after 1 year).

5 Conclusion

PV degradation rates for the six PV modules are calculated for a field exposure of five years in Oman. The results have shown higher degradation rates than the manufacturer proposed values. Possible reasons are due to high temperatures and salty humid climate affecting the PV module encapsulant, thereby causing the reduction in short circuit current (Isc). Further, a loss factor equation is developed to incorporate the losses that occurred due to visual defects. The total output current of the PV module is modified and simulated using MATLAB to validate the measured data. The results have shown a satisfactory match with an error percentage of 0.3%. However, this equation modeling is dependent on factors arising from the local environmental conditions. The estimation of PV degradation will help the industry utilities to select the proper PV technology for better performance under local climatic conditions and also improvise the quality of the installation methods. The other major advantage is to accurately estimate the return of investments (ROI) for the stakeholders.

References

1. Umar T, Egbu C (2018) Global commitment towards sustainable energy. In: Proceedings of the institution of civil engineers-engineering sustainability, Vol 172, No 6, September 2019, pp 315–323 In Focus: SEB16. Thomas Telford Ltd. https://doi.org/10.1680/jensu.17.00059
2. Umar T, Wamuziri S (2016) Briefing: Conventional, wind and solar energy resources in Oman. In: Proceedings of the institution of civil engineers-energy, Vol 169, No 4, pp 143-147. https://doi.org/10.1680/jener.16.00011
3. Umar T (2017) Briefing: Towards a sustainable energy: the potential of biomass for electricity generation in Oman. In: Proceedings of the institution of civil engineers-engineering Sustainability, Vol 171, No 7, Thomas Telford Ltd., pp 329–333. https://doi.org/10.1680/jensu.17.00001
4. Umar T (2018) Geothermal energy resources in Oman. In: Proceedings of the Institution of Civil Engineers-Energy, Vol 171, No 1, pp 37–43. https://doi.org/10.1680/jener.17.00001
5. Umar T, Egbu C, Ofori G, Honnurvali MS, Saidani M, Opoku, A (2019) Challenges towards renewable energy: an exploratory study from the Arabian Gulf region. In: Proceedings of the institution of civil engineers-energy, pp 1–34. https://doi.org/10.1680/jener.19.00034
6. Umar T (2017) Rooftop solar photovoltaic energy: a case study of India. Nanomaterials and Ener-gy. 6(1):17–22. https://doi.org/10.1680/jnaen.16.00002
7. Jäger-Waldau A (2018) PV status report 2018. European Commission
8. Malik K, Rahman SM, Khondaker AN, Abubakar IR, Aina YA, Hasan MA (2019) Renewable energy utilization to promote sustainability in GCC countries: policies, drivers, and barriers. Environ. Sci. Pollut. Res. 1–17
9. Honnurvali MS, Gupta N, Goh K, Umar T, Kabbani A, Nazecma N (2019) January. Can Future Smart Cities Powered by 100% Renewables and Made Cyber Secured-A Analytical Approach. In: 2019 IEEE 12th International conference on global security, safety and sustainability (ICGS3) pp. 206–212 IEEE
10. Busquet S, Kobayashi J, Rocheleau RE (2017) Operation and performance assessment of grid-connected PV systems in operation in Maui, Hawaii. In: 2017 IEEE 44th Photovoltaic specialist conference (PVSC), IEEE, pp 1061–1066)

11. Dhoke A, Mengede A (2017) Degradation analysis of PV modules operating in Australian environment. In: 2017 Australasian universities power engineering conference (AUPEC), IEEE, pp 1–5
12. Honnurvali MS, Gupta N (2017) PV electrical parameters degradation analysis-Oman perspective. In: 2017 8th International renewable energy congress (IREC), IEEE, pp 1–5
13. Honnurvali MS, Gupta N, Goh K, Umar T, Kabbani A, Nazeema N (2018) Case study of PV output power degradation rates in Oman. IET Renew Power Gener 13(2):352–360
14. Honnurvali MS, Gupta N, Goh K, Umar T, Kabbani A, Nazeema N (2017) Study of Photovoltaics (PV) Performance Degradation Analysis in Oman. Int J Sustain Energ 6(2):18
15. Jordan DC, Silverman TJ, Wohlgemuth JH, Kurtz SR, VanSant KT (2017) Photovoltaic failure and degradation modes. Prog Photovoltaics Res Appl 25(4):318–326
16. Umar T, Egbu C, (2018) Heat stress, a hidden cause of accidents in construction. In: Proceedings of the institution of civil engineers-municipal engineer, Thomas Telford Ltd., pp. 1–12. https://doi.org/10.1680/jmuen.18.00004
17. Umar T, Egbu C, Honnurvali MS, Saidani M, Al-Bayati AJ (2019) Briefing: Status of occupational safety and health in GCC construction. In: Proceedings of the institution of civil engineers-management, procurement and law, pp 1–5. https://doi.org/10.1680/jmapl.18.00053
18. Lopez-Garcia J, Sample T (2018) Evolution of measured module characteristics versus labelled module characteristics of crystalline silicon-based PV modules. Sol Energy 160:252–259
19. Kazem HA, Chaichan MT (2019) The effect of dust accumulation and cleaning methods on PV panels' outcomes based on an experimental study of six locations in Northern Oman. Sol Energy 187:30–38
20. Kuitche JM, Pan R, TamizhMani G (2014) Investigation of dominant failure mode (s) for field-aged crystalline silicon PV modules under desert climatic conditions. IEEE J. Photovoltaics 4(3):814–826
21. Sánchez-Friera P, Piliougine M, Pelaez J, Carretero J, Sidrach de Cardona M (2011) Analysis of degradation mechanisms of crystalline silicon PV modules after 12 years of operation in Southern Europe. Prog Photovoltaics Res Appl 19(6):658–666
22. Jordan DC, Kurtz SR (2013) Photovoltaic degradation rates—an analytical review. Prog Photovoltaics Res Appl 21(1):12–29
23. Hejri M, Mokhtari H, Azizian MR, Ghandhari M, Söder L (2014) On the parameter extraction of a five-parameter double-diode model of photovoltaic cells and modules. IEEE J. Photovoltaics 4(3):915–923
24. Siddique, H.A.B., Xu, P. and De Doncker, R.W., 2013, June. Parameter extraction algorithm for one-diode model of PV panels based on datasheet values. In 2013 International Conference on Clean Electrical Power (ICCEP), IEEE, pp 7–13
25. Kumar R, Singh SK (2018) Solar photovoltaic modeling and simulation: As a renewable energy solution. Energy R. 4:701–712

Converter Efficiency Improvement of Islanded DC Microgrid with Converter Array

S. K. Rai[1(✉)], H. D. Mathur[1], and Shazia Hasan[2]

[1] Birla Institute of Technology and Science, Pilani Campus, Pilani, Rajasthan, India
{p2016503,mathurhd}@pilani.bits-pilani.ac.in
[2] Birla Institute of Technology and Science Pilani, Dubai Campus, Dubai, UAE
shazia.hasan@dubai.bits-pilani.ac.in

Abstract. Photovoltaic (PV) solar energy is growing rapidly in energy supplying for residential buildings. Since solar energy directly generates DC power, a DC microgrid is a better choice particularly in islanded mode of operation. The DC–DC converter is the most essential part of DC microgrid, and therefore, overall efficiency of microgrid largely depends upon the converter's efficiency. The efficiency of converters depends upon the controller along with load conditions. The converter has typically lower efficiency in both the cases of heavy and light load conditions. This paper presents the analysis of converter efficiency improvement of DC microgrid using converter arrays at the place of centralized converters. The data of solar power generation and load demand have been used in the study, and it is found that converter array improves the efficiency by maximum 2.587% than centralized converter architecture.

Keywords: DC microgrid · DC-DC converters · Photo-Voltaic system · Converter array · Converter efficiency

1 Introduction

DC microgrid is getting popular due to generation of DC power by the renewable energy sources and availability of storage units [1]. It also overcomes the major disadvantages of the AC microgrid like transformer inrush current, frequency synchronization, reactive power flow, power quality issues, phase unbalance, etc. [2–4].

Converters (DC–DC/DC–AC) are the main and essential part of the DC microgrid. The efficiency of the overall system is improved with the help of enhancement in control technologies. There are a number of literature available on the control aspect of the microgrid to optimize the system [5–7]. Apart from the control aspect, the converter efficiency also depends upon its utilization. In both the cases of heavy and light load condition, the efficiency of converters decreases.

This paper investigates the DC microgrid converter efficiency with respect to centralized converter and converter array. Photovoltaic (PV) solar power generation-based DC microgrid with battery storage system has been considered for the study.

© Springer Nature Singapore Pte Ltd. 2020
N. Goel et al. (eds.), *Modelling, Simulation and Intelligent Computing*,
Lecture Notes in Electrical Engineering 659,
https://doi.org/10.1007/978-981-15-4775-1_8

Section 2 explains the system architecture. Section 3 covers the converter array for DC microgrid. Simulation setup and results are covered in Sects. 4 and 5, respectively. Section 6 includes the conclusion of the presented work.

2 System Architecture

The proposed system architecture considers a 2500 sq.ft institutional office building. The electrical energy consumption of the building is in the range of 15–40 units (15–40 kWh) depending upon the weather conditions and need.

The DC microgrid architecture used in the building has been shown in Fig. 1 which consists of PV generation unit, power converters, different loads and battery energy storage system.

The twelve 500 W_P (YS500M-96) solar panels designed by Yangtze Solar Power Co. Ltd are used for power generation which generate maximum 6 kW power. The specifications of the used solar panel have been given in Table 1.

A 20 kWh battery storage system (BSS) is used for backup during office working hours which uses 12 V, 150 Ah batteries in series and parallel combinations to meet desired capacity and voltage level. As shown in Fig. 1, the DC microgrid system has two DC voltage bus, high voltage DC (HVDC) bus of 380 V and low voltage DC (LVDC) bus of 48 V. The BSS is connected to the 48 V DC bus along with all other

Table 1 YS500M-96 SPECIFICATIONS AT STC

Parameters	Values
Maximum power (Pmax)	500 WP
Voltage at maximum power (Vmpp)	48.35 V
Open circuit voltage (Voc)	58.89 V
Short circuit current (Isc)	10.04 A
Panel efficiency	19.51%
Power tolerance (Positive)	+3%
Surface area	2.56 m^2

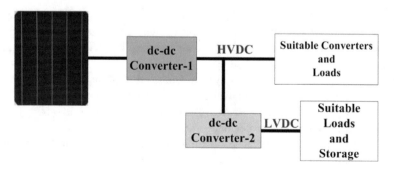

Fig. 1 A typical DC microgrid

DC load. The converter preceding to BSS is bidirectional in nature to charge and discharge the batteries. The heavy loads are connected with the 380 V DC bus which includes single-phase AC loads also.

3   Converter Array for DC Microgrid

Converter array conversion is a way to employ more than one converter at the place of single converter-based conversion system. It has switching scheme in its architecture to improve overall efficiency for power conversion/inversion in the microgrid. The most common conversion system in DC microgrid is the centralized conversion system.

3.1   Architecture of DC Microgrid

Figure 2a shows a DC microgrid with centralized converter system. In this case, both high voltage DC bus and low voltage DC bus get its input directly from PV junction box. The architecture shown in Fig. 2b is similar to the architecture of Fig. 2a apart from the input supply for the low voltage DC bus converter. Here, the low voltage DC bus converter is fed from high voltage DC bus, and it does not require a separate converter for BSS. Figure 2c shows a DC microgrid with input-series-output-parallel conversion system [8]. This architecture has a number of series converters, and its outputs are paralleled to a junction box. High voltage dc bus feeds input to the low voltage dc bus converter. Since each series converter handles a set of PV panels, it improves the reliability and system balance.

The considered architecture for analysis in the paper is shown in Fig. 2d. The main converter of this architecture has three boost converters at the place of one boost converter. This main converter array gets its input from junction box, and its DC output is 380 V. The architecture has one more converter array which is bidirectional in nature and converts high voltage DC (380 V) to low voltage DC (48 V) and vice versa. The low voltage DC bus is connected to both low voltage DC load and battery storage system. The bidirectional converter array is used to both sink and source power to the high voltage DC bus.

3.2   Switching Scheme for Converter Array

The efficiency of converter varies with the load. At light and heavy load conditions, the converter gives poor efficiency which is independent from used control techniques. A typical efficiency curve of the DC–DC converter has been shown in Fig. 3, and the same is utilized for the presented study.

In the considered DC microgrid system, the rated total input is 6 Kw, and rated power (RP) of each converter in converter array is 2 kW. The maximum efficiency of DC–DC converter lies between 40% and 80% of RP, and the maximum power capacity is 1.2 times RP. On the basis of this data, the switching scheme is developed as shown in Fig. 4 and 5. The first converter works till required output power of 0.8RP (1.6 kW). The second converter starts at 0.8RP, and the third converter starts at 1.6RP (3.2 kW). When all three converters are working and if power requirement decreases, then the

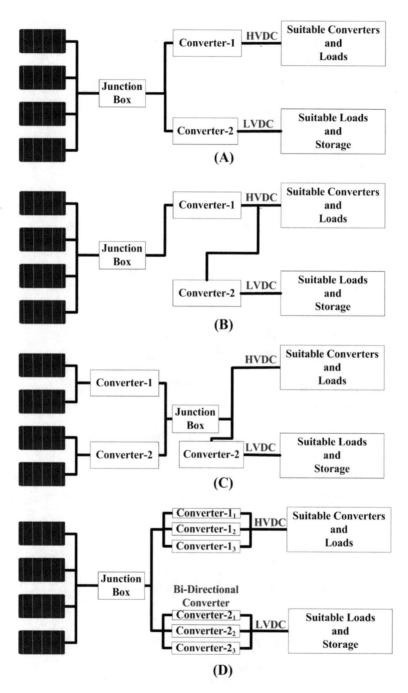

Fig. 2 Power converter architecture for DC microgrid

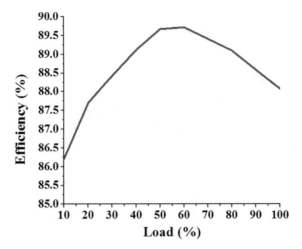

Fig. 3 Typical DC–DC converter efficiency

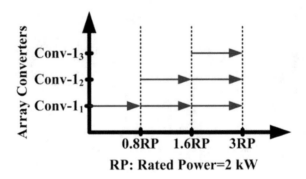

Fig. 4 Switching Scheme of converters for increasing power

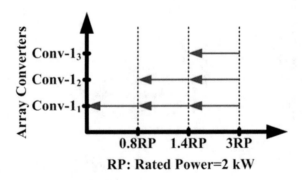

Fig. 5 Switching Scheme of converters for decreasing power

third converter stops at 1.4RP (2.8 KW), and if further required power decreases, then the second converter stops at 0.8RP. The switching scheme keeps the converters into maximum efficiency region as per the load power requirement.

4 Simulation Setup

The operation of considered DC microgrid is simulated using MATLAB^{TM} software. The hourly data of 24-hour are used for simulation purpose. The solar irradiance and temperature data of a particular day are taken from [9]. Figure 6 and 7 show the plot of irradiance and temperature, respectively. The power produced by solar panel is calculated using the Eq. (1), and its plot is shown in Fig. 8.

$$P_t =_P *S_P * \emptyset_t(1 - 0.005(T_t - 25)) \tag{1}$$

where

$$P_t = \text{Total Generated Power by Solar Array}$$

$$_P = \text{Panel Efficiency}$$

$$S_P = \text{Total Surface Area of Solar Array}$$

$$\emptyset_t = Irradiance\left(kW/m^2\right)$$

$$T_t = \text{Temperature } \left(^0C\right)$$

The analysis of considered DC microgrid is performed with a typical day load shown in Fig. 9. It is hourly measured data.

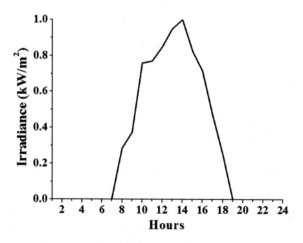

Fig. 6 Hourly irradiance data

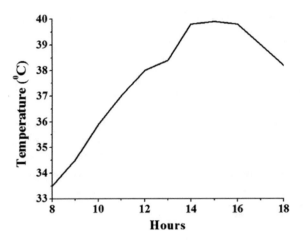

Fig. 7 Hourly temperature data

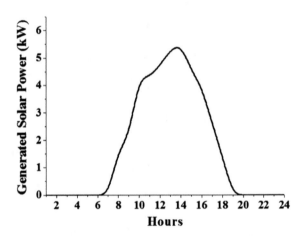

Fig. 8 Hourly generated solar PV power

5   Simulation Results

The analysis is performed with consideration of equal sharing of power among working converters in converter array architecture.

5.1   Simulation with Gradual Increasing and Decreasing Load Power Demand

At the beginning, simulation is performed with increasing and decreasing load power demand with both centralized converter (Fig. 2b) and converter array (Fig. 2d). The simulation results of increasing power (0–6 kW) and decreasing power (6–0 kW) are

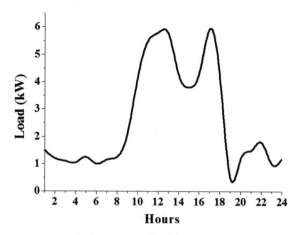

Fig. 9 Hourly load demand

shown in Figs. 10 and 11, respectively. The switching scheme for increasing/ decreasing power of converter array is considered in the study. The thin line curve is of centralized converter architecture, whereas thick line (red) curve is of converter array. As per the obtained graph, the converter array enhances the efficiency by maximum 2.587% with respect to centralized converters in both the cases of increasing and decreasing power demand, whereas the average enhancements in efficiency during completed span of output power due to converter array are 0.67% (increasing power demand) and 0.47% (decreasing power demand).

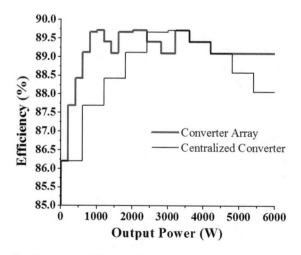

Fig. 10 Converter efficiency with increasing output power (0–6 kW)

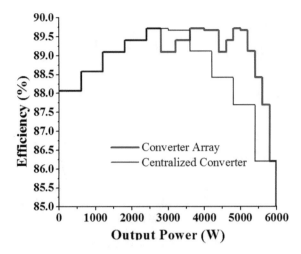

Fig. 11 Converter efficiency with decreasing output power (6–0 kW)

5.2   Simulation with Solar Power Generation and Hourly Load Demand

At the second stage, the analysis is performed as per generated solar power and load data into consideration. The variation in the load profile is the mixture of both increasing and decreasing power demand nature. The study is performed during availability of solar power, and therefore, data between 8 and 16 h have been used. The simulated result has been shown in Fig. 12. The graph is reflecting that the converter array enhances the efficiency by maximum 1.938% with respect to centralized converters, whereas the average enhancements in efficiency during complete span of analysis duration due to converter array are 0.43%.

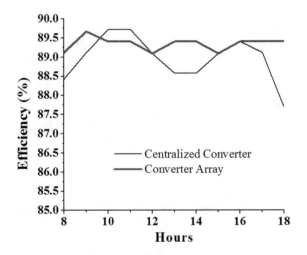

Fig. 12 Converter efficiency with hourly load demand and PV generation

6   Conclusion

In this paper, a study has been performed to enhance the converter efficiency by using converter array at the place of centralized converter. The analysis is done with increasing/decreasing output power pattern as per switching scheme, and it is found that the conversion efficiency enhances maximum by 2.587% with average enhancement through whole simulation span is 0.67% (for increasing output power) and 0.47% (for decreasing output power). The considered architecture is also tested with the hourly load demand data along with generated PV solar power, and it is found that the conversion efficiency enhances by maximum 1.938% with average enhancement of 0.43% throughout the simulation span.

Acknowledgements. This research work is supported by the Department of Science and Technology, New Delhi, under the ICPS Scheme through letter no. DST/CPS/CLUSTER/IoT/2018/General.

References

1. Fregosi D, Sharmila R, Dusan B, John S, Stephen F, Eric B, Jennifer S, Eric W (2015) A comparative study of DC and AC microgrids in commercial buildings across different climates and operating profiles. In: 2015 IEEE First international conference on DC microgrids (ICDCM), IEEE, pp 159–164
2. Kwasinski A (2010) Quantitative evaluation of DC microgrids availability: effects of system architecture and converter topology design choices. IEEE Trans Power Electron 26(3):835–851
3. Justo JJ, Mwasilu F, Lee J, Jung JW (2013) AC-microgrids versus DC-microgrids with distributed energy resources: a review. Renewable sustainable energy Rev. 24:387–405
4. Dragičević T, Lu X, Vasquez JC, Guerrero JM (2015) DC microgrids—Part II: a review of power architectures, applications, and standardization issues. IEEE Trans Power Electron 31 (5):3528–3549
5. Liu X, Peng W, Poh CL (2011) A hybrid AC/DC microgrid and its coordination control. IEEE Trans smart grid 2(2):278–286
6. Guerrero JM, Vasquez JC, Matas J, De Vicuña LG, Castilla M (2010) Hierarchical control of droop-controlled AC and DC microgrids—A general approach toward standardization. IEEE Trans Industr Electron 58(1):158–172
7. Salomonsson D, Lennart S, Sannino A (2007) An adaptive control system for a DC microgrid for data centers. In: 2007 IEEE Industry applications annual meeting, IEEE, pp 2414–2421
8. Chaudhary P, Suvendu S, Parthasarathi S (2014) Input-series–output-parallel-connected buck rectifiers for high-voltage applications. IEEE Trans Industr Electron 62(1):193–202
9. Jakhar S, Soni MS, Gakkhar N (2016) Parametric modeling and simulation of photovoltaic panels with earth water heat exchanger cooling. Geothermal energy 4(1)

Memristive Computational Amplifiers and Equation Solvers

P. Michael Preetam Raj, Amlan Ranjan Kalita, and Souvik Kundu$^{(\boxtimes)}$

Department of Electrical and Electronics Engineering, Birla Institute of
Technology and Science Pilani, Hyderabad Campus, Hyderabad 500078, India
{p20160017, souvik.kundu}@hyderabad.bits-pilani.ac.in

Abstract. Computational amplifiers are extremely useful for generating
waveforms and solving numerous equations. In this work, memristive computational amplifier circuits were developed on spice simulator platform for the
generation of step, pulse, exponential, and parabolic signals. On one side,
decaying characteristics have been obtained based on the controlled decrement
(or increment) in the memristance when the memristor is connected in the output
(or input) loop of the amplifier. On the other side, rising characteristics were
generated through exchanging the polarity of the input applied signal. These
characteristics were further employed to solve exponential, linear, and parabolic
equations. External voltage signals and internal circuit resistances were utilized
to control the signal parameters such as rise time, fall time, delay, and amplitude.
In the proposed circuits, the extension or reduction in the range of the generated
signals was made possible through adjusting the external bias voltages. This
work paves the way for futuristic low power, improved latency, and reduced
on-chip area-based computational memristive amplifiers.

Keywords: Memristor · Computational amplifiers · Equation solvers

1 Introduction

Since the inception of the memristor technology, the fields of nonvolatile memory and
computing have experienced enormous research advancements. Thirty years post its
theoretical postulation [1], the first physical memristor had been fabricated by HP
laboratories in 2008 [2, 3]. It is widely regarded as the fourth basic electronic component besides resistor, capacitor, and inductor [1]. It is noteworthy to mention that the
memristance is controllable through utilizing the external voltage signal [4]. When this
external signal is disconnected, the memristor continues to maintain its previously
attained resistance until its applied voltage is reconnected; hence the name memristor,
which means memory resistor [5]. Recent advancements have proved that the hysterical
current–voltage relationship exhibited by memristors accounts for numerous electronic
applications [2, 6, 7]. Prior studies have also proven that, within memristive systems,
variations in individual memristor characteristics can affect the overall circuit and
system performances [8]. In order to address this issue, algorithms have been developed to mimic the high precision switching behavior of memristive systems [9].

© Springer Nature Singapore Pte Ltd. 2020
N. Goel et al. (eds.), *Modelling, Simulation and Intelligent Computing*,
Lecture Notes in Electrical Engineering 659,
https://doi.org/10.1007/978-981-15-4775-1_9

Importantly, memristors have shown laudable resolution in programmable resistance with reduced effect of parasitic components [10].

Although the primary focus of memristive research has been nonvolatile memory [2, 3], various important applications have also been realized in the fields of artificial neural networks, image processing, digital logic systems, and analog circuit design [7, 11–13]. In order to facilitate the development of advanced electronic applications, SPICE models for memristors were proposed in the simulation environment [14–16]. It is important to mention that memristors have shown tremendous potential to act as operational and computing elements. These devices have been utilized in the development of programmable threshold comparator, gain amplifier, etc. [13, 17]. Furthermore, memristor-based circuits have closely mimicked the transfer characteristics of operational amplifier circuits [18]. Interestingly, memristors were utilized for mathematical operations such as multiplication and division, through the employment of memristive crossbar array and its analog characteristics [19, 20]. These analog computation-based operations are extremely essential in the fields of signal processing, and control systems [21]. Memristors along with capacitors and inductors have also been utilized to design systems which can be tuned to provide under-damped or over-damped response [22]. Variable first-order and second-order filters using memristors were modeled, which are inexpensive and multifunctional when compared to the conventional transistor-based systems [23]. In addition, adaptive filters with improved quality factors have also been developed [24]. These important advancements in memristive technology were accompanied with improvement in terms of on-chip area ($\sim 40\%$ improvement), power consumption (hundredth times), and latency (<1 ns), when compared to the widely used transistor technology [25–28]. Despite all the above-mentioned technological advancements in the fields of memristive circuit and systems, some of the computational capabilities of memristors have been less explored. Importantly, to the best of our knowledge, no one has yet developed the memristive characteristics-based improved computational amplifiers, signal generators, and equation solvers.

This work proposes memristor-based analog circuits for computational applications. Step, pulse, exponential, and parabolic signals were generated with tunable parameters. Within the signals, the decrement of voltage with respect to time was obtained based on the memristor location in the amplifier circuit. On the other side, rising characteristics were generated through the exchange of the polarity of input external signal. Based on these characteristics, exponential, linear, and parabolic equations were solved for real-time applications.

2 Simulation Details

LTspice XVII software has been utilized to simulate the computational amplifier circuits. The simulation model of memristor is based on a bilayered 10 nm TiO_2 thin film, which was electronically accessed through utilizing two platinum electrodes [2]. This device was first fabricated by Hewlett–Packard laboratories, and the same was incorporated into LTspice simulator [2, 15].

3   Results and Discussions

A memristive amplifier circuit for the generation of step signal is shown in Fig. 1a. The voltage–time characteristics at the output terminal (V(out)) were considered for three different resistance values of 200, 500, and 700 Ω as depicted in Fig. 1b. Interestingly, the delay time and rise time of the step signal varied with the value of the resistance R1. This is owing to the fact that increase in the value of R1 causes the reduction in current through the memristor (same current flows through R1 and memristor which are placed in input and output loops, respectively), which in turn leads to the slower drift of memristance. One can observe that the general shape of these characteristics follows the voltage–time relation as given in the below equation.

$$v = k(1 - e^{-at-n}) - m \tag{1}$$

where v represents the voltage, t denotes the time, k, a, m and n are variables which fit the equation with the graphs as shown in Fig. 1b. The value of the k can be tuned through adjusting the voltage source V2 as the maximum step height is limited by the applied external signal. Furthermore, an external positive voltage can be utilized to shift the voltage range (shown in Fig. 1b) into the positive polarity. The variables a, m, and n are controllable through adjusting the value of R1 since memristance varies with the current flowing through it, which is ultimately responsible for the obtained characteristics. One can utilize the circuit in Fig. 1a as a step signal generator for analog or

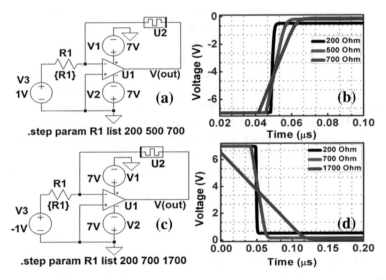

Fig. 1 a Memristive step signal generator with variable delay time and rise time, and **b** its voltage–time characteristics. These characteristics were utilized to solve positive sloped linear and exponentially rising equations. **c** Pulse signal generation with variable ON durations and fall times. **d** Generated pulse signal. The signal features were employed to solve positive sloped linear and exponentially decaying relations

digital electronic applications. On the other side, during the rise time, the voltage–time characteristics (Fig. 1b) can be approximated as straight lines with tunable slopes. When one needed to solve the equations of straight lines, it was made possible through these characteristics. Firstly, the x–y relation of the given straight line was matched with the said voltage–time characteristics, through adjusting the values of voltage sources $V1$, $V2$, and $V3$ as well as the resistor $R1$. Thereafter, the x value was considered as time and the corresponding measured voltage represented the required y value. In this manner, the equation of the line was solved to find y value from the given x value. It is notable to mention that the generation of step signal and the straight line solving were possible owing to the movement of oxygen vacancies within the memristor [2].

Modifications to the circuit in Fig. 1a yield different computational applications. Such an attempt shown in Fig. 1c comprises of a memristive computational amplifier-based pulse signal generator and its waveform. It can be seen that different values of the resistor $R1$ yielded different ON durations and fall times owing to the varied rates of drifts in memristance. It is noteworthy to mention that the decrement or increment in the value of the memristance depends on the polarity of its connection toward the external voltage source. In Fig. 1d, the shape of the characteristics is owing to the decrement of memristance with respect to time. The value of the pulse height is adjustable through utilizing the voltage source $V1$. This circuit was utilized to solve equations with delayed exponentially decaying response with variable rate of decay. One can note that the inversion in the polarity of the output voltage when compared to the input voltage (in Fig. 1a, c) owes to the inverting nature of the amplifier circuit. In Fig. 2a, a memristive inverting amplifier circuit was utilized to generate pulse signals of variable pulse height. The height of the pulse was tunable by adjusting the resistance $R1$. The obtained characteristics (Fig. 2b) are owing to the fact that the gain of the amplifier is limited by the value of $R1$. Once the drift in memristance has reached its termination, the output depends only on the value of $R1$. Hence, three different lines parallel to the time axis were obtained corresponding to the three values of $R1$. Importantly, the decrease in memristance with respect to time was responsible for the signal characteristics. Interestingly, one can utilize a positive external offset voltage in order to shift the signal into the positive voltage domain. The range of the pulse height was tunable through adjusting the value of $V2$. When one needed to solve an x–y relation, which follows the shape of a pulse, it was achieved through utilizing the circuit in Fig. 2a. The y value was obtained by considering the input x value as the corresponding time. The ON duration of the pulse is controllable by utilizing the value of $V1$. This is owing to the fact that low voltages cause slower drift of memristance, whereas high-speed memristance switching occurs at larger voltages.

In a different case, two memristances have been connected in an inverting amplifier (shown in Fig. 2c), and the voltage–time characteristics are depicted in Fig. 2d. In this circuit, memristance $U2$ decreases with respect to time while $U3$ increases based on the external bias signal. The net result led to an increase in the value of the voltage $V(\text{out})$ with respect to time. Analogous to the case in Fig. 1a, this circuit was utilized to solve exponential equations. However, the parameters of the generated signal were adjustable through utilizing the voltage sources $V2$ and $V3$ (depicted in Fig. 2c). It is noteworthy to mention that the characteristics saturate to a constant value, once the drift in memristance

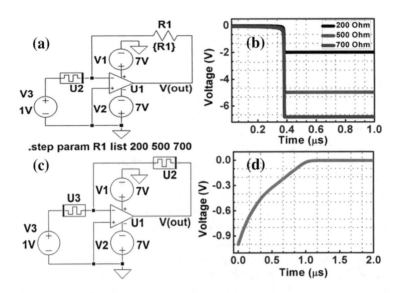

Fig. 2 Generation of pulse signals with variable amplitude. **a** Pulse generator circuit, and **b** its output signals. The pulse height was controlled through utilizing the value of the resistance in the output loop. **c** Memristive exponentially rising signal generator, and **d** its output voltage. These characteristics were utilized to solve $v = k(1 - e^{-at}) + n$

has terminated. This common phenomenon was observed in all the graphs reported in this work. In contrary to the previous cases, a parabolic relation between the voltage and time was obtained for the circuit shown in Fig. 3a. The equation is as given below:

$$v = kt^2 \tag{2}$$

In this circuit, the memristance $U2$ has a decrementing tendency, while $U3$ was increased with respect to time. Owing to the ratio of the two oppositely acting memristive drifts, the net effect resulted in the voltage being incremented proportional to the square of the time. When one needed to perform square of a number, it was achieved by tuning $k = 1$, and measuring voltage V(out) at the time value for which the square operation has to be performed. It is important to mention that k value is adjustable through utilizing $V3$. On the other side, $V1$ (external bias) limits the maximum voltage for which the square operation can be performed. Beyond this range, the output voltage V(out) saturates to a constant value as shown in Fig. 3b. Importantly, the termination of the drift in memristance is responsible for the saturation of the value, as discussed in the previous case. Based on the modifications to the circuit in Fig. 3a, exponentially decaying signal was obtained by utilizing the circuit in Fig. 3c. In this case, $U3$ tends to increase when $U2$ reduces with time, which results in the signal shown in Fig. 3d. It is important to mention that the curve in Fig. 3d follows the relation:

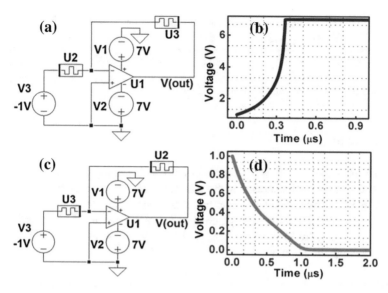

Fig. 3 **a** Step signal generator, and **b** its parabolic rise time characteristics. These features have been utilized to solve $v = kt^2$. **c** Circuit for generation of an exponentially decaying signal, and **d** the obtained voltage–time characteristics. This behavior was employed to solve $v = ke^{-at}$

$$v = ke^{-at} \tag{3}$$

The signal parameters k and a are controllable through adjusting the voltage sources shown in Fig. 3c. The saturation of the voltage signal is due to the same reason mentioned in the earlier cases.

The memristive circuits proposed in this work are advantageous over their conventional transistor-based counterparts (as given in Table 1) owing to the memristive high circuit density ($\sim 40\%$ improvement), reduced latency (<1 ns), and low power dissipation ($\sim 1\%$ of transistor-based technology) [25–27]. Interestingly, the faster switching within the practical memristors does not create any notable delay in the overall circuit performances. While different practical memristors operate in dissimilar resistance ranges [25–27], the common phenomena of the high-speed drift in the value of memristance conveniently supports there employment in the proposed circuits. When one utilizes different memristors, the corresponding scaling along the x and y-axes needs to be considered. Although researchers have identified that memristors exhibit high variability in its device performances, the characteristics of the practical circuits shall perform reliably through the employment of randomness compensation circuits. Hence, the circuits proposed in this work are highly reliable.

Table 1 Comparison between memristor- and transistor-based circuits

Feature	Transistor	Memristor
Circuit density	1400 devices/unit area	1000 devices/unit area
Latency	In μs	<1 ns
Power consumption	A fraction of Watt	Few milliwatt

4 Conclusions

This work establishes memristive amplifier circuits for the generation of step, pulse, exponential, and parabolic signals, with variable parameters. The circuit characteristics were utilized to solve exponential, parabolic, and linear equations. These applications were made possible primarily through controlling memristance. The ideas proposed in this work pave way for low power, improved latency, and reduced on-chip area based futuristic computational amplifiers.

Acknowledgements. One of the authors P. M. P. Raj acknowledges UGC, Govt. of India, for Ph.D. fellowship support through NET JRF (3509/(OBC) (NET-JAN 2017)).

References

1. Chua LO (1971) Memristor—the missing circuit element. IEEE Trans Circ Theory 18 (5):507–519. https://doi.org/10.1109/TCT.1971.1083337
2. Strukov DB, Snider GS, Stewart DR, Williams RS (2008) The missing memristor found. Nature 453(7191):80–83. https://doi.org/10.1038/nature06932
3. Williams RS (2014) How we found the missing memristor. In: Memristors and memristive systems. Springer, New York, pp 3–16. https://doi.org/10.1007/978-1-4614-9068-5_1
4. Raj PMP, Subramaniam A, Priya S, Banerjee S, Kundu S (2019) Programming of memristive artificial synaptic crossbar network using PWM techniques. J Circ Syst Comput 28(12):S0218126619502013. https://doi.org/10.1142/S0218126619502013
5. Chua L (2011) Resistance switching memories are memristors. Appl Phys A 102(4):765–783. https://doi.org/10.1007/s00339-011-6264-9
6. Raj PMP, Ranjan Kalita A, Hudait MK, Priya S, Kundu S (2018) Nonlinear DC equivalent circuits for ferroelectric memristor and its FSM application. Integr Ferroelectr 192(1):16–27. https://doi.org/10.1080/10584587.2018.1521667
7. Kvatinsky S, Belousov D, Liman S, Satat G, Wald N, Friedman EG, Kolodny A, Weiser UC (2014) MAGIC—memristor-aided logic. IEEE Trans Circ Syst II Express Briefs 61 (11):895–899. https://doi.org/10.1109/TCSII.2014.2357292
8. Yakopcic C, Taha TM, Subramanyam G, Pino RE (2013) Generalized memristive device SPICE model and its application in circuit design. IEEE Trans Comput Des Integr Circ Syst 32(8):1201–1214. https://doi.org/10.1109/TCAD.2013.2252057
9. Alibart F, Gao L, Hoskins BD, Strukov DB (2012) High precision tuning of state for memristive devices by adaptable variation-tolerant algorithm. Nanotechnology 23 (7):075201. https://doi.org/10.1088/0957-4484/23/7/075201

10. Shin S, Kim K, Kang S-M (2009) Memristor-based fine resolution programmable resistance and its applications. In: 2009 International conference on communications, circuits and systems. IEEE, pp 948–951. https://doi.org/10.1109/ICCCAS.2009.5250376

11. Zamarreño-Ramos C, Camuñas-Mesa LA, Pérez-Carrasco JA, Masquelier T, Serrano-Gotarredona T, Linares-Barranco B (2011) On spike-timing-dependent-plasticity, memristive devices, and building a self-learning visual cortex. Front Neurosci 5:26. https://doi.org/10.3389/fnins.2011.00026

12. Merrikh-Bayat F, Bagheri Shouraki S, Merrikh-Bayat F (2014) Memristive fuzzy edge detector. J Real-Time Image Process 9(3):479–489. https://doi.org/10.1007/s11554-012-0254-9

13. Pershin YV, Di Ventra M (2010) Practical approach to programmable analog circuits with memristors. IEEE Trans Circ Syst I Regul Pap 57(8):1857–1864. https://doi.org/10.1109/TCSI.2009.2038539

14. Mahvash M, Parker, AC (2010) A memristor SPICE model for designing memristor circuits. In: 2010 53rd IEEE international midwest symposium on circuits and systems. IEEE, pp 989–992. https://doi.org/10.1109/MWSCAS.2010.5548803

15. Biolek Z, Biolek D, Biolková V (2009) Spice model of memristor with nonlinear dopant drift. Radioengineering 18(2):210–214

16. Kvatinsky S, Ramadan M, Friedman EG, Kolodny A (2015) VTEAM: a general model for voltage-controlled memristors. IEEE Trans Circ Syst II Express Briefs 62(8):786–790. https://doi.org/10.1109/TCSII.2015.2433536

17. Shin S, Kim K, Kang S-M (2011) Memristor applications for programmable analog ICs. IEEE Trans Nanotechnol 10(2):266–274. https://doi.org/10.1109/TNANO.2009.2038610

18. Yu Q, Qin Z, Yu J, Mao Y (2009) Transmission characteristics study of memristors based Op-Amp circuits. In: 2009 International conference on communications, circuits and systems. IEEE, pp 974–977. https://doi.org/10.1109/ICCCAS.2009.5250356

19. Bickerstaff K, Swartzlander EE (2010) Memristor-based arithmetic. In: 2010 Conference record of the forty fourth Asilomar conference on signals, systems and computers. IEEE, pp 1173–1177. https://doi.org/10.1109/ACSSC.2010.5757715

20. Merrikh-Bayat F, Shouraki SB (2011) Memristor-based circuits for performing basic arithmetic operations. Procedia Comput Sci 3:128–132. https://doi.org/10.1016/J.PROCS.2010.12.022

21. Mouttet B (2009) Proposal for memristors in signal processing. Presented at the nano-net, pp 11–13. https://doi.org/10.1007/978-3-642-02427-6_3

22. Xu K-D, Zhang YH, Wang L, Yuan MQ, Fan Y, Joines WT, Liu QH (2014) Two memristor SPICE models and their applications in microwave devices. IEEE Trans Nanotechnol 13(3):607–616. https://doi.org/10.1109/TNANO.2014.2314126

23. Li L, Chew ZJ (2012) Printed circuit board based memristor in adaptive lowpass filter. Electron Lett 48(25):1610–1611. https://doi.org/10.1049/el.2012.2918

24. Driscoll T, Quinn J, Klein S, Kim HT, Kim BJ, Pershin YV, Di Ventra M, Basov DN (2010) Memristive adaptive filters. Appl Phys Lett 97(9):093502. https://doi.org/10.1063/1.3485060

25. Knag P, Lu W, Zhang Z (2014) A native stochastic computing architecture enabled by memristors. IEEE Trans Nanotechnol 13(2):283–293. https://doi.org/10.1109/TNANO.2014.2300342

26. Yilmaz Y, Mazumder P (2017) A drift-tolerant read/write scheme for multilevel memristor memory. IEEE Trans Nanotechnol 16(6):1016–1027. https://doi.org/10.1109/TNANO.2017.2741504

27. Zidan MA, Fahmy HAH, Hussain MM, Salama KN (2013) Memristor-based memory: the sneak paths problem and solutions. Microelectr J 44(2):176–183. https://doi.org/10.1016/j.mejo.2012.10.001
28. Jo K-H, Jung C-M, Min K-S, Kang S-M (2010) Self-adaptive write circuit for low-power and variation-tolerant memristors. IEEE Trans Nanotechnol 9(6):675–678. https://doi.org/10.1109/TNANO.2010.2052108

Low Complexity DCT Approximation Algorithm for HEVC Encoder

Sravan K. Vittapu, Uppugunduru Anil Kumar,
and Sumit K. Chatterjee[✉]

EEE Department, Birla Institute of Technology and Science Pilani, Hyderabad
Campus, Jawahar Nagar, Shameerpet, Hyderabad 500078, India
vsravan9l@gmail.com, sumit2702@hyderabad.bits-pilani.
ac.in

Abstract. The Discrete Cosine Transform (DCT) plays a major role in many
video coding standards such as High Efficiency Video Coding (HEVC). In this
paper, a new algorithm that generates low complexity DCT approximation
matrices with minimum number of low-frequency coefficients for all transform
sizes 8, 16 and 32, is proposed. This algorithm accelerates the HEVC encoder in
terms of total encoding time (ΔET) by 39.93% with a small (0.6104%) increase
in bitrate and a small (0.5489 dB) decrease in Peak Signal to Noise Ratio.

Keywords: Discrete Cosine Transform (DCT) · High Efficiency Video Coding
(HEVC) · Quad-Tree · Rate Distortion Optimization (RDO)

1 Introduction

High Efficiency Video Coding (HEVC) is the newest video coding standard authorized
by Joint Collaborative Team on Video Coding (JCT-VC) [1]. HEVC provides twice the
coding efficiency and 50% reduction in bitrate using computationally complex coding
tools such as new asymmetric frame partitions, large transform sizes and new prediction
modes, compared to its successor, the H.264 standard [2]. The frame partitioning hier-
archy of HEVC is a Quad-Tree like structure. Coding Tree Unit (CTU) of size $M \times M$ is
the starting block of the Quad-Tree, where M takes values 64, 32 or 16. Each CTU is split
into Coding Units (CUs) and each CU is recursively partitioned into Prediction Units
(PUs) and Transform Units (TUs) and thus forms the Quad-Tree structure. Each node
(CTUs, CUs, PUs and TUs) in the tree contains one luma and two chroma components.
In HEVC, high compression ratio is achieved for larger blocks for high resolution video
sequences. The block size at each depth in Quad-Tree depends on the Rate Distortion
Optimization (RDO), which is performed for all nodes of the tree. Discrete Cosine
Transform (DCT) [3] plays an important role in several image/video compression
standards. In HEVC, the integer DCT or its inverse (IDCT) supports TUs of sizes 4×4,
8×8, 16×16 and 32×32 [4]. It also supports Discrete Sine Transform (DST) and its
inverse (IDST) for intra mode to further reduce bitrate by 1% [5].

In HEVC, one-fourth of the total Encoding Time (ET) is required for Transform
and Quantization (TcomTrQuant) class [6]. As the DCT satisfies kernel separable

© Springer Nature Singapore Pte Ltd. 2020
N. Goel et al. (eds.), *Modelling, Simulation and Intelligent Computing*,
Lecture Notes in Electrical Engineering 659,
https://doi.org/10.1007/978-981-15-4775-1_10

property [7], 2D DCT in HEVC is implemented by executing two 1D DCTs in a row-column approach [6]. In literature, several DCT approximations have been proposed to reduce the arithmetic complexity [8]. In [9], two low complexity DCT multiplication-free eight-point approximations were proposed, angular similarity based DCT approximation were proposed in [10], and in both papers, the eight-point DCT is scaled to 16 and 32-point using Jridi-Alfalou-Meher (JAM) algorithm [11] for the implementation in HEVC encoder. Suzuki et al. [12] proposed fast multiplication-free transforms for image processing applications. In [13], hardware efficient HEVC transform was implemented. One-bit transform-based low complexity fast search algorithm was proposed in [14]. Adaptive multiple transform-based approximations were proposed in [15] for post HEVC encoding standard. Multiplier-less 2D DCT was implemented in [16]. In [17], DCT architectures with scalable approximations were proposed. Power-efficient 2D DCT was proposed in [18]. In [19], area-efficient IDCT/IDST architecture was introduced for Ultra High Definition (UHD) video sequences. Field Programmable Gate Array (FPGA) implementation of 1D transform for HEVC using DSP slices was proposed in [20]. DCT kernels with reduced bit depths and efficient hardware architecture were proposed in [21]. Hierarchical multiplier-free HEVC transform and its optimized hardware architecture were presented in [22]. High throughput architecture using single 1D IDCT for 32×32 transform unit was proposed in [23]. Efficient hardware architecture of variable block size HEVC 2D DCT for FPGA platforms was designed in [24].

The 1D DCT for an N-point input vector $\mathbf{x} = [x_0, x_1, \ldots, x_{N-1}]^{\mathrm{T}}$ is a linear transformation of $\mathbf{x}$ into transformed output vector $\mathbf{X} = [X_0, X_1, \ldots, X_{N-1}]$, mathematically represented as [7]

$$X_k = \alpha_k \cdot \sqrt{\frac{2}{N}} \cdot \sum_{n=0}^{N-1} x_n \cdot \cos\left\{\frac{(n+1/2)k\pi}{N}\right\} \tag{1}$$

where $k = 0, 1, \ldots, N - 1$. The values for α are given as: $\alpha_0 = \frac{1}{\sqrt{2}}$ and for $k > 0$ $\alpha_k = 1$. In matrix form (1) is given as:

$$\mathbf{X} = \mathbf{C} \cdot x \tag{2}$$

where $\mathbf{C}$ is an N-point DCT matrix and its elements are mathematically entered as:

$$c_{m,n} = \alpha_m \cdot \sqrt{\frac{2}{N}} \cdot \cos\{(n+1/2)m\pi/N\}, \quad m, n = 0, 1, \ldots, N - 1 \tag{3}$$

As the DCT obeys orthogonal property, the IDCT is given as

$$\mathbf{x} = \mathbf{C}^{\mathrm{T}} \cdot \mathbf{X} \tag{4}$$

The 2D DCT is mathematically expressed in terms of 1D DCTs as

$$X = C x \cdot C^T \tag{5}$$

and the 2D IDCT is represented as

$$x = C^T X \cdot C \tag{6}$$

The rest of the paper is organized as follows: in Sect. 2 an overview of eight-point DCT approximation based on angular similarity is provided. The proposed algorithm is discussed in Sect. 3. Section 4 presents the experimental results of the described work using HM16.18 reference software. Conclusions are finally drawn in Sect. 5.

2 Overview of DCT Approximation Based on Angular Similarity

In this method, a low complexity approximate matrix T of size 8×8 is obtained from the row entries from the search space D, where $D = P^8$ [10]. The search space D is a collection of vectors of size 1×8 and each vector element is taken from the set P. In this method, the row vectors of T are entered by searching in two search spaces D_1 and D_2, where $D_1 = P_1^8$, $D_2 = P_2^8$, $P_1 = \{0, \pm 1\}$ and $P_2 = \{0, \pm 1, \pm 2\}$: therefore, the search spaces D_1 and D_2 have 6561 and 390,625 elements, respectively. This method is restricted to eight-point since the generation of T depends on approximation order. As there are eight elements in each vector, $8! = 40,320$ possible permutations are considered. Therefore, this method is not suitable for 16 and 32-points to implement in HEVC encoder as it drastically increases the computational complexity. This method is embedded in HEVC encoder by generating the 16 and 32-point versions of T using JAM scalable algorithm. This algorithm reduces the computational complexity of HEVC Test Model (HM) encoder in terms of additions and bit-shift operations.

3 Proposed Algorithm

The proposed algorithm decreases the computational complexity by reducing the number of coefficients required to compute N-point DCT, where N takes the values 8, 16 and 32. In this algorithm, the low complexity matrix T to compute eight-point DCT is generated using angular similarity [10] and then the 16 and 32 point versions of T are generated using JAM scalable algorithm as shown below:

$$T_{(N)} = \frac{1}{\sqrt{2}} M_N^{per} \begin{bmatrix} T_{(N/2)} & Z_{(N/2)} \\ Z_{(N/2)} & T_{(N/2)} \end{bmatrix} M_N^{add}, \tag{7}$$

where $Z_{(N/2)}$ is a null matrix of order $N/2$, the scaling factor $\frac{1}{\sqrt{2}}$ is merged into the step size of quantization in HM software and the matrices M_N^{per} and M_N^{add} are represented as:

$$\mathbf{M}_N^{\text{add}} = \begin{bmatrix} \mathbf{I}_{(N/2)} & \bar{\mathbf{I}}_{(N/2)} \\ \bar{\mathbf{I}}_{(N/2)} & \mathbf{I}_{(N/2)} \end{bmatrix} \tag{8}$$

and

$$\mathbf{M}_N^{\text{per}} = \begin{bmatrix} \mathbf{P}_{N-1,N/2} & \mathbf{Z}_{1,N/2} \\ \mathbf{Z}_{1,N/2} & \mathbf{P}_{N-1,N/2} \end{bmatrix} \tag{9}$$

where $\mathbf{I}_{(N/2)}$ and $\bar{\mathbf{I}}_{(N/2)}$ are the identity and counter identity matrices of order $N/2$ respectively and $\mathbf{P}_{N-1,N/2}$ is a matrix of order $(N-1) \times (N/2)$ with row vectors given as:

$$\mathbf{P}_{N-1,N/2}^{(i)} = \begin{cases} \mathbf{Z}_{1,N/2}, & \text{if } i = 1, 3, 5, \ldots, N-1 \\ \mathbf{I}_{N/2}^{\left(\frac{i}{2}\right)}, & \text{if } i = 0, 2, 4, \ldots, N-2 \end{cases} \tag{10}$$

After generating the low complexity approximation matrices of transform size 8, 16 and 32, the minimum number of coefficients ($N_{\min}$) algorithm [25] is applied to further reduce the computational complexity of HEVC encoder. The $N_{\min}$ required for computation is evaluated by this algorithm in HM16.18 encoder, the obtained Peak Signal to Noise Ratio (PSNR) is 0.5489 dB smaller and the average bitrate is increased by 0.6104% as compared to the HM reference software algorithm. The coefficients are varied from minimum ($N_{\min}$) to maximum ($N_{\max}$) value in the proposed algorithm. The P_{TH}, R_{TH} are the threshold values of PSNR (P) and bitrate (R), respectively, whose values are P_{TH} is 0.284 dB smaller than minimum PSNR ($P_{\min}$) and R_{TH} take the value 0.6104 times maximum bitrate ($R_{\max}$). By applying this algorithm for different video sequences at Quantization Parameter (QP) values 22, 27, 32 and 37, the average number of coefficients to be computed are $N_{4\min} = 4$, $N_{8\min} = 3$, $N_{16\min} = 6$ and $N_{32\min} = 7$. The functions $G_P(S_n)$ and $G_R(S_n)$ are the PSNR and bitrate values for N-point DCT.

The proposed algorithm is given as follows:

Step 1: Generate the eight-point low complexity matrix $\mathbf{T}$ using angular similarity algorithm.

Step 2: Taking $\mathbf{T}$ as a basic matrix, generate the matrices of large transform sizes (S) = 16 and 32 using JAM scalable algorithm.

Step 3: Apply the $N_{\min}$ algorithm for the generated matrices from Step 1 and Step 2, as shown below:

$P_{\text{TH}} = P_{\max} - 0.5489$;

$R_{\text{TH}} = 0.6104 * R_{\min}$;

$M = P_{max}/R_{min}$, $T = 0$; //initialize

    for $(S = 0, S < 4, S^{++})$
    {

    for $(S = 0, S < 8, S^{++})$
    {
    for $(S = 0, S < 16, S^{++})$
    {
    for $(S = 0, S < 32, S^{++})$

{

P = mean $\{G_P(S_4), G_P(S_8), G_P(S_{16}), G_P(S_{32})\}$ //overall PSNR obtained

R = mean $\{G_R(S_4), G_R(S_8), G_R(S_{16}), G_R(S_{32})\}$ //overall bitrate obtained

Temp = P/R;

If $(T < M$ && $P > P_{TH}$ && $R < R_{TH})$ then

    $N_{4min} = S_4$;

    $N_{8min} = S_8$;

    $N_{16min} = S_{16}$;

    $N_{32min} = S_{32}$;

    end; //end of if statement

    end; //end of 4-point for loop

    end; //end of 8-point for loop

    end; //end of 16-point for loop

    end; //end of 32-point for loop

4   Experimental Results

The proposed algorithm is implemented in HM16.18 reference software using standard test video sequences of class B to E [26]. The encoding process is done at QP values 22, 24, 32 and 37. The performance metrics shown in Table 1 and are mathematically represented as:

$$\text{Average } \Delta\text{Bitrate} = \frac{\text{Bitrate}_{proposed} - \text{Bitrate}_{HM}}{\text{Bitrate}_{HM}} \times 100\% \tag{11}$$

$$\text{Average } \Delta\text{PSNR} = \frac{\text{PSNR}_{proposed} - \text{PSNR}_{HM}}{\text{PSNR}_{HM}} \times 100\% \tag{12}$$

$$\Delta\text{ET} = \frac{\text{ET}_{HM} - \text{ET}_{Proposed}}{\text{ET}_{HM}} \times 100\% \tag{13}$$

The BD-Rate and BD-PSNR are the average change in rate for a fixed PSNR and average change in PSNR for a fixed rate, respectively. The pursuance of the proposed algorithm is tested in Random Access (RA) encoder configuration, as shown in

Table 1 Performance comparison of proposed algorithm with HM16.18 in RA configuration

Class	Video sequence	BD-rate	BD-PSNR	Average ΔBitrate (%)	Average ΔPSNR (%)	Average ΔET (%)
Class B	Kimono	−6.9410	0.1093	0.0743	−0.053	4.987
	Basketball Drive	−7.9525	0.1141	0.0508	−0.071	5.120
Class C	Party Scene	−5.6185	0.1620	0.0114	−0.024	15.32
	BQMall	−3.8775	0.1586	0.0143	−0.016	11.09
	Basketball Drill	−5.6195	0.1434	0.0120	−0.033	13.17
Class D	RaceHorses	−7.9279	0.4915	0.0176	−0.021	10.44
	Blowingbubbles	−3.3043	0.0352	0.0290	−0.003	12.08
	BQSquare	−2.2551	0.1576	0.0203	−0.010	11.63
Class E	FourPeople	−6.3513	0.1582	0.0127	−0.017	10.55
	Johnny	−6.3720	0.1471	0.0335	−0.035	7.981
	KristnAndSara	−4.6743	0.1049	0.0244	−0.011	7.339
	Average	−4.3495	0.1273	0.0214	−0.021	7.863

Table 2 Required computational cost of proposed algorithm DCT Algorithm

N	Algorithm [6]			Algorithm [10]		Algorithm [25]		Proposed algorithm	
	Mul.	Add	Shift	Add	Shift	Add	Shift	Add	Shift
8	20	28	4	24	6	17	7	15	5
16	84	100	8	64	12	54	31	46	10
32	340	372	16	160	24	121	69	97	14

Table 1. The proposed algorithm decreases the computational load of HM encoder in terms of total Encoding Time (ET) by 39.93%, with a small decrement in PSNR of 0.5489 dB and an average bitrate increment of 0.6104%. The proposed algorithm involves less computational cost than the HM encoder [6, 10, 25], as shown in Table 2. Therefore, through the observation of Tables 1, 2 and 3 it may be inferred that the proposed algorithm outperforms the algorithm [25] as compared to HM16.18 in terms of a small decrement in average PSNR of −0.008% and a small increment in the average bitrate of 0.0268% in LDP configuration as shown in Table 3. The RDO

curves of the proposed algorithm, HM and [10] for the video sequences of classes B to D are shown in Fig. 1. The frame of BQSquare video sequence is taken for comparison among the proposed algorithm, HM [6, 10] at QP = 32 and is shown in Fig. 2.

Table 3 Performance comparison of proposed algorithm and algorithm [25] with HM16.18 in LDP configuration

Video sequence	Class	Algorithm [25]		Proposed algorithm	
		Average ΔBitrate (%)	Average ΔPSNR (%)	Average ΔBitrate (%)	Average ΔPSNR (%)
Party Scene	C	0.180	−0.308	0.0150	−0.005
BQMall	C	0.285	−0.216	0.0206	−0.001
Basketball Drill	C	−0.595	−0.245	0.0505	−0.012
Blowingbubbles	D	0.270	−0.163	0.0150	−0.010
BQSquare	D	0.206	−0.212	0.0025	−0.003
FourPeople	E	−0.130	−0.206	0.0383	−0.015
KristnAndSara	E	−0.535	−0.137	0.0457	−0.014
Average		−0.045	−0.212	0.0268	−0.008

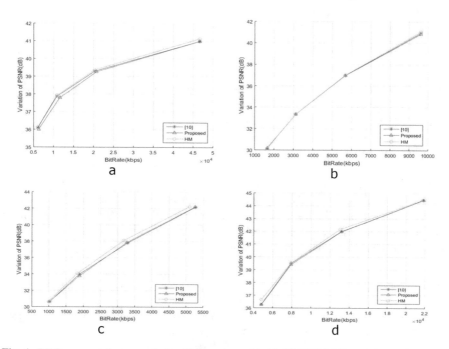

Fig. 1 RDO curves of proposed algorithm compared with HM16.18 [6] encoder and [10] for class B–E video sequences: **a** Basket Ball Drive, **b** Party Scene, **c** Race Horses, **d** KristenAndSara

Fig. 2 Frame comparison of BQSquare sequence: **a** original, **b** encoded output of HM [6], **c** decoded output of HM [6], **d** encoded output of proposed algorithm, **e** decoded output of the proposed algorithm, **f** encoded output of [10], **g** decoded output of [10]

5 Conclusion

In this paper, an algorithm is proposed, which reduces the computational cost and speeds up the HM encoder, thereby outperforms existing algorithms [10, 25]. From the experimental results, it is shown that the proposed algorithm reduces the computational load in terms of total encoding time with a small increment in bitrate and a very small decrease in PSNR compared to the HM encoder.

References

1. Sullivan GJ et al (2012) Overview of the high efficiency video coding (HEVC) standard. IEEE Trans Circ Syst Video Technol 22(12):1649–1668
2. Wiegand T et al (2003) Overview of the H. 264/AVC video coding standard. IEEE Trans Circ Syst Video Technol 13(7):560–576
3. Britanak V, Yip PC, Rao KR (2010) Discrete cosine and sine transforms: general properties, fast algorithms and integer approximations. Elsevier, Amsterdam

4. Bossen F et al (2012) HEVC complexity and implementation analysis. IEEE Trans Circ Syst Video Technol 22(12):1685–1696
5. Sze V, Budagavi M, Sullivan GJ (2014) High efficiency video coding (HEVC). In: Integrated circuit and systems, algorithms and architectures. Springer, Berlin
6. Bossen F, Flynn D, Sharman K, Shring K (2018) HM software manual. Document: JCTVC-Software Manual [Online]. Available: https://hevc.hhi.fraunhofer.de
7. Cintra RJ et al (2016) Energy-efficient 8-point DCT approximations: theory and hardware architectures. Circ Syst Sig Process 35(11):4009–4029
8. Rao KR, Yip P (2014) Discrete cosine transform: algorithms, advantages, applications. Academic Press, Cambridge
9. Tablada CJ et al (2017) DCT approximations based on Chen's factorization. Sig Process Image Commun 58:14–23
10. Oliveira RS et al (2019) Low-complexity 8-point DCT approximation based on angle similarity for image and video coding. Multidimension Syst Sig Process 30(3):1363–1394
11. Jridi M, Alfalou A, Meher PK (2014) A generalized algorithm and reconfigurable architecture for efficient and scalable orthogonal approximation of DCT. IEEE Trans Circ Syst I Regul Pap 62(2):449–457
12. Suzuki T et al (2018) Multiplierless lifting-based fast X transforms derived from fast Hartley transform factorization. Multidimension Syst Sig Process 29(1):99–118
13. Masera M et al (2017) Analysis of HEVC transform throughput requirements for hardware implementations. Sig Process Image Commun 57:173–182
14. Vittapu SK, Chatterjee SK (2018) Complexity reduction for HEVC encoder using multiplication free one-bit transformation. J Electron Imaging 27(6):063028
15. Jdidia SB et al (2018) Low-complexity algorithm using DCT approximation for POST-HEVC standard. In: Pattern recognition and tracking XXIX, Proceedings of SPIE, vol 10649
16. Kulasekera SC et al (2014) Low-complexity multiplierless DCT approximations for low-power HEVC digital IP cores. In: Geospatial infofusion and video analytics IV; and motion imagery for ISR and situational awareness II, Proceedings of SPIE, vol 9089
17. Jridi M, Meher PK (2016) Scalable approximate DCT architectures for efficient HEVC-compliant video coding. IEEE Trans Circ Syst Video Technol 27(8):1815–1825
18. Kalali E, Mert AC, Hamzaoglu I (2016) A computation and energy reduction technique for HEVC discrete cosine transform. IEEE Trans Consum Electron 62(2):166–174
19. Liang H et al (2016) Area-efficient HEVC IDCT/IDST architecture for 8K × 4K video decoding. IEICE Electron Express 13(6):20160019–20160019
20. Arayacheeppreecha P, Pumrin S, Supmonchai B (2015) 1-D integer transform for HEVC encoder using DSP slices on FPGA. Appl Mech Mater 781:151–154
21. Harize S et al (2018) HEVC transforms with reduced elements bit depth. Electron Lett 54 (22):1278–1280
22. Fan C et al (2017) A hierarchical multiplier-free architecture for HEVC transform. Multimedia Tools Appl 76(1):997–1015
23. Chen Y-H, Ko Y-F (2017) High-throughput IDCT architecture for high-efficiency video coding (HEVC). Int J Circuit Theory Appl 45(12):2260–2269
24. Chen M, Zhang Y, Lu C (2017) Efficient architecture of variable size HEVC 2D-DCT for FPGA platforms. AEU-Int J Electron Commun 73:1–8
25. Singhadia A, Bante P, Chakrabarti I (2019) A novel algorithmic approach for efficient realization of 2-D-DCT architecture for HEVC. IEEE Trans Consum Electron 65(3):264–273
26. Test sequences (2012). ftp://tnt.uni-hannover.de

An Area and Power-Efficient Serial Commutator FFT with Recursive LUT Multiplier

Jinti Hazarika$^{(\boxtimes)}$ [ID], Mohd Tasleem Khan [ID], Shaik Rafi Ahamed [ID],
and Harshal B. Nemade

IIT Guwahati, Guwahati, Assam, India
`jinti@iitg.ac.in`

Abstract. This paper presents an area and power-efficient architecture for serial commutator real-valued fast Fourier transform (FFT) using recursive look-up table (LUT). FFT computation consists of butterfly operations and twiddles factor multiplications. The area and power performance of FFT architectures are mainly limited by the multipliers. To address this, a new multiplier is proposed which stores the partial products in LUT. Moreover, by adding the shifted version of twiddle coefficients, the stored partial products gain symmetry, and thus the size of LUT can be reduced to half. Further symmetry is achieved by adding another shifted version of twiddle coefficients and so on. This makes the proposed LUT multiplier recursive in nature. A new data management scheme is suggested for the proposed architecture. To validate the proposed architecture, application-specific integrated circuit (ASIC) synthesis and field-programmable gate array (FPGA) implementation are carried out for different symmetry factor. For instance, the proposed architecture for 1024-point with symmetry factor of two achieves 39.11% less area, 42.29% less power, 33.27% less sliced LUT (SLUT) and 29.18% less flip-flop (FF) as compared to the best existing design.

Keywords: Fast Fourier transform · Recursive LUT multiplier

1 Introduction

Fast Fourier transform (FFT) is widely used in transforming a signal from time-domain to frequency-domain which has key applications in bio-medical, orthogonal frequency division multiplexing (OFDM), ultrawide-band (UWB) [1–3], etc. Many practical applications such as echo cancellation, etc. require high-point FFT computations to provide better performance. However, this comes under huge amount of computational resources. FFT processors are desired to occupy low-area, consume low-power and provide high-speed for real-time applications. In this context, pipelined FFT architectures are preferred as they are suitable for real-time processing and can simultaneously meet the architecture efficiency in terms of area, power and throughput.

Pipelined FFT architectures are divided into two categories: feedforward and feedback architectures. Depending on the number of samples processed per clock cycle, they can again be divided into serial (or single-path) and parallel (or multi-path)

© Springer Nature Singapore Pte Ltd. 2020
N. Goel et al. (eds.), *Modelling, Simulation and Intelligent Computing*,
Lecture Notes in Electrical Engineering 659,
https://doi.org/10.1007/978-981-15-4775-1_11

architectures. This leads to four types of FFT architectures: single-path delay feedback (SDF), single-path delay commutator (SDC), multi-path delay feedback (MDF), and multi-path delay commutator (MDC) [4–6]. Parallel FFT architectures are often employed when the throughput desired is beyond that of serial FFT architectures at a given clock frequency but they require high area-power due to large number of multipliers involved. Therefore, serial architectures are suitable in certain applications which have area-power constraints. However, when compared with parallel architectures, serial architectures have low hardware utilization. In this work, we focus on the SDC FFT architecture to increase its hardware efficiency.

For low-power applications such as energy-aware systems, IoT, etc., serial FFT architectures are desired [7, 8]. The performance of such architectures is mainly limited by the size of multipliers. High-radix FFT is a suitable approach to reduce the multipliers as the number of twiddle factor coefficients decreases with increase in radix size, however it increases the overall system complexity. Conventional concepts such as distributed arithmetic (DA) are commonly used to reduce the complexity of multipliers in the FFT architectures. DA based implementations generally have low-area and low-power consumption at the cost of additional memory units. In [9], the exponential growth of look-up table (LUT) size in DA based designs has been reduced to linear. This will reduce the number of transistors for the implementation of any digital design. According to semiconductor roadmap [10], LUT-based design is an attractive solution. This is due to advancements in semiconductor memories which results in shorter LUT access time, high-throughput and reduced-latency. This motivates us to design a multiplier using LUT for serial FFT architecture to process real-valued signals.

The rest of this paper is organized as follows: Sect. 2 presents details of the proposed scheme. Section 3 compares the proposed architecture with existing architectures and validates the proposed FFT through implementation results. Section 4 gives the conclusions drawn from the presented work.

2   Proposed Scheme

An N-point DFT $X(k)$ of a discrete-time sequence $x(m)$ can be expressed as

$$X(k) = \sum_{m=0}^{N-1} x(m) W_N^{mk}; \quad 0 \leq m \leq N-1 \text{ and } 0 \leq k \leq N-1 \qquad (1)$$

where W_N^{mk} is the twiddle factor. The computation of FFT involves several adders and multipliers. Among them, the twiddle factor multipliers are the main source of power consumption and chip-area and pose as a bottleneck in the design of FFT processors. It is therefore important to develop a multiplier that could result in low-area and low-power for efficient realization of FFT.

2.1   Formulation of the Proposed Multiplier

Consider a B-bit intermediate input $x = b_{B-1}b_{B-2}\ldots b_0$ and A_n be the twiddle coefficient of nth stage with $n = \log_2 N$ and N denotes the number of FFT points. By writing x in binary form, we get

$$x = \sum_{i=0}^{B-1} b_i 2^i = 2 \sum_{i=0}^{B-2} b_{i+1} 2^i + b_0 \tag{2}$$

By multiplying x with coefficient A_n, the intermediate output I_n can be computed as

$$I_n = \left[2 \sum_{i=0}^{B-2} b_{i+1} 2^i + b_0 \right] A_n \tag{3}$$

From (3), it can be observed that I_n takes the value of all multiples $\{0, A_n, 2A_n, \ldots, (2^B - 1)A_n\}$. Assuming $r_{B-1} = \sum_{i=0}^{B-2} b_{i+1} 2^i$ and $b_0 = 1$ which means I_n would take odd multiples $\{A_n, 3A_n, \ldots, (2^{B-1} - 1)A_n\}$. Even multiplies can be obtained by left-shifting selected odd multiples, for instance, $\{1 \ll A_n = 2A_n, 2 \ll A_n = 4A_n, \ldots, 1 \ll (2^{B-1} - 1)A_n = 2 \cdot (2^{B-1} - 1)A_n\}$. Pre-computing and storing only the odd multiples in LUT would be an advantage. This is because each of them requires a separate memory location, unlike [9] where a separate adder is needed for the generation of each odd-multiple. However, in such case the size of LUT can be a performance bottleneck for large values of B. To address this issue, $2^{B-2}A_n$ is added to and subtracted from (3) which leads to

$$I_n = [2r_{B-1} + 1]A_n - 2^{B-2}A_n + 2^{B-2}A_n \tag{4}$$

Combining the first two terms of (4) would result in mirror symmetries between the odd multiples with an offset $+2^{B-2}A_n$. Thus, (4) becomes

$$I_n = [2r_{B-2} + 1]A_n + 2^{B-2}A_n = I_n^1 + 2^{B-2}A_n \tag{5}$$

where $r_{B-2} = r_{B-1} \oplus b_{B-2} = \sum_{i=0}^{B-3} b_i 2^i$ and $b_i = b_{i+1} \oplus b_{B-2}$ and '$\oplus$' is XOR operation. It is clear from (5) that the size of LUT is reduced to 2^{B-2} with one adder and few XOR gates. The odd multiples become $\{A_n, 3A_n, \ldots, (2^{B-2} - 1)A_n\}$. Similarly, when $2^{B-3}A_n$ is added to and subtracted from (5), we get

$$I_n = I_n^2 + 2^{B-3}A_n + 2^{B-2}A_n \tag{6}$$

Again, there are mirror symmetries between odd multiples with an offset $+2^{B-3}A_n$. Now, I_n would take values from the set $\{A_n, 3A_n, \ldots, (2^{B-3} - 1)A_n\}$. Thus, for any recursive factor γ, (6) can be written as

$$I_n = I_n^{\gamma} + 2^{B-\gamma-1}A_n + 2^{B-\gamma-2}A_n \cdots + 2^{B-2}A_n \tag{7}$$

where $I_n^{\gamma} = [2r_{B-1-\gamma} + 1]A_n, r_{B-2-\gamma} = \{\ldots\{\{r_{B-1} \oplus b_{B-2}\} \oplus b_{B-3}\}\ldots \oplus b_{B-1-\gamma}\} = \sum_{i=0}^{B-2-\gamma} b_{i+1}2^i$ and $b_i = \{\ldots\{\{b_{i+1} \oplus b_{B-2}\} \oplus b_{B-3}\}\ldots \oplus b_{B-1-\gamma}\}$. In general, the odd multiples would take values from the set $\{A_n, 3A_n, \ldots, (2^{B-1-\gamma} - 1)A_n\}$. Two approaches can be used to realize (7). First, the odd multiples can be stored in LUT at the cost of single adder delay. This gives the advantage of reduced LUT size. Second, A_n can be pre-computed and stored with the odd-multiples. But, this requires A_n to be translated through multi-input XOR operation $\{\ldots\{\{A_n \oplus b_{B-2}\} \oplus b_{B-3}\}\ldots \oplus b_{B-1-\gamma}\}$. However, this would increase the LUT size but solves the delay problem. In this paper, first approach is considered where the delay of the proposed multiplier is reduced by the method of pipelining in serial FFT architecture implementation.

The comparison of different LUT multipliers storing all multiples, odd multiples [11], and recursive (R) odd multiples in terms of number of transistors and access delay is shown in Fig. 1. It is found that the proposed R-odd LUT multiplier offers significant savings in number of transistors with diminishing effect on LUT access delays which has less effect for higher values of γ. We have synthesized all the multipliers for different input width by Cadence RTL Compiler using TSMC 90 nm library. The corresponding synthesized results are listed in Table 1. The savings in the area in the proposed multiplier results from lower storage requirements. Moreover, on increasing the symmetry factor γ, the savings in the area can be significant. For instance, the savings in the area for the proposed multiplier when $B = 16$, $\gamma = 2$ and $B = 16$, $\gamma = 4$ are 55.25% and 81.08% with respect to all odd-multiple LUT design [11] respectively.

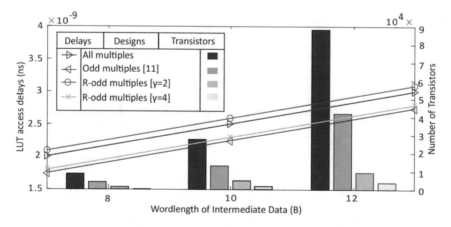

Fig. 1 Access delays and number of transistors for all multiples, odd multiples and R-odd multiples pre-computed and stored in LUT

Table 1 Area comparison of different LUT multipliers

Input width B	All multiples	Odd multiples [11]	Proposed design	
			$\gamma = 2$	$\gamma = 4$
8	956.8	538.4	209.2	87.7
16	1611.3	858.7	384.3	162.5
24	2213.2	1172.0	543.9	225.4
32	2836.2	1488.1	723.5	331.9

2.2 Data Management Scheme (DMS)

The proposed DMS for serial computation of 16-point real FFT with normal order input-output is shown in Fig. 2. The re-ordering and computation of input data, intermediate data and output data at different stages along with their time indices 't' are explicitly shown. It is clear that most of the FFT stages require data re-ordering (DR), butterfly (BF) operation and quarter complex multiplication (QCM). However, last two stages require only DR and BF operations.

2.3 FFT Architecture

The architecture for 16-point FFT consists of eight DR units, two types of BF units-BFI, BFII, two QCM units and a BF-cum-DR unit. The DR and BFI units employed in the proposed architecture are identical to that of [8]. The detailed explanation of each of these units is described below.

Architectural Details of DR Unit DR circuits are required to perform DR operations shown in Fig. 2. DR unit shown in Fig. 3a basically contains two 2-to-1 multiplexers separated by buffers for data shuffling. If the control of multiplexer 's' is set to 1, then the sample is passed through the buffer, whereas if it is set to 0, then the input sample at that instant is interchanged with the sample 'L' clock cycles apart, where 'L' is the stage-dependent buffer length (indicated by 'Lat' in Fig. 2).

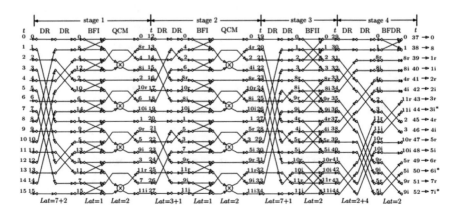

Fig. 2 Data management scheme for a 16-point real-valued FFT

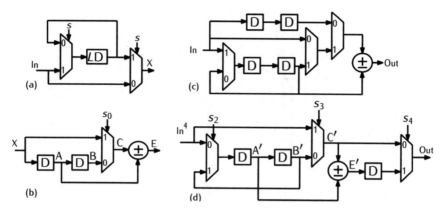

Fig. 3 **a** Circuit for DR unit with buffer length L [8]. **b** Circuit for BFI unit [8]. **c** Circuit for BFII unit. **d** Circuit for stage 4 with combined BF and DR units

Architectural Details of BF Unit Two types of BF units—BFI, BFII are employed in the proposed architecture. BFI unit shown in Fig. 3b carries out addition or subtraction in a given clock cycle in stage 1 and 2. BFII shown in Fig. 3c is newly proposed to carry out two BF operations in an interleaved manner in stage 3 as per the DMS in Fig. 2. Unlike BFI which computes the BF operation with its adjacent data, the BFII unit computes the BF operation with a data two clock cycles apart. This reduces the latency in the DR units of the subsequent stage. Interestingly, the latency savings in DR units increase with N at the cost of two registers.

Architectural Details of QCM As the name suggests, QCM unit performs the quarter operation of complex multiplication in a given clock cycle and is shown in Fig. 4. It requires four 2-to-1 multiplexers, five registers, one decoder, three XOR gates, one barrel shifter (BS) and two adders. The real multiplier consists of R-LUT with word size $(W + B - \gamma)$-bits and external logic (decoder, BS unit and XOR gates), where W is the word length of A_n. The output of decoder becomes the address lines of R-LUT storing the odd multiples. To generate even-multiples, BS provides variable left-shifts on the selected contents of R-LUT. The size of BS can be determined by the maximum number of left-shifts required on the odd-multiple. Note the AND gate before BS unit introduces '0' when all the bits of intermediate input x of a particular stage become zero.

Architectural Details of BF-cum-DR Unit In the last stage, a combined unit for BF and DR is employed to perform any one of the three operations in one clock cycle: a half-BF computation, an interchange of samples two clock cycles apart or pass a sample through the buffer, as depicted in Fig. 3d. A timing diagram is provided in Table 2 to understand its operation in the last stage. The samples with indices 0 and 1 undergo BF computation in the BF-cum-DR unit after two stages of DR with $L = 2$ followed by $L = 4$. After the BF operation is over, the rest of the samples are either passed through the buffer with a latency of 2 or they are interchanged with a sample two clock cycles apart to achieve a normal order FFT output.

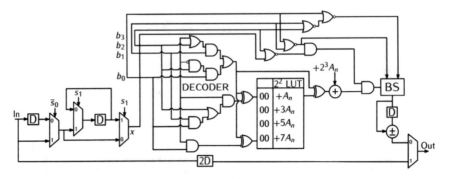

Fig. 4 Circuit for QCM unit with R-LUT multiplier. BS Barrel shifter

Table 2 Timing diagram of BF-cum-DR unit in stage 4

Clk	I_n^4	s_2	A'	B'	s_3	C'	E'	s_4	Out
35	x_0^3	0	–	–	0	–	–	–	–
36	x_1^3	0	x_0^3	–	1	x_1^3	x_0^4	1	–
37	x_{8r}^3	0	x_1^3	x_0^3	0	x_0^3	x_1^4	1	x_0^4
38	x_{8i}^3	0	x_{8r}^3	x_1^3	0	x_1^3	–	1	x_1^4
39	x_{4r}^3	0	x_{8i}^3	x_{8r}^3	0	x_{8r}^3	–	0	x_{8r}^4
40	x_{4i}^3	0	x_{4r}^3	x_{8i}^3	0	x_{8i}^3	–	0	x_{8i}^4

x_m^n: output of stage n with sample index m

3 Results and Discussion

3.1 Computational Complexities

The hardware complexities of the proposed and existing designs are listed in Table 3. It is important to note that when the word length of input becomes high, the recursive LUT size is still high. To avoid this issue, the input word length B is split into K groups of size B_K such that $B = B_K K$. In general, a shift-adder tree of depth $\log_2 K$ would result due to splitting of B into K sub-LUT units. Each B_K is individually processed in sub-LUT units, where the external addition of shifted twiddle coefficients is controlled by γ.

3.2 Implementation Results

ASIC Synthesis In order to validate the proposed design, it is coded in Verilog along with the existing designs. ASIC synthesis is carried out by Cadence RTL Compiler at 1.0 V and clock frequency of 100 MHz. The obtained results in terms of area, power, minimum clock period (MCP), area-delay product (ADP) and power-delay product (PDP) are listed in Table 4. The main advantage of the proposed design is that the LUT size can be reduced depending on the value of γ. Hence, the savings obtained in area and power can be even higher for larger γ values for a given B and W. For instance, the

Table 3 Hardware complexities of various serial pipelined real-valued FFT

Design	Real MULT	LUT	Real ADD	Real REG	TP
FFT$_1$ [7]	$4\log_2 N - 12$	0	$6\log_2 N - 10$	$5N/4 - 2$	1
FFT$_2$ [7]	$4\log_2 N - 12$	0	$4\log_2 N - 6$	$3N/2 - 5$	1
FFT$_3$ [8]	$\log_2 N - 2$	0	$2\log_2 N - 2$	$9\log_2 N + 2N - 24$**	1
Proposed $(\gamma = 2)$[a]	0	$8K$ $(\log_2 N - 1)$	$2\log_2 N - 2$	$7\log_2 N + 2N - 20$	1
Proposed $(\gamma = 4)$[a]	0	$2K$ $(\log_2 N - 1)$	$2\log_2 N$	$7\log_2 N + 2N - 20$	1

MULT multipliers, *LUT* look-up table, *ADD* adders, *REG* registers, *TP* throughput
**For fair comparison, the $N - 5$ registers in output re-ordering circuits are included in addition to the reported $N + 9 \log_2 N - 19$ registers
[a]Complexity of post-multiplication register in QCM is half of that in [8]

Table 4 Performance comparison of ASIC synthesis results using TSMC 90 nm CMOS library and FPGA implementation on Xilinx ZYNQ (XC7Z020-1CLG84C) for $W = B = 8$ and $N = 1024$

Design	Area (mm^2)	Power (mW)	MCP (ns)	ADP (mm^2 ns)	PDP (mW ns)	MCP	NOS	SLUT ($\times 1000$)	FF ($\times 1000$)	SDP
FFT$_1$ [7]	1.572	5.04	7.15	11.24	36.04	9.73	16285	59.13	52.56	158453.05
FFT$_2$ [7]	1.133	4.12	6.56	7.43	27.03	8.98	11093	46.84	42.66	99615.14
FFT$_3$ [8]	1.043	3.12	6.14	6.40	19.16	8.32	9982	42.89	39.87	83050.24
Proposed $(\gamma = 2)$	0.635	1.81	6.49	4.12	11.75	8.78	6127	28.62	28.24	53795.06
Proposed $(\gamma = 4)$	0.415	1.24	6.98	2.89	8.65	9.36	4016	19.31	17.89	37589.76

proposed design when implemented for 1024-point with $\gamma = 2$ occupies nearly 39.11% less area, 42.29% less power, 35.62% less ADP than [8] at the cost of marginal increase in MCP with respect to design in [8]. This is because the delay due to XOR gates in the addressing logic is mainly compensated by the reduction in LUT access time.

FPGA Implementation In order to obtain the logic utilization, FPGA implementation is performed using Xilinx ZYNQ device (XC7Z020-1CLG84C). The corresponding results in terms of number of slices (NOS), slice look-up tables (SLUT), flipflops (FF) and slice-delay-product (SDP) are expressed in Table 4 at 100 MHz system clock. It is found that the proposed design for 1024-point FFT with $\gamma = 2$ utilizes 33.27% less SLUT and 29.18% less FF as compared to the design in [8].

4   Conclusion

In this paper, an area and power-efficient architecture for serial commutator real-valued fast Fourier transform (FFT) using recursive look-up table (LUT) multiplier has been presented. In general, serial commutator real-valued FFT architectures have latency issues while multipliers are the most computationally expensive elements. In the proposed scheme, the multiplier has been realized with recursive LUT which has symmetry in LUT contents obtained by appropriately adding the shifted versions of twiddle coefficient. Depending on the requirement, appropriate symmetry factor can be used to reduce the LUT size. The proposed architecture is validated on ASIC synthesis and FPGA implementation.

References

1. Elamaran V, Arunkumar N, Hussein AF, Solarte M, Ramirez-Gonzalez G (2018) Spectral fault recovery analysis revisited with normal and abnormal heart sound signals. IEEE Access 6:62874–62879
2. Liu S, Liu D (2018) A high-flexible low-latency memory-based FFT processor for 4G, WLAN, and future 5G. IEEE Trans Very Large Scale Integr (VLSI) Syst 27(3):511–523
3. Kaveh S, Norouzi Y (2016) Non-uniform sampling and super-resolution method to increase the accuracy of tank gauging radar. IET Radar Sonar Navig 11(5):788–796
4. Garrido M, Andersson R, Qureshi F, Gustafsson O (2016) Multiplierless unity-gain SDF FFTs. IEEE Trans Very Large Scale Integr (VLSI) Syst 24(9):3003–3007
5. Wang Z, Liu X, He B, Yu F (2015) A combined SDC-SDF architecture for normal I/O pipelined radix-2 FFT. IEEE Trans Very Large Scale Integr (VLSI) Syst 23(5):973–977
6. Chen SG, Huang SJ, Garrido M, Jou SJ (2014) Continuous-flow parallel bit-reversal circuit for MDF and MDC FFT architectures. IEEE Trans Circ Syst I Regul Pap 61(10):2869–2877
7. Chinnapalanichamy A, Parhi KK (2015) Serial and interleaved architectures for computing real FFT. In: 2015 IEEE international conference on acoustics, speech and signal processing (ICASSP). IEEE, pp 1066–1070
8. Garrido M, Unnikrishnan NK, Parhi KK (2017) A serial commutator fast Fourier transform architecture for real-valued signals. IEEE Trans Circ Syst II Express Briefs (2017)
9. Khan MT, Shaik RA (2019) Optimal complexity architectures for pipelined distributed arithmetic-based LMS adaptive filter. IEEE Trans Circ Syst I Regul Pap 66(2):630–642
10. Arden WM (2002) The international technology roadmap for semiconductors-perspectives and challenges for the next 15 years. Curr Opin Solid State Mater Sci 6(5):371–377
11. Meher PK (2009) New approach to look-up-table design and memory-based realization of FIR digital filter. IEEE Trans Circ Syst I Regul Pap 57(3):592–603; Author F (2016) Article title. Journal 2(5):99–110

Neuro-fuzzy Classifier for Identification of Stator Winding Inter-turn Fault for Industrial Machine

Amar Kumar Verma$^{(\boxtimes)}$ (iD), Aakruti Jain, and Sudha Radhika

Birla Institute of Technology & Science Pilani, Hyderabad Campus, Hyderabad, Telangana 500078, India
amarverma710@gmail.com

Abstract. Induction machines have extensive demand in industries as they are used for large-scale production and therefore vulnerable to both electrical and mechanical faults. Over the past decade, online condition monitoring of industrial machinery has become one of the major research areas in fault detection and diagnosis. There are different types of stator winding insulation faults, of which the current work is focused on the identification of stator winding inter-turn fault as it accounts for 37% of the overall machine failures. Also, this fault, if identified at its incipient stage, can predominantly improvise machine downtime and maintenance cost. The proposed work uses acquired experimental data from both healthy and faulty three-phase induction motor to train the neuro-fuzzy classifier for fault severity evaluation. It has been observed that AI-based neuro-fuzzy classifier is capable of generating rules and membership functions on its own with a given set of experimental data whereas fuzzy classifier requires manual intervention for defining rules and membership functions. A comparison of fuzzy and neuro-fuzzy based fault identification is made, and the efficiency of both classifiers was compared.

Keywords: Neuro-fuzzy logic (NFL) · Stator inter-turn fault (SITF) · Fault identification · Neural network (NN) · Artificial intelligence techniques (AI)

1 Introduction

Induction machines are highly adapted for commercial and industrial applications. Being robust and reliable, these electro-mechanical devices find their wider application in ranging from the simplest of the task such as pumps, centrifugal machines, elevators, etc. to the most dangerous and difficult surroundings like mines, marine applications, petroleum industry, etc. As induction machines are inevitable for large-scale production industries and are used profusely, they are vulnerable to different kinds of uncontrollable, electrical and mechanical faults such as bearing fault, stator insulation failure, rotor bar breaking, and other related faults [1, 2]. Voltage sag and swell, failure of electrical connections, damage of insulation, stator inter-turn fault, stator coil to coil fault, stator phase to ground fault, stator phase to phase fault, and stator open circuit faults, etc., fall into electrical fault category. Bearing damage, broken bars, rotor misalignment, rotor mechanical unbalanced, etc., are mechanical faults [3–6]. These

© Springer Nature Singapore Pte Ltd. 2020
N. Goel et al. (eds.), *Modelling, Simulation and Intelligent Computing*,
Lecture Notes in Electrical Engineering 659,
https://doi.org/10.1007/978-981-15-4775-1_12

faults may lead to shut down of the machines unexpectedly during the working hours, adversely affecting the production time, which in turn results in substantial production losses. These faults may have only minor symptoms, but if not identified at an early stage may result in lower efficiency hence high energy consumption and long term degradation [4].

At present industries are focused on reactive, planned, and proactive maintenance strategies. However, there are certain limitations with such strategies, such as unplanned downtime, the potentially greater damage to machine beyond failed part, need for additional spare parts inventory, etc. [5]. Thus, there comes a need for predictive maintenance (PdM). PdM can break the trade-offs of the older strategies by using early stage detection techniques of these faults and thereby maximizing the useful life of machine parts. There are various fault identification techniques for condition monitoring of induction machine using different signals measured from the motor such as voltage, flux and instantaneous power [6], vibration [7], temperature [8], motor current signature analysis [9, 10], electromagnetic or mechanical torque [11]. A diagnosis process based on the AI technique and multiple signatures should be more reliable in stator fault detection and severity evaluation [12–15].

Fuzzy logic is another emerging method in the field of fault prediction due to its capability to mimic the human brain. However, due to the lack of manual intervention in this method, it may lead to a high percentage of misclassification and thereby yielding poor identification accuracy [16, 17]. The neuro-fuzzy logic model is a combination of artificial neural network and fuzzy logic which generates rules and membership functions of output on its own with provided set of inputs. Neuro-fuzzy systems (NFS) has a wide range of applications in electrical and electronics system, manufacturing and system modeling, image processing and feature extraction, etc. but most importantly in forecasting and predictions. Thus, the current research focuses on an early-stage detection of stator winding fault at the incipient stage in a three-phase induction motor both y using fuzzy logic technique and the latest neuro-fuzzy logic technique. A comparison of both fuzzy, as well as neuro-fuzzy identification results, is also performed in the current research. It has been observed that neuro-fuzzy based logic is capable of identifying the healthy as well as the faulty with 93.3% accuracy, whereas fuzzy logic-based algorithm is capable of only 86.3%.

2   Methodology

Stator current signal data acquired from the experimental set-up of three-phase induction motor under healthy and faulty conditions are used to train the fuzzy inference system of neuro-fuzzy for fault severity evaluation. It has been observed that artificial intelligence based neuro-fuzzy is capable of generating rules and membership functions on its own with a given set of experimental data whereas fuzzy requires manual intervention for defining rules and membership functions. A numerical model of the same with healthy as well as stator inter-turn fault is also designed, and the same has been validated using the neuro-fuzzy logic which was trained using the experimental data.

Fig. 1 Experimental setups for both healthy and faulty induction motor (Power Electronics Lab of BITS Pilani, Hyderabad)

2.1 Experimental Setup

The experimental setup for both healthy and faulty induction motor has been setup is shown in Fig. 1. Stator inter-turn fault has been intentionally created by tapping winding at 25%, and 50% in one of the phases in three-phase induction motor. Further, the stator current signal data has been acquired in the time-domain.

Data Collection Three-phase stator current I_a, I_b, and I_c from both healthy and faulty induction motor have been acquired using a current transducer (hall effect), data acquisition system, and LabView software platform. The acquired data from both machines is shown in Fig. 2.

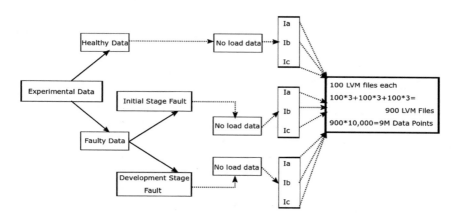

Fig. 2 Data acquisition

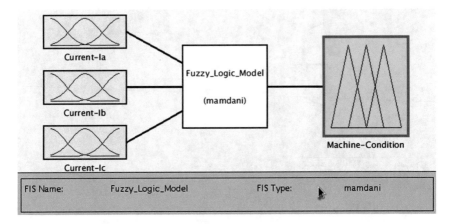

Fig. 3 Fuzzy logic controller

Data Pre-processing The experimental results obtained from both healthy and faulty induction motor under different loading conditions have been sampled at a 10 kHz rate. The conversion ratio (K_N) of the sensor is 1:1000. The resistance of 100 Ω is connected across each of the sensors to give current out in terms of voltage.

2.2   Fuzzy Logic Designer

The acquired data has been processed to create a membership function for a given set of input and output, as shown in Fig. 3. The current input is commonly divided into three ranges as small, medium, and high. Similarly, the output current is classified as good, initial, and development stage.

Finally, based on membership function and created rules, a fuzzy inference system (FIS) is generated to test on the numerically modeled induction motor with both healthy as well as faulty conditions. The performance of the fuzzy logic controller has been quantified and observed as 86.3%. However, the fuzzy logic designer requires manual observation and setup of membership functions and rules, which affected the accuracy of identification.

2.3   Neuro-fuzzy Logic Designer

As it has been observed that fuzzy logic requires manual intervention for defining rules and membership functions whereas the AI-based neuro-fuzzy logic is capable of generating rules and membership functions on its own with a given set of experimental data, in the current research identification of faulty data is also performed by designing a suitable neuro-fuzzy system. In fuzzy, membership functions and rules should be provided for it to make decisions and predict the result, but the fusion of neural network (NN) and fuzzy logic will give the perfect blend in decision making and least manual intervention, automatizing the complete process as shown in Fig. 4. The quick learning ability of neural network helps in implementing and automatizing the membership

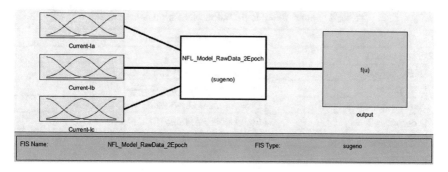

Fig. 4 Neuro-fuzzy logic controller

functions and rules, while fuzzy logic principles detect the fault using manually provided before mentioned set of rules and membership functions.

Since NFL facilitates the training of data and generation of membership functions and rules without manual intervention, the experimental data is loaded into the neuro-fuzzy designer by defining performance parameter of NFL such as membership function (MF) type for both input and output, optimization algorithm of neural network, tolerance error, and number of epochs before training NFL model. The adaptive network-based fuzzy inference system (ANFIS) structure of the neuro-fuzzy model for the given set of input and output is shown in Fig. 5.

The membership functions generated for input current and the output shows machine condition, i.e., faulty or healthy conditions. It is observed that the membership

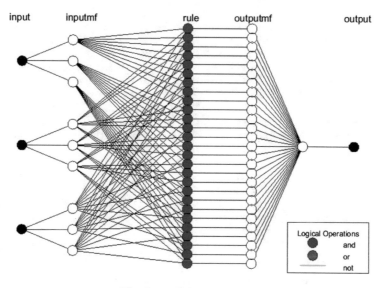

Fig. 5 ANFIS structure plot

Table 1 Neuro-fuzzy logic output versus machine condition

NFL output	Machine condition
NFL output = 0	Healthy
0 < NFL output < 1	Initial stage fault; prior to SITF
NFL output = 1	Stator inter-turn fault (SITF)
1 < NFL output < 2	Development stage fault; prior to SCCF
NFL output = 2	Stator coil-coil fault (SCCF)
2 < NFL output < 4	Stator phase neutral fault

function for each phase current is similar but has a different range based on the fault severity in the experimental data used for training. The output range varies from 0 to 4 as tabulated in Table 1.

In a neuro-fuzzy system, instead of membership function for output, constant values are generated for the combination of inputs, known as rules. Thus, with three input currents, 27 rules are devised giving 27 outputs. The combination of input currents and output can be viewed through rule viewer is shown in Fig. 6. The first 3 columns are the three-phase current I_a, I_b, and I_c from both healthy and faulty induction motor while the last column is the output (severity of faults). The numbers at the beginning of each row indicate the rule number, last block at the end of last column gives the aggregate output using all the rules for a given set of input currents. If the current is in the transition state, aggregate outputs are generated using both the values

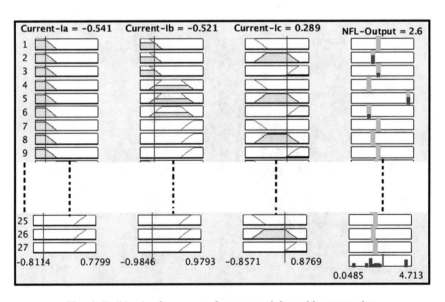

Fig. 6 Build rules from neuro-fuzzy on training with current data

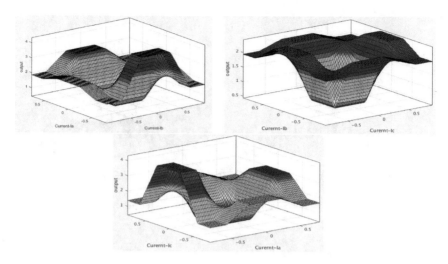

Fig. 7 Neuro-fuzzy output corresponding to current I_a, I_b, and I_c

in transition and the final output is the weighted mean of all the aggregated outputs, where weight depends on the position of input current. Similarly, the fault severity is quantified and shown in Fig. 7.

3   Numerical Model

The numerical model of an induction motor under both healthy and faulty condition have been modeled and implemented to validate the results obtained from the experimental setup under the same operating condition.

3.1   Induction Motor with Stator Inter-turn Fault

The stator winding configuration for both healthy and faulty induction motors with stator inter-turn short is represented in Fig. 8. In the case of faulty motor, an additional short-circuit winding has to be considered and the corresponding numerical model is derived [18]. The θ_f is the angle between phase a and turn short phase and the value can be 0, 2π or 4π correspond to fault on stator winding phases a, b or c, respectively.

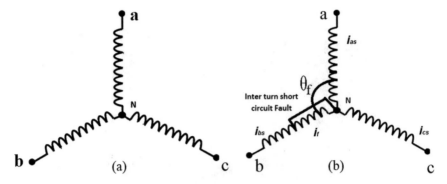

Fig. 8 Stator winding without and with inter-turn short

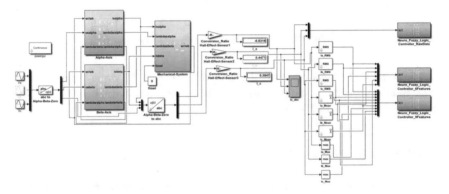

Fig. 9 Induction motor numerical model with stator inter-turn short

3.2   Numerical Implementation

The derived numerical model for both healthy and faulty induction motors has been implemented by transforming abc-to-$\alpha\beta$ coordinate is shown in Fig. 9. There are mainly three subsystems in the implemented numerical model; α, β and mechanical. The α and β axis values are used to calculate developed electromagnetic torque, stator current, and angular speed of the induction motor in the mechanical subsystem block. Finally, neuro-fuzzy classifier is integrated with the numerical model in order to validate the results.

4   Results and Discussion

The generated fuzzy inference system (FIS) file from neuro-fuzzy logic (NFL) controller is integrated with numerical model of faulty induction motor (IM). Stator winding inter-turn fault (SITF) can be created in the numerical model in any of the phases by assigning corresponding θ_f values. In the experimental setup, stator winding

Table 2 Machine current under different loading conditions

Load (in %)	Torque (in N-m)	Normalized stator current (A)	
		Experimental	Numerical
0	15	0.19	0.185
25	9	0.33	0.322
50	3	0.57	0.563

fault is intentionally created in the b-phase, therefore θ_f is $2\pi/3$ and has been considered for validation of results.

Three-phase output stator current I_a, I_b, and I_c from the numerical model are input to the neuro-fuzzy classifier block, which are then analyzed with the help of FIS file generated by training data obtained from the experimental setup. Thus, output of NFL shows the machine fault severity. The final output is mean of the instantaneous values generated, thus predicting the fault in numerical model. The instantaneous output values are further classified in different ranges and the count of values in those ranges helps in identifying the machine condition. The developing stage fault, i.e., stator coil-coil fault is intentionally created in the induction motor and made to run for 0.5 s to propagate it into a complete fault stage, i.e., stator phase neutral fault. The neuro-fuzzy logic was able to classify 293 samples out of 314 whereas, fuzzy logic was able to classify only 271 samples out of 314. The numerical results obtained from healthy induction motor under different loading condition is tabulated in Table 2.

5 Conclusion

Fuzzy and neuro-fuzzy techniques for fault identification of stator winding inter-turn fault (SITF) have been implemented. It has been observed that AI technique-based neuro-fuzzy logic system was capable of identifying the healthy as well as the faulty with 93.3% accuracy, whereas fuzzy logic-based algorithm with 86.3%. The numerical model of induction motor with and without stator inter-turn fault has been implemented, and the same has been validated using experimental results. A correlation factor of 0.94 is obtained, thus validating the numerical model.

References

1. Verma AK, Radhika S, Padmanabhan S (2018) Wavelet based fault detection and diagnosis using online MCSA of stator winding faults due to insulation failure in industrial induction machine. In: 2018 IEEE recent advances in intelligent computational systems (RAICS). IEEE, pp 204–208
2. Vamsi IV, Abhinav N, Verma AK, Radhika S (2018) Random forest based real time fault monitoring system for industries. In: 2018 4th International conference on computing communication and automation (ICCCA). IEEE, pp 1–6
3. Radhika S, Sabareesh G, Jagadanand G, Sugumaran V (2010) Precise wavelet for current signature in 3φ IM. Expert Syst Appl 37(1):450–455

4. Verma AK, Spandana P, Padmanabhan SV, Radhika S (2019) Quantitative modeling and simulation for stator inter-turn fault detection in industrial machine. In: International conference on intelligent computing and communication, pp 87–97
5. Ranjan GS, Verma AK, Radhika S (2019) K-nearest neighbors and grid search cv based real time fault monitoring system for industries. In: 2019 IEEE 5th international conference for convergence in technology (I2CT), pp 1–5. IEEE
6. Verma AK, Akkulu P, Padmanabhan SV, Radhika S (2020) Automatic condition monitoring of industrial machines using FSA-based hall-effect transducer. IEEE Sens. J
7. Alameh K, Cité N, Hoblos G, Barakat G (2015) Vibration-based fault diagnosis approach for permanent magnet synchronous motors. IFAC Pap OnLine 48(21):1444–1450
8. Singh G, Kumar TCA, Naikan V (2016) Induction motor inter turn fault detection using infrared thermographic analysis. Infrared Phys Technol 77:277–282
9. Seera M, Lim CP, Ishak D, Singh H (2012) Fault detection and diagnosis of induction motors using motor current signature analysis and a hybrid FMM–CART model. IEEE Trans Neural Netw Learn Syst 23(1):97–108
10. Shukla S, Jha M, Qureshi M (2014) Motor current signature analysis for fault diagnosis and condition monitoring of induction motors using interval type-2 fuzzy logic. Int J Innov Sci Eng Technol 1(5):84–94
11. Haroun S, Seghir AN, Touati S, Hamdani S (2015) Misalignment fault detection and diagnosis using AR model of torque signal. In: 2015 IEEE 10th international symposium on diagnostics for electrical machines, power electronics and drives (SDEMPED). IEEE, pp 322–326
12. Seera M, Lim CP, Nahavandi S, Loo CK (2014) Condition monitoring of induction motors: a review and an application of an ensemble of hybrid intelligent models. Expert Syst Appl 41 (10):4891–4903
13. Sharifi R, Ebrahimi M (2011) Detection of stator winding faults in induction motors using three-phase current monitoring. ISA Trans 50(1):14–20
14. Palácios RHC, da Silva IN, Goedtel A, Godoy WF (2015) A comprehensive evaluation of intelligent classifiers for fault identification in three-phase induction motors. Electr Power Syst Res 127:249–258
15. Duan F, Zivanović R (2015) Condition monitoring of an induction motor stator windings via global optimization based on the hyperbolic cross points. IEEE Trans Industr Electron 62 (3):1826–1834
16. Rodríguez PVJ, Arkkio A (2008) Detection of stator winding fault in induction motor using fuzzy logic. Appl Soft Comput 8(2):1112–1120
17. Azgomi HF, Poshtan J (2013) Induction motor stator fault detection via fuzzy logic. In: 2013 21st Iranian conference on electrical engineering (ICEE). IEEE, pp 1–5
18. Duan F, Zivanovic R (2012) A model for induction motor with stator faults. In: 2012 22nd Australasian universities power engineering conference (AUPEC). IEEE, pp 1–5

Quantification of the Extent of Multiple Node Charge Collection in 14 nm Bulk FinFETs

Nanditha P. Rao[1(✉)] and Madhav P. Desai[2]

[1] International Institute of Information Technology Bangalore, Bengaluru, India
nanditha.rao@iiitb.ac.in
[2] Indian Institute of Technology Bombay, Mumbai, India
madhav@ee.iitb.ac.in

Abstract. A key issue to consider in the case of FinFET-based circuits is their susceptibility to multiple transients as a result of a neutron-induced particle strike. In this paper, we perform a device simulation-based characterization study on representative layouts of 14 nm bulk FinFETs to get insights into the charge collection efficiency and the extent to which multiple transistors are affected. We find that multiple transistors do get affected and the impact can last up to five transistors away (~ 200 nm). We show that the likelihood of two adjacent FinFETs getting affected (collecting maximum charge) is high, when their source/drain regions are biased high. This observation could be used as an indicator to identify vulnerable parts of the layout by looking at regions which have adjacent signal rails and possibly reduce such areas. In the case of two nearby multi-fin FinFETs, the charge collected per fin is seen to reduce as the number of fins increase. Thus, smaller FinFETs are more susceptible to high amounts of charge collection. A careful placement of small vulnerable gates may be necessary to reduce the likelihood of multiple transients.

Keywords: Multiple transients · FinFET · Charge collection

1 Introduction

One of the key factors contributing to better performance of computing systems, has been, the scaling of process technology and Moore's law. Scaling comes with its own set of challenges, such as sub-threshold leakage currents, short channel effects, increased fabrication costs and so on. In this paper, we look at the device and soft-error reliability aspects of the FinFET technology which has replaced the planar MOSFETs for sub-22 nm technology node. When high energy particles such as alpha particles, protons or neutrons strike a semiconductor integrated circuit, they generate electron-hole pairs (charge) in the substrate. This charge can either recombine or get collected in the source/drain regions of a transistor through diffusion or other mechanisms, resulting in a glitch or a transient current known as single event transient (SET). SET is modeled as a current injection into the drain of a single transistor [1]. If multiple transients and multiple flips were to occur, these models will not accurately represent the reality and will result in optimistic reliability estimates. Therefore, it is important to quantify the

© Springer Nature Singapore Pte Ltd. 2020
N. Goel et al. (eds.), *Modelling, Simulation and Intelligent Computing*,
Lecture Notes in Electrical Engineering 659,
https://doi.org/10.1007/978-981-15-4775-1_13

extent to which a circuit/layout is susceptible to multiple transients especially in the latest technology.

There is a need to study the extent to which device layouts with FinFETs, which have replaced the planar MOSFETs for sub-22 nm technologies, are susceptible to SETs. Most existing studies on soft errors in FinFETs focus on memories and show that the radiation sensitivity of FinFET-based SRAMs is better than that of planar SRAMs [2]. This is mainly attributed to the fact that the volume of the source/drain region (the fin) that connects to the substrate is small as compared to planar devices, resulting in reduced charge collection. With technology scaling [3, 4], a radiation-induced strike can have a large region of influence and can affect multiple transistors and logic gates. The phenomenon of a radiation-induced strike affecting multiple transistors has been studied to some extent in [5] for planar MOSFETs but has not been understood to the same extent in FinFETs. Some studies with respect to FinFETs are performed in [6] and they report that multiple cell upsets do occur in FinFET-based SRAMs. In this paper, we study the extent to which multiple transistors are affected, by performing a device simulation-based characterization study of the impact of a particle strike on layouts of 14 nm bulk FinFETs. We quantify the range of impact of a particle strike, collection efficiency of multi-fin FinFETs and the role of potentials on charge collection. To the best of our knowledge, such a study on 14 nm FinFETs has not been done before.

In an array of six-FinFETs, we find that a strike can have an impact up to five transistors away (nearly 200 nm) from the strike location. However, the nearest two transistors are the ones that are most affected. We find that the likelihood of two adjacent FinFETs getting affected (collecting maximum charge) is high, when their corresponding source/drain regions are biased to a high potential. This implies that, if there are adjacent logic gates, it is best to do the layout such that their signal rails are not adjacent to each other (signal rails which could be potentially high) or reduce the number of adjacent signal rails. This could help in restricting the SET to a single logic gate and prevent one of the two logic gates from getting affected. We can also use this fact as an indicator to identify vulnerable parts of the layout by looking at regions which have adjacent signal rails. In the case of a single multi-fin FinFET, we find that the key contributing factor to charge collection is the proximity of the strike to the drain region rather than the number of fins. However in the case of two adjacent multi-fin FinFETs, the charge collected per fin reduces as the number of fins increase. Thus, FinFETs with smaller widths (smaller logic gates) are more susceptible to high amounts of charge collection. This implies that a careful placement of small vulnerable gates in the layout may be necessary to mitigate the effects of multiple SETs.

The rest of this paper is organized as follows. In Sects. 2 and 3, we describe the device construction and simulation setup respectively. We explain the role of potentials on multiple node charge collection in Sect. 4. In Sect. 5, we present the quantification of the range of impact of a particle strike and collection efficiency of multi-fin FinFETs. We summarize and conclude the paper in Sect. 6.

2 Device Characterization

In Fig. 1, we show a bulk n-FinFET built based on the information available in [8]. The geometry of the device and doping concentrations are shown in Table 1. The gate stack is constructed with TiN/HfO$_2$. Parameters such as the fin height, STI depth, source/drain doping, channel doping and fin doping are calibrated to meet the on/off current (I_{on} and I_{off}) and sub-threshold slope requirements of the 14 nm bulk FinFET data [7]. The upper fin has a Gaussian doping profile toward the channel. The stress in the channel region is modeled as a Gaussian profile as per the data available in [9]. The device had a total of 78 k elements after meshing. In Fig. 1c, we show a plot of the drain current versus gate to source voltage (V_{gs}) of our device ('Sim') for a drain-source voltage (V_{ds}) of 0.7 and 0.05 V, as compared with that in the specification ('Spec') [7].

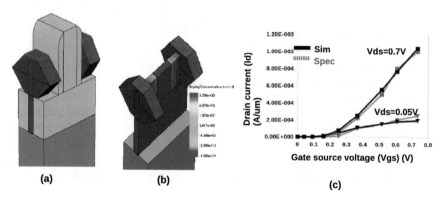

(a)   **(b)**   **(c)**

Fig. 1 a Structure of the 14 nm bulk n-FinFET. **b** The fin is shown explicitly. **c** Drain current (A/μm) versus gate source voltage (V_{gs}). The plot from our device is denoted by 'Sim' and the one in the specification [7] is denoted by 'Spec'

Table 1 Geometry and doping concentrations of the 14 nm bulk n-FinFET

Device parameter	Value
Gate length	20 nm
Effective oxide thickness	1.2 nm
Fin height	45 nm
Fin width	10 nm
Fin pitch	42 nm
Gate pitch	70 nm
STI depth	60 nm
Total substrate depth	400 nm
Lower fin doping (Boron)	1e19 cm^{-3}
Channel doping (Boron)	3e18 cm^{-3}
Source/drain doping (Arsenic)	1e20 cm^{-3}

3   Simulation Setup

When a neutron strikes the silicon substrate, several secondary ions such as, 28 Al and 25 Mg [3, 5] get generated. The energy of these secondary ions is expressed in terms of linear energy transfer or LET (MeV-cm^2/mg). It is the energy that the particle transfers to Silicon per unit length. There are several nuclear libraries and codes available. However, we use the ENDF/B-VI (Evaluated Nuclear Data File) library to determine the energy of the secondary ions for a given neutron energy. In Silicon, 3.6 eV of energy is required to create one electron hole pair or charge (1 MeV LET = 0.01 pC/μm). The secondary ion energy obtained from the nuclear code is provided as an input to a tool called TRIM (Transport Ions in Matter) to obtain the length and radius of the charge track in Silicon. The energy of secondary ions is known to range from 1 to 11 MeV [5]. We create the device layout in Sentaurus Structure Editor. We simulate the charge track on the generated device layout using Sentaurus Device. We are mainly interested in measuring the charge collected at the source/drain regions of the FinFET.

We model the particle strike as a cylindrical column of charge with a Gaussian radial track (sigma = 0.1–0.4 for energies of 1–11 MeV) and is simulated using the HeavyIon module in Sentaurus Device. The physics models used in the simulation are as follows. Mobility degradation effects due to impurity scattering, carrier-carrier scattering, high electric fields and mobility degradation at the silicon-insulator interface are specified using the following models: PhuMob, CarrierCarrierScattering, HighFieldsaturation, inversion and accumulation layer model (IALMob) and Enormal (Lombardi) respectively. Generation and recombination processes of electron-hole pairs are modeled using Auger, Band2Band and SRH recombination models. Quantum effects at the semiconductor–insulator interface are modeled using the eMultiValley modified local-density approximation (MLDA) model. These physics models are consistent with the models used in [10].

4   The Role of Potentials on Multiple Node Charge Collection

We construct a layout of two bulk n-FinFETs with a 11 MeV particle strike in between the two devices, as shown in Fig. 2a. We vary the voltages of all the source/drain regions and perform particle strike simulations for all possible voltage combinations. The charge collected in each source/drain region (A1, B1, A2 or B2) is plotted against the corresponding voltages in Fig. 2b. We notice that the charge collected in adjacent FinFETs is high, when their corresponding source/drain regions (B1, A2 for instance) is high, and it reduces by nearly 50% when the terminal is biased low. Similarly, when all nodes are biased low the charge collected in all the source/drain regions is minimum. Thus, the likelihood of two adjacent FinFETs getting affected depends heavily on the potentials of their source/drain regions being biased high. This implies that, if there are adjacent logic gates, it is best to do the layout such that their signal rails are not adjacent to each other or reduce the number of such adjacent signal rails, so that we can prevent one of the two logic gates from getting affected.

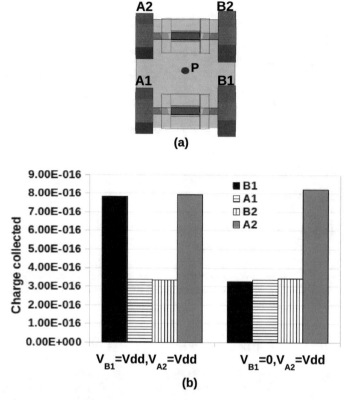

Fig. 2 a Layout of two bulk n-FinFETs showing the particle strike location 'P'. The source/drain regions are marked as A1, A2, B1 and B2. **b** Charge collected in B1 and A2 is high when the respective nodes are biased high

5 Collection Efficiency and Multiple Node Charge Collection

5.1 Multiple Node Charge Collection

We construct a layout of six bulk n-FinFETs as shown in Fig. 3a, to study the extent to which a particle strike can affect multiple transistors. Layouts with three device separations are simulated: 4λ, 7λ and 14λ (where $2\lambda = 14$ nm). We assume a particle to be incident in between the first two devices as shown in the figure and measure the charge collected in the drains (biased high to measure maximum charge). For a 11 MeV strike, we see that up to five devices (collect at least 2fC) can be affected due to a single particle strike. The overall part of the layout that collects 2fC is marked in the figure and the range of impact is nearly 200 nm. However, the nearest two devices collect maximum amount of charge (8–10fC). In the case of 5 MeV strike, two devices get affected on an average. Thus, a single particle strike can affect multiple transistors and the region of the layout which is affected by the strike is substantial. We performed a similar experiment on an array of six bulk p-FinFETs. However, the number of devices affected in the case of p-FinFETs is less than that affected in a layout of n-FinFETs.

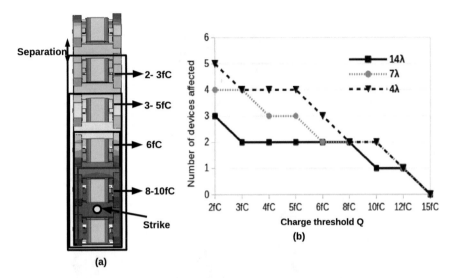

Fig. 3 **a** Layout of 6 bulk n-FinFETs showing the strike location and charge collection map for a 11 MeV particle strike. **b** Graph showing the number of devices collecting more than a certain charge Q (x-axis)

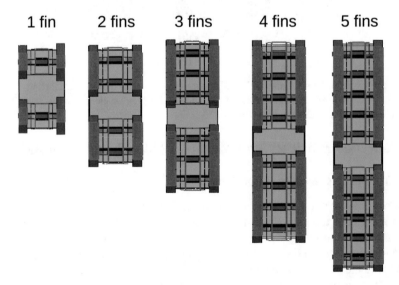

Fig. 4 Layout of two adjacent multi-fin bulk n-FinFETs. The number of fins varies from one to five

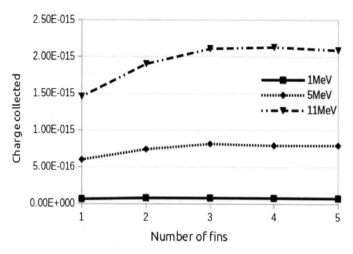

Fig. 5 The minimum charge collected in "both" the FinFETs is plotted as the number of fins is varied

5.2 Multiple Node Charge Collection in Multi-fin FinFETs

In this section, we study multiple node charge collection in a layout of multi-fin FinFETs as the number of fins is varied. We construct five different layouts of two adjacent multi-fin FinFETs as shown in Fig. 4. The number of fins in these five layouts varies from one to five respectively as shown in the figure. We simulate particle energies of 1, 5 and 11 MeV and assume it to be incident in between the two devices. We want to understand the extent to which both the FinFETs are affected as a result of the particle strike. We measure the charge collected in the source/drain regions of both the devices. In Fig. 5, we plot the minimum charge collected in "both" the devices as the number of fins is increased. For example, in the layout of two 1-fin FinFETs, a 11 MeV particle strike generates a charge of at least 1.5fC in "both" the devices. We see that a 2 fin FinFET does not collect twice as much charge as a single fin FinFET; it collects less. Similarly, a 5 fin FinFET does not collect five times as much charge as a single fin FinFET. Further, we see that the charge collected per fin reduces as the number of fins is increased as indicated by Fig. 5. Thus, FinFETs with smaller widths (and hence, smaller logic gates) are more susceptible to high amounts of charge collection. This implies that a careful placement of small vulnerable gates in the layout may be necessary to mitigate the effects of multiple SETs. Avoiding placement of small gates next to each other may reduce the likelihood of multiple SETs.

6   Conclusion

We studied the extent to which multiple transistors are affected in a layout of 14 nm bulk FinFET devices, by performing a device simulation-based characterization study of the impact of a particle strike. We find that multiple transistors are affected and the impact can last up to five transistors away ($\sim$200 nm). However, the nearest two transistors are the ones that are most affected. The likelihood of two adjacent FinFETs getting affected is high, when their corresponding source/drain regions are biased high. So, it is best to do the layout such that the number of adjacent signal rails is less, so that we can restrict the SET to a single logic gate. In the case of two adjacent multi-fin FinFETs, the charge collected per fin reduces as the number of fins increase. Thus, smaller FinFETs are more susceptible to high amounts of charge collection. A careful placement of smaller gates is needed to reduce the likelihood of multiple SETs.

References

1. Zhou Q, Mohanram K (2006) Gate sizing to radiation harden combinational logic. IEEE Trans Comput Aided Des Integr Circ Syst 25(1):155–166
2. Fang Y-P, Oates AS (2011) Neutron-induced charge collection simulation of bulk FinFET SRAMs compared with conventional planar SRAMs. IEEE Trans Device Mater Reliab 11 (4):551–554
3. Narasimham B, Gupta S, Reed D, Wang JK, Hendrickson N, Taufique H (2018) Scaling trends and bias dependence of the soft error rate of 16 nm and 7 nm FinFET SRAMs. In: 2018 IEEE international reliability physics symposium (IRPS), Burlingame, CA, pp 4C.1-1–4C.1-4
4. Uemura T, Lee S, Monga U, Choi J, Lee S, Pae S (2018) Technology scaling trend of soft error rate in flip-flops in $1 \times$ nm bulk FinFET technology. IEEE Trans Nucl Sci 65(6):1255–1263
5. Amusan OA, Steinberg AL, Witulski AF, Bhuva BL, Black JD, Baze MP, Massengill LW (2007) Single event upsets in a 130 nm hardened latch design due to charge sharing. In: Proceedings of the 45th annual IEEE international reliability physics symposium, 2007, pp 306–311
6. Tam N, Bhuva BL, Massengill LW, Ball D, McCurdy M, Alles ML, Chatterjee I (2015) Multi-cell soft errors at the 16-nm FinFET technology node. In: 2015 IEEE international reliability physics symposium (IRPS), pp 4B.3.1–4B.3.5
7. Natarajan S, Agostinelli M et al (2014) A 14 nm logic technology featuring 2nd-generation FinFET, air-gapped interconnects, self-aligned double patterning and a $0.0588 \ \mu m^2$ SRAM cell size. In: 2014 IEEE international electron devices meeting, pp 3.7.1–3.7.3
8. Kedzierski J, Ieong M, Nowak E, Kanarsky TS, Zhang Y, Roy R, Boyd D, Fried D, Wong HSP (2003) Extension and source/drain design for high-performance FinFET devices. IEEE Trans Electron Devices 50(4):952–958
9. Serra N, Esseni D (2010) Mobility enhancement in strained n-FinFETs: basic insight and stress engineering. IEEE Trans Electron Devices 57(2):482–490
10. Kumar US, Rao VR (2016) A thermal-aware device design considerations for nanoscale SOI and bulk FinFETs. IEEE Trans Electron Devices 63(1):280–287

A Comparative Assessment of Genetic and Golden Search Algorithm for Loss Minimization of Induction Motor Drive

Keerti Rai[(⊠)] 🔟, S. B. L. Seksena, and A. N. Thakur

Electrical Engineering Department, National Institute of Technology
Jamshedpur, Jamshedpur, Jharkhand, India
raikeerti07@gmail.com

Abstract. This paper presents a comparative performance assessment for loss minimization of vector controlled induction motor (IM) drive based on two different algorithms, namely Genetic Algorithm (GA) and Golden Search (GS). Here the features of GA and GS both for estimation and recalculation of optimized flux component of current have been utilized for a better optimal efficiency operation of the IM drive. The GA- and GS-based algorithms greatly improve efficiency by reducing the core loss of the drive system. Moreover, both the approaches have no effect on parameter variation and also need no additional hardware for hardware implementation. However, GA-based loss minimization scheme proves its edge over GS-based scheme for the IM drive. The simulation results for different operating conditions are presented here. Stability study of the whole drive system is also carried out utilizing both the schemes. The performance of the proposed drive is validated experimentally on dSPACE-1104 based laboratory prototype.

Keywords: Induction motor drive · Loss minimization · Genetic algorithm · Golden search algorithm · Stability · Performance

1 Introduction

In the last three decades, huge efforts have been forward by both developed and developing countries for energy-saving [1]. The electric motors are the largest consumer of total electricity produced globally (nearly 65–70%), [2]. Precisely, out of this share of percentage, the three-phase induction motors (IMs) of different ratings utilize approximately 85–90% of this electricity. Further, the IM drives have an annual expansion rate of 1.5% in the industrial sector and 2.2% in tertiary sector. Hence, an improvement in efficiency by minimizing the losses will significant impact on saving of revenue, fuel consumption and other associated negative factors [3]. According to the literature available, for each and every 1% improvement in the motor efficiency might lead to savings of over $1 billion per annum in energy prices, the consumption of coal is reduced by 5.4–9.1 million tons per annum and further bringing down greenhouse emission by nearly 13.6–18.1 million tons per annum [4].

© Springer Nature Singapore Pte Ltd. 2020
N. Goel et al. (eds.), *Modelling, Simulation and Intelligent Computing,*
Lecture Notes in Electrical Engineering 659,
https://doi.org/10.1007/978-981-15-4775-1_14

Minimization of loss in an induction motor is directly related to the choice of the flux level. But the extreme minimization causes a high copper loss. For constant speed operation, provided torque is variable the flux has to vary, so as to improve the drive efficiency. A number of energy optimization strategies such as simple state control [5], search control [6], and loss model-based control [7] for IM drive have been reported in the literature. The loss model-based control consists of computing losses by using the machine model and selecting a flux level that can be used to minimize the losses. The second category is the power-measure-based approach, also known as search controllers (SCs), in which the flux is decreased until the electrical input power settles down to the lowest value for a given torque and speed [8, 9, 10]. The genetic algorithm (GA) based method mainly works on the principle of optimization of a significant parameter (e.g., DC link power or DC link current or stator current or drive losses) by trial and error method [11, 10]. Unlike other control strategies, the method does not depend upon the motor or converter parameters. However, the method suffers from the torque ripples and slow convergence rate. The golden search algorithm is generally used to minimize IM drive loss. This also has a close relation to the Fibonacci search method [12]. However, the golden search technique has a definite edge over the Fibonacci search algorithm as the later needs to know a priori the number of evaluations in the minimum searching process, which is totally eliminated in the former one [13]. In the present work, GA and GS algorithms are used to minimize the loss in the IM drive. A comparative performance analysis of the IM drive system is carried out for both GA- and GS-based algorithm in different operating conditions to generate optimal value of i_{ds}^* and hence optimal rotor flux.

2 Loss Minimization Mechanism

The loss minimization algorithm is based on searching optimal value of the flux component of stator current (i_{ds}), for which the input power of the system can be minimal. The input power is calculated as the product of the measured DC voltage (V_{dc}) and DC current (I_{dc}) as follow:

$$P_{in} = V_{dc} \cdot I_{dc} \tag{1}$$

Therefore, to increase the efficiency of a motor, the electrical input power must be optimized to reduced value by minimizing the total losses. Thus, the loss minimization is accomplished by adjusting the flux level through the set flux component of current (i_{ds}^*) for the generation of reference flux current, the technique for the loss minimization is divided into two categories. The first one based on loss model-based approach [14–16]. In this method, the loss is computed by using the machine model and selecting the flux level that minimizes the losses. The second one is power-based approach, known as Search Controllers (SC), in which the flux is decreased until the electrical input power comes to the optimal lowest level [17–19].

3 Genetic Algorithm and Efficiency Optimization of IM Drive

Genetic algorithm (GA) is inspired by Darwin's theory about evolution—"the Survival of the fittest". In GA, the optimization problem is resolved by imitating the practices through natural uses, i.e. selection, crossover, mutation and accepting. It is intelligent exploitation of random search and exploits historical information to direct the search into the region of better performance within the search space [20, 21]. In GA method applied for optimization, the chromosomes are the solution for problems involved in optimization [22]. The mechanism of selection, crossover and mutation leads to the birth of superior quantity offspring from the previous generation (parents). However, throughout the genetic evolution, only the stronger chromosomes survive. At the end-stage near-optimal or optimal solutions can be achieved. Generation of random population of '*popsize*' chromosomes. Evaluation of the fitness function $f(i_{ds})$ of each chromosome 'i_{ds}^*' in the population ($f(i_{ds}^*)$). Minimize ($f(i_{ds}^*)$), where $i_{ds}^* = i_{ds1}^*$, $i_{ds2}^*, i_{ds3}^*, \ldots, i_{dsn}^*$. Creation of a new population by iterating the selection, crossover, mutation and accepting process until the new population is complete. Use of new generated population for a further run of the algorithm, and if the end condition is satisfied, stop and return the best solution in the existing population.

4 Golden Search Based Controller

In a golden search based algorithm, the maximum value of the flux current equals to its rated value $(i_{ds\,max} = i_{ds})$ is defined and the minimal value equals to $\alpha * i_{ds}$, where $\alpha = \sim 0.5$ depending on load levels. The algorithm calculates the flux current in the interval between $(i_{ds\,min}, i_{ds\,max})$, which is fed to the control system to reduce the total input power of the drive. The input power is calculated in Eq. (1).

The new values of the reference flux current (i_{ds}) are calculated using two golden search sections; F_1 and F_2 as below;

$$F_1 = \frac{\sqrt{5} - 1}{2} = 0.618 \text{ and } F_2 = \frac{3 - \sqrt{5}}{2} = 0.382 \tag{2}$$

The algorithm calculates two values of the reference flux currents (i_{ds1}^*, i_{ds2}^*) in the interval $(i_{ds\,min}, i_{ds\,max})$ using the golden sections. The input power corresponding to these two current levels is calculated as depicted in Eq. (1). The search procedure repeats itself until the desired accuracy is achieved.

$$\left| i_{ds2}^* - i_{ds1}^* \right| < \varepsilon_i \tag{3}$$

where ε_i is the flux current tolerance. The final value of reference flux current is calculated by averaging the values of i_{ds1}^*, i_{ds2}^*

$$i_{ds}^* = \frac{i_{ds1}^* + i_{ds2}^*}{2} \tag{4}$$

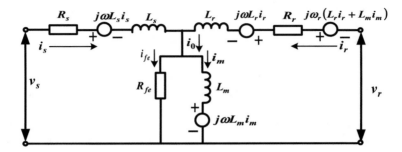

Fig. 1 T-model of an equivalent circuit model of induction motor showing iron loss

The technique comprises of all the losses including the loss in inverter, since the power entering the system is measured and used in the optimization algorithm.

5 Loss Model of Induction Motor

Losses in the IM corresponding to model under consideration as in Fig. 1 may be given as:

$$P_{\text{loss}} = \frac{3}{2}\left[R_s\left(i_{ds}^{s\,2} + i_{qs}^{s\,2}\right) + R_r\left(i_{dr}^{s\,2} + i_{qr}^{s\,2}\right) + \frac{1}{R_{fe}}\left(v_{di}^{s\,2} + v_{qi}^{s\,2}\right)\right] \tag{5}$$

or,

$$P_{\text{loss}} = P_{js} + P_{jr} + P_{fe} \tag{6}$$

6 Input Power Measurement

The drive is equipped with DC voltage and current sensors to evaluate P_{in}. The motor is driven in closed-loop speed control without load, such that the contribution of the rotor copper losses is neglecting rotor copper loss. The currents, voltages and input powers are recorded when steady-state conditions are reached.

7 Loss Minimization of IM Drive

The improved dynamic performance with minimum loss is the important requirement for the IM drive. Therefore, loss minimization schemes using GA and GS algorithms [23–25] have been incorporated separately in the outer loop of the control scheme. The vector control not only has the advantage of providing excellent dynamic performance

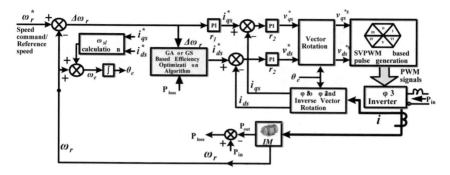

Fig. 2 Schematic diagram of vector controlled IM drive and efficiency optimization scheme

but also enables decoupled control of torque and flux through d-axis (flux-producing) and q-axis (torque-producing) currents in the steady-state. This makes the inclusion of the loss minimization algorithm simple. However, the present energy optimization algorithm using GS technique for the IM drive is based on powerless of the drive system. The drive loss is calculated from the difference between the power input to the inverter and the shaft power output. The reference flux current i_{ds}^* is generated by the optimization algorithm, while the torque component of current is acquired from the speed control loop. In the transient state, when either the speed command or the load torque is changed, the nominal value of the i_{ds}^* comes into play. The transient speed is easily detected when the speed error signal ($\Delta\omega_e$) reaches the maximum value 0.5 rad/s and the energy optimization algorithm starts settling the i_{ds}^* to the required optimal value. The optimal value of i_{ds}^* generates the optimized required flux without affecting the output power. The optimal value of flux reduces the power loss of the drive system thus fulfilling the objective of the proposed work. The detailed schematic diagram for estimation of speed with loss minimization algorithm is depicted in Fig. 2.

8 System Stability Analysis

To carry out the stability analysis of any system, the variables must be time-invariant [4]. In the state space domain, equations IM model are represented as

$$\dot{x} = Ax + Bu \tag{7}$$

$$y = Cx + Du \tag{8}$$

where $x = \begin{bmatrix} i_{ds} & i_{qs} & \psi_{dr} & \psi_{qr} \end{bmatrix}^{\mathrm{T}}$, $u = \begin{bmatrix} v_{ds} & v_{qs} \end{bmatrix}^{\mathrm{T}}$ and $y = \begin{bmatrix} i_{ds} & i_{qs} \end{bmatrix}^{\mathrm{T}}$.
   The following expressions have been derived from Fig. 2.

$$v_{ds}^* = \left(k_{p3} + \frac{k_{i3}}{s} \right) \times \left(i_{ds}^* - i_{ds} \right) = r_2 \left(i_{ds}^* - i_{ds} \right) \tag{9}$$

124　K. Rai et al.

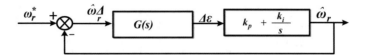

Fig. 3 Closed-loop representation of speed estimator

$$v_{qs}^* = \left(k_{p2} + \frac{k_{i2}}{s}\right)\left\{\left(k_{p1} + \frac{k_{i1}}{s}\right) \times (\omega_r^* - \omega_r) - i_{qs}\right\} = r_2\{r_1(\omega_r^* - \omega_r) - i_{qs}\}$$

(10)

where $r_1 = \left(k_{p1} + \frac{k_{i1}}{s}\right)$ = transfer function of the speed PI controller, $r_2 = \left(k_{p2} + \frac{k_{i2}}{s}\right)$ = transfer function of current loop PI controllers, as in Fig. 2.

To check the feasibility of the algorithm for speed estimation, From the small-signal error equation, $\Delta\varepsilon/\Delta\hat{\omega}_r$ is represented as:

$$\frac{\Delta\varepsilon}{\Delta\hat{\omega}_r} = k_2\frac{\Delta i_{ds}}{\Delta\omega_r} + k_3\frac{\Delta i_{ds}}{\Delta\omega_r} + (k_5 - k_4) = G(s)$$

(11)

The closed-loop transfer function representation (Fig. 3) of the speed estimator is obtained as:

$$\frac{\hat{\omega}_r}{\omega_r^*} = G(s)\left(k_p + \frac{k_i}{s}\right)/\left(1 + G(s)\left(k_p + \frac{k_i}{s}\right)\right)$$

(12)

where $\left(k_p + \frac{k_i}{s}\right)$ = is the transfer function of the PI controller (Fig. 4).

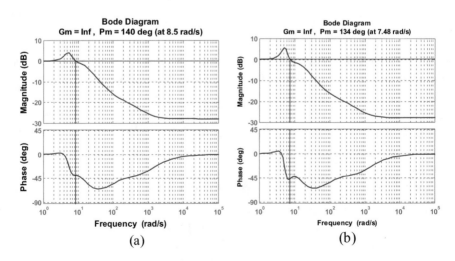

Fig. 4 Bode plot using **a** GA algorithm and **b** using GS algorithm for motoring mode

Making use of Eq. (12), the stability analysis, using linearized machine equations for the complete drive system based on GS and GA algorithms are investigated for both motoring as well as regenerating modes of operations. The drive system based on these two algorithms is found to be stable in both modes of operations confirmed by Bode Plot.

9 Simulation Results

The performance of the proposed GA and GS algorithms for loss minimization of the vector controlled IM drive is verified in Matlab/Simulink for various test conditions as described further in this section. The simulation results show the speed and losses of the drive, before and after the optimization schemes are invoked. The detailed specification of the 3-phase, 1.5 kW, IM is given in Appendix.

9.1 Step Change in Rotor Speed

The performance of the IM drive is studied in the motoring mode by a step change in the reference speed as shown in Figs. 5a and 6a for GA and GS algorithms. An increasing step change in the speed command is applied at every 10 s and the reference and actual speeds. Throughout the operation, a constant torque of 5 N m is maintained. After every speed step of change period, i_{ds} is adjusted to the optimal value by the GA and GS algorithms. The flux orientation is also altered as the i_{ds} changes. The loss of the IM drive system is shown in Figs. 5b and 6b.

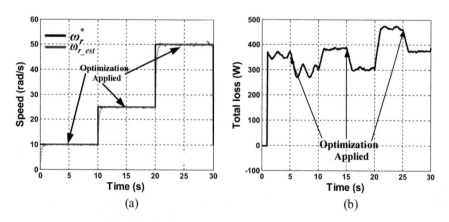

(a)   (b)

Fig. 5 Simulation results of the IM drive with GA scheme for step change in rotor speed at 5 N m load torque: **a** reference, estimated and actual speeds, **b** total loss of the drive

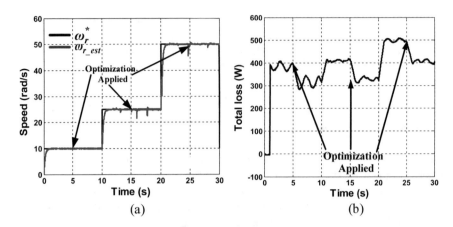

Fig. 6 Simulation results of the IM drive with GS scheme for step change in rotor speed at 5 N m load torque: **a** reference, estimated and actual speeds, **b** total loss of the drive

9.2 Regenerating Mode of Operation

The performance of the IM drive system in regenerating mode is shown in Figs. 7 and 8. The transition of the estimator from motoring mode and back keeping the load torque constant at 5 N m. The estimated speed follows the actual speed satisfactorily (Figs. 7a and 8a) for both GA and GS loss optimization scheme.

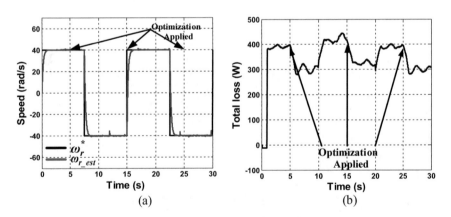

Fig. 7 Simulation results of the IM drive with GA scheme for regenerating mode of operation at 5 N m load torque: **a** reference, estimated and actual speeds, **b** total loss of the drive

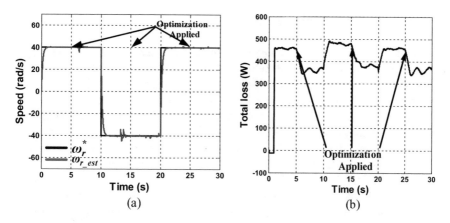

Fig. 8 Simulation results of the IM drive with GS scheme for regenerating mode of operation at 5 N m load torque: **a** reference, estimated and actual speeds, **b** total loss of the drive

The optimization algorithm comes into play at 5, 15 and 25 s. The decrement in loss when the GA optimization is working is shown in Fig. 7b, whereas for the GS optimization, the minimization in loss is shown in Fig. 8b. It can be observed from the results that GA-based optimization scheme is more efficient in reducing the loss than the GS-based scheme. The comparative diagram of total loss of the IM drive with the GA and GS optimization schemes is shown in Fig. 9.

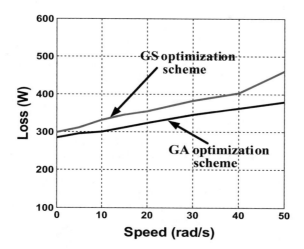

Fig. 9 Comparison of total drive's loss for IM drive with GA and GS optimization scheme

10   Experimental Validation

A laboratory prototype is developed to validate the performance of the proposed induction motor drive using dSPACE-1104. The code of the proposed estimator for dSPACE-1104 controller is generated with 10-kHz sampling frequency at 4.7-kHz inverter switching rate using MATLAB/Simulink interface. The stator windings are fed through a space-vector pulse width modulated (SVPWM) inverter operating with a switching frequency of 4.7-kHz for control purposes. The control pulses for the inverter are generated according to the proposed algorithm and field orientation. Two current sensors are employed to sense the line currents for the execution of the proposed control scheme in dSPACE-1104 controller board which has a built-in analog-to-digital converter (ADC), digital-to-analog converter (DAC) and dedicated Input/Output (I/O) ports. The results of the operation of the drive are presented in the following subsections.

10.1   Step Change of Rotor Speed

The experimental results corresponding to the simulation test are shown in Figs. 10 and 11 for GA- and GS-based loss minimization schemes, respectively. Here again, the step change in the reference speed is considered as shown in Figs. 10a and 11a. The loss of the IM drive shown in Fig. 10b for GA-based scheme and in Fig. 11b, for GS-based scheme.

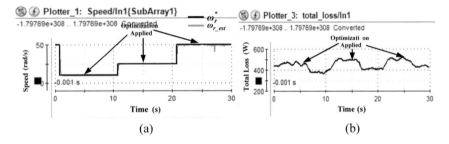

Fig. 10 Experimental results of IM drive with GA scheme for step change in rotor speed at 5 N m load torque: **a** reference, estimated and actual speeds, **b** total loss of the drive

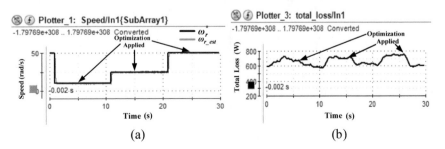

Fig. 11 Experimental results of IM drive with GS scheme for step change in rotor speed at 5 N m load torque: **a** reference, estimated and actual speeds, **b** total loss of the drive

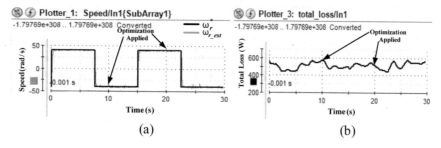

Fig. 12 Experimental results of IM drive with GA scheme for second quadrant operation at 5 N m load torque: **a** reference, estimated and actual speeds, **b** total loss of the drive

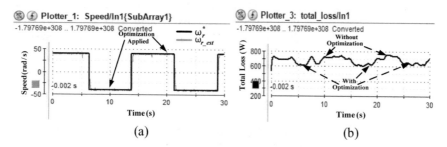

Fig. 13 Experimental results of IM drive with GS scheme for second quadrant operation at 5 N m load torque: **a** reference, estimated and actual speeds, **b** total loss of the drive

10.2 Regenerating Mode Operation

The experimental results for both GA- and GS-based loss minimization schemes in both the motoring and regenerating modes of operation are presented in Figs. 12 and 13. The load torque is kept constant at 5 N m while the reference speed is reversed alternatively at +40 rad/s and −40 rad/s every after 10 s (Figs. 12a and 13a).

The estimated and actual speeds track the reference speed satisfactorily in these conditions. The total losses of the IM drive before and after optimization algorithm applied are shown in Figs. 12b and 13b.

It can be easily perceived from the results that the performance of GA-based loss minimization scheme for IM drive surpasses that of GS-based algorithm.

11 Conclusion

This paper presents a comparison between two different loss minimization schemes for IM drive known as genetic algorithm (GA) and golden search (GS) method. The loss minimization algorithms are developed considering the drive's loss for both the schemes. The performance of the IM drive system considering GA and GS algorithms have been tested in both simulation and experiment at different operating conditions. Moreover, the stability analysis of the IM drive system in the motoring and regenerating modes of operation confirms a stable drive operation for both the schemes. GA- and GS-based schemes for IM drive have been implemented in real-time for a laboratory 1.3 kW motor using dSPACE-1104. The efficiency optimization algorithm generates the optimal value of i_{ds}^*. Hence, core loss of the drive system is minimized for optimize value of the flux level is optimized. It is found from the results that the performance of the IM drive with the GA-based optimization algorithm remarkably reduces the loss of drive system as compared to results obtained with GS-based optimization scheme.

Appendix

Machine Parameters: Rated shaft power (P_r) = 1.3 kW, Line-to-line voltage $(V_{LL\,rms})$ = 415 V, 50 Hz, Poles (P) = 4, Magnetizing inductance (L_m) = 0.7 H, Stator and rotor leakage inductance $(L_{ls} = L_{lr})$ = 0.0352 H, Stator resistance (R_s) = 5.205 Ω.

References

1. Rai K, Seksena SBL, Thakur AN (2016) On some aspects of energy conservation in industries. J Inst Eng India Ser B 97(2):233–237
2. Yatim YHM, Utomo WM (2007) To develop an efficient variable speed compressor motor system. Universiti Teknologi Malaysia (UTM)
3. Yakhelet Y (2007) Energy efficiency optimization of induction motors. Boumerdes University, Boumerdes, Algeria
4. Bose BK (2002) Modern power electronics and AC drives. Prentice Hall, New Delhi
5. Benbouzid MEH, Said NSN (1998) An efficiency-optimization controller for induction motor drives. IEEE Power Eng Rev 18(5):63–64
6. Kirschen DS, Novotny DW, Suwanwissot W (1984) Minimizing induction motor losses by excitation control in variable frequency drives. IEEE Trans Ind Appl IA-20(5):1244–1250
7. Vieira RP, Gastaldini CC, Azzolin RZ, Grundling H (2014) Sensorless sliding-mode rotor speed observer of induction machines based on magnetizing current estimation. IEEE Trans Ind Electron 61(9):4573–4582
8. Li J, Ren H, Huang Q, Zhong Y (2010) A novel on-line MRAS rotor resistance identification method insensitive to stator resistance for vector control systems of induction machines. In: IEEE international symposium on industrial electronics, Bari
9. Wei R, Gensior A (2016) A model-based loss-reduction scheme for transient operation of induction machines. In: 18th European conference on power electronics and applications (EPE'16 ECCE Europe), Sept 2016
10. Vaez-Zadeh S, John VI, Rahman MA (1999) An on-line loss minimization controller for the Interior Magnet motor drives. IEEE Trans Energy Convers 14(4):1435–1440
11. Sousa GCD, Bose BK, Cleland JG (1995) Fuzzy logic based on-line efficiency optimization control of an indirect vector-controlled induction motor drive. IEEE Trans Ind Electron 42:92–198
12. Ta CM, Hori Y (2001) Convergence improvement of efficiency-optimization control of induction motor drives. IEEE Trans Ind Appl 37(6):1746–1753

13. Chakraborty C, Hori Y (2003) Fast efficiency optimization techniques for the indirect vector-controlled induction motor drives. IEEE Trans Ind Appl 39(4):1070–1076
14. Kusko A, Galler D (1983) Control means for minimization of losses in AC and DC drives. IEEE Trans Ind Appl IA-19(4):561–570
15. Lorenz RD, Yang SM (1992) Efficiency-optimized flux trajectories for closed-cycle operation of field-orientation induction machine drives. IEEE Trans Ind Appl 28(3):574–580
16. Garcia GO, Luis JCM, Stephan RM, Watanabe EH (1994) An efficient controller for an adjustable speed induction motor drive. IEEE Trans Ind Electron 41(5):533–539
17. Kirschen DS, Novotny DW, Lipo TA (1985) On-line efficiency optimization of a variable frequency induction motor drive. IEEE Trans Ind Appl 21:610–615
18. Zidani F, Benbouzid MEH, Diallo D (2000) Fuzzy efficient-optimization controller for induction motor drive. IEEE Power Eng Rev 20(10):43–44
19. Moreira JC, Lipo TA, Blasko V (1991) Simple efficiency maximizer for an adjustable frequency induction motor drive. IEEE Trans Ind Appl 27:940–946
20. Odofin S, Gao Z, Liu X, Sun K (2016) Robust actuator fault detection for an induction motor via genetic-algorithm optimisation. In: IEEE 11th conference on industrial electronics and applications (ICIEA), pp 468–473
21. Maitre J, Bouchard B, Bouzouane A, Gaboury S (2015) 9 Parameters estimation of an extended induction machine model using genetic algorithms. In: 9th International conference on electrical and electronics engineering (ELECO), pp 608–612
22. Rahman OAA, Munetomo M, Akama K (2013) An adaptive parameter binary-real coded genetic algorithm for constraint optimization problems: performance analysis and estimation of optimal control parameters. Inf Sci 233(1):54–86
23. Fortes MZ, Ferreira VH, Coelho APF (2013) The induction motor parameter estimation. IEEE Trans Lat Am 11(5):1273–1278
24. Rai K, Seksena SBL, Thakur AN (2018) A comparative performance analysis for loss minimization of induction motor drive based on soft computing techniques. Int J Appl Eng Res 13(1):210–225
25. Rai K, Seksena SBL, Thakur AN (2018) Loss minimization of vector control induction motor drive using genetic algorithm. J Adv Res Dyn Control Syst 10(1):152–164

Space Vector Controlled Single-Phase to Three-Phase Direct Matrix Converter Drive

Manoj A. Waghmare[ID], B. S. Umre[(⊠)], M. V. Aware[ID], and Anup Kumar[ID]

Visvesvaraya National Institute of Technology, Nagpur 440010, India
{manoj10waghmare, anup4321kumar}@gmail.com,
{bsumre, mvaware}@eee.vnit.ac.in

Abstract. This paper presents a study of space vector modulated single-phase to three-phase (1 × 3) direct matrix converter (DMC). In the direct matrix converters, DC link is not present; therefore, switching ripples get reflected in the output of matrix converter, and to attenuate these ripples, input as well as output filters are essential for better voltage regulation and performance of the converter. Filter design which is either of input source side or load side is based on some important operating parameters in the induction motor drive control. The design of filters considering source side as well as load side parameters is also described in the paper. MATLAB/Simulink environment is used to simulate the control strategy of the single-phase to three-phase direct matrix converter with a three-phase induction motor model; feasibility and validity of the converter are also discussed.

Keywords: Indirect matrix converter · Direct matrix converter · Rectifier + inverter · AC–AC · SVM · Voltage conversion ratio · Unity power factor (UPF)

1 Introduction

Nowadays, various electrical applications like traction motors and some home appliances are connected with single-phase AC supply and drive the three-phase induction motors. To control and maintain constant frequency at the load side, a front-end rectifier connected with an inverter is generally employed in single-phase to three-phase conversion system. In rectifier to inverter scheme, it is very necessary to compensate for the difference in instantaneous power, as instantaneous power in single-phase system varies with time, while in three-phase system, it does not vary with time [1]. Therefore, a large volume power decoupling capacitor is needed as intermediate just after the rectifier. But, this capacitor makes the system more bulky, for example, in traction system, the weight of power decoupling capacitor of the rectifier reaches approximately 10% of the whole converter (rectifier + inverter), and hence to reduce the weight of the traction supply, system becomes difficult. Additionally, overall efficiency gets reduced in back-to-back conversion system as DC-link capacitor causes high distortion in input current with total harmonic distortion (THD) which can cause up to 140%.

© Springer Nature Singapore Pte Ltd. 2020
N. Goel et al. (eds.), *Modelling, Simulation and Intelligent Computing*,
Lecture Notes in Electrical Engineering 659,
https://doi.org/10.1007/978-981-15-4775-1_15

To overcome all such problems, matrix converters come with number of advantages such as sinusoidal input and output waveforms, compact size, good power-to-weight ratio, regenerative capabilities and many more. In the early 80 s, Alensia and Venturini explained the basic principle for working of matrix converters [2]. Then, matrix converters are classified into two major categories, namely direct matrix converters (DMCs) and indirect matrix converters (IMCs), and after that, a full fledge research in MCs begins. Indirect matrix converter composed of a current source rectifier (CSR) and voltage source inverter (VSI) is joined together at their common fictitious DC-link point without any passive component, which leads to a compact design. On the other hand, DMCs consist of bidirectional switches, connecting m-input paths to the n-output paths to convert m-phase power supply into n-phase power supply. DMCs configuration enable the adjustment of input side power factor; in spite of the nature of the load connected to the output side, it makes easier to maintain the unity power factor at the source side.

The single-phase to three-phase direct matrix converter design is based on six bidirectional switches with three legs for each phase of the converter as shown in Fig. 1. Each leg of the converter is designed using a pair of bidirectional switches for the forward as well as reverse power flow. It is a forced commutated converter which performs power conversion directly from AC power source of one phase to another without any DC link [3]. His topology has several advantages like the absence of bulky DC-link capacitor, which makes DMC as a compact power electronics circuit which improves efficiency of the system; also one can operate three-phase motors using only a single-phase supply. DMC adopts direct AC–AC conversion technique which converts single-phase supply voltage into balance three-phase voltage that differs by 120° to each other [4].

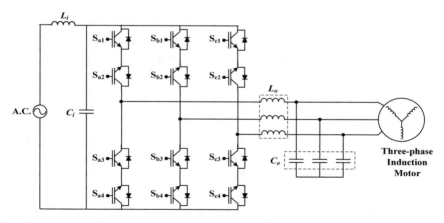

Fig. 1 Proposed single-phase to three-phase converter based on bidirectional power semiconductor switches

2 Principle of Working

The basic power circuit for single-phase to three-phase direct matrix converter can be regarded as two converters connected in series as shown in Fig. 2. According to the supply voltage, in one complete cycle of power frequency, the operation of 1×3 DMC can be divided into two different cycles. One cycle is explained above the zero reference which is for positive half cycle and another one is for negative $(-)$ cycle.

During positive period of phase 'a', in order to avoid short circuit in source side switches, S_{a1} and S_{a3} should be switched complementary to each other, and also to avoid open circuit on the load side, at least one of the switch from S_{a1} and S_{a3} should be ON. Now, if we consider an inductive load at the output of the converter, if S_{a1} is ON and S_{a3} if OFF, diode of the switch S_{a2} must provide a path for the flow of current from load side to source side. Similarly, diode of S_{a4} should also provide flow of energy from load to source when the switches S_{a3} is ON and S_{a1} is OFF. For phase 'b' and phase 'c', similar switching approach can be applied. According to the working discussed here, in positive half cycle of the source switches, S_{a1}, S_{a3}, S_{b1}, S_{b3}, S_{c1} and S_{c3} should be kept ON to have a flow of current from source side to load side, and diodes of switches S_{a2}, S_{a4}, S_{b2}, S_{b4}, S_{c2} and S_{c4} are conducting to give the flow of current from load side to source side. Similarly, in negative half cycle of the source voltage, switches S_{a2}, S_{a4}, S_{b2}, S_{b4}, S_{c2} and S_{c4} should be switched to make the flow of energy from the AC source to load, and diodes of the switches S_{a1}, S_{a3}, S_{b1}, S_{b3}, S_{c1} and S_{c3} conduct to flow the energy from load to source side.

Considering the discussed modes of operation, there are a total of eight operating states in each of positive and negative half cycles, respectively. Let us assume a switching function $S = \pm(S_a, S_b, S_c)$ where '+' denotes positive period of the source and '−' denotes negative period of the source. S_a, S_b and S_c are controlled switching states defined as in (1).

$$S_i = a, b, c = \begin{cases} P, \text{upper switch is ON} \\ N, \text{lower switch is ON} \end{cases} \tag{1}$$

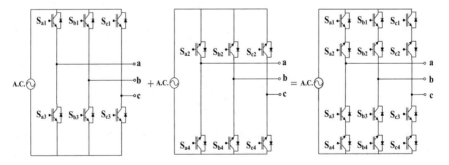

Fig. 2 Basic configuration (power circuit) of single-phase to three-phase direct matrix converter

For example, the switching state +(PPN) shows that during positive half cycle of the supply, switches S_{a1}, S_{b1} and S_{c3} are ON, while the complementary switches S_{a3}, S_{b3} and S_{c1} are OFF. All the modes of operation for (1×3) DMC for the current flow in a three-phase balance load in positive half cycle of the load are shown in Fig. 3.

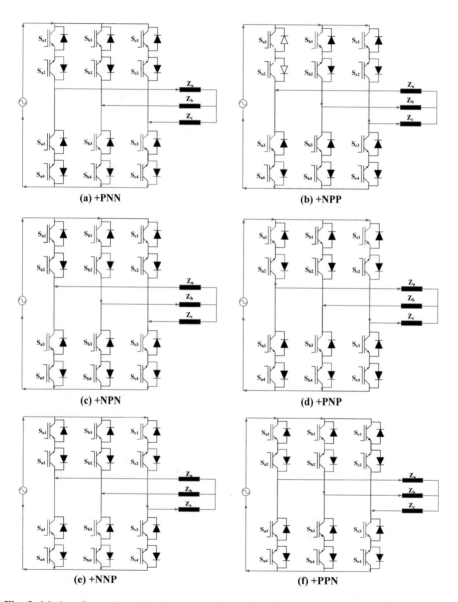

Fig. 3 Modes of operation of single-phase to three-phase direct matrix converter during positive half cycle of the source

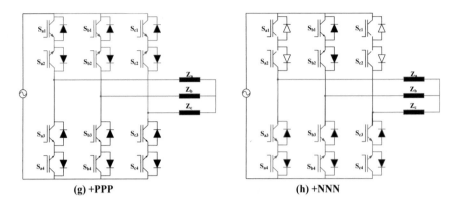

(g) +PPP   (h) +NNN

Fig. 3 (*continued*)

3   Input and Output Filter Design

The direct matrix converter for supplying electric loads acts as current source converter. This makes necessary to install input and output side LC filters for the DMCs. Ripple attenuation, internal losses reactive current loading and voltage regulation are some points of consideration to design any filter; apart from these, there are some other constraints like commutation reliability and dynamics of converter operation that need to be considered for designing of filters [5].

3.1   Input Filter Design

Many researches are going on for the input filter design of controlled rectifiers for the decades. An allowable distortion level in the voltage and current to ensure good power quality in an industrial environment is specified in IEEE519 document [6]. However, the filter complying with IEEE519 may be of active or passive type, add cost and occupy space in the system. As the direct matrix converter operates at higher switching frequency, so the input filter design of DMC becomes very easy as the input current harmonics contents are near to the switching frequency.

In order to remove the harmonics and electromagnetic interference (EMI) problems in the supply current, input side filter is generally needed in matrix converters. Since matrix converters are considered to be the current source from the source side, an LC circuit is usually present in source side [7]. A matrix converter with a small size input filter can eliminate the current spikes, and also, it avoids high-frequency interferences on the converters in the power system.

To design an input filter, several constraints like cost, volume and switching noise attenuation must be taken into account. In this paper, a single-stage damped LC filter design is considered which aims to maximize the input displacement factor (IDF). The capacitor is the main element to reduce IDF [8]. Therefore, the value of the capacitor to be optimized is given as (2).

$$C_i = \frac{I_p}{\omega V_p} \tan \left[\cos^{-1}(\text{IDF}) \right] \tag{2}$$

where

I_p peak value of input current.
V_p peak value of input voltage.
ω angular frequency of the supply voltage.

From the value of C_i and the natural angular frequency ω_n of the filter, value of inductor L_i can be calculated as (3).

$$L_i = \frac{1}{\omega_n^2 C_i} \tag{3}$$

3.2 Output Filter Design

Usually, the specified voltage regulation restricts the use of inductor as output filter component. Hence, due to high ripple content at the zero crossing of the load current, commutation based on the output current direction becomes difficult. Also, the presence of large capacitor in output side introduces large delay in the detection of current direction based on switching voltage [9].

In [10], SVM is used with proper zero vector placement so that in spite of inaccuracies in voltage measurement, safe commutation can be achieved. But, this method has the restriction to operate with input displacement angle within the voltage zero crossing, which can be decided by voltage ripples also. A complete analysis of ripple voltage therefore provides analytical design of output filters.

The output side filter is required to eliminate higher order harmonics and to maintain proper voltage regulation across the load side. If Z is considered to be total load impedance, then value of inductance L_o and capacitance C_o of the filter can be given as in (4) and (5), respectively.

$$L_o = \frac{Z}{\pi f} \tag{4}$$

$$C_o = \frac{1}{\pi f Z} \tag{5}$$

4 Implementation of Space Vector Modulation (SVM) Technique

Direct matrix converters face major difficulty in switching because of alternating and bidirectional nature of the input supply. But, in the last few decades, various modulation strategies are discussed, and many new methods of modulations are developed

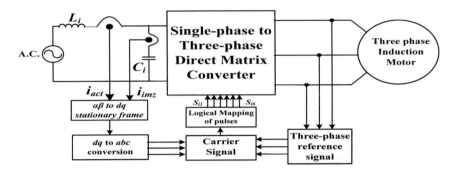

Fig. 4 Switching control block for single-phase to three-phase direct matrix converter

[11, 12]. However, in these literatures, power factor at the source side is not analyzed. Figure 4 shows switching control block for presented 1×3 DMC drive. The switching of the converter can be categorized into two different controls; the first one is switching the converter as a current source rectifier (CSR) from source side control and secondly switching the converter as a voltage source inverter (VSI) from load side control. Logical mapping of both side control pulses can give switching pulses for the 1×3 DMC.\

4.1   Source Side Current Control Strategy

The presented 1×3 DMC requires a proper modulation strategy at the source side which is needed in order to maintain unity power factor (UPF). As only one phase is present in supply side, it is necessary to convert single-phase waveform into three-phase reference waveform. Therefore, single-phase to dq frame conversion is needed, and thereafter, dq to three-phase (abc) conversion is needed to have switching pulses for CSR.

The capacitor current of the input side can be extracted as missing orthogonal component [13]. This current leads the source voltage by 90°. Now, we can consider the actual current signal as the real current (i_{act}) and capacitor current as imaginary current (i_{imz}) which corresponds to α and β components, respectively. These currents in sinusoidal form can be expressed as (6) and (7), respectively.

$$i_{act} = A \sin(\omega t) \tag{6}$$

$$i_{imz} = A \sin\left(\omega t - \frac{\pi}{2}\right) \tag{7}$$

where $\omega = 2\pi f$, f = supply frequency.

The linear transform from i_{act} and i_{imz} to dq frame is given as in (8).

$$\begin{bmatrix} i_d \\ i_q \end{bmatrix} = \begin{bmatrix} \cos\phi & \sin\phi \\ -\sin\phi & \cos\phi \end{bmatrix} \begin{bmatrix} i_{act} \\ i_{imz} \end{bmatrix} \tag{8}$$

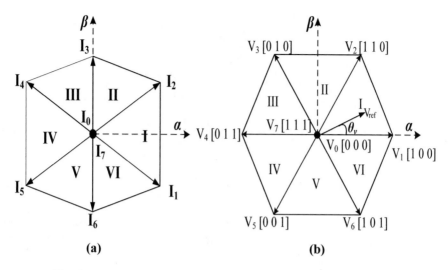

Fig. 5 Space vector-based available **a** current vectors, **b** voltage vectors

where ϕ = angle between i_{act}–i_{imz} frame and dq reference frame.

Here, the DC component of the input supply current is mapped with original signal at the power frequency in the dq frame. These dq components are then converted to the three-phase 120° shifted frame by using (9).

$$\begin{bmatrix} i_a \\ i_b \\ i_c \end{bmatrix} = \begin{bmatrix} \cos\phi & -\sin\phi \\ \cos(\phi - 120°) & -\sin(\phi - 120°) \\ \cos(\phi + 120°) & -\sin(\phi + 120°) \end{bmatrix} \begin{bmatrix} i_d \\ i_q \end{bmatrix} \tag{9}$$

These three-phase current waveforms are utilized as the reference wave to generate switching current vector for a CSR as shown in Fig. 5(a).

4.2   Load Side Voltage Control Strategy

In this control algorithm, three-phase load side voltages are needed to be sensed and used as reference waveform to generate switching pulses for a voltage source inverter (VSI). In the inverter stage, there are a total of six active vectors, and two null vectors are possible as shown in Fig. 5b.

4.3   Mapping of Voltage and Current Vectors

The load side voltage vectors and source side current vectors at any given instance are called as reference quantities. By controlling phase angle of the source side, the current vectors can be controlled. Now, by considering duty cycles, both input side current vectors and output side voltage vectors can be synthesized. From the phase angle difference between the voltage and current vectors, duty cycles can be calculated as given in equations below.

$$\delta_1 = -1^{(k_v + k_i + 1)} \frac{2}{\sqrt{3}} m_i \frac{\cos\left(\theta'_v - \frac{\pi}{2}\right)\cos\left(\theta'_i - \frac{\pi}{2}\right)}{\cos(\Delta\theta)} \tag{10}$$

$$\delta_2 = -1^{(k_v + k_i)} \frac{2}{\sqrt{3}} m_i \frac{\cos\left(\theta'_v - \frac{\pi}{2}\right)\cos\left(\theta'_i + \frac{\pi}{6}\right)}{\cos(\Delta\theta)} \tag{11}$$

$$\delta_3 = -1^{(k_v + k_i)} \frac{2}{\sqrt{3}} \frac{\cos\left(\theta'_v + \frac{\pi}{6}\right)\cos\left(\theta'_i - \frac{\pi}{2}\right)}{\cos(\Delta\theta)} \tag{12}$$

$$\delta_4 = -1^{(k_v + k_i + 1)} \frac{2}{\sqrt{3}} \frac{\cos\left(\theta'_v + \frac{\pi}{6}\right)\cos\left(\theta'_i + \frac{\pi}{6}\right)}{\cos(\Delta\theta)} \tag{13}$$

where m_i is modulation index, $\Delta\theta$ angular displacement between supply voltage and input current reference i_{ref}, k_v and k_i are voltage and current sectors, respectively, and

$$m_i = \frac{V_0}{V_i}; \theta'_v = \theta_v - (k_v - 1)\frac{\pi}{6}; \theta'_i = \theta_i - (k_i - 1) \tag{14}$$

For the fix switching frequency, duty cycle of the zero vector (δ_0) is such that sum of all the duty cycles should be unity.

$$\delta_0 = 1 - (\delta_1 + \delta_2 + \delta_3 + \delta_4) \tag{15}$$

4.4   Voltage Conversion Ratio (VCR)

For modulation index given in (14) and constraint that all the duty ratios should never be zero, it is found that at the modulation index of 0.58, there is a changeover of linear modulation to overmodulation. Hence, for single-phase to three-phase direct matrix converter, maximum output voltage is 0.58 times of the supply voltage.

5   Simulation Parameters and Results

The validation of the presented (1×3) DMC is done by simulating the complete system in MATLAB/Simulink environment with the converter parameters as given in Table 1. Modeling of 50 Hz, three-phase, four-pole induction motor has been done,

Table 1 Single-phase to three-phase DMC parameters

Parameters	Symbol	Values
Supply voltage	V_s	230 V
Supply frequency	f	50 Hz
Input filter inductor	L_s	5 mH
Input filter capacitance	C_s	50 μf
Switching frequency	f_{sw}	5 kHz

Table 2 Three-phase induction motor parameters

Parameters	Symbol	Values
Rated power	P	750 W
Stator resistance	Rs	10 Ω
Stator inductance	L_s	0.05 H
Rotor resistance	R_s	6 Ω
Rotor inductance	L_r	0.03 H

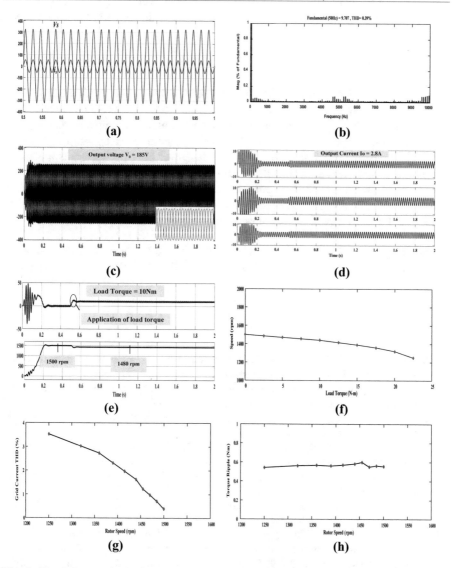

Fig. 6 Simulation results for implemented 1×3 DMC **a** input source voltage and current ($\times 5$ times the actual current), **b** THD analysis of source current (**c**) three-phase output line voltage (**d**) three-phase stator current (**e**) motor load torque (N-M) and rotor speed (Rpm), (**f**) load torque versus rotor speed (**g**) rotor speed versus grid current THD (**h**) rotor speed versus torque ripple curves

and it is fed with the converter at the load torque of 10 N-m. The modeled induction motor parameters are presented in Table 2.

Here, 1×3 DMC is simulated on different load torque conditions by the variation of mechanical torque. A total of six bidirectional switches are controlled by SVPWM technique, and results of filtered signals for the given specifications are taken along supply side and motor side as shown in Fig. 6. From the waveform, it can be seen that RMS value of three-phase balanced output line voltage (V_o) and steady-state stator current (I_o) of the converter is 185 V and 2.8 A, respectively. The FFT analysis of the supply current is done, which gives 0.39% of THD for the given system parameter.

6 Conclusion

The single-phase to three-phase DMC has promising features like stability of operation, reliability and efficiency. It replaces the classical (AC–DC–AC) converter for the generation of different phases of voltage signals without any interlinked DC capacitor. Here, the problem of commutation of switches is eliminated by proper switching sequence by using SVPWM techniques. The design of filter parameters is a comprehensive review of the earlier literatures. To avoid distortion in source current during switching of DMC, LC filter is provided in source side, and good performance is exhibited. By simulating three-phase induction motor driven by 1×3 DMC, it can be stated that the converter is having strong capability of practical applications.

Acknowledgements. This work is sponsored and supported by Department of Science and Technology, Science and Engineering Research Board (DST-SERB), India. Bearing project reference number EMR/2016/007657.

References

1. Saito M, Matsui N (2008) A single to three-phase matrix converter for a vector-controlled induction motor. In: IEEE Industry Application Society Annual Meeting, pp 1–6
2. Alesina A, Venturini M (1989) Analysis and design of optimum amplitude nine switch direct ac–ac converters. IEEE Trans Power Electron 4(1):101–112
3. Filho M, Filho A, Ruppert E (2012) A three-phase to three-phase matrix converter prototype. Revista controle & Automacao 23(3):247–262
4. Velu V, Marium N, Amran M, Farzilah N (2017) Realization of single phase to three phase matrix converter using SVPWM algorithm. Automatica 57(1):129–140
5. Dasgupta A, Sensarma P (2014) Filter design of direct matrix converter for synchronous applications. IEEE Trans Ind Electron 61(12):6483–6493
6. IEEE recommended practices and requirement for harmonic control in Electrical Power System, IEEE standard 519-1992
7. Huber L, Borojevic D (1995) Space vector modulated three-phase to three-phase matrix converter with input power factor correction. IEEE Trans Ind Appl 31(6):1234–1246
8. Zargari N, Joos G, Ziogas P (1994) Input filter design for PWM current-source rectifier. IEEE Trans Ind Appl 30(6):1573–1579

9. Gopinath D, Ramnarayanan V (2007) Implementation of bi-directional switch commutation scheme for matrix converter. In: Proceedings of National Power Electronics Conference, IISc Banglore, India
10. Shi H, Lin H, He B, Wang X, Yu L, Au X (2012) Implementation of voltage based commutation in space vector modulated matrix converter. IEEE Trans Ind Electron 59 (1):154–166
11. Vadillo J, Echeverria JM, Galarza A, Fontan L (2008) Modelling and simulation of space vector modulation techniques for matrix converters: analysis of different switching strategies. In: International Conference, Wuhan, China, pp 1299–1304
12. Nakata Y, Itoh JI (2012) Pulse density modulation control using space vector modulation for a single-phase to three-phase indirect matrix converter. In: Proceedings of IEEE Energy Conversion Congress and Exposition
13. Ryan MJ, Lorenz RD (1997) A synchronous frame controller for a single phase sin wave inverter. In: Proceedings of applied power elctronics conference, vol 2, pp 813–819

A Power- and Area-Efficient CMOS Bandgap Reference Circuit with an Integrated Voltage-Reference Branch

Santunu Sarangi, Dhananjaya Tripathy, Subhra Sutapa Mahapatra, and Saroj Rout[✉]

Silicon Institute of Technology, Bhubaneswar 751024, India
saroj.rout@silicon.ac.in

Abstract. This work presents a compact and low-power bandgap voltage-reference design using self-biased current mirror circuit. This design eliminates the standard complementary-to-absolute-temperature (CTAT) bipolar device in the voltage-reference branch, reducing the bipolar area by 20%. Instead, the design shares the same bipolar device in the main CTAT branch for generating the reference voltage. An additional benefit of eliminating the voltage-reference branch is the reduction of total power consumption by approximately 30%. This novel topology reduces power and area of the core bandgap reference circuit without compromising temperature drift performance. Designed, fabricated and functionally tested in a 0.6um CMOS process. The simulation result shows the temperature coefficient of this design is 6.3 ppm/°C for a temperature range of −40 to 125 °C. This bandgap reference design occupies a silicon area of 0.018 mm^2 and draws an average quiescent current of 2 μA from a supply voltage of 3.3–5 V. The simulated flicker voltage noise is 4.34 μV/√Hz at 10 Hz.

Keywords: Bandgap · Voltage-Reference · Temperature coefficient · Self-Biased current mirror · CMOS

1 Introduction

Voltage-reference circuits are an essential block in most applications from a simple integrated circuit (IC) to a large System-on-Chip (SoC) ranging from purely digital circuits to mixed-signal applications such as Analog to Digital converters (ADCs), Digital to Analog converters (DACs), phase-locked loops (PLLs), low noise amplifiers (LNAs), digital multimeters, battery chargers, low-power IoT sensor nodes, portable data acquisition systems and so on. Since the first bandgap reference (BGR) circuit introduced by Robert Widlar in 1971 [1], BGR has been widely used since it provides a well-defined voltage-reference with a very weak dependence on process, voltage and temperature. Most analog and mixed-signal circuits also require a current reference that sets the internal bias current for the circuits. The BGR can also provide a reference current directly which has positive temperature coefficient (PTC). For most bias

© Springer Nature Singapore Pte Ltd. 2020
N. Goel et al. (eds.), *Modelling, Simulation and Intelligent Computing*,
Lecture Notes in Electrical Engineering 659,
https://doi.org/10.1007/978-981-15-4775-1_16

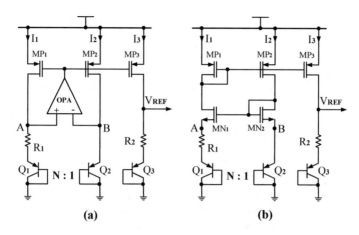

Fig. 1 Traditional BGR circuits; **a** using Op-Amp, **b** using self-biased current mirror

currents, a PTC reference current is sufficient. For circuits demanding more stable current reference can achieve so with some additional circuits [2].

Figure 1 shows two types of traditional BGR circuit: (a) one using operational amplifier (Op-Amp) and (b) using a self-biased current mirror circuit [3]. The principle of operations is the same in both cases where the nodes 'A' and 'B' are forced to be the same by (a) the Op-Amp or (b) the self-biased current mirror. Forcing same node voltages makes the voltage drop across R_1 be exactly difference between the base-to-emitter voltage (V_{BE}) of the two bipolar transistor provided that, the size of the transistor $Q_1 = N \cdot Q_2$. The voltage across the resistor R_1 produces a proportional-to-absolute-temperature (PTAT) voltage, which is multiplied with a suitable constant and added to V_{BE} of Q_3 to generate a stable voltage [4] as follows:

$$V_{REF} = V_{BE3} + V_T \cdot \frac{R_2}{R_1} \cdot \ln(N) \tag{1}$$

where V_{REF} is the output reference voltage and V_T is the thermal voltage of the semiconductor.

Typically, Op-Amp based BGR is preferred over self-biased for better power supply rejection (PSR) performance and lower supply requirement. Although the self-biased BGR may have a lower performance in those two metrics, it is a simpler design consuming less area and power while achieving almost similar temperature drift performance. In most IC or SoC designs the self-biased BGR performance may suffice to allow less design time, lower risk, lower area and power which is always desirable for any SoC design. Moreover, the PSR in a self-biased BGR can be improved by using cascade current mirrors [3] or symmetric biasing of both the branches [5].

In this paper an improved self-biased based bandgap reference circuit has been proposed which further lowers the area and power of the reference circuit while preserving the temperature coefficient performance. The improved circuit generates the reference voltage without using the separate *reference-voltage* branch as in the traditional self-biased BGR.

This paper is organized as follows: Sect. 2 describes the proposed architecture of the BGR along with its design procedure and circuit implementation. Simulation and measurement results are presented in Sect. 3, followed by a conclusion in Sect. 4.

2 Proposed Bandgap Reference

Figure 2 shows the core part of the proposed bandgap reference circuit. As evident from the figure, this modified circuit avoids a bipolar device in the reference branch. Here the BJT Q_2 used for a dual purpose; firstly, it helps for generating a PTAT voltage across resistor R_1 and secondly, voltage across this adds with voltage across R_2 for generating reference voltage (V_{REF}) at the output node. This elegant modification in the traditional self-biased current mirror-based BGR provides some great advantages particularly in power consumption and silicon area of the core circuit. These advantages are:

- Since we eliminate the standard voltage-reference branch, the bipolar device area reduces by approximately 20% and the PMOS current mirror area reduces by approximately 30%. Note that, bipolar devices and the current mirrors are a significant portion of the core BGR area.
- One-third of the total current is reduced in the core BGR and therefore one-third reduction in power consumption in the core BGR circuit as well.

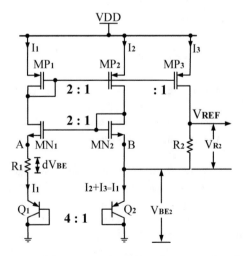

Fig. 2 Core part of the proposed self-biased current mirror-based BGR

The self-biased current mirror uses two PMOS transistor MP_1, MP_2 and two NMOS transistor MN_1, MN_2. These four transistor forms the self-biased feedback loop which makes the node voltages at 'A' and 'B' equal. The second branch of the circuit uses a single bipolar device Q_2, which produces a CTAT voltage V_{BE2} across the BJT Q_2, whereas, in the first branch, four parallel BJTs are connected with a resistor R_1 in series. As both the node voltages at 'A' and 'B' are the same and current flowing through both the BJTs are same, a PTAT voltage dV_{BE} produced across resistor R_1.

$$dV_{BE} = V_{BE2} - V_{BE1} \tag{2}$$

where V_{BE1} is voltage across the four parallel BJTs Q_1.

As V_{BE2} is a CTAT voltage and dV_{BE} is a PTAT voltage, so the addition of CTAT voltage with some appropriate constant multiplication of the PTAT voltage will generate a reference voltage which will be zero temperature coefficients at a reference temperature.

The power supply rejection (PSR) performance does not change significantly from the traditional self-biased BGR. The PSR can be improved by using cascode current mirrors [3] or symmetric biasing of both the branches [5]. Our proposed integration of reference branches will also work with symmetric biasing as shown in [5].

2.1   Design Procedure of Improved BGR

In this section, the expressions to calculate the resistance values of the core BGR circuit for a current value will be shown. For a low-power BGR, Q_1 and Q_2 were each biased with 1 µA. Given the bias current, R_1 can be expressed as:

$$R_1 = V_T \cdot \ln(4)/I_1 \tag{3}$$

where V_T is the thermal voltage of the semiconductor and its value at room temperature is approximately 25.8 mV. Applying the values of V_T and I_1 in Eq. (3), R_1 evaluates to 35.76 kΩ.

The reference voltage can be calculated by combining the voltage across the BJT Q_2 (CTAT in nature) and the voltage across the resistor R_2 (PTAT in nature) as;

$$V_{REF} = V_{BE2} + V_{R2} \tag{4}$$

where V_{R2} is the PTAT voltage across the resistor R_2 and can be expressed as:

$$V_{R2} = V_T \cdot \frac{R_2}{R_1} \cdot \ln(4) \tag{5}$$

Equation (4) can be rewritten as;

$$V_{REF} = V_{BE2} + \alpha \cdot V_T \tag{6}$$

where $\alpha = \frac{R_2}{R_1} \cdot \ln(4)$ is a constant.

For calculating zero temperature coefficient reference voltage at the reference temperature, the derivative of V_{REF} should be zero.

$$\frac{\partial V_{REF}}{\partial T} = \frac{\partial (V_{BE2} + \alpha \cdot V_T)}{\partial T} = 0 \tag{7}$$

Using $\frac{\partial V_{BE2}}{\partial T} = -1.6 \text{ mV/°C}$ and $\frac{\partial V_T}{\partial T} = 85 \text{ μV/°C}$ [4] in Eq. (7), α evaluates to 18.82.

Now we can calculate V_{REF} value from the Eq. (6) as:

$$V_{REF} = 0.67 \text{ V} + 18.82 \times 25.8 \text{ μV} = 1.155 \text{ V}$$

For this modified architecture the current flowing through the resistor R_2 is half of that current flowing in the resistor R_1. So, the constant α for this circuit will be;

$$\alpha = \frac{R_2}{2R_1} \cdot \ln(4) \tag{8}$$

Applying α and R_1 values in Eq. (8), R_2 evaluates to 971 kΩ.

2.2   Design Procedure of Improved BGR

Figure 3 shows the complete implementation of the proposed BGR. MP_{1-3}, MN_{1-2}, R_{1-2}, and Q_{1-2} forms the core part of the bandgap and the value of the resistors are calculated in the previous sub-section. MPS_{1-2} and MNS_1 form the startup circuitry since there are two stable states. MP_B transistors are the PTAT current sources for biasing internal circuits. MN_{1-2} are biased in the deep-sub-threshold (weak inversion) region to provide the maximum g_m/I_D for a given bias current [6], which ensures the

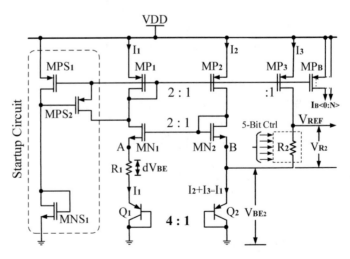

Fig. 3 Complete schematic diagram of the proposed BGR circuit

voltages of node 'A' and 'B' are only offset by the V_T mismatch of the MN_{1-2} and the systematic offset of I_1 and I_2. Typically, $(g_m/I_D) > 20$ ensures deep-sub-threshold operation. Please note that this offset is similar to an offset in an Op-Amp based BGR where the input-referred offset of the Op-Amp is dominated by the V_T mismatch of the input pair of the differential amplifier which also biased in deep-subthreshold region for low-power application.

PMOS current mirrors MP_{1-3} and MP_B are biased in the saturation region where g_m/I_D is typically less than 10 [6], to ensure the minimum systematic offset in I_1 and I_2. As mentioned before, this systematic offset can be minimized by using cascode current mirrors [3] or symmetric biasing of both the branches [5]. The unit sizes of Q_1 and Q_2 are chosen to be the minimum allowable in the implemented technology and the ratio between them is chosen such that the area of Q_{1-2} and R_{1-2} is minimized. For the implemented technology, the BJT ratio 4:1 was found to be optimum. A high-sheet-rho poly resistor ($R_{sheet} = 3.76$ kΩ/sq) was chosen to minimize the resistor area. In order to trim the output temperature coefficient (TC) of the BGR after fabrication, R_2 is a five-bit programmable resistor is used as shown in Fig. 4, which is programmed through an Inter-IC Communication (I^2C) protocol with a range of 890–940 kΩ. Each of the programmable resistor in R_2 is made of series-parallel combination of unit resistors of 20 kΩ. R_1 is also constructed from combination of same unit resistors so they can be matched in layout with R_2. During startup, MPS_2 ensures that the current mirror is pulled out of the $zero - V_{gs}$ state and once the circuit is operating normally ($V_{REF} \approx 1.155$), the voltage drop across MNS_1 should be high enough that it shuts OFF MPS_2. MNS_1 needs to be sized such that there is no leakage current during normal operation. MPS_1 provides a trickle current for MNS_1 and MNS_1 is sized with a very long length transistor to provide a large voltage drop for the minimum amount of current. For layout, special care is taken to match MP_{1-3}, MN_{1-2}, R_{1-2}, and Q_{1-2} which affects the TC directly.

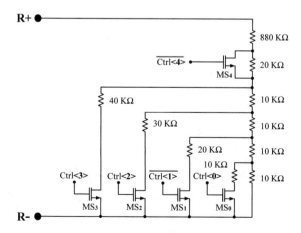

Fig. 4 Implementation of the 5-bit trimmable resistor R_2

3   Simulation and Measurement Results

3.1   Simulation Results

The improved self-biased bandgap reference has been simulated with a commercially available Spectre simulator using the Process Design Kit (PDK) from the foundry. The first-order temperature drift performance is simulated over the entire temperature range of −40 to 125 °C. A simulated reference voltage (V_{REF}) versus temperature curve is shown in Fig. 5. The calculated temperature coefficient (TC) from the figure is 6.3 ppm/°C. Figure 6, shows the parametric plot of V_{REF} versus temperature at all 32 (5-bit) trimming resistance values. The simulated PSR performance at room temperature for the improved BGR circuit is about 40 dB at DC and 35 dB at 1 kHz. The noise performance at room temperature is 4.34 $\mu V/\sqrt{Hz}$ at 10 Hz and 1.47 $\mu V/\sqrt{Hz}$ at 100 Hz which is dominated by the flicker noise of current mirrors, MN_{1-2} (46%) and MP_{1-3} (52%). The simulated average quiescent current is about 2 μA over the temperature range of −40 to 125 °C. Table 1 summarizes the simulation parameters and their corresponding simulated values.

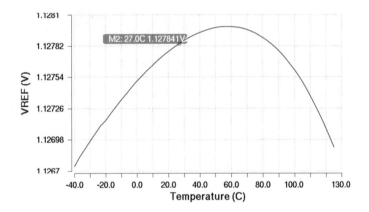

Fig. 5   Simulation result: V_{REF} versus temperature (tempco)

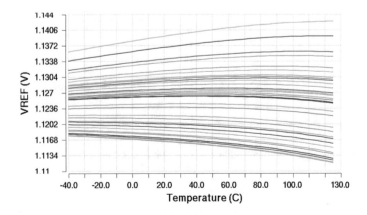

Fig. 6   Simulation result: V_{REF} versus temperature for different R_2 trims

Table 1 Summary of the simulation results

Technology	0.6 μm CMOS	Temperature coefficient	6.3 ppm/°C
Power supply voltage	3.3–5 V	Line regulation	16 mV/V
Temperature range	−40 to 125 °C	Flicker Noise@10 Hz	4.34 μV/√Hz
Reference voltage (V_{REF})	1.15 V	PSR @ DC/1 kHz	40 dB/35 dB
Quiescent current	2.08 μA		

3.2 Test Setup and Measurement Results

This work has been fabricated in a commercially available 0.6 μm CMOS technology. The proposed work has been integrated to provide a bias voltage to other blocks inside the chip. Figure 7 shows the chip micrograph with highlighting the proposed BGR and its corresponding layout view. The whole BGR consumes 0.018 mm^2 of silicon area inside the chip.

At the time of this writing, the ability to do a full temperature characterization using an environmental chamber along with R_2 trimming through I^2C was unavailable. A functional test of the fabricated BGR was done using the test setup as shown in Fig. 8 with the R_2 set to the default value. For the functional test, the packaged silicon chip is mounted on a temporary prototype board to test the functionality. We used a buffer (OP-90) at the output of the chip to avoid loading from the low-impedance measurement device. A hot air stream was used to heat the device to temperatures between 25 and 100 °C from the top side of the chip.

The temperature was changed by changing the distance between the source of the hot air stream and the device. The output of the BGR was measured using high precision (6-1/2 digit) voltage meter (Keysight 34461A). After each temperature value settled, the temperature of the device was measured using a mounted laser-guided infrared thermometer. The device was powered using a programmable power supply (Keysight E3631A).

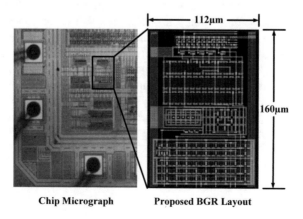

Chip Micrograph   Proposed BGR Layout

Fig. 7 Chip micrograph and layout of proposed BGR

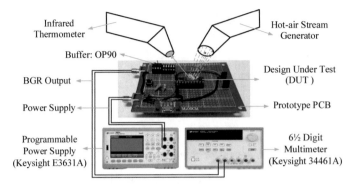

Fig. 8 Test setup for the functional verification of the fabricated BGR

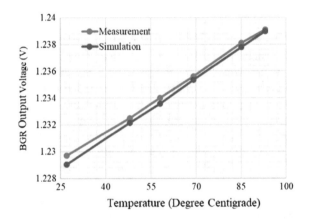

Fig. 9 Temperature coefficient (tempco) plot

Figure 9 shows the measurement result of output voltage versus temperature. As seen from the result, the untrimmed temperature coefficient is strongly PTAT in nature (115 ppm/°C). Some of the random mismatch pairs that could contribute to this are MP_{2-3}, R_1/R_2 ratio, Q_{1-2} and MN_{1-2} as well. In the simulation, when R_2 is increased by 6.5% and V_T offset value of $\sigma V_T/\sqrt{A}$ is applied between MP_{2-3} and MN_{1-2} the simulation results match the test result as shown in Fig. 9.

For the same offsets added as for the tempco simulation, the line regulation in both simulations and measurements match closely showing a line regulation of 16 mV/V as shown in Fig. 10. On availability of an environmental chamber, we will be able to get to the root of the tempco response by doing accurate temperature characterization for different R_2 trim value.

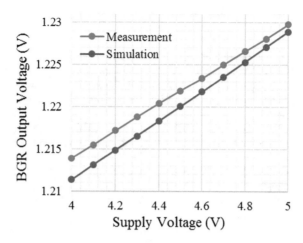

Fig. 10 Line Regulation plot

4 Conclusion

In this paper, a self-biased based BGR was improved for area and power by eliminating the reference voltage branch and integrating it in the main core without compromising temperature drift performance. By using the CTAT voltage in the core of the BGR to generate the reference voltage (V_{REF}), the power consumption of the core and area of the BJTs reduces by 33% and 20%, respectively. The BGR is implemented in a 0.6-μm CMOS process with an area of 0.018 mm^2mm^2 that includes the core bandgap and bias currents. This architecture greatly simplifies the design complexity with a temperature coefficient of 6.3 ppm/°C for a temperature range of −40 to 125 °C from the simulation. The simulated PSR is 35 dB at 1 kHz which can be improved by using the cascode self-biased current mirror. This architecture gives a spot noise of 4.34 μV/√Hz dominated by the flicker noise of NMOS and PMOS current mirrors. The flicker noise can be reduced by increasing the area of those devices or chopping the current mirror. Table 2 shows a comparison of the core performances with previously published work.

Table 2 Comparison of core performance specifications with previously published work

	Unit	[5]	[3] First- order	[3] Second- order	This work (conventional)	This work (improved)
Tech		0.35-μm CMOS			0.6-μm CMOS	
Supply	V	3.5	3.3	3.3	3.3–5	3.3–5
Current	μA	9.8	260	260	3.12	2.08
Area	mm^2	0.0432	–	–	0.0216	0.018
Tempco	ppm/°C	38.3	22	7	6.3	6.3

References

1. Widlar R (1970) New developments in IC voltage regulators. In: 1970 IEEE international solid-state circuits conference. Digest of Technical Papers, vol XIII, pp 158–159
2. Ji Y, Jeon C, Son H, Kim B, Park H, Sim J (2017) 5.8 A 9.3 nW all-in-one bandgap voltage and current reference circuit. In: 2017 IEEE International Solid-State Circuits Conference (ISSCC), pp 100–101
3. Wu W, Zhiping W, Yongxue Z (2007) An improved CMOS bandgap reference with self-biased cascoded current mirrors. In: 2007 IEEE Conference on Electron Devices and Solid-State Circuits, pp 945–948
4. Allen PE, Holberg DR (2012) CMOS analog circuit design. OUP USA (2012)
5. Lam Y, Ki W (2010) CMOS Bandgap references with self-biased symmetrically matched current–voltage mirror and extension of sub-1-V design. IEEE Trans Very Large Scale Integr (VLSI) Syst 18(6):857–865
6. Harrison RR, Charles C (2003) A low-power low-noise CMOS amplifier for neural recording applications. IEEE J Solid-State Circ 38(6):958–965

A Dynamic Base Data Compression Technique for the Last-Level Cache

Shreya Jayateerth Joshi, Prashant Mata$^{(\boxtimes)}$, and Nanditha Rao

International Institute of Information Technology, Bengaluru, India
prashant.mata@iiitb.org, nanditha.rao@iiitb.ac.in

Abstract. Cache compression improves the efficiency of a cache by increasing the effective cache size through compression and compaction of data blocks. In this paper, we propose a data compression technique which determines the base value of a cache line dynamically and stores the deltas with respect to this base, the base could be 2 bytes (B2), 4 bytes (B4) or 8 bytes (B8) in size. The dynamic base is chosen such that it maximizes the total number of compressed blocks in a cache line. We implement two types of dynamic base techniques which we call the B2B4 (combines B2 and B4) and B4 techniques. These dynamic base techniques are tested on image workloads and the results are compared against the fixed base compression technique. We see a 52.31% improvement in the number of compressed bytes over the fixed base method on an average, for B2B4 technique, which translates to an average improvement of 3.95% and a maximum improvement of 10.5% in compression factor. We also proposed a cache compaction scheme which utilizes the B2B4 compression technique and finds that such a scheme saves 8.2% of the cache area. We implemented the proposed scheme on an FPGA to analyze the performance and hardware overhead.

Keywords: Cache compression · Dynamic base · Image application · Compression factor · Compaction · FPGA

1 Introduction

Cache compression increases the effective size of a cache, through compression and compaction (layout/placement) of data blocks. Cache compression technique when implemented should have low hardware complexity, low latency, and high compression factor. Chen et al. [1] proposed a compression algorithm (C-pack) which compresses data based on the data patterns. It compresses multiple data simultaneously using a single dictionary and has a high decompression latency. Another technique called Base-Delta-Immediate (BDI) compression [2] takes advantage of the low dynamic range found in data patterns and represents the data in a compact form using a base and relative differences with respect to the base (hence called B + Δ). A dictionary-based compression scheme is proposed in [3] where the authors improve upon the previous methods. This compression scheme is aware of the data that is present in multiple neighboring blocks.

© Springer Nature Singapore Pte Ltd. 2020
N. Goel et al. (eds.), *Modelling, Simulation and Intelligent Computing*,
Lecture Notes in Electrical Engineering 659,
https://doi.org/10.1007/978-981-15-4775-1_17

A frequent pattern compression (FPC) technique [4] predefines seven patterns and encodes them into 3 bits each. If the block matches any predefined pattern, then it is encoded with the three bits of that pattern, otherwise, it is stored as it is. Adaptive cache compression [5] technique dynamically selects between the uncompressed and compressed cache line based on whether the decompressor overhead is high or low. Zhang et al. [6] proposed a method called Frequent value compression (FVC), based on frequent value locality, that is some values occur very frequently in memory.

In this paper, we propose a compression scheme which inherits few features from BDI, but uses a dynamic base instead of a fixed base to compress the data. Our algorithm assumes that the size of the data block to be compressed is 32 bytes, and the base can be either 2 bytes, 4 bytes or 8 bytes wide (we call this B2, B4, and B8 respectively). The algorithm searches through the cache line for a base which when chosen, will result in maximum compression. Further, in BDI, a cache line is considered for compression only if "all" the blocks are compressed with respect to the selected base. In our scheme, we store the deltas with respect to the chosen base, even if not "all" blocks are compressible. This scheme was motivated by analyzing a set of image data, which shows that having a dynamic base improves compression. Hence, all our analysis in this paper is presented with respect to image workloads. To the best of our knowledge, this is the first attempt at performing a detailed analysis of the cache compression technique on hardware.

The rest of the paper is organized as follows. In Sect. 2, we describe the motivation behind our proposed approach. In Sect. 3, we describe the dynamic base algorithm and its implementation. We present our results in Sect. 4. Section 5 concludes the paper.

2 Motivation

The BDI compression scheme [2] compresses a cache line with fixed base values, that is, the base can be either the first 2-bytes, 4-bytes, or 8-bytes in the cache line. For example, consider the cache line shown in Fig. 1a. If the fixed base BDI compression technique is adopted, the base will be the "first set" of 4 bytes, that is "29272f25", and the remaining groups of data will result in 4 bytes of delta with respect to the base. This is called the Base 4 BDI technique. The cache line cannot be compressed as shown in Fig. 1b. So, a large number of cache lines can remain uncompressed since the base is always 'fixed'. With our proposed dynamic base compression method, we choose "07070707" as the base and store the deltas with respect to it as shown in Fig. 1c. The deltas with respect to the first four sets of blocks are four bytes in length indicating that they are uncompressed. The remaining deltas are just one byte in length. This scheme with a 4-byte base and mostly 1-byte deltas is referred to as the Dynamic Base 4–Delta 1 (B4D1) scheme. Such a method is not allowed in the BDI algorithm, however, we note that this scheme clearly results in better compression.

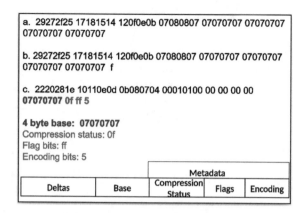

Fig. 1 a Uncompressed cache line of 32 bytes. **b** Fixed base BDI technique applied on the cache line. **c** Cache line being compressed after applying dynamic base technique. This includes four-byte deltas for the first four groups of data, one byte deltas for the remaining groups of data. Structure of the compressed cache line is also shown

3   Implementation and Simulation

3.1   Dynamic Base Algorithm

We implement the following two types of algorithms.

Dynamic B4 method. This method chooses between the following bases: an 8-byte base (called Base 8) which is fixed (the first cache block), a 2-byte base (called Base 2) which is fixed, and a 4-byte base (called Base 4) which is dynamic. Hence, the name Dynamic B4.

Dynamic B2B4 method. This method chooses between the following bases: A fixed 8-byte base, a dynamic 4-byte base and a dynamic 2-byte base.

The details of the algorithm are now described.

Choosing the base. The dynamic base is chosen as follows. For a base of 4 bytes, a cache line of 32 bytes is divided into eight blocks, which is eight possible bases. For every base, the delta with respect to all the other bases is calculated. For a base of 4 bytes, the possible deltas are 1 or 2 bytes in length. We check whether the deltas are consistently 1 byte or 2 bytes in length. The total length of the respective compressed cache line (CCL) is stored for all bases. The base that is able to compress a minimum of four blocks for delta of 1 byte and 3 blocks for a delta of 2 bytes is chosen as the final base.

Selecting the Compression scheme. Table 1 summarizes all possible compression schemes and the minimum number of blocks that should be compressible in order to adopt a particular technique. For example, there are nine ways of compressing a 32 bytes cache line using bases of size 8, 4, and 2 bytes as shown in Fig. 2. All the nine ways are employed on the same cache line and the resulting compressed cache line (CCL) is then fed to the priority-based selector. In this unit, the technique (Table 1) which gives minimum length is given the higher priority. In case there are two CCL of the same length then the decompressor complexity is considered to set the priority. For example, base 4 delta 2 and base 2 delta 1 both give a length of 212 bits. In this case,

Table 1 A summary of the size of compressed lines and the compression algorithms that lead to the compression size

Compression technique (B = Bytes)	Minimum compressible blocks	Possible lengths (compressed cache line + metadata) (B = Bytes, b = bits)	Encoding (4 bits) or CoN
All zeros	–	12b (8b + 4b)	0000
Base8B Repeated	4	68 (8B + 4b)	0001
Base8B Delta1B	4	104 (12B + 8b)	0010
Base8B Delta2B	4	136 (16B + 8b)	0011
Base8B Delta4B	4	200 (24B + 8b)	0100
Base4B Delta1B	4	116(12B + 20b), 140 (15B + 20b), 164(18B + 20b) 188(21B + 20b), 212 (24B + 20b), 236(27B + 20b)	0101
Base4B Delta2B	3	180(20B + 20b), 196 (22B + 20b), 212(24B + 20b) 228(26B + 20b), 244 (28B + 20b)	0110
Base4B Repeated	8	36 (4B + 4b)	1001
Base2B Delta1B	10	180(18B + 36b), 188 (19B + 36b), 196(20B + 36b) 204(21B + 36b), 212 (22B + 36b), 220(23B + 36b) 228(24B + 36b)	0111
No Compression	–	260 (32B + 4b)	1111

base 4 delta 2 is considered as it would require operation on only eight cache blocks as opposed to sixteen cache blocks in base 2 delta 1 scheme. This is how an efficient and shortest length CCL is selected.

Metadata. Metadata consists of compression status bits, flag bits, and encoding bits. Compression status bits indicate whether a block is compressed using the selected base or not. For example, in Fig. 1c, the first four blocks of data are uncompressed and the next four blocks are compressed. Hence, we set the compression status bits as "00001111" (0x0F). Flag bits indicate whether the cache block is larger (1) or smaller (0) than the base. In the example in Fig. 1c, all the eight blocks are either equal to or greater than the base in magnitude. Hence the bits are set to "11111111" (0xFF). The encoding bits are 4 bits in length which identify the technique chosen to compress the data. For example, in Fig. 1c, "0101" (Table 1) are the encoding bits since the scheme we used is Base 4 delta 1.

3.2   Decompression

For decompressing (see Fig. 2) a compressed cache line, the compression number (CoN) indicates the technique used for compression (for example Base 4 delta 1, Base 4 delta 2, etc.) and also tells whether the base chosen is dynamic or fixed. If the base is dynamic, then it will first check for the compression status of each block which informs the deltas corresponding to each uncompressed block. The next step is to check the flag bits and determine whether the base was smaller or greater than the uncompressed block. The cache line can be finally decompressed using the base and deltas.

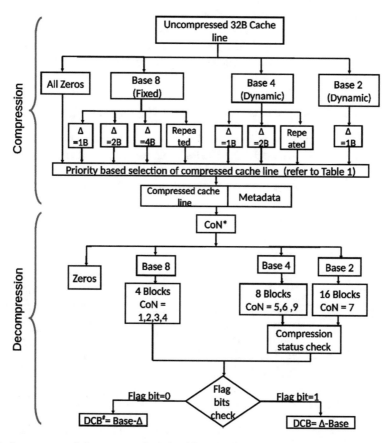

Fig. 2 A summary of the compression algorithms implemented. CoN stands for compression number. DCB is decompressed cache block

4 Simulation Results

The compression algorithms are implemented in Verilog and tested on a set of standard images chosen from a wide variety of sources at [7, 8].

4.1 Compression Factor

We find that a large number of cache lines remain uncompressed with the fixed base method as compared to the dynamic base methods. In Table 2, we present the total number of bits, before and after compression. The ratio of the number of uncompressed bits to the compressed bits is the compression factor. A ratio of less than 1 indicates that the compressed image had more bits, which is due to the added metadata. This essentially means that either the image did not get compressed or very few lines were compressed. We find that there is an improvement in the compression factor by 3.45% on an average across all workloads for the Dynamic B4 method and 3.95% on an average for the Dynamic B2B4 method. The improvement in bits reduction for Dynamic B2B4 method came out to be 52.31%.

4.2 Classification and Analysis of Size of the Compressed Lines

We report the percentage of compressed lines that got compressed into different sizes with B2B4 algorithm in Table 3. The data for compression sizes such as 196, 192, 176, 132, 112, 100, 64, and 8 are not reported since the percentage of lines with such sizes were either significantly less or zero. We classify all such lines as "Others" and sum up

Table 2 Comparison of the number of bits before and after compression

Image	Total number of bits				Compression factor		
	Uncompressed	Compressed fixed base	Compressed B4	Compressed B2B4	Compressed fixed base	Compressed B4	Compressed B2B4
Goldhill	2,097,152	2,124,932	2,081,708	2,063,176	0.9869	1.0074	1.0165
Mountain	245,7600	2,483,296	2,428,400	2,408,148	0.9897	1.0120	1.0205
Lena	2097152	2,128,508	2,082,156	2,065,740	0.9853	1.0072	1.0152
Barbara	2,097,152	2,129,040	2091,232	2,076,264	0.9850	1.0028	1.0101
Boy	3145728	3,161,200	3,143,152	3,131,400	0.9951	1.0008	1.0046
Earth	4,601,856	3,210,688	3,148,732	3,145,224	1.4333	1.4615	1.4631
Sun	29,920,512	30,386,318	30,138,900	2,983,9016	0.9847	0.9928	1.0027
Brain	5,094,144	4,858,576	4,639,616	4,629,964	1.0487	1.0980	1.1003
Lungs	1,444,864	1,426,456	1,390,040	1,386,832	1.0129	1.0394	1.0418
Black hole	404,480	213,836	193,960	193,500	1.8915	2.0854	2.0903
California fire	7,313,152	7,426,060	7,263,672	7,203,252	0.9848	1.0068	1.0153
Single atom	3,206,400	3,230,312	3,131,176	3,124,804	0.9926	1.0240	1.0261
Arctic	3,417,600	3,397,900	3,250,520	3,244,264	1.0058	1.0514	1.0534

Table 3 This table shows the data after applying the B2B4 algorithm. Each cell in the table shows the percentage of cache lines belonging to a classification. 256,240 etc. indicate the size of the cache line in bits after compression

Image	256	240	232	224	216	208	200	184	160	136	32	Others
Goldhill	80.70	1.46	4.00	4.11	2.48	3.14	0.95	1.25	0.62	0.28	0.01	1.00
Mountain	81.86	0.67	2.71	3.21	2.38	3.88	1.30	1.49	0.80	0.26	0.44	1.00
Lena	79.50	1.61	5.58	4.11	2.12	3.78	1.07	1.16	0.32	0.01	0.01	0.73
Barbara	82.74	1.34	4.58	3.59	2.09	3.14	0.79	0.90	0.28	0.05	0.00	0.50
Boy	92.46	0.57	1.28	1.79	0.76	0.83	0.24	0.26	0.27	0.15	1.11	0.28
Earth	57.79	0.07	1.10	0.21	0.07	1.16	0.06	1.04	1.13	0.97	35.97	0.43
Sun	86.81	1.69	1.84	4.15	2.27	1.89	0.70	0.25	0.04	0.01	0.00	0.35
Brain	74.58	0.40	2.93	0.78	0.31	4.18	0.18	3.58	4.13	1.71	6.57	0.65
Lungs	82.74	0.62	4.06	0.97	0.53	3.17	0.16	1.91	1.67	0.82	3.03	0.32
Black hole	23.73	0.70	3.67	0.32	0.06	5.13	0.00	3.92	4.11	2.34	55.06	0.96
California fire	78.13	1.92	6.86	4.65	2.03	4.37	0.47	1.00	0.31	0.05	0.01	0.20
Single atom	82.32	0.67	4.25	1.11	0.46	5.85	0.10	1.14	2.87	0.31	0.90	0.02
Arctic	75.14	0.76	5.78	1.16	0.34	7.64	0.12	2.29	3.60	0.75	2.30	0.12

their percentages. We find that the dynamic B2B4 algorithm results in more compressed lines as compared to the B4 algorithm.

4.3   Cache Compaction Scheme to Utilize the B2B4 algorithm

We utilize the data presented in Table 3 (B2B4 scheme), to design a cache compaction scheme for a 4-way set associative cache shown in Fig. 3. We assign two ways - way 1 and way 2 to store compressed data of size 256, 240, and 232 bits since they are the maximum occurring instances. We reserve way 3 to store the next maximum occurring instances such as 224, 216, 184,160, and 32 bits etc. The rest of the compressed data is directed to way 4. The most interesting activity happens in way 3. For instance, compressed lines of size 184 and 32 bits can be placed in the cache block adjacent to each other if their index bits are the same, thus saving the space. Extra bits will be needed along with the tag bits to reflect the compression factor. Such an ordered arrangement of compressed blocks will not only save space but also reduce the access latency since we need to search in a limited number of ways for a particular compressed length.

4.4   FPGA Resource Utilization

In Table 4, we report the FPGA resource utilization for the compressor. From the table, it can be noticed that hardware utilization by B4 algorithm is almost twice as compared to fixed base technique. The algorithms are implemented through Xilinx Vivado 2017.4 on the Xilinx Basys3 Artix-7 FPGA.

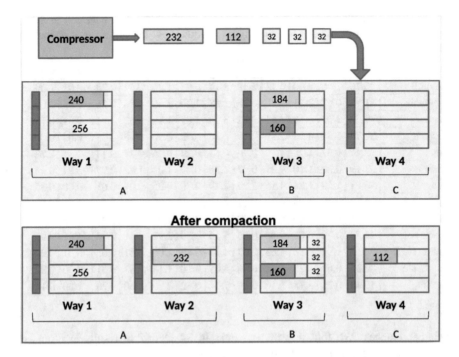

Fig. 3 The proposed compaction scheme for a 4-way set associative cache

Table 4 Compressor resource utilization in the Xilinx Basys3 Artix-7 FPGA

Algorithm	Look up tables (LUTs)	Flip-flops (FF)	Block RAMs
Fixed Base	3967	4637	19
B4	9099	4410	19
B2B4	18,541	4414	19

5 Conclusions

In this paper, we proposed a data compression technique that determines the base value of a cache line dynamically and stores the deltas with respect to this base. We tested these algorithms on image workloads and observed a 52.31% improvement in bits reduction over the fixed base method on an average. In terms of compression factor, this translated to a 3.95% improvement on an average and a maximum of 10.5% over the fixed base approach. We find that such a scheme saves 8.2% of the cache area. Overall, we have shown an improvised compression method that can be used for efficient cache compaction.

References

1. Chen X, Yang L, Dick RP, Shang L, Lekatsas H (2010) C-pack: a high-performance microprocessor cache compression algorithm. IEEE Trans Very Large Scale Integr (VLSI) Syst 18:1196–1208
2. Pekhimenko G, Seshadri V, Mutlu O, Kozuch MA, Gibbons PB, Mowry TC (2012) Base-delta-immediate compression: Practical data compression for on-chip caches. In: 2012 21st international conference on parallel architectures and compilation techniques (PACT), Sept 2012, pp 377–388
3. Panda B, Seznec A (2016) Dictionary sharing: an efficient cache compression scheme for compressed caches. In: 2016 49th Annual IEEE/ACM international symposium on microarchitecture (MICRO), Oct 2016, pp 1–12
4. Alameldeen A, Wood D (2004) Frequent pattern compression: a significance-based compression scheme for l2 caches. Technical report, University of Wisconsin-Madison Department of Computer Sciences
5. Alameldeen R, Wood DA (2004) Adaptive cache compression for high-performance processors. In: Proceedings of the 31st annual international symposium on computer architecture (ISCA '04), Washington, DC, USA. IEEE Computer Society, p 212
6. Yang J, Zhang Y, Gupta R (2000) Frequent value compression in data caches. In: Proceedings 33rd annual IEEE/ACM international symposium on microarchitecture (MICRO-33), Dec 2000, pp 258–265
7. Image/video processing/compression. https://www.hlevkin.com/06testimages.htm
8. Test images. https://homepages.cae.wisc.edu/~ece533/images/

Spatiospectral Feature Extraction and Classification of Hyperspectral Images Using 3D-CNN + ConvLSTM Model

Alkha Mohan[✉] and M. Venkatesan

Department of Computer Science & Engineering, National Institute of
Technology Karnataka, Surathkal, Manglore 575025, India
{mohan.alkha, venkisakthi77}@gmail.com

Abstract. Hyperspectral images (HSIs) are contiguous bands captured beyond
the visible spectrum. The evolution of deep learning techniques places a massive
impact on hyperspectral image classification. Curse of dimensionality is one of
the significant issues of hyperspectral image analysis. Therefore, most of the
existing classification models perform principal component analysis (PCA) as
the dimensionality reduction (DR) technique. Since hyperspectral images are
nonlinear, linear DR techniques fail to reserve the nonlinear features. The usage
of both spatial and spectral features together improves the classification accuracy of the model. 3D-convolutional neural networks (CNN) extract the spatiospectral features for classification, whereas it is not considering the
dependencies in features. This research work proposes a new model for HSI
classification using 3D-CNN and convolutional long short-term memory
(ConvLSTM). The optimal band extraction is performed by a hybrid DR
technique, which is the combination of Gaussian random projection (GRP) and
Kernel PCA (KPCA). The proposed deep learning model extracts spatiospectral
features using 3D-CNN and dependent spatial features using 2D-ConvLSTM in
parallel. Combination of extracted features is fed into a fully connected network
for classification. The experiment is performed on three widely used datasets,
and the proposed model is compared against the various state-of-the-art techniques and found better classification accuracy.

Keywords: Hyperspectral images · Convolutional neural network ·
ConvLSTM

1 Introduction

Remote sensing image analysis has become an emerging area nowadays. Various types
of remote sensing images such as radar images, multispectral images and hyperspectral
images are used for different applications based on their spatial resolution, spectral
resolution, data collection techniques, etc. The reflected energy from any material has a
unique footprint, which called spectral signature of that material [1]. This spectral
signature is used to discriminate various objects on the earth's surface. Multispectral
images are having around 3–10 number of spectral components which are captured
beyond the visible spectrum. When the number of bands in multispectral images

© Springer Nature Singapore Pte Ltd. 2020
N. Goel et al. (eds.), *Modelling, Simulation and Intelligent Computing*,
Lecture Notes in Electrical Engineering 659,
https://doi.org/10.1007/978-981-15-4775-1_18

increases, then the bandwidth decreases, and reconstruction of the complete spectral signature becomes easy. Therefore, researchers started the use of hyperspectral images for crop monitoring [2], change detection, weather prediction, etc., due to its spectral–spatial features and a large number of bands.

Hyperspectral images are contiguous band images that look like a 3D image cube. Significant challenges in hyperspectral image classification are lack of ground truth images and limited spectral library. Even though a large number of bands improve the material discrimination, it leads to the curse of dimensionality due to the smaller number of labeled samples and redundancy in bands. Hyperspectral images are nonlinear, and the application of linear reduction techniques fails in their processing [3]. Kernel PCA (KPCA), local linear embedding (LLE) and Isomap are different nonlinear DR techniques, and they are highly computational complex for large volume images.

Various HSI classification techniques work by extracting features from image either manually or through self-learning classifiers. Traditional classifiers such as SVM [4] and random forest consider only the spectral features of HSIs for classification, and thus, the classification accuracy is poor. The evolution of deep learning made a huge impact on hyperspectral image classification. Deep learning models learn the features automatically during the training phase and use them for classification.

Chen et al. [5] introduce the concept of deep learning for hyperspectral image classification at first. Many CNN-based approaches consider both spectral and spatial features [6–8]. Lee et al. [9] exploit the spatiospectral information for HSI classification using multiscale filter banks. In 2018, Chen et al. [10] proposed another spatial–spectral-based hyperspectral image classification. Different LSTM-based classification techniques are evolved for hyperspectral images. Zhou et al. [11] designed spectral–spatial LSTM for HSI classification. It is experienced that LSTM model found most relevant dependencies in spatial and spectral dimension of hyperspectral images. Since the HSIs having higher spatial resolution, neighborhood image patches will be correlated. The usage of convolution technique in LSTM helps to retain the dependencies and avoid this unfolding technique in standard LSTM. Inspired from bidirectional convolutional LSTM [12] and spatial–spectral convolutional LSTM [13], a hybrid 3D-CNN + 2D-ConvLSTM model is designed for hyperspectral image classification. With the help of 3D-CNN and convolutional LSTM, most significant features are extracted and improve the classification performance. The main contribution of the paper is: A hybrid DR technique is used for optimal band extraction, which preserves nonlinear features of hyperspectral images. Prior usage of Gaussian random projection reduces the computational complexity of KPCA. Concatenation of ConvLSTM and 3D-CNN features improves the classification accuracy.

The rest of the paper is arranged as follows. The detailed architecture of the proposed hybrid model is discussed in Sect. 2. Section 3 contains the dataset description and experimental setup. Section 4 compares the experimental results of the proposed technique with state-of-the-art techniques. Finally, the work is summarized in Sect. 5 with major findings.

2   Proposed Hybrid CNN + ConvLSTM Model

The proposed methodology is divided into two subparts, namely band extraction and classification. In classification technique, the proposed model uses the combination of two models and extracts spatiospectral features. The detailed algorithm and architecture of band extraction and classification are explained in the following subsections.

2.1   Hybrid Band Extraction

Hyperspectral image X is denoted as a 3D cube of size $M \times N \times D$, where M and N are the spatial width and height of the image and D denotes the number of spectral bands. Ground truth of the input image Y is converted using one-hot encoding and represented as $Y = (y_1, y_2, \ldots, y_C)$, where C denotes the number of classes present in the input image. In hyperspectral images, there is a possibility of redundancy in the spectral band, and this redundancy reduces the intraclass similarity. Band selection and band extraction are the primary treatments to cure the curse of dimensionality, in which dimensionality reduction through band extraction is the most popular and less computational task. The proposed classification model uses a hybrid DR technique, which is the combination of linear Gaussian random projection (GRP) and a nonlinear form of PCA (Kernel PCA).

Random projection [14] is a dimensionality reduction technique, which maps the high-dimensional data into a lower dimension by preserving the distance between the data points. Suppose the input data X having size $n \times d$ random projection maps X into a lower dimensional data Y of size $n \times k$ using a random projection matrix $R^{d \times k}$. The computational complexity of random projection is $O(knd)$.

$$Y^{n \times k} = X^{n \times d} D^{d \times k} \tag{1}$$

The idea behind random projection is Johnson–Lindenstrauss lemma; it states that while converting input data from d dimension to k dimension, its Euclidean distance is preserved with a factor of $1 \pm \varepsilon$. The computational complexity of the nonlinear technique diminished by applying GRP on input image X, thereby reducing the number of bands from D to $D/2$. The desired dimension for the nonlinear technique is not predefined. The experiment evaluation is carried out using different numbers of bands ranging from 2 to 30. The proposed technique chooses KPCA as nonlinear DR and finds the desired number of bands using cumulative eigenvalues of KPCA matrix. Here, the spatial dimension of input image retained after DR, while the spectral dimension reduced from D to B, minimum band number having cumulative eigenvalue more than 95%.

2.2   Classification Model

The input Y for the classification deep learning model is changed to $M \times N \times B$ after hybrid dimensionality reduction. The first step of the classification model is to extract the features. The usage of both spectral and spatial features improves the classification of accuracy of hyperspectral images rather than the usage of single feature alone. 3D-CNN helps to extract both the features, while it lost the dependency between nearby image patches. The proposed model is a hybridization of two deep learning models. Most relevant dependent features from HSI are extracted using ConvLSTM. The proposed model uses a window size w and generates $M * N$ number of image patches for classification. The summary of the proposed 3D-CNN model is shown in Table 1. The time step in ConvLSTM is set to B, the output dimension of band extracted data. The 3D input is converted into B number of 2D components. The summary of the proposed 2D-ConvLSTM is represented in Table 2. The classification is performed using fully connected neural networks. The features extracted from model 1 and model 2 (Output 1 and Output 2 in Tables 1 and 2) are concatenated. The summary of classification network is represented in Table 3.

Table 1 Summary of the proposed 3D-CNN model

Layer (type)	Output shape	Kernel size
Input	$15 \times 15 \times 15 \times 1$	NA
Conv3D	$13 \times 13 \times 9 \times 16$	$3 \times 3 \times 7$
Batch norm	$13 \times 13 \times 9 \times 16$	NA
Conv3D	$11 \times 11 \times 5 \times 32$	$3 \times 3 \times 5$
Batch norm	$11 \times 11 \times 5 \times 32$	NA
Conv3D	$9 \times 9 \times 3 \times 64$	$3 \times 3 \times 3$
Output1 (reshape)	$9 \times 9 \times 192$	

Table 2 Summary of the proposed 2D-ConvLSTM model

Layer (type)	Output shape	Kernel size
Input	$15 \times 15 \times 15 \times 1$	NA
ConvLSTM2D	$15 \times 13 \times 13 \times 16$	3×3
ConvLSTM2D	$15 \times 11 \times 11 \times 32$	3×3
ConvLSTM2D	$15 \times 9 \times 9 \times 64$	3×3
Output2 (reshape)	$9 \times 9 \times 960$	

Table 3 Summary of classification of fully connected network

Layer (type)	Output shape	Kernel size
Input	$9 \times 9 \times 192, 9 \times 9 \times 960$	NA
Concatenate	$9 \times 9 \times 1152$	NA
Flatten	93312	NA
Dense	16	NA
Dropout (0.2)	16	0
Dense	32	NA
Dropout (0.2)	32	0
Dense	C	

Rectified linear unit (ReLU) activation function is used in all convolution layers, and softmax is used for classification. The optimization of the proposed model is done using adam optimizer with categorical-crossentropy loss function having learning rate 0.0001 and decay 1e−06. The training process repeats for 100 epochs of batch size 100 without any augmentation on input samples.

3   Dataset Description and Experimental Setup

The proposed method is evaluated by experiments in universally available hyperspectral image datasets. All executions are done on Intel(R) Xeon(R) Silver 4114 CPU @ 2.24 GHz with a RAM of 196 GB under CentOS Linux release 7.4.1708 (Core) using Python3 programming implementation. Three hyperspectral datasets are used to evaluate the performance of the proposed method and state-of-the-art techniques. They are Indian pines (IP), Pavia University (PU) and Salinas (SA) collected from the website http://www.ehu.eus/ccwint\\co/index.php. Dimensionality reduction process sets the desired dimension D for random projection to 100, 51 and 102 (D/2) for IP, PU and SA datasets to maintain the experimental similarity. The cumulative eigenvalue of principal components in KPCA for three datasets reaches more than 95% in the band that ranges in between 12 and 18. Thus, the proposed model sets the desired dimension for hybrid dimensionality reduction to 15 for all the datasets, i.e., $B = 15$.

4   Results and Discussions

The entire dataset is randomly divided into 20% training set, 10% validation set and remaining 70% for testing. The proposed method is compared against six state-of-the-art techniques which start from conventional SVM [4] classifier to new deep learning techniques such as 2D-CNN [15], 3D-CNN [15], SSLSTM [11] and SSCL2DNN [13]. The classification performance of each method is compared against the proposed method using the evaluation parameters, overall accuracy (OA), kappa statistics (K) and average accuracy (AA). The classification result of IP, PU and SA dataset is shown in Tables 4, 5 and 6, respectively. The classification map of three datasets is shown in Figs. 1, 2 and 3. Analyzing the classification accuracy and classification map, it is found that the proposed model performs better than that of the existing state-of-the-art techniques.

Table 4 Classification result of Indian pine dataset

Class	SVM	2D-CNN	3D-CNN	SSLSTM	SSCL2DNN	Proposed
1	70.73	90.24	99.39	79.88	92.68	97.17
2	89.32	95.47	97.86	94.24	97.78	99.52
3	91.93	95.21	97.05	90.70	95.65	96.15
4	86.50	92.84	96.95	88.85	96.13	99.71
5	90.57	94.02	96.55	91.09	95.86	98.07
6	97.72	97.45	99.20	96.58	98.82	98.09
7	58.00	75.00	94.00	77.00	93.00	98.00
8	98.72	99.07	100.00	97.09	99.53	100.00
9	44.44	61.11	76.39	63.89	62.50	70.89
10	77.20	94.86	97.31	91.49	97.09	99.74
11	95.02	98.03	98.95	95.21	98.71	99.33
12	84.27	91.06	97.94	86.42	96.40	99.78
13	89.13	93.75	97.01	91.98	95.24	99.73
14	97.41	98.79	99.56	98.51	99.65	99.69
15	93.73	97.33	99.06	94.42	98.20	99.42
16	68.75	93.15	94.35	79.46	88.99	99.56
OA	91.20	96.21	98.28	93.69	97.72	99.19
AA	83.34	91.71	96.35	88.56	94.14	97.17
K	89.91	95.67	98.04	92.79	97.40	98.62

Table 5 Classification result of Pavia University dataset

Class	SVM	2D-CNN	3D-CNN	SSLSTM	SSCL2DNN	Proposed
1	52.21	79.64	82.91	70.57	90.32	97.62
2	89.78	93.79	96.25	89.90	98.72	99.95
3	17.73	44.26	66.25	32.20	55.37	82.96
4	55.23	77.30	90.61	69.21	78.87	96.52
5	68.05	77.76	93.83	77.69	87.42	98.51
6	39.14	70.75	85.72	50.47	87.04	97.75
7	15.19	38.22	77.80	35.20	35.91	93.61
8	60.56	84.19	88.48	68.19	91.34	97.41
9	53.39	79.13	68.64	85.37	75.59	88.60
OA	65.67	81.89	89.14	73.90	89.04	97.10
AA	50.14	71.67	83.39	64.31	77.84	95.44
K	51.52	75.86	85.59	64.42	85.16	96.14

Table 6 Classification result of Salinas dataset

Class	SVM	2D-CNN	3D-CNN	SSLSTM	SSCL2DNN	Proposed
1	74.26	81.35	98.59	82.02	91.64	99.91
2	85.23	86.21	99.96	82.15	97.80	100.00
3	59.56	78.14	96.22	54.62	98.77	100.00
4	98.19	97.66	99.81	95.53	99.40	100.00
5	94.33	93.45	99.66	93.05	98.83	99.99
6	94.40	94.61	99.98	95.74	100.00	100.00
7	78.97	88.90	98.74	82.30	96.23	100.00
8	85.36	93.66	89.17	85.36	93.05	99.36
9	68.44	84.14	99.84	77.75	99.89	99.97
10	71.65	87.81	99.04	70.77	98.44	99.76
11	93.98	94.54	97.45	95.90	98.08	99.68
12	88.38	87.47	99.09	86.60	99.06	100.00
13	80.04	82.21	97.46	68.72	97.02	99.45
14	84.04	88.10	99.84	91.34	99.69	99.75
15	90.36	94.59	99.49	88.93	92.78	99.48
16	64.92	78.96	99.85	61.13	93.39	100.00
OA	82.39	89.65	97.17	83.10	96.30	99.29
AA	82.01	88.24	98.39	81.99	97.13	99.83
K	80.30	88.47	96.85	81.15	95.88	99.21

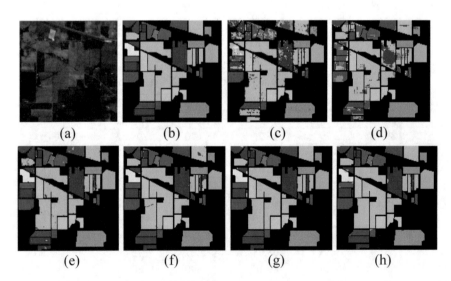

Fig. 1 Classification maps for the Indian pines dataset. **a** Sample band. **b** Ground truth. **c** SVM.
d 2D-CNN. **e** 3D-CNN. **f** SSLSTMs. **g** SSCL2DNN. **h** Proposed

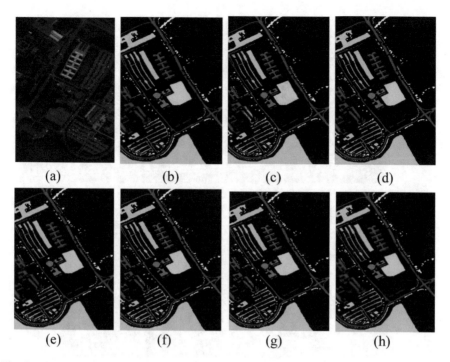

Fig. 2 Classification maps for the Pavia University dataset. **a** Sample band. **b** Ground truth. **c** SVM. **d** 2D-CNN. **e** 3D-CNN. **f** SSLSTMs. **g** SSCL2DNN. **h** Proposed

5   Conclusions

This work proposes a novel hybrid CNN model for hyperspectral image classification considering spatial and spectral features. Dimensionality reduction technique used in this model is a combination of linear and nonlinear methods. The proposed method is compared against various state-of-the-art techniques on three widely used datasets. It is found that the proposed technique produces a better result for classification. The usage of ConvLSTM helps to find the dependency between features, and thus, it leads to improve classification accuracy. The introduction of 3D-ConvLSTM model will increase the classification accuracy in the future.

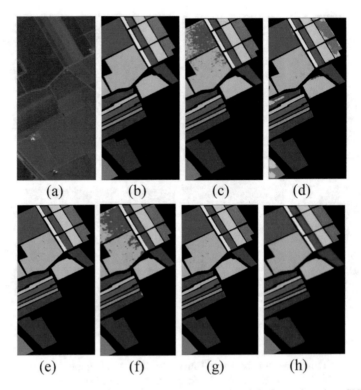

Fig. 3 Classification maps for Salinas dataset. **a** Sample band. **b** Ground truth. **c** SVM. **d** 2D-CNN. **e** 3D-CNN. **f** SSLSTMs. **g** SSCL2DNN. **h** Proposed

References

1. Jensen JR (2007) Remote sensing of the environment: an earth resource perspective, 2 edn. Prentice-Hall, Englewood Cliffs
2. Thenkabail PS, Smith RB, De Pauw E (2000) Hyperspectral vegetation indices and their relationships with agricultural crop characteristics. Remote Sens Environ 71:158–182. https://doi.org/10.1016/S0034-4257(99)00067-X
3. Huang H, Luo F, Liu J, Yang Y (2015) Dimensionality reduction of hyperspectral images based on sparse discriminant manifold embedding. ISPRS J Photogramm Remote Sens 106:42–54. https://doi.org/10.1016/j.isprsjprs.2015.04.015
4. Melgani F, Bruzzone L (2004) Classification of hyperspectral remote sensing. 42:1778–1790
5. Chen Y, Lin Z, Zhao X, Wang G, Gu Y (2014) Deep learning-based classification of hyperspectral data. IEEE J Sel Top Appl EARTH Obs. Remote Sens 7:2094–2107. https://doi.org/10.1109/jstars.2014.2329330
6. Pan B, Shi Z, Xu X (2017) MugNet: deep learning for hyperspectral image classification using limited samples. ISPRS J Photogramm Remote Sens. https://doi.org/10.1016/j.isprsjprs.2017.11.003
7. Yu S, Jia S, Xu C (2017) Convolutional neural networks for hyperspectral image classification. Neurocomputing 219:88–98. https://doi.org/10.1016/j.neucom.2016.09.010

8. Chen Y, Zhao X, Jia X (2015) Spectral-spatial classification of hyperspectral data based on deep belief network. IEEE J Sel Top Appl Earth Obs Remote Sens. https://doi.org/10.1109/jstars.2015.2388577

9. Lee H, Kwon H, Member S (2017) Going deeper with contextual CNN for hyperspectral image classification. IEEE Trans Image Process 26:4843–4855. https://doi.org/10.1109/TIP.2017.2725580

10. Chen C, Jiang F, Yang C, Rho S, Shen W, Liu S, Liu Z (2018) Hyperspectral classification based on spectral–spatial convolutional neural networks. Eng Appl Artif Intell 68:165–171. https://doi.org/10.1016/j.engappai.2017.10.015

11. Zhou F, Hang R, Liu Q, Yuan X (2019) Hyperspectral image classification using spectral-spatial LSTMs. Neurocomputing 328:39–47. https://doi.org/10.1016/j.neucom.2018.02.105

12. Liu Q, Zhou F, Hang R, Yuan X (2017) Bidirectional-convolutional LSTM based spectral-spatial feature learning for hyperspectral image classification. Remote Sens 9:1330. https://doi.org/10.3390/rs9121330

13. W. Hu, H. Li, S. Member, L. Pan, W. Li, S. Member, Feature Extraction and Classification Based on Spatial-Spectral ConvLSTM Neural Network for Hyperspectral Images, IEEE Trans. Geosci. Remote Sens. (2019) 1–15

14. Johnson WB, Lindenstrauss J (1984) Extensions of lipschitz mappings into a Hilbert space, pp 189–206. https://doi.org/10.1090/conm/026/737400

15. Chen Y, Jiang H, Li C, Jia X, Ghamisi P (2016) Deep feature extraction and classification of hyperspectral images based on convolutional neural networks. IEEE Trans Geosci Remote Sens 54:6232–6251. https://doi.org/10.1109/TGRS.2016.2584107

A Comparative Analysis of Community Detection Methods in Massive Datasets

B. S. A. S. Rajita$^{(\boxtimes)}$ ⓘ, Deepa Kumari, and Subhrakanta Panda ⓘ

CSIS Department, Birla Institute of Technology and Science Pilani,
Hyderabad Campus, Pilani, India
{p20150409, p20190020, spanda}@hyderabad.bits-pilani.
ac.in

Abstract. Nowadays there is a boom in social network data streaming from various fields of interest related to finance, engineering, medicine, and general sciences. All these data are modeled as graphs for better analysis. Community detection is one such mechanism for the analysis of such massive data. Many community detection algorithms exist in literature. The existing algorithms are compared by using either real-world or artificial networks (modeled as graphs) but not both. This paper aims to make a comparative study of two popular existing community detection algorithms both on real-world and synthetic data and verify their performance. The approach in this paper makes good use of recent advances in graphical modeling of different social networks. We generated a random graph that represents most of the observed properties of a real-world dataset. The experimental results are tabulated and the computed metrics help in inferring the suitability or scalability of an algorithm for small or massive datasets.

Keywords: Community detection · Social networks · Evaluation metrics

1 Introduction

A Social Network (SN) modeled as a graph is a social structure that represents interaction among the social entities and their relationship. The relationship among the social entities forms interactions in the SN. Social entities are represented as nodes and the interactions are represented as edges in the graphical representation [1]. One important aspect of such graphical representation is to identify a set of nodes that are densely connected among themselves but are sparsely connected to the remaining nodes of the graph. These sets of nodes are called, *community* [2]. Hence, *Community Detection (CD)* can be used to find users that behave similarly, detect groups of interests, and or cluster users in an e-commerce application. The increasing size of social networks like *Facebook* (approx. 1.1 billion users), *Twitter* (approx. 300 million users), *LinkedIn* (approx. 500 million users), etc. [3] has made CD more difficult. The graph of such massive data can reach up to billions of nodes and edges.

As we know that some methods often tend to perform exceptionally well or poorly on different kinds of graphs. Moreover, one might prefer to choose a specific method depending on the target application or the properties of the graph that is to be analyzed.

© Springer Nature Singapore Pte Ltd. 2020
N. Goel et al. (eds.), *Modelling, Simulation and Intelligent Computing*,
Lecture Notes in Electrical Engineering 659,
https://doi.org/10.1007/978-981-15-4775-1_19

Thus, it is necessary to analyze the performance metrics of the community detection algorithms on large size datasets. According to our study, most of the researchers [17, 18] have selected *Louvain* [4] (modularity based approach) and *Label Propagation* (*LP*) [5] (diffusion-based approach) for a comparative study on synthetic datasets. Hence, in this paper, we analyze the performance of two very popular community detection algorithms in [4, 5] to explore their efficiency in the detection of communities in massive real and synthetic datasets (randomly generated graphs).

The structure of the remaining paper is as follows: Sect. 2 gives a brief overview of two community detection algorithms. Section 3 discussed the proposed work, Sect. 4 presents the experimental setup and performance analysis of the algorithms, and Sect. 5 concludes the work with some insights into our future work.

2   Background

This section briefly describes community detection algorithms and their and working.

2.1   Community Detection

Community detection is a method to find the sets of densely connected nodes in a graphically represented social network [6]. These social networks can either be *static* or *dynamic*. It is an unsupervised learning approach similar to the clustering technique in data mining [7]. Existing community detection algorithms implement the following steps to detect the communities. First, every node is considered as a single cluster. Second, identify a set of high-density nodes as a community-based on a criterion. This criterion (such as a *set of densely connected nodes* [8], *similar topological properties* [9], *similar structural properties* [10], etc.) may vary for each algorithm. In this section, we discuss the aforementioned two well-known community detection algorithms.

2.2   Louvain Algorithm

Louvain algorithm is a hierarchical bottom-up approach [4] to find a set of communities in a given graph. It works on the principle of *modularity* to detect the existence of any community in a given graph. This approach uses the *greedy optimization* strategy for optimizing the modularity values of the detected communities. Initially, this algorithm considers each node of the input graph as a single community (singleton set). In every iteration, it considers each node and its connected neighboring nodes. When that node is included in the neighboring community then the modularity gain of that community is calculated by using the density of links inside communities. If the inclusion of that node maximizes the modularity, then that node is permanently added to the community. Otherwise, the node is added to other neighboring communities. This procedure is repeated until there is no further improvement in the modularity of that community.

2.3   Label Propagation (LP)

LP was introduced by Raghavan et al. [5]. It is a *diffusion-based* community detection approach. The concept behind this algorithm is that similar type of nodes exchanges more information and hence form communities. Initially, this algorithm assigns a unique label to each node of the graph. The next steps are finding a similar set of nodes and assign the same label to make them as one community. We can get a similar set of nodes by applying a *random walk*. It is used for finding all the shortest distance from a given node. They can be considered similar sets of nodes. The shortest distance between *node i* and *node j* is calculated by averaging the number of edges that a random walker has to traverse from *node i* to *node j*. To get this shortest distance among all the nodes, the distances are stored in a distance matrix P, wherein similar value nodes are grouped as communities. So, in this method modularity calculation is not required. Only finding a similar set and assigning the same label to those nodes is enough to detect the communities. This algorithm terminates when all the nodes are visited.

3   Proposed Work

In this section, we show the experimental setup of our implementation of CD algorithms on two different types of datasets. Then we compare their results in terms of the following parameters: *detected communities, modularity, clustering coefficient,* and *performance speed.*

We considered two well-known datasets: *Zachary Karate club* and *Enron email* datasets. Along with these standard datasets, we randomly generated three graphs for comparison purposes. The graphical model of the Zachary Karate club consists of 34 nodes and 78 edges. Whereas the graph of the Enron dataset consists of 36,692 nodes and 367,662 edges. Based on our literature review, we generated three random graphs following uniform distribution between edges and power-law distribution between nodes. These random graphs are generated by using the ***Erdos–Rényi*** [11] model by fixing the node size and generating random edges with the given uniform probability for fixing the number of connecting edges between any two nodes. These generated graphs represent datasets of varying size. The first random graph consists of 1000 nodes to represent a small size dataset, the second random graph consists of 50,000 nodes to represent a medium size dataset and the third graph consists of 100,000 nodes to represent a large dataset. The required steps for our experimentation can be summarized as given below:

1. Process the data to get the node and edge information.
2. Generate the graphical model of the corresponding datasets based on the node and edge information.
3. Generate synthetic graphs.
4. Implement community detection algorithms on all the constructed graphs.
5. Compute the metrics and compare the results.

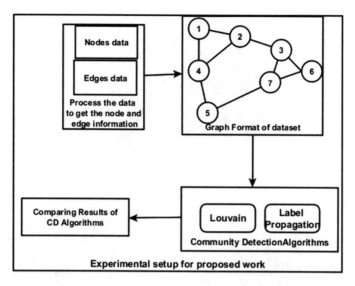

Fig. 1 Block diagram for proposed comparative work

Figure 1 shows the block diagram of the steps followed for the proposed comparative work. The generated graphs are given as input to the CD algorithms (Louvain and Label Propagation). The communities detected by these algorithms are compared based on some evaluation metrics mentioned in Sect. 4.

3.1 Detected Communities Visualization

Visualization provides an accessible way to see and understand the trends, outliers, and patterns in data. In Fig. 2, we show the communities detected by the CD algorithms on a synthetic dataset. Figure 2a shows the graphical model of the Synthetic dataset that is given as input to Louvain and LP algorithms. Figure 2b shows the communities detected by the Louvain algorithm. It detected 137 communities with the smallest and largest communities consisting of 5 and 63 nodes, respectively. Similarly, Fig. 2c shows that LP detected 116 communities with the smallest and largest community consisting of 5 and 72 nodes, respectively. This kind of visualization gives an idea of the number of detected communities. Such visualization neither help in interpreting the importance of statistical measures nor speak about the performance of the algorithms. Hence, it is important to analyze and answer the following *research questions (RQs)*:

RQ1. Can random graphs and real-world datasets form the basis of comparing community properties?
RQ2. Can evaluation metrics throw some insight into the performance of different community detection algorithms?

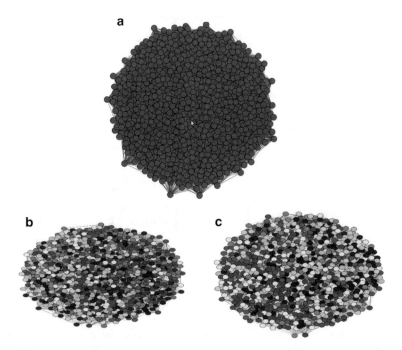

Fig. 2 a Input to the CD algorithms [synthetic dataset(1000 nodes)]. **b** The output of Louvain. **c** The output of Label Propagation

4   Experimental Results

In this section, we discuss the obtained results, and comparative analysis of evaluation metrics using illustrative methods (Table and Figures). All the experiments are conducted on a computer server which is equipped with 4Quad-CoreAMD Opteron CPU, 32 GB memory, and 12 T disk storage. The *mint OS 6.4* is installed as the operating system. Other software environments include *python* 3.6 and *pycharm* 4.5.3. All the graphs used in the experiments are generated using the *igraph* package in python.

RQ1. Can Random Graphs and Real-World Datasets form the Basis of Comparing Community Properties?
We implemented the CD algorithms (Louvain and Label Propagation) on the graphical representation of the two real-world datasets and generated random graphs. We executed each algorithm 500 times to avoid any variation or biasness in the obtained results. We have shown the average values of the results obtained from each execution. The algorithms could process 34 to 367,662 nodes within an acceptable time duration. Figure 3 represents the relation between the number of communities detected and the size of the graph (in terms of nodes). In Fig. 3, *X*-axis corresponds to the number of nodes and *Y*-axis shows the number of detected community values for the two algorithms. Our results confirm that these two CD algorithms are detecting communities for two types of datasets in a linear distribution [12]. In Fig. 3, the line named Louvain

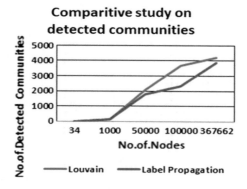

Fig. 3 Comparative study on detected communities

(blue color) shows the performance of Louvain. This method detects communities in a linear distribution with respect to the size of the graph. The *average number of detected communities* is 2023 for a graph with an average of 103,739 number of nodes. So each detected community consists of 51 nodes on average. The overall density of the detected communities is the ratio of the *average number of detected communities* and the *average number of nodes* in each detected community, which is 2023/51 = 40 (39.66). From the results, it can be inferred that the algorithm is able to detect communities uniformly for all types of data.

Similarly, in Fig. 3, the line named Label Propagation (in red color) shows the performance of LP. This method detects communities in a linear distribution with regards to the size of the graph. We can interpret that the *average number of detected communities* is 1617 for a graph with an average 103,739 number of nodes. So the *average number of nodes* in each detected community is 64. The overall density of each detected community is 1617/64 = 25. As the Louvain algorithm detects communities with higher density, therefore unlike LP, Louvain is suitable to detect communities uniformly for all the sizes of the dataset. Therefore, we can infer that random graphs and real-world datasets form a good basis for comparing community detection algorithms.

RQ2. Can Evaluation Metrics Throw Some Insight into the Performance of Different Community Detection Algorithms?

The detected communities (output of two algorithms) are compared on the basis of some evaluation metrics. These evaluation metrics are the *number of communities detected (C)*, *Modularity (M)*, and *Clustering Coefficient (CC)* [12]. Table 1 displays the performance of each algorithm expressed in terms of these evaluation metrics. The first and the last row correspond to the results on Zachary and Enron datasets, respectively. Whereas the middle rows correspond to the results on the synthetic dataset (randomly generated). In total 518,696 nodes and 87,725 edges are processed. The average time taken by the Louvain algorithm is approximately 0.14 s compared to 0.37 s taken by the LP algorithm. Modularity is one measure that signifies the strength of a graph to undergo division for forming the communities. Graphs with high modularity have dense connections between the intra-modular nodes and sparse

Table 1 Comparative study of Two CD algorithms

#Edges of graphs	#Nodes of graphs	Louvain_C	Louvain_M	Louvain_CC	Louvain_T ()	LP_C	LP_M	LP_CC	LP_T (s)
78	34	4	0.41	0.35	0.01	3	0.62	0.25	0.05
398	1000	137	0.29	0.48	0.03	116	0.57	0.07	0.09
10167	50,000	2085	0.59	0.78	0.15	1786	0.43	0.08	0.17
40390	10,0000	3675	0.79	0.75	0.47	2316	0.37	0.08	0.57
36692	367,662	4216	0.92	0.78	0.59	3865	0.31	0.08	0.69

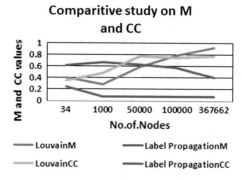

Fig. 4 Comparative study on *M* and CC

connections between inter-modular nodes. It is considered to be a good performance measure for community detection and has been used in several studies [13–15]. The clustering coefficient (CC) is a measure of the degree to which nodes in a graph tend to cluster together. It indicates that in most real-world graphs, nodes tend to create tightly knit groups characterized by a relatively high density of ties [16].

In Fig. 4, *X*-axis corresponds to the number of nodes and *Y*-axis shows the M and CC values for the two algorithms. Initially, Louvain is tested on a small size dataset having 34 nodes. It detected four communities with *M* = 0.41. Then this experiment is tested with different sizes (1000, 50,000, 100,000, etc.) of nodes (in increasing order of their node size) to find the number of detected communities. This experimental result is plotted in Fig. 4, labeled LouvainM (blue color line). It exhibits linear growth with respect to the number of detected communities versus the size of the graph. The average value of *M* is 0.55 for the synthetic graphs with an average node size of 50,333. And the average value of *M* is 0.66 for the real-world graphs with an average node size of 183,848. So, its average is greater than the mean modularity of 0.5, where modularity ranges between 0 to 1. LP tested on a small size dataset of 34 nodes, detected 3 communities with *M* = 0.35. Then this experiment is tested with different sizes (1000, 50,000, 100,000, etc.) of nodes (in increasing order of their node size) to find the number of detected communities. This experimental result is plotted in Fig. 4, named Label PropagationM (red color). Its *M* value decreases, with the increase in graph size (in terms of number nodes). The average value of *M* is 0.45 for synthetic graphs with an average node size of 50,333, and *M* = 0.46 for real-world graphs with an average node size of 183,848. So, its average is less than the mean modularity of 0.5. Any CD algorithm that detects communities with modularity greater than the mean value indicates that those algorithms are efficient in detecting the communities [12]. So, we can infer that Louvain gives better *M* value for real datasets as well as random graphs.

CC of the communities detected by the Louvain algorithm is shown in Fig. 4 and labeled as Louvain CC (green color line). It exhibits a linear growth until the size of the graph is the average of the node size. Once the number of nodes is greater than average, this line is maintained at a consistent value of CC = 0.78. Hence, we can infer that CC values of the communities are consistent in Louvain for large datasets. CC of the communities detected by the LP algorithm is shown in Fig. 4 and named Label

Propagation CC (violet color line). It exhibits a linear decrease in CC for graphs with up to 1000 nodes. Once the node size of the graph is greater than 1000, CC is consistent at 0.08. We can interpret that it is not maintaining consistent CC and M. So, our observation is that Label Propagation is suitable for small size dataset. Thus, evaluation metrics throw good insight into the performance of different community detection algorithms.

5   Conclusion and Future Work

In this paper, we mined the hidden patterns of social networks to answer two important research questions. We studied community properties by applying CD algorithms on two different types of datasets of considerable size. We observed that Label Propagation performed well for smaller datasets. Louvain algorithm performed better for massive datasets. This work would help researchers to understand the ideas of community detection methods better, and help them in selecting the appropriate method in their practical applications.

In the future, we aim to develop our own parallel community detection algorithm and parallelize the existing algorithms to achieve faster computational results. We also plan to apply meta-heuristic techniques to optimize the modularity of algorithms in the detection of communities.

References

1. Girvan M, Newman MEJ (2002) Community structure in social and biological networks. Proc Nat Acad Sci 99(12):7821–7826
2. Fortunato S (2010) Community detection in graphs. Phys Rep 486(3):75–174
3. Sharma R, Oliveira S (2017) Community detection algorithm for big social networks using hybrid architecture. Big Data Res 10(2):44–522
4. Blondel VD, Guillaume JL, Lambiotte R, Lefebvre E (2008) Fast unfolding of communities in large networks. J Statistical Mech Theory Experim 2008(10):10008
5. Raghavan UN, Albert R, Kumara S (2007) Near linear time algorithm to detect community structures in large-scale networks. Phys Rev E 76(3):036106
6. Chakrabarti D, Kumar R, Tomkins A (2006) Evolutionary clustering. In: Proceedings of the 12th ACM SIGKDD international conference on knowledge discovery and data mining, Aug 2006, pp 554–560
7. Li X, Tan Y, Zhang Z, Tong Q (2016) Community detection in large social networks based on relationship density. In: IEEE 40th Annual computer software and applications conference (COMPSAC 2016), pp 524–533
8. Alamuri M, Surampudi BR, Negi A (2014) A survey of distance/similarity measures for categorical data. In: International joint conference on neural networks (IJCNN), vol 9(3), pp 1907–1914
9. Leskovec J, Kleinberg J, Faloutsos C (2005) Graphs over time: densification laws, shrinking diameters and possible explanations. In: Proceedings of the 11th ACM SIGKDD international conference on knowledge discovery in data mining, pp 177–187, 21 Aug 2005
10. Sumith N, Annappa B, Bhattacharya SA (2018) Holistic approach to influence maximization in social networks: STORIE. Appl Soft Comput 66(2):533–547

11. Clauset A, Newman ME, Moore C (2004) Finding community structure in very large networks. Phys Rev E 70(6):066111
12. Lancichinetti A, Kivelä M, Saramäki J, Fortunato S (2010) Characterizing the community structure of complex networks. PloS One 5(8):e11976
13. Leskovec J, Lang KJ, Mahoney M (2010) Empirical comparison of algorithms for network community detection. In: Proceedings of the 19th international conference on world wide web, pp 631–640
14. Tyler JR, Wilkinson DM, Huberman BA (2005) E-mail as spectroscopy: automated discovery of community structure within organizations. Inform Soc 21(2):143–153
15. Palla G, Derényi I, Farkas I, Vicsek T (2005) Uncovering the overlapping community structure of complex networks in nature and society. Nature 435(70):793–814
16. Pons P, Latapy M (2005) Computing communities in large networks using random walks. In: International symposium on computer and information sciences, pp 284–293, Oct 2005
17. Wagenseller P, Wang F, Wu W (2018) Size matters: a comparative analysis of community detection algorithms. IEEE Trans Comput Soc Syst 5(4):951–960
18. Zhao Z, Zheng S, Li C, Sun J, Chang L, Chiclana F (2018) A comparative study on community detection methods in complex networks. J Intell Fuzzy Syst 35(1):1077–1086

Improving Performance of Relay-Assisted Molecular Communication Systems Using Network Coding

Prabhat K. Upadhyay$^{(\boxtimes)}$ ⓘ

Discipline of Electrical Engineering, Indian Institute of Technology Indore,
Indore, Madhya Pradesh 453552, India
pkupadhyay@iiti.ac.in

Abstract. The performance of a molecular communication system depends on various parameters like diffusion coefficients of the messenger molecules, the distance between the two communicating nanomachines, the volume of the nanomachines, the time taken for molecules to reach the receiver, and concentrations of molecules for different signals. Even after optimizing the adjustable parameters, sometimes nanomachines need to communicate with distances between them being comparably large than their optimal distances, making the communication to become unreliable. Hereby, in this paper, an intermediate nanomachine called relay is incorporated to assist the molecular communication process. Especially, network coding strategy is employed to improve the performance of the system by reducing the time taken by the signals and minimizing the error probability of detection. The numerical results will help to choose reliable parameters for the considered relay-based model of the molecular communication system.

Keywords: Diffusion · Molecular communication · Network coding · Error performance · Detection threshold optimization

1 Introduction

Molecular communication (MC) is an emerging communication paradigm that occurs at nano- and micro-scales [1, 2]. It is highly efficient and can be used for in vivo biomedical applications [3]. As the name suggests, in MC, the signal transmission is achieved by using molecules as the information carriers. These signal carrying molecules can be transmitted and received by microscopic devices called nanomachines.

Based on how the molecules reach the receiver, MC can be divided into two types [1]. If the signal carrying molecules travel to the receiver simply by diffusion, then it is called as passive transport, and if they travel to the receiver using some directional chemical energy, then it is called as active transport. In passive type of communication, the molecules simply diffuse in all possible directions making them dynamic and unpredictable; hence, it will be the best choice to use them in environments where we cannot use a properly connected infrastructure. However, the time taken by molecules to reach the receiver varies with square of the communication distance. As such,

© Springer Nature Singapore Pte Ltd. 2020
N. Goel et al. (eds.), *Modelling, Simulation and Intelligent Computing*,
Lecture Notes in Electrical Engineering 659,
https://doi.org/10.1007/978-981-15-4775-1_20

diffusion to the longer destinations would take longer time and large number of signaling molecules would be needed for the proper detection of the signal. As a consequence, passive transport is only effective for small distances. In active transport, the signal molecules can travel over large distances by use of external means like microtubules guiding it to the receiver and molecular motors which carry the signaling molecules to the receiver. Thus, active transport needs smaller number of molecules to be sent for a particular signal. However, the energy required to transport the molecules must be replenished continuously. The attenuation of molecular concentration in MC system limits the communication range along with the reduction in the transmission rate and fidelity. Moreover, the propagation time increases with the square of the distance [2]. Hence, it necessitates the use of a nanorelay for MC with distant receivers.

Several works have studied relaying and network coding in MC system to alleviate the above problem. For instance, see [4–7] for decode and forward relaying, [4] for sense and forward relaying, [5] for amplify and forward relaying, and [6, 7] for network-coded MC. Different detection techniques have been covered in the literature for MC systems such as amplitude and energy detection in [8], maximum likelihood detection in [9], and weighted sum detection in [2, 5, 9, 10]. In [6], the authors have calculated probability of error using energy detection for network-coded MC system with an arbitrary threshold. However, in this paper, we calculate the probability of error for optimal network-coded MC system with weighted sum detector since it is physically more reliable. The proposed error model explicitly considers diffusion noise and the inter-symbol interference (ISI). The performance of the network-coded MC is presented through numerical results along with the investigation highlighting the impact of different parameters on error performance. The performance analysis for optimal system is also examined with various key parameters.

2 System Model

We consider a network-coded MC system in which a nanomachine R (nanorelay) is placed in between and equidistant from the two end nanomachines A and B. The nanomachines A and B release different types of signaling molecules. We assume a three-dimensional medium having a fixed diffusion coefficient D. The released number of molecules corresponding to information bits 0 and 1 are N_0 and N_1, respectively. The distance between nanomachines A and R is 'r', which is also the distance between nanomachines R and B. Further, the radius of nanomachines is taken as ρ. Moreover, T denotes the time taken by molecular signal to arrive at the nanorelay. More importantly, signals from A and B take the same time to reach the nanorelay and vice versa.

3 Network Coding

In network coding, nanomachines A and B use a nanorelay [11] as the signal repeater. In such a case, the following four steps will be involved in the communication process.

(a) Nanorelay receiving and analyzing the signal from A.
(b) Nanorelay transmitting the signal from A into the medium to B.
(c) Nanorelay receiving and analyzing the signal from B.
(d) Nanorelay transmitting signal from B into the medium for A.

Now, one can realize how the nanorelay responds to the incoming signals and make it send a desired output. With network coding technique, instead of sending the messages to A and B separately, it sends a message that has the information of both the signals A and B, and the combined signal is transmitted into the medium. The nanomachines A and B are coded in such a way that they can extract the required signal from the combined signal of the nanorelay.

Hereby, the number of steps involved would be three.

(a) Nanorelay receiving and analyzing the signal from A, i.e., x_A.
(b) Nanorelay receiving and analyzing the signal from B, i.e., x_B.
(c) Nanorelay transmitting a combined signal to be received by both A and B.

As such, we use XOR logic to reduce the number of communication time slots. The nanorelay is coded to apply XOR function on the incoming signals and transmit the resulting signal. Hence, the combined transmitted signal would be $x_A \oplus x_B$. Now, the nanomachines A and B are coded to apply XOR function on the received signal and their own signal to get the intended one as follows.

Received signal at A,

$$x_A \oplus (x_A \oplus x_B) = x_B \tag{1}$$

Received signal at B,

$$x_B \oplus (x_A \oplus x_B) = x_A \tag{2}$$

Hence, by using network coding, intended signal can be decoded at the respective nanomachines in less time slots. Let $\hat{x}_m$, $m \in \{A, B, R\}$, denote the decoded symbol corresponding to x_m at the receiving nanomachine.

4   Error Performance Analysis

The molecular concentration corresponding to bit 0 at time t and distance r can be expressed as

$$C^0(r, t) = \frac{N_0}{(4\pi Dt)^{\frac{3}{2}}} . e^{\frac{-r^2}{4Dt}}. \tag{3}$$

Consequently, the received number of molecules for information bit 0 can be given as

$$N^0(r, t) = C^0(r, t).V, \tag{4}$$

where

$$V = \frac{4}{3}.\pi\rho^3,$$ (5)

is the volume of spherical reception region.

Since $C^0(r,t)$ and $N^0(r,t)$ give instantaneous values, receiving nanomachine takes L number of samples and adds them to be compared with the detection threshold τ, i.e.,

$$T_s = \frac{T}{L},$$ (6)

where T_s is the sampling duration. Thus, the observed number of molecules, when information bit 0 was transmitted, can be written as

$$N_{obs}^0 = \sum_{k=1}^{L} N^0(kT_s).$$ (7)

Similarly, the observed signal, when information bit 1 was transmitted, can be calculated as

$$N_{obs}^1 = \sum_{k=1}^{L} N^1(kT_s),$$ (8)

where

$$N^1(r,t) = C^1(r,t).V$$ (9)

is the instantaneous received number of molecules at time t and distance r when information bit 1 was transmitted. Herein, $C^1(r,t)$ is the molecular concentration corresponding to information bit 1 and can be represented as

$$C^1(r,t) = \frac{N_1}{(4\pi Dt)^{\frac{3}{2}}}.e^{\frac{-r^2}{4Dt}}.$$ (10)

Now, we calculate the statistics of Brownian or counting noise for large N_1, N_0. As such, variance of counting noise, $n^0(t)$, corresponding to information bit 0, can be expressed as

$$\sigma_0^2 = \sum_{k=1}^{L} N^0(kT_s).$$ (11)

Likewise, variance of counting noise, $n^1(t)$, corresponding to information bit 1, can be expressed as

$$\sigma_1^2 = \sum_{k=1}^{L} N^1(kT_s). \tag{12}$$

Hereby, average probability of error when $\widehat{x}_A$ is in error can be written as

$$P_e^A = P_E^{x_A=0} P(x_A = 0) + P_E^{x_A=1} P(x_A = 1)$$
$$= 0.5(P_E^{x_A=0} + P_E^{x_A=1}),$$

for equally likely symbols.

$$P_E^{x_A=0} = P[N_{obs}^0 + n_{ob}^0 > \tau], \text{ where } n_{ob}^0 = \sum_{k=1}^{L} n^0(kT_s).$$

As $P_E^{x_A=0} = P[n_{ob}^0 > \tau - N_{obs}^0] = \dfrac{1}{\sqrt{2\pi\sigma_0^2}} \int\limits_{\tau-N_{obs}^0}^{\infty} e^{\frac{-\left(n_{ob}^0\right)^2}{2\sigma_0^2}} d\left(n_{ob}^0\right),$

hence,

$$P_E^{x_A=0} = \frac{1}{2} \text{erfc}\left(\frac{\tau - N_{obs}^0}{\sqrt{2\sigma_0^2}}\right). \tag{13}$$

Similarly, $P_E^{x_A=1} = P\left[n_{ob}^1 \leq \tau - N_{obs}^1\right]$, where $n_{ob}^1 = \sum_{k=1}^{L} n^1(kT_s).$

Therefore,

$$P_E^{x_A=1} = \frac{1}{2} \text{erf}\left(\frac{\tau - N_{obs}^1}{\sqrt{2\sigma_1^2}}\right). \tag{14}$$

Eventually,

$$P_e^A = \frac{1}{4}\left[\text{erfc}\left(\frac{\tau - N_{obs}^0}{\sqrt{2\sigma_0^2}}\right) + \text{erf}\left(\frac{\tau - N_{obs}^1}{\sqrt{2\sigma_1^2}}\right)\right]. \tag{15}$$

Now, we calculate the probability of observing ISI causing molecules as

$$p(t) = \frac{V}{(4\pi D(T+t))^{\frac{3}{2}}} \cdot e^{\frac{-r^2}{4D(T+t)}}. \tag{16}$$

Then, we calculate ISI statistics as

$$n_{\text{ISI}} = N\left(\mu, \sigma_{\text{ISI}}^2\right),$$

where

$$\mu = 0.5 \sum_{k=L+1}^{2L} \left\{ N^1(kT_s) + N^0(kT_s) \right\},$$

and

$$\sigma_{\text{ISI}}^2 = 0.5\left(N^0 + N^1\right). \sum_{k=L+1}^{2L} p(kT_s).(1 - p(kT_s)). \tag{17}$$

Further, average probability of error when $\widehat{x}_B$ is in error can be written as

$$P_e^B = P_E^{x_B=0} P(x_B = 0) + P_E^{x_B=1} P(x_B = 1)$$
$$P_E^{x_B=0} = P\left[n_{\text{ob}}^0 + n_{\text{ISI}} > \tau - N_{\text{obs}}^0\right].$$

$$= \frac{1}{\sqrt{2\pi\sigma_{t0}^2}} \int_{\tau-N_{\text{obs}}^0}^{\infty} e^{-\frac{(n_{t0}-\mu_{t0})^2}{2\sigma_{t0}^2}} d(n_{t0}), \text{ where } n_{t0} \sim N\left(\mu_{t0}, \sigma_{t0}^2\right),$$

with $\sigma_{t0}^2 = \sigma_0^2 + \sigma_{\text{ISI}}^2$ and $\mu_{t0} = \mu$.

$$P_E^{x_B=0} = \frac{1}{2} \text{erfc}\left(\frac{\tau - N_{\text{obs}}^0 - \mu}{\sqrt{2}\sigma_{t0}}\right). \tag{18}$$

Similarly, $P_E^{x_B=1} = \frac{1}{\sqrt{2\pi\sigma_{t1}^2}} \int_0^{\tau-N_{\text{obs}}^1} e^{-\frac{(n_1-\mu_{t1})^2}{2\sigma_{t1}^2}} d(n_{t1}),$

where $n_{t1} \sim N\left(\mu_{t1}, \sigma_{t1}^2\right)$, $\sigma_{t1}^2 = \sigma_1^2 + \sigma_{\text{ISI}}^2$, and $\mu_{t1} = \mu$.

$$P_E^{x_B=1} = \frac{1}{2}\left[\text{erf}\left(\frac{\mu}{\sqrt{2}\sigma_{t1}}\right) + \text{erf}\left(\frac{\tau - N_{\text{obs}}^1 - \mu}{\sqrt{2}\sigma_{t1}}\right) \right]. \tag{19}$$

Now, $P_e^R = P_e^A + P_e^B$, when x_R is in error.
$P_e^{\hat{R}} = \frac{1}{2}\left(P_E^{\hat{x}_R=0} + P_E^{\hat{x}_R=1}\right)$, when $\hat{x}_R$ is in error.
$P_e^{\hat{R}} = P_e^A$, since $\hat{x}_R$ is not affected by ISI.
Finally, the overall probability of error in network-coded MC can be given as

$$P_e = P_e^R + 2P_e^{\hat{R}} = 3P_e^A + P_e^B \tag{20}$$

$$
\begin{aligned}
P_e = \frac{3}{4} &\left[\mathrm{erfc}\left(\frac{\tau - N^0_{\mathrm{obs}}}{\sqrt{2}\sigma_0} \right) + \mathrm{erf}\left(\frac{\tau - N^1_{\mathrm{obs}}}{\sqrt{2}\sigma_1} \right) \right] \\
+ \frac{1}{4} &\left[\mathrm{erfc}\left(\frac{\tau - N^0_{\mathrm{obs}} - \mu}{\sqrt{2}\sigma_{t0}} \right) + \mathrm{erf}\left(\frac{\mu}{\sqrt{2}\sigma_{t1}} \right) + \mathrm{erf}\left(\frac{\tau - N^1_{\mathrm{obs}} - \mu}{\sqrt{2}\sigma_{t1}} \right) \right].
\end{aligned}
\tag{21}
$$

5   Detection Threshold Optimization

Increasing detection threshold τ enhances the probability of miss detection. However, decreasing τ increases the probability of false alarm. Hence, we are interested in finding the optimal detection threshold, τ_{opt}, that minimizes P_e. Since τ is a continuous function, the minimum value of the function P_e would be at certain value of τ such that

$$
\frac{\partial P_e}{\partial \tau} = 0.
$$

Now, using $\mathrm{erfc}(x) = 1 - \mathrm{erfc}(x)$ and $\frac{\mathrm{d}}{\mathrm{d}x}\mathrm{erfc}(x) = \frac{-2}{\sqrt{\pi}}\mathrm{e}^{-x^2}$, one can arrive at

$$
\begin{aligned}
&\frac{3}{4}\left[\frac{-\sqrt{2}}{\sqrt{\pi}}\mathrm{e}^{-\left(\frac{\tau-N}{\sqrt{2}\sigma}\right)^2} \right] - \frac{3}{4}\left[\frac{-\sqrt{2}}{\sqrt{\pi}}\mathrm{e}^{-\left(\frac{\tau-N}{\sqrt{2}\sigma}\right)^2} \right] + \frac{1}{4}\left[\frac{-\sqrt{2}}{\sqrt{\pi}\sigma}\mathrm{e}^{-\left(\frac{\tau-N-\mu}{\sqrt{2}\sigma}\right)^2} \right] \\
&- \frac{1}{4}\left[\frac{-\sqrt{2}}{\sqrt{\pi}\sigma}\mathrm{e}^{-\left(\frac{\tau-N-\mu}{\sqrt{2}\sigma}\right)^2} \right] \\
&= 0
\end{aligned}
\tag{22}
$$

Finally, one can numerically get $\tau_{\mathrm{opt}} = \tau$, such that $N^0_{\mathrm{obs}} < \tau < N^1_{\mathrm{obs}}$.

6   Numerical Results

In this section, we conduct numerical investigations to evaluate the performance of the proposed system using MATLAB. The simulation parameters are listed in Table 1. In Fig. 1, the error probability of a network-coded diffusive molecular communication system is evaluated as a function of diffusion coefficient D, with $\tau = (N^0_{\mathrm{obs}} + N^1_{\mathrm{obs}})/2$ and τ_{opt}, for system parameters $N_0 = 500$, $r = 350$ nm, and $\rho = 45$ nm. Evidently, one can observe that the error probability increases as diffusion coefficient increases. This is because the molecular pulse decays more quickly as value of diffusion coefficient increases. Moreover, one can see that the optimal detection threshold τ_{opt} minimizes the error probability significantly.

Table 1 Simulation Parameters

Parameters	Notation	Values
# molecules for sending bit 0	N_0	{500, 600}
# molecules for sending bit 1	N_1	1000
Diffusion coefficient	D	$\{1, 5\text{–}12\} \times 10^{-10}$ m^2/s
Distance between R and $\{A, B\}$	r	{325–350} nm
Radius of A, R, B	ρ	{45, 50} nm
# samples	L	10
Sampling duration	T_s	20 μs

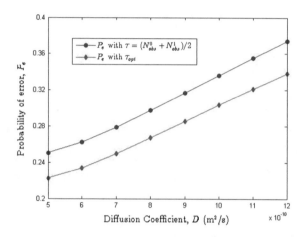

Fig. 1 Probability of error as a function of diffusion coefficient

In Fig. 2, the error probability of the considered system is evaluated as a function of communication distance r, with $\tau = (N_{obs}^0 + N_{obs}^1)/2$, for system parameters $N_0 = \{500, 600\}$, $\rho = \{45, 50\}$ nm, and $D = 1 \times 10^{-10}$ m^2/s. One can notice that the error probability reduces as the radius of receiver nanomachine increases due to reduced variance of diffusion noise. Intuitively, the error probability increases as the distance between nanomachines and the nanorelay increases. Moreover, for a fixed value of N_1, error probability increases with increasing value of N_0 due to the reduced constellation distance.

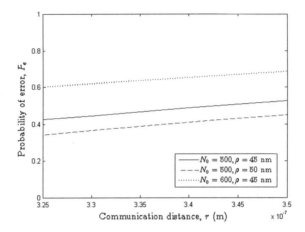

Fig. 2 Probability of error as a function of communication distance r

7 Conclusions

In this paper, the probability of error for a network-coded molecular communication system with physically more realizable weighted sum detector is calculated. Moreover, the detection threshold of the considered system is optimized to minimize the error probability. Eventually, the effect of several parameters on the error performance is demonstrated.

Acknowledgements. This research work was supported by the Council of Scientific and Industrial Research (CSIR), Government of India, under Grant Number 22(0763)/18/EMR-II.

References

1. Farsad N, Yilmaz HB, Eckford A, Chae C-B, Guo W (2016) A comprehensive survey of recent advancements in molecular communication. IEEE Commun Surv Tutorials 18 (3):1887–1919
2. Ahmadzadeh A, Noel A, Burkovski A, Schober R (2015) Amplify-and-forward relaying in two-hop diffusion-based molecular communication networks. In: IEEE GLOBECOM, pp 1–7, San Diego, CA, USA (2015)
3. Huang S, Lin L, Yan H, Xu J, Liu F (2019) Statistical analysis of received signal and error performance for mobile molecular communication. IEEE Trans Nanobiosci 18(3):415–427
4. Einolghozati A, Sardari M, Fekri F (2013) Relaying in diffusion-based molecular communication. In: IEEE ISIT, pp 1844–1848
5. Ahmadzadeh A, Noel A, Schober R (2014) Analysis and design of two-hop diffusion-based molecular communication networks. In: IEEE GLOBECOM, pp 2820–2825, Austin, TX, USA (2014)
6. Aijaz A, Aghvami AH, Nakhai MR (2014) On error performance of network coding in diffusion-based molecular nanonetworks. IEEE Trans Nanotechnol 13(5):871–874

7. Unluturk B, Malak D, Akan O (2013) Rate-delay tradeoff with network coding in molecular nanonetworks. IEEE Trans Nanotechnol 12(2):120–128
8. Llatser I, Cabellos-Aparicio A, Pierobon M et al (2013) Detection techniques for diffusion-based molecular communication. IEEE J Sel Areas Commun 31(12):726–734
9. Noel A, Cheung KC, Schober R (2014) Optimal receiver design for diffusive molecular communication with flow and additive noise. IEEE Trans Nanobiosci 13(3):350–362
10. Ahmadzadeh A, Noel A, Schober R (2015) Analysis and design of multihop diffusion-based molecular communication networks. IEEE Trans Mol Biol Multi-Scale Commun 1 (2):144–157
11. Bergel I (2019) Detection and amplification of molecular signals using cooperating nano-devices. In: IEEE ICASSP, Brighton, UK (2019)

Compact Yagi–Uda-Shaped Patch Antenna for 5 GHz WLAN Applications

Doondi Kumar Janapala⑩, M. Nesasudha$^{(\boxtimes)}$⑩, and Sam Prince Tensing

Karunya Institute of Technology and Sciences, Coimbatore, Tamil Nadu 641114, India
jdoondikumarece@gmail.com, nesasudha@karunya.edu

Abstract. In this article, the design of a compact Yagi–Uda-shaped patch antenna is presented for 5 GHz WLAN applications. The ground plane is maintained partial with slots etched symmetrical on both sides of the strip feed, the slots in ground plane are made, and the dimensions are tuned to achieve compact size. FR4 material is used as substrate with a compact size of 15 mm × 15 mm × 1.6 mm. Various dimensions of the radiator can be varied to tune the frequency. This tuning produces different channels around the 5 GHz to suite 802.11a/h/j/n/ac/ax protocols. The antenna performance is presented with the help of reflection coefficient, radiation pattern, and other antenna parameters.

Keywords: Compact antenna · Yagi–uda · WLAN · 5 GHz

1 Introduction

Wireless communication systems are grown vastly in the past few decades due to the technical advancements. The growing demand for new communication systems with improved data rate, low cost, low power consumptive, and enhanced performance leads to creation of different new antennas that are capable of handling the needs. Over the years, several researchers invented different new antennas for various applications like Wi-Fi, WLAN, and Bluetooth. One of the main concerns when developing the antennas for these communication systems is that the size of the antenna should be maintained short.

WLAN consists of majority of communication systems. WLAN follows IEEE 802.11 protocols which consist of different frequency bands around 900 MHz, 2.4 GHz, 3.65 GHz, 5 GHz, 5.8 GHz, and 60 GHz. Several antennas are developed to operate at these bands, and some antennas are developed to have capabilities to cover more than one frequency bands of WLAN. These WLAN systems need antennas which are small in size and easy to realize. Different techniques have been invented to reduce the size of the antenna. Fractal geometry or loops are most commonly used methods in miniaturizing the antenna size. In our work, we concentrate on developing compact antenna to work at 5 GHz for WLAN applications. In [1], a miniaturized monopole antenna is developed to serve at 5.2–5.8 GHz, where the defective ground surface is used to reduce the size of the antenna. A compact-sized double-dollar symbol antenna

© Springer Nature Singapore Pte Ltd. 2020
N. Goel et al. (eds.), *Modelling, Simulation and Intelligent Computing*,
Lecture Notes in Electrical Engineering 659,
https://doi.org/10.1007/978-981-15-4775-1_21

is designed for WLAN 5.1–6 GHz applications in [2]. In a similar way, these compact antennas can be developed to serve dual- or multi-band applications. Fractals are used mostly to miniaturize antenna size in several cases, and in [3], a dual-band elevated slotted patch antenna is designed to cover 2.4–2.5 and 4.9–5.9 GHz bands; here dual-reverse-arrow fractal geometry is used in reducing the size of the antenna. A compact multi-band antenna for WLAN/WiMAX/ITU applications is developed in [4] using metamaterial-based structures. Compact-sized dual-polarized multi-band antenna is developed using slots for IEEE 802.11a/b/g/n/ac/ax applications in [5]. In a similar way, the MIMO antennas are also developed for WLAN applications to increase the throughput of the system. MIMO antenna is developed to cover frequency band from 5.3 to 6.7 GHz using meander line structure as radiator [6]. A dual-band MIMO antenna is designed using metamaterial structures for LTE and WLAN applications [7].

Yagi class antennas can get compact-sized designs, and in [8], a Yagi class antenna is designed for GSM, WLAN, and WiMAX applications. Where inverted L- and inverted F-shaped wires are used as active and parasitic elements in the array, here the antenna is not planar. In our current work, a planar Yagi class antenna is proposed. The proposed antenna consists of an array having successive loop and straight-line combinations. The partial ground with slots helped in reducing the size of the antenna. The variation in the dimensions of the Yagi class radiator can tune the antenna to different bands around 5 GHz to suite the WLAN applications.

2 Design of the Proposed Antenna

The proposed antenna geometry is presented in the following Fig. 1. And its dimensions are listed in Table 1. The proposed antenna used FR4 material with relative permittivity 4.4 and dielectric loss tangent 0.02. Yagi class radiator is having different shapes on top and bottom of the array to match impedance, and in between, it consists of the loop and straight-line combinations. The ground is maintained partial with slots etched to reduce the size of the overall antenna.

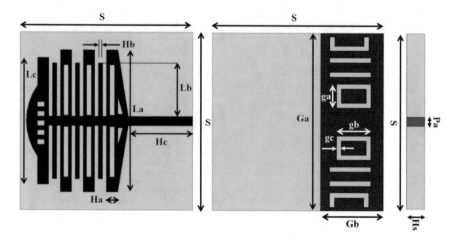

Fig. 1 Fractal antenna design. **a** Top view, **b** bottom view, and **c** front view

Table 1 Dimensions of the proposed Yagi class antenna in mm

S. No.	Symbol	Value (mm)
1	S	15
2	Hs	1.6
3	La	12
4	Lb	4.5
5	Lc	10.7
6	Ld	0.8
7	Ga	15
8	Gb	5.5

2.1 Description of the Design

The proposed antenna is designed using ANSYS HFSS 19.2v software. The substrate with dimensions S, used for this design, is FR4 epoxy with substrate thickness Hs. The overall dimensions, i.e., length and breadth of the radiating surface, are Hd and La, respectively. The length and width of the feed-line strip are Hc and Ld, respectively. Partial ground shown in Fig. 1b dimensions is Ga and Gb. Slots were made in the ground plane to achieve minimum return loss and compact size. The slots in the ground Fig. 1b have dimensions Ga, Gb, Gc, Gd, Ge, and Gf with values 3 mm, 2 mm, 3.6 mm, 3.4 mm, 0.4 mm, and 0.8 mm, respectively.

2.2 Parametric Study and Comparative Analysis

The dimensions of the loop structures and slots in the antenna have been varied to get the required target frequency with high radiation efficiency.

Lc Variation. In the patch, the first strip as shown in Fig. 1a is analyzed through parametric analysis which has dimensions Lc and Ha in which Lc is varied with values Lc = 8.7 mm, Lc = 9.7 mm, Lc = 10.7 mm, and comparative analysis is shown in Fig. 2. From the graph, it has been observed that the variation "Lc = 10.7 mm" reflection coefficient at 5 GHz and return loss with −22 dB has a bandwidth from 4.9 to 5.2 GHz. While other dimensions also cover bands in between 5 and 5.3 GHz.

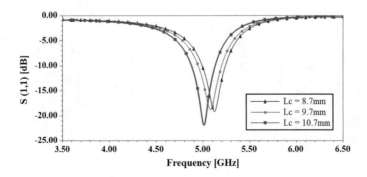

Fig. 2 Reflection coefficient comparison for Lc variation

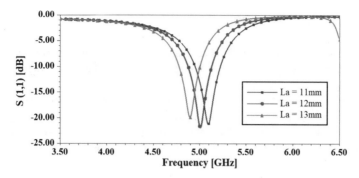

Fig. 3 Reflection coefficient comparison for La variation

La Variation. In this next analysis, the dimensions of the loop structure are subjected to parametric analysis, i.e., length La is varied and analyzed for the following values La = 11 mm, La = 12 mm, and La = 13 mm. The reflection coefficient comparison is presented in the following Fig. 3. Here, La at 11 mm it operates below 5 GHz at 4.8 GHz, at 12 mm it operates at 5 GHz, and at 13 mm it operates at 5.1 GHz.

Lb Variation. In this analysis, the length of the strip (Lb), which is present in between the loop structures, is varied with different values, and comparative analysis is presented in Fig. 4.

From the above graph Fig. 4, it is observed that after the parametric analysis with various values of Lb, i.e., Lb = 2.5 mm, Lb = 3.5 mm, Lb = 4.5 mm, Lb = 5.5 mm, Lb = 6.5 mm, it shows that for the Lb variation, the proposed antenna operates at 5 GHz and below 5 GHz range. The Lb variation shifts the resonant frequency to lower range from 5 GHz.

By taking parametric analysis into consideration, the proposed antenna dimensions are finalized as follows: "Lc = 10.7 mm," "La = 12 mm," and "Lb = 4.5 mm." With these dimensions, the antenna is fabricated and the reflection coefficient is measured using FieldFox Microwave Analyzer N9915. The simulated and measured analysis is presented in the following Results and Discussion section.

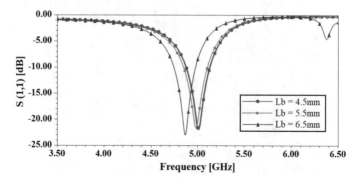

Fig. 4 Reflection coefficient comparison for Lb variation

3   Results and Discussion

The fabricated antenna along with its measurement setup is presented in the following Fig. 5. The reflection coefficient comparison for the simulated and measured is presented in Fig. 6.

From the above graph Fig. 6, it can be seen that the simulated and measured results were both covering the 5 GHz band. The slight deviation could be the result of unavoidable fabrication issues. The fabricated antenna works in the required target frequency with minimum return loss of −30 dB along with the radiation efficiency of 86%.

The simulated radiation gain is presented in the following Fig. 7.

Figure 7 represents the gain plot obtained which is omni-directional and also covers most of its areas by being sensitive to signals from all the directions.

The VSWR curve for the proposed Yagi class antenna is presented in the following Fig. 8.

Figure 8 represents the voltage standing wave ratio (VSWR). The standard value for VSWR has been given from 0 to 2, and the value which we obtain is 1.669. Therefore, the achieved VSWR is good for operation.

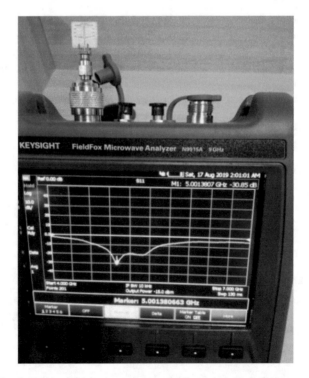

Fig. 5 Fabricated antenna and its reflection coefficient measurement

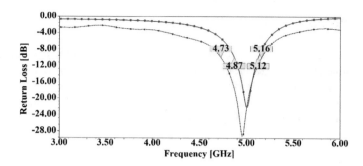

Fig. 6 Simulated and measured reflection coefficient comparison

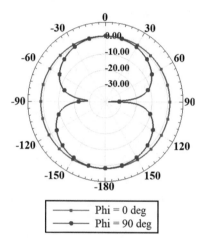

Fig. 7 Simulated radiation pattern at 5 GHz, Phi = 0° (red) and Phi = 90° (blue)

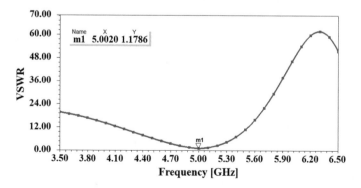

Fig. 8 VSWR curve for the proposed Yagi antenna

4   Conclusion

The proposed antenna is designed using ANSYS HFSS, and dimensions of the patch were determined by parametric analysis and after the comparative study. The simulated and measured reflection coefficients are compared and found to be operating at 5 GHz. The simulated radiation efficiency is 86%. The proposed antenna is suitable candidate for WLAN applications at 5 GHz.

References

1. Chetouah F, Aidel S, Bouzit N, Messaoudene I (2018) A miniaturized printed monopole antenna for 5.2–5.8 GHz WLAN applications. Int J RF Microwave Comput-Aided Eng 28(5): e21250
2. Sabbar N, Hati K, Asselman H, Hajjaji AE (2018) A new monopole antenna in the form of double dollar-symbol for WLAN (5.1–6 GHz) applications. Procedia Manuf 22:539–544
3. Soleimani H, Oraizi H (2017) Miniaturization and dual-banding of an elevated slotted patch antenna using the novel dual-reverse-arrow fractal. Int J RF Microwave Comput-Aided Eng 27(5):e21085
4. Rajkumar R, Usha Kiran K (2016) A compact metamaterial multiband antenna for WLAN/WiMAX/ITU band applications. AEU Int J Electron Commun 70(5):599–604
5. Saxena S, Kanaujia BK, Dwari S, Kumar S, Tiwari R (2017) A compact microstrip fed dual polarised multiband antenna for IEEE 802.11 a/b/g/n/ac/ax applications. AEU Int J Electron Commun 72:95–103
6. Chouhan S, Panda DK, Gupta M, Singhal S (2017) Meander line MIMO antenna for 5.8 GHz WLAN application. Int J RF Microwave Comput-Aided Eng: e21222
7. Rezvani M, Zehforoosh Y (2018) A dual-band multiple-input multiple-output microstrip antenna with metamaterial structure for LTE and WLAN applications. AEU Int J Electron Commun 93:277–282
8. Chen G, Chan KK-M, Rambabu K (2010) Miniaturized yagi class of antennas for GSM, WLAN, and WiMax applications. IEEE Trans Consum Electron 56(3):1235–1240

Hybrid Green Energy Systems for Uninterrupted Electrification

S. Lavanya Devi$^{(\boxtimes)}$ (ID) and S. Nagarajan (ID)

Department of EEE, Jerusalem College of Engineering, Chennai, India
lavanyadevi@jerusalemengg.ac.in, nagu_shola@gmail.com

Abstract. In this paper, a new multiport, hybrid green-fed DC-DC bidirectional converter is designed and fed into the standalone system. This model is developed and integrated to study the concept of generating systems. In this model, a PV panel of 300 watts is designed and integrated to a bidirectional converter with a battery backup to extract power, while the wind power is harnessed with a transformer coupled a dual-half-bridge converter. The model proposed is designed with a meritorious objective of sustainable, cost-effective, less component count reduced losses, and good efficiency with a good reliability. The system works day and night to produce output with good efficiency. The simulation results are obtained using MATLAB software 2014a, and hence, the performance is analyzed.

Keywords: Bidirectional converter (BDC) · Photovoltaic (PV) · Wind · Dual half bridge converter

1 Introduction

Due to the fast depletion of fossil fuels, we are in a situation to extract power from the renewable energy sources. Sun and wind are the chief sources of energy for planet Earth, which are supportable and unlimited. Fortunately, our India is highly enriched with solar insolation levels, and wind has its own impact throughout the year, making it wise to harvest the energy from them. Among the many renewable energy sources available, solar photovoltaic and wind energy have the highest potential to generate green power clean and for all our future needs. Calculating of RES varies according to availability, season, and their considerable instability [2]. Hence, this proposed system is introduced and used intensively. This topology allows energy source diversification. The buildings using this concept are transformed to independent energy system.

- Usage of RES in a useful way
- High efficiency
- Prolonged existence time
- Low cost and reliability
- Good power quality
- Low maintenance
- Low acoustic noises and losses
- Enhancement in battery charging efficiency

© Springer Nature Singapore Pte Ltd. 2020
N. Goel et al. (eds.), *Modelling, Simulation and Intelligent Computing*,
Lecture Notes in Electrical Engineering 659,
https://doi.org/10.1007/978-981-15-4775-1_22

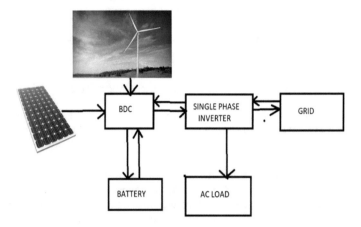

Fig. 1 Proposed system using green energy systems

The proposed system shown in Fig. 1 is the complementary behavior of solar insolation and wind velocity with battery led to its usage in household applications. This hybrid system works in either standalone mode or grid-connected mode [3]. This use of multi-input converter for hybrid power system is attracting increasingly its attention because of its numerous advantages with centralized control.

2 Converter Configuration

The basic circuit diagram of a bidirectional DC-DC converter is shown in Fig. 2. It consists of two switches. It acts as a boost converter in a forward direction (mode 1) and acts as a back converter in the reverse direction (mode 2). The complete analysis and design of a bidirectional DC-DC converter are explained. The converter is designed using these formulas.

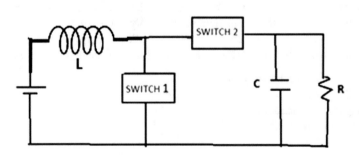

Fig. 2 Bidirectional DC-DC converter

$$I_o = \frac{P_o}{V_o}$$

$$R = \frac{V_o}{I_o}$$

$$\Delta I_o = 0.2$$

$$AV_o = 1\% \text{ of } V_o$$

$$L = \frac{D(1-D)^2 . V_o}{2.f_s . AI_o}$$

$$D = 1 - \frac{V_{in}}{V_{out}}$$

$$C_{out} = \frac{D.I_o}{f \times \Delta V_o}$$

Thus, the values of inductors, capacitors, and gate pulses are designed to get the good efficient output.

3 Proposed Topology for Standalone Systems

The proposed converter consists of PV-fed DC-DC converter and wind with a transformer-coupled bidirectional converters fused with a single-phase inverter connected to a RL load. This proposed system has reduced power-converting stages with less component count and high efficiency. The system is described in Fig. 3.

There exists two DC links on both sides of high-frequency transformers. The control of voltage on both sides is facilitated by these two DC links. Thus, simple control strategy is followed. The boosting capacity is incorporated by pv coupled with a converter and wind with a DC-DC converter [1]. This transformer provides the

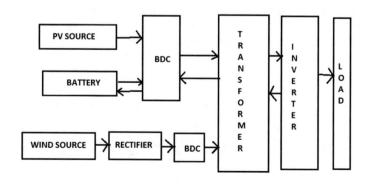

Fig. 3 Schematic diagram of the proposed green energy system

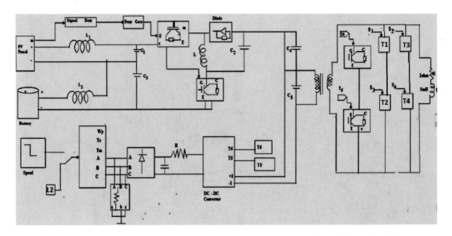

Fig. 4 Circuit diagram of the proposed green energy system fed to bidirectional converter

isolation to load from sources and battery. The extraordinary feature of the converter is maximum power tracking using P&O algorithm of both pv and wind along with the charging and discharging control of the battery. The solar- and series-connected batteries are controlled by MPPT algorithm, and pulses are generated to control the switches T_1 and T_2 as shown in Fig. 4. When the voltage level of the pv is greater than the battery, the battery is charged and in the absence of pv, battery comes into role. That is, when the voltage level of the battery is high when compared to solar, inductor current reverses. The battery discharges. Thus, the wind fed to DC-DC converters is coupled with BDC fed by PV through dc link connected to the high frequency transformer then to the inverter to the load for domestic applications.

4   Simulation Results

The simulation studies are done with MATLAB 2014a platform [5]. The results obtained for the proposed topology using green energy are fed to a bidirectional converter to a standalone system. The PV panel and wind turbine are designed to generate a voltage of 300 V when fed to a bidirectional converter to provide a smooth DC link when connected to a multiport transformer. The simulation results obtained from the PV panel are shown below in Figs. 5, 6, and 7.

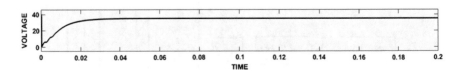

Fig. 5 Output voltage from the PV source

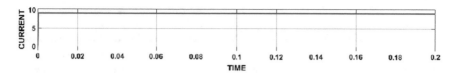

Fig. 6 Output current from the PV source

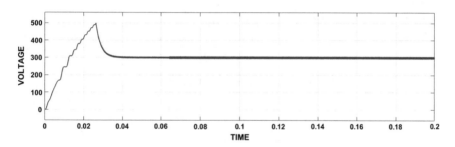

Fig. 7 Output power from the PV source

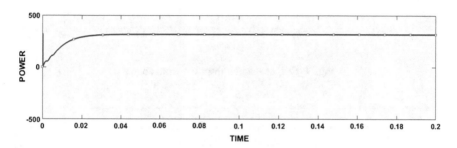

Fig. 8 Output voltage from the bidirectional converter fed with PV source

Thus, the output voltage from the converter fed with PV source is obtained as 300 V shown below in Fig. 8.

The simulated results of wind source are designed to effectively integrate with bidirectional converter as shown below in Figs. 9, 10, and 11.

The output voltage from the bidirectional converter fed with wind source is obtained as 300 V in order to have a smooth coupling of the DC link at the primary side of the transformer shown below in Fig. 12.

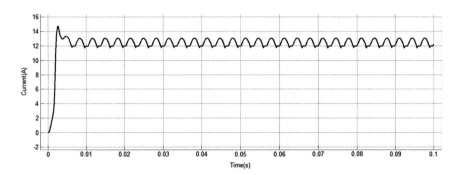

Fig. 9 Output voltage from the wind source

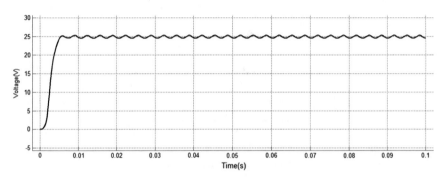

Fig. 10 Output current from the wind source

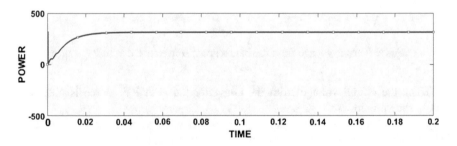

Fig. 11 Output power from wind source

The pulses generated are shown below in Fig. 13 to control the switches of the single-phase inverter fed to a RL load.

Thus, the output voltage and current are obtained from the proposed system to justify the viability as shown in Figs. 14 and 15, respectively.

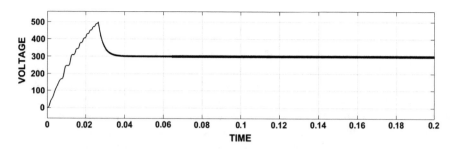

Fig. 12 Output voltage from the bidirectional converter fed with wind source

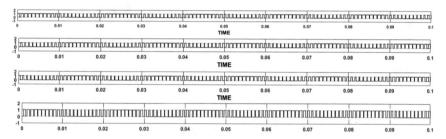

Fig. 13 Pulses generated to the switches of inverter

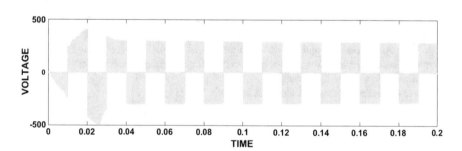

Fig. 14 Output voltage of the proposed green energy systems fed with bidirectional converter to the standalone systems

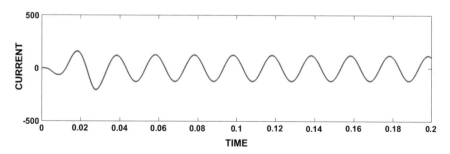

Fig. 15 Output current of the proposed green energy systems fed with bidirectional converter to the standalone systems

5   Conclusion

A power emigration scheme using a standalone hybrid PV, wind, and battery for domiciliary applications is proposed. The planned green energy hybrid system provides efficient integration of PV and wind with battery backup to extract maximum power. It is realized by a new multiport transformer-coupled bidirectional converter followed by a full-bridge converter with an efficient control scheme, which utilizes PV, wind power, and battery capacity for the application. Detailed simulation studies are carried out to justify the viability of the scheme. Thus, in future the experimental studies can be carried out and also the plan is to be carried out in grid-connected mode for better results in future. Thus, the proposed model is highly efficient to provide uninterrupted electrification all around the clock to ensure high reliability, and the values are ascertained.

References

1. Mangu B, Fernandes BG (2014) Multi-input transformer coupled DC-DC converter for PV-wind based stand-alone single-phase power generating system. In: Proceedings of IEEE Energy Conversation Congress and Exposition (ECCE), pp 5288–5295, Pittsburgh, PA, USA, Sep 2014
2. Valenciaga F, Puleston PF (2005) Supervisor control for a stand-alone hybrid generation system using wind and photovoltaic energy. IEEE Trans Energy Convers 20(2):398–405
3. Qi W, Liu J, Chen X, Christofides PD (2011) Supervisory predictive control of standalone wind/solar energy generation systems. IEEE Trans. Control Syst Tech 19(1):199–207
4. Na W, Gou B Analysis and control of bidirectional DC/DC converter for PEM fuel cell applications. In: IEEE Power and energy society general meeting—conversion and delivery of electrical energy in the 21st century
5. http://www.mathworks.com/products/Simulink

Wear Debris Shape Classification

Mohammad Shakeel Laghari$^{(\boxtimes)}$, Ahmed Hassan,
and Mubashir Noman

U. A. E. University, Al Ain, United Arab Emirates
mslaghari@uaeu.ac.ae

Abstract. Wear debris is produced in all machines containing moving parts. Wear debris or particles separate from these moving parts because of close contacts and friction and are contained in oil in an oil-wetted system. Analysis of wear debris provides important information about the condition of a machine. The produced particles come in different shapes, sizes, colors, and surface texture. This paper describes the morphological analysis of wear particles by using computer vision and image processing techniques. The aim is to classify these particles according to their shape attributes. Four particle shapes are classified by using Histogram of Oriented Gradients (HOG) and shape attributes including eccentricity, extent, major and minor axis length, equiv-diameter, and centroid distance. The shape classification can be used to identify origin of particle generation and thus predict wear failure modes in engines and other machinery. The objective of particle classification obviates reliance on visual inspection techniques and the need for specialists in the field.

Keywords: Computer vision · Image processing · Wear debris · Particle shape classification · Histogram of oriented gradients

1 Introduction

Computer vision is being used in diverse fields of applications. One of the key fields is the automation of visual inspection systems that facilitates the manufacturing industries to improve economy, production, and quality. These visual inspection systems include domain of microscopic applications such as wear debris analysis. The wear debris present in the lubrication oil contains important information about the machine condition. An early detection and recognition of the wear debris could help to prevent the loss of expensive equipment thereby saving cost and time.

A few examples of wear particles are *fatigue wear, rubbing wear, cutting wear, abrasive wear, adhesive wear, severe sliding wear*, etc. Research in the field has suggested approximately 29 different types of wear particles [1].

Several methods are used to monitor machine wear such as X-rays and ultrasound. Particles can be separated from oil for examination by using several methods. One of the methods is the use of filters according to the particle size. Another method permits particles to be deposited on glass slides. *Ferrography* is yet another method to separate wear particles from oil. The particles are arranged according to their size on a transparent substrate slide for examination and analysis. Another method to extract particles

© Springer Nature Singapore Pte Ltd. 2020
N. Goel et al. (eds.), *Modelling, Simulation and Intelligent Computing*,
Lecture Notes in Electrical Engineering 659,
https://doi.org/10.1007/978-981-15-4775-1_23

is *Magnetic Chip Detector*. These detectors (Mag. Plugs) are fitted with a powerful permanent magnet and are located at suitable positions in the machine. Metal particles attach to the plug that is later spread on a slide [2].

The number of types of wear particles depends on a relationship between the condition under which they are formed and their properties. Each type relates to a different machine condition and performance. Wear particle properties are segregated in terms of their *morphology, size, quantity,* and *composition*. These four properties exhibit the severity and rate of generation (quantity), the source of the particle generation (composition), the source, type, and rate of generation (morphology), and the rate, type, and severity (size) [3].

Morphology is further divided into six attributes of size, shape, edge details, thickness ratio, color, and surface texture. These attributes can support in predicting wear failure modes [4].

Wear particles come in irregular and arbitrary shapes. Many of these particles have similarity in shapes which make it difficult to classify them. For example, rubbing wear particles are somehow similar in shape to severe sliding particles. This paper focuses on automated classification of four different types of wear particles (including cutting, rubbing, severe sliding, and spherical) by using shape-based features and Support Vector Machine (SVM). Shape-based features like HOG, eccentricity, major/minor axis length, extent, centroid distance are some of the salient attributes that can be used to differentiate the wear particles. In addition to morphological features, some other significant features are texture and color of the wear debris. Using a combination of these features, i.e., color, texture, and morphology tends to improve better accuracy results.

2   Literature Review

Enough research is being performed on the classification of wear particles using its morphology, texture, and color. Li et al. proposed an extreme learning machine (ELM) based technique to classify the wear particles. They used particle morphology, color, and texture as features to train a single-hidden-layer feed forward neural network by using ELM [5]. Stachowiak et al. used texture and shape features to do automated wear particles classification by using linear SVM [6]. Peng et al. tried to find some correlation between wear debris and vibration analysis for machine fault diagnosis [7]. The authors used Fourier descriptors to recognize the wear particle shapes [8, 9]. Yuan et al. have used a particle boundary signal to analyze wear particle features in conjunction with a new radial concave deviation (RCD) method [10]. Peng et al. used a three-level search tree model approach to distinguish between six types of wear particles. Their approach uses multiclass Support Vector Data Description (SVDD) to classify red-oxide, black-oxide, and other debris. It is followed by k-means clustering to separate out cutting and spherical particles. Finally, an SVM classifier differentiates between fatigue and sliding debris [11]. Authors in this paper have devised an interactive image analysis system to process and store quantitative information of particle shape and edge details. Analysis of the stored data is described that allows systematic morphological analysis of wear particles [12–15].

3 Proposed Methodology

This section describes different techniques used to classify wear particles based on their shape. The main steps of these techniques are pre-processing, feature extraction, and classification that is shown as a block diagram in Fig. 1.

3.1 Pre-processing

The images of wear particles have granular noise that is needed to be removed and also edges are needed to be blurred to make the boundary smooth. Therefore, a median filter is used to remove the granular dots in the image followed by Gaussian blur to make the edges smooth. Mathematically, Gaussian smoothing is expressed as:

$$I_{\text{new}}[i,j] = \sum_{u=-k}^{k} \sum_{v=-k}^{k} I[i-u, j-v] H[u,v] \tag{1}$$

$$H[u,v] = \frac{1}{2\pi\sigma^2} e^{\frac{-(u^2+v^2)}{2\sigma^2}} \tag{2}$$

Then, global thresholding is applied to extract the boundary of the particle. Thresholding introduced some unwanted regions, therefore the images are filtered based on the area of the connected components (regions) and only that region is preserved that has the largest area (boundary of the particle). After extracting the

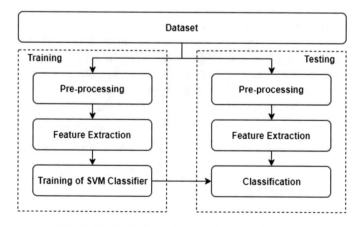

Fig. 1 Block diagram of the proposed methodology

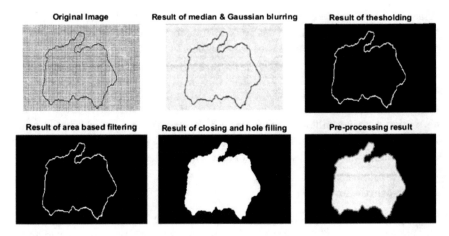

Fig. 2 Pre-processing steps

boundary of the particle, morphological closing is performed with a circular structuring element to fill any small gaps in the contour. Then, the interior region of the particle is filled with foreground color. Finally, a clean foreground region of the particle is achieved as shown in Fig. 2.

3.2 Feature Extraction

To classify the particles correctly, it is necessary to choose the essential features in the image. Therefore, different types of features are used to represent the shape of the particles. These include HOG and other shape features like centroid—the center of mass of object, eccentricity—the ratio of the distance between the foci of the ellipse and its major axis length, equiv-diameter—the diameter of a circle with the same area as the region, extent—the ratio of pixels in the region to pixels in the total bounding box, major axis length and minor axis length.

HOG uses the distribution of local intensity gradients that makes it a good prospect to describe the structural shape and texture of an object. The HOG descriptors are computed using default parameters [16], i.e., cell size of 8×8, block size of 16×16, 50% block overlap, and nine unsigned orientation bins. In HOG descriptors, first of all, gradients of the image are computed using one-dimensional kernel $[-1, 0, 1]$ in both x and y directions. Then magnitude and angle of the gradients are calculated by using $G = \sqrt{G_x^2 + G_y^2}$ and $\theta = \tan^{-1}\left(\frac{G_y}{G_x}\right)$. After calculating magnitude and phase of the gradients, orientation binning is performed by using nine evenly spaced bins. Then, the block normalization is performed by using L2-norm, i.e., $\sqrt{||v||_2^2 + \epsilon^2}$. Finally, all block features are converted into a vector.

3.3 Classification

A multiclass SVM classifier is used to classify the earlier mentioned four wears types because SVM is more accurate, avoids overfitting, and provides better separability between classes due to the optimal margin gap between hyperplanes. This classifier uses a one-versus-one Error-Correcting Output Codes (ECOC) scheme by using binary SVM learners. It assigns the class $k^\wedge$ to the new observation that minimizes the aggregation of L binary learner's loss, i.e.,

$$k^\wedge = \text{argmin}_k \frac{\sum_{l=1}^{L} |m_{kl}| g(m_{kl}, s_l)}{\sum_{l=1}^{L} |m_{kl}|} \tag{3}$$

where g is the loss function expressed as

$$g(m_{kl}, s_k) = \max(0, 1 - m_{kl}s_k)/2 \tag{4}$$

The classifier is trained by using computed features as described in the previous step. Finally, the trained model is validated by using the test data.

4 Experimentation Results

This section describes the detailed experiments performed. The algorithm is implemented and tested in Matlab 2015a.

4.1 Dataset

The dataset is gathered from the authors of this paper [17]. The dataset contains six different types of wear particles however only four types are used for this investigation. Eight images of each particle type of rubbing, severe sliding, cutting, and spherical are used to reach a total count of 32 images. The size of each image is 900×600 pixels. Some of the dataset images are shown in Fig. 3.

4.2 Testing Procedure

All the images are loaded and pre-processed as discussed in Sect. 3.1. Then, different feature extraction techniques are used to extract features and an SVM model is built as described below.

In the first technique, at least 100 strongest corner points are extracted from the image by using Harris corner detector. At each interest point, HOG features are computed and feature vector is extracted. A multiclass SVM is trained by using these features and cross-validated on the dataset. The cross-validation accuracy achieved by using this procedure is about 25%. So, it was decided to experiment with less and more interest points for feature computation. As the number of interest points is increased, cross-validation accuracy also increased, and vice versa.

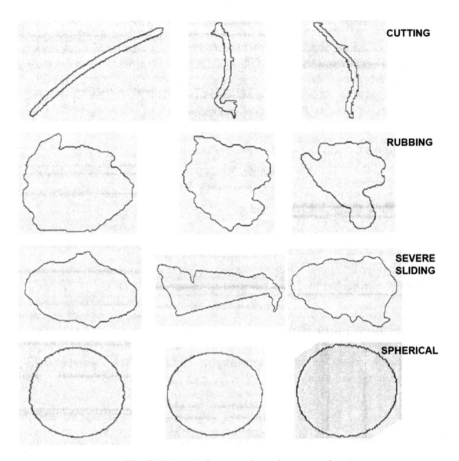

CUTTING

RUBBING

SEVERE
SLIDING

SPHERICAL

Fig. 3 Dataset after granular noise removal

Observing the above results, interest points computation is removed and HOG features of a complete image are computed. Now, by using these new features, a multiclass SVM model is trained and cross-validated. It is observed that the accuracy of the model is increased to about 69%. Therefore, the use of a complete image for HOG features is better for shape classification instead of using only interest points.

The accuracy by using features described in the first technique was not achieved as perceived. Therefore, it was necessary that some other features of eccentricity, extent, equiv-diameter, major axis length, minor axis length, and distance of each boundary point from the centroid are selected and are computed.

A maximum of 2000 points is used for centroid distance features. Those shapes with points less than 2000 are padded with zeros. When classifier is trained with these features, the accuracy is decreased to about 15%. These results were really discouraging. As contribution of boundary points was very large in this feature set, therefore, it is decided to train the classifier by removing the distance of boundary points from feature vectors. The model is again trained and validated by using only five shape features as described above. The results improved and an accuracy of about 90% is achieved by using only these five features.

4.3   Discussion

For the corner points and HOG descriptors' technique, accuracy is very low. This is obviously due to not using the other boundary points of the shape which are contributing to the shape separation. It is confirmed when HOG descriptors are computed for a complete image, accuracy is improved. The confusion matrix for the technique using a complete image for HOG descriptors is shown in Fig. 4.

The figure shows that more rubbing particles are classified as either spherical or severe sliding which is due to some shape similarity in these particles. Similarly, some severe sliding particles are classified as rubbing particles because the dataset has some rubbing and severe sliding particles to be very similar in shape. The accuracy of this technique can be improved by using a larger dataset that will certainly provide some more distinguished features that will help in classifying the particles correctly. As this dataset does not contain the surface texture of the particles, therefore, another option to increase the accuracy could be the use of that dataset which contains particle images with surface texture.

Alternatively, the technique which uses the above-mentioned shape features gives better accuracy and its confusion matrix is shown in Fig. 5. However, a careful analysis shows that the features like major axis length, minor axis length, and equiv-diameter are scale-dependent and change in size of the particles will certainly affect the accuracy of these features classifier.

Confusion Matrix

	Cutting	Rubbing	Severe Sliding	Spherical
Cutting	8 25.0 %	0 0.0 %	0 0.0 %	0 0.0 %
Rubbing	0 0.0 %	3 9.4 %	3 9.4 %	2 6.3 %
Severe Sliding	0 0.0 %	3 9.4 %	5 15.6 %	0 0.0 %
Spherical	0 0.0 %	1 3.1 %	0 0.0 %	7 21.9 %

Fig. 4 Confusion matrix using HOG features classification

Confusion Matrix

	Cutting	Rubbing	Severe Sliding	Spherical
Cutting	8 25.0 %	0 0.0 %	0 0.0 %	0 0.0 %
Rubbing	0 0.0 %	6 18.8 %	0 0.0 %	2 6.3 %
Severe Sliding	0 0.0 %	0 0.0 %	8 25.0 %	0 0.0 %
Spherical	0 0.0 %	1 3.1 %	0 0.0 %	7 21.9 %
	Cutting	Rubbing	Severe Sliding	Spherical

Fig. 5 Confusion matrix using shape-based features classification

The above figure depicts that these features are able to distinguish rubbing and severe sliding particles correctly which the HOG based classifier is unable to distinguish but some rubbing particles are still classified wrongly as spherical particles.

5 Conclusion

In this paper, wear particle classification is discussed by using shape-based features and multiclass SVM classifiers. It is observed that HOG features have less accuracy whereas other features like eccentricity, extent, equiv-diameter, major/minor axis length give better accuracy. However, by using more attributes such as size, color, and surface texture can certainly help in improving accuracy.

References

1. Anderson DP (1991) Wear particle atlas (Revised). In: 4th print, prepared for the NavalAirEngineeringCenter, Lakehurst, NJ
2. Cumming AC (1989) Condition monitoring today and tomorrow—an airline perspective. In: proceedings of 1st international conference on COMADEN 89. Birmingham, UK (1989)
3. Khuwaja GA, Laghari MS (2002) Computer vision techniques for wear debris analysis. Int J Comp App Tech 5(1/2/3): 70–78
4. Laghari MS, Memon QA, Khuwaja GA (2004) Knowledge based wear particle analysis. Int J Inf Technol 1(1–4):31–37
5. Li Q, Zhao T, Zhang L, Sun W, Zhao X (2017) Ferrography wear particles image recognition based on extreme learning machine. J Electr Comput Eng 2017(2):1–6
6. Stachowiak GP, Stachowiak GW, Podsiadlo P (2008) Automated classification of wear particles based on their surface texture and shape features. Tribol Int 41(1):34–43

7. Peng Z, Kessissoglou NJ, Cox M (2005) A study of the effect of contaminant particles in lubricants using wear debris and vibration condition monitoring techniques. Wear 258(11–12):1651–1662

8. Laghari MS, Soomro TR, Khuwaja GA (2011) The use of Fourier descriptors to recognize particle profile. In: Proceedings of 4th IEEE international conference on modeling, simulation, and applied optimization. Kuala Lumpur, pp 1–6

9. Laghari MS (2008) Wear particle profile analysis by using Fourier analyses. In: Proceedings of 5th IEEE international workshop on signal processing and its applications, Sharjah

10. Yuan W, Chin KS, Hua M, Dong G, Wang C (2016) Shape classification of wear particles by image boundary analysis using machine learning algorithms. Mech Syst Signal Process 72–73:346–358

11. Peng Y, Wu T, Cao G, Huang S, Wu H, Kwok N, Peng H (2017) A hybrid search-tree discriminant technique for multivariate wear debris classification. Wear 392–393:152–158

12. Laghari MS, Ahmed F, Aziz J (2010) Wear particle shape and edge detail analysis. In: Proceedings of 2nd IEEE international conference on computer and automation engineering. IEEE Xplore, Singapore, pp 122–125

13. Laghari MS (2003) Shape and edge detail analysis for wear debris identification. Int J Comput Appl 10(4):271–279

14. Laghari MS, Ahmed F (2009) Wear particle profile analysis. In: Proceedings of IEEE international conference on computer design and applications, Singapore, pp 546–550

15. Gaidhane V, Hote YV (2018) An efficient edge extraction approach for flame image analysis. In: Proceedings of pattern analysis and applications, pp 1139–1150

16. Dalal N, Triggs B (2005) Histogram of oriented gradients for human detection. In: Proceedings of IEEE computer society conference on computer vision and pattern recognition. IEEE Xplore, San Diego, CA, pp 886–89

17. Albidewi IA (1993) The application of computer vision to the classification of wear particles in oil. In: Ph.D. thesis, University of Wales, Swansea

On the Physical Layer Security for Land Mobile Satellite Systems

Vinay Bankey and Prabhat K. Upadhyay$^{(\boxtimes)}$

Discipline of Electrical Engineering, Indian Institute of Technology Indore,
Indore, Madhya Pradesh 453552, India
{phd1501202007,pkupadhyay}@iiti.ac.in

Abstract. Land mobile satellite (LMS) systems have become prominent in the fifth-generation broadband wireless communications by providing high quality-of-services to terrestrial mobile users at low cost. However, with the increasing smart technologies, wiretapping and security threats are becoming a major concern in such systems. In this paper, we investigate the secrecy performance of a downlink LMS system by employing a friendly jammer in the presence of an eavesdropper on the ground. Specifically, we derive the secrecy outage probability (SOP) and the probability of strictly positive secrecy capacity (SPSC) expressions of the considered LMS system under the pertinent hetero-geneous fading models for the satellite channels and terrestrial jamming chan-nels. We validate our analytical hypothesis through simulations and reveal the impact of jamming and key parameters on the secrecy performance of LMS systems.

Keywords: Jamming · Land mobile satellite systems · Physical layer security · Secrecy capacity · Secrecy outage probability · Shadowed-Rician fading

1 Introduction

Land mobile satellite (LMS) systems have emerged in the fifth-generation (5G) wireless communication networks with a great promise owing to its advantages of providing seamless connectivity to remote mobile users with a high transmission rate [1]. The importance of such systems is continuously growing for a variety of applications such as broadcasting, navigation, disaster relief, military, etc. However, the inherent broadcasting nature of LMS communication systems always gives an open call to the wiretappers, and thereby, such systems are more vulnerable to wiretapping attacks. The security concerns have rapidly increased and posed a challenge in LMS systems. Traditionally, cryptographic methods have been utilized to achieve secure communication in satellite systems [2, 3]. Recently, information-theoretic based physical layer security (PLS) technique excelled cryptographic methods and has become a leading candidate to ensure overall security [4]. The PLS technique basically exploits the physical characteristics of wireless channels to strengthen communication against security attacks [5]. Based on this pioneering technique, few works [6, 7] have analyzed the PLS in LMS systems. Particularly, the authors have studied the secrecy

© Springer Nature Singapore Pte Ltd. 2020
N. Goel et al. (eds.), *Modelling, Simulation and Intelligent Computing*,
Lecture Notes in Electrical Engineering 659,
https://doi.org/10.1007/978-981-15-4775-1_24

outage probability (SOP) performance of an interference-limited LMS system in [6], and average secrecy capacity performance of a basic LMS system in [7].

PLS performance can be further enhanced by employing a friendly jamming technique where the jammer is utilized to disturb the eavesdropper by emitting the artificial noise and to prevent it from wiretapping the information of a legitimate node [8]. The authors in [9–11] have analyzed the secrecy performance using friendly jammer for terrestrial communication. In [9], authors have employed friendly jammer to improve the secrecy performance of a multi-user wireless network. A cooperative jamming relay has been introduced to degrade the reception at eavesdropper in [10]. Further, the authors in [11] have optimized the secrecy rate in a wireless network using full-duplex jamming receivers.

Although the above-said works have analyzed the PLS performance incorporating friendly jamming, however, they are only limited to the terrestrial communication scenarios. To the extent of the authors' awareness, no results regarding the PLS performance analysis of LMS systems with cooperative jamming have reported so far. Note that the jammer, by sending an artificial noise signal, can confuse the eavesdropper and thereby enhance the secrecy performance.

With the above motivation, we investigate the secrecy performance of a downlink LMS system considering a friendly jammer on ground to protect the communication against an eavesdropper. For this system, we consider that the satellite channels experience shadowed-Rician fading and terrestrial jamming channels follow the Nakagami-m fading. With these heterogeneous channel models, we derive the novel expressions for SOP and probability of strictly positive secrecy capacity (SPSC) of the considered LMS system. Numerical and Monte-Carlo simulation results are provided to corroborate the analytical findings and to highlight the impact of various key parameters on the PLS performance of the considered LMS system.

2   System and Channel Model Description

2.1   System Model

As shown in Fig. 1, we consider a downlink LMS system which consists of a geostationary satellite S, a legitimate terrestrial user U, an eavesdropper E, and a friendly jammer J. In this system, S communicates with U in the presence of E at ground while an outside friendly J is utilized to send a jamming signal to interfere with E. It is assumed that each node, including the S, is equipped with a single antenna and operates in a half-duplex mode. Further, we consider that the satellite channels i.e., $S \rightarrow U$ and $S \rightarrow E$, are assumed to experience shadowed-Rician fading distribution which accurately characterizes the statistical behavior of the LMS communication channel [1]. Besides, the jamming channels, i.e., $J \rightarrow U$ and $J \rightarrow E$ follow Nakagami-m fading. Both receiving terrestrial nodes, i.e., U and E are assumed to be inflicted by additive white Gaussian noise (AWGN) with zero mean and variances σ_u^2 and σ_e^2, respectively. Moreover, throughout this paper, we represent $S \rightarrow U$ and $S \rightarrow E$ channels as main and wiretap channels, respectively, and use subscripts s, u, e, and j for denoting nodes S, U, E, and J, respectively.

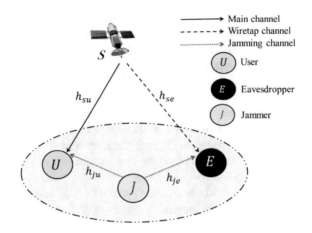

Fig. 1 System model of LMS communication with a friendly jammer

We assume that the user and jammer cooperate with each other such that the impact of jamming signal can be nulled out at the user, thus, jammer would not interfere with the cooperative user [8]. Whereas, E receives both jamming signal and signal from the satellite which keep E unclear, and hence, E needs extra effort to extract the confidential information.

Let x_s and x_j denote the signal transmitted by satellite S and friendly jammer J, respectively. Then, the received signals at the user U and eavesdropper E can be expressed, respectively, as

$$y_{su} = \sqrt{P_s}h_{su} + v_u \tag{1}$$

and

$$y_{se} = \sqrt{P_s}h_{se} + \sqrt{P_J}h_{je}x_j + v_e, \tag{2}$$

where P_s and P_J are the transmit powers at satellite S and jammer J, respectively, h_{su} is the channel coefficient between S and U, h_{se} is the channel coefficient between S and E, and h_{je} denotes the channel coefficient between J and E. Herein, $v_u \sim \mathcal{CN}(0, \sigma_u^2)$ and $v_e \sim \mathcal{CN}(0, \sigma_e^2)$ show the AWGN at U and E, respectively, with $\mathcal{CN}(\cdot, \cdot)$ representing the complex normal distribution. Based on (1), the instantaneous signal-to-noise ratio (SNR) at user U can be given as

$$\Gamma_U = \rho_u|h_{su}|^2 = \gamma_{su}, \tag{3}$$

where $\rho_u = P_s/\sigma_u^2$. Similarly, from (2), the instantaneous signal-to-interference-plus-noise ratio (SINR) at eavesdropper E can be given as

$$\Gamma_E = \frac{\gamma_{se}}{\gamma_J + 1},\tag{4}$$

where $\gamma_{se} = \rho_e |h_{se}|^2$ and $\gamma_J = \rho_J |h_{je}|^2$ with $\rho_e = P_s/\sigma_e^2$ and $\rho_J = P_J/\sigma_e^2$.

Now, we formulate the secrecy capacity of the considered LMS system which is defined as the non-negative difference between the channel capacity of main channel and that of wiretap channel [12]. In this way, we first express the channel capacity of the main and wiretap channels, using (3) and (4), as

$$C_{\Gamma_U} = \log_2(1 + \Gamma_U)\tag{5}$$

and

$$C_{\Gamma_E} = \log_2(1 + \Gamma_E).\tag{6}$$

Hence, the secrecy capacity for the considered LMS system can be expressed as

$$C_{sec} = [C_{\Gamma_U} - C_{\Gamma_E}]^+,\tag{7}$$

where $[z]^+ = \max(z, 0)$. Note that the secrecy capacity is an important metric which is used to evaluate secrecy performance such as SOP and probability of SPSC in wireless communication systems. We now characterize the statistics of the satellite and terrestrial channels in succeeding subsection.

2.2   Channel Model

As stated before, we assume that the satellite channels follow a shadowed-Rician fading distribution, the probability density function (PDF) of $|h_{st}|^2$ between S and terrestrial node l, for $\iota \in \{u, e\}$, is given by [1]

$$f_{|h_{st}|^2}(x) = \alpha_\iota e_1^{-\beta_\iota x} F_1(m_\iota; 1; \delta_\iota x), \quad x \geq 0,\tag{8}$$

where $\alpha_\iota = ((2b_\iota m_\iota)/(2b_\iota m_\iota + \Omega_\iota))^{m_\iota}/2b_\iota, \delta_\iota = \Omega_\iota/(2b_\iota(2b_\iota m_\iota + \Omega_\iota)), \beta_\iota = 1/2b_\iota$, and $_1F_1(\cdot; \cdot; \cdot)$ is the confluent hypergeometric function of the first kind [13, Eq. 9.210.1]. Herein, Ω_ι and $2b_\iota$ represent the average power of line-of-sight (LOS) and multipath components, respectively, and m_ι denotes the fading severity parameter of the pertinent channel. Now, we can simplify (8) for integer-valued fading parameter m_ι and express $f_{|h_{st}|^2}(x)$ as [12]

$$f_{|h_{st}|^2}(x) = \alpha_\iota \sum_{\kappa=0}^{m_\iota - 1} \Xi_\iota(\kappa) x^\kappa e^{-(\beta_\iota - \delta_\iota)x},\tag{9}$$

where $\Xi_\iota(\kappa) = (-1)^\kappa (1 - m_\iota)_\kappa \delta_\iota^\kappa/(\kappa!)^2$. Herein, $(\cdot)_n$ represents the Pochhammer symbol [13, p. xliii]. Further, by making a transformation of variable, the PDFs of γ_{su} and γ_{se} can be derived, respectively, as

$$f_{\gamma_{su}}(x) = \alpha_u \sum_{\kappa=0}^{m_u-1} \frac{\Xi_u(\kappa)}{\rho_u^{\kappa+1}} x^\kappa e^{-\chi_u x} \tag{10}$$

and

$$f_{\gamma_{se}}(x) = \alpha_e \sum_{r=0}^{m_e-1} \frac{\Xi_e(r)}{\rho_e^{r+1}} x^r e^{-\chi_e x}, \tag{11}$$

where $\chi_u = (\beta_u - \delta_u)/\rho_u$ and $\chi_e = (\beta_e - \delta_e)/\rho_e$. We can further obtain the corresponding cumulative distribution functions (CDFs) $F_{\gamma_{su}}(x)$ and $F_{\gamma_{se}}(x)$ by integrating the respective PDFs, with the aid of [13, Eq. 3.351.2], as

$$F_{\gamma_{su}}(x) = 1 - \alpha_u \sum_{\kappa=0}^{m_u-1} \frac{\Xi_u(\kappa)}{(\rho_u)^{\kappa+1}} \sum_{p=0}^{\kappa} \frac{\kappa!}{p!} \chi_u^{-(\kappa+1-p)} x^p e^{-\chi_u x} \tag{12}$$

and

$$F_{\gamma_{se}}(x) = 1 - \alpha_e \sum_{r=0}^{m_e-1} \frac{\Xi_e(r)}{(\rho_e)^{r+1}} \sum_{s=0}^{r} \frac{r!}{s!} \chi_e^{-(r+1-s)} x^s e^{-\chi_e x}. \tag{13}$$

On the other hand, assuming Nakagami-m fading distribution for the terrestrial jamming channels, the PDF of the channel gain γ_J is given as $f_{\gamma_J}(x) = (x^{m_j-1}/\Gamma(m_j))(m_j/\eta_j)^{m_j} e^{-(m_j/\eta_j)x}$, with average power Ω_j and fading severity m_j of the pertinent channel, where $\eta_j = \rho_J \Omega_j$. Now, we concentrate on the secrecy performance analysis of the considered LMS system in the next section.

3   Secrecy Performance Analysis

In this section, we study the secrecy performance analysis by investigating the SOP and the probability of SPSC for the considered LMS system.

3.1   SOP

The SOP is defined as the probability of the event when the secrecy capacity drops below a predefined secrecy rate R_s and can be given as

$$\mathcal{P}_{sec} = \Pr[C_{sec} < R_s] = \Pr\left[\frac{1+\Gamma_U}{1+\Gamma_E} < \gamma_s\right], \tag{14}$$

where $\gamma_s = 2^{R_s}$. We can further write $\mathcal{P}_{sec}$ as

$$\mathcal{P}_{\text{sec}} = \int_0^\infty F_{\Gamma_U}(x\gamma_s + \gamma_s - 1)f_{\Gamma_E}(x)dx. \tag{15}$$

To solve (15), we require CDF of Γ_U which can be computed directly from (12) using (3). Next, we can evaluate PDF of Γ_E as $f_{\Gamma_E}(x) = dF_{\Gamma_E}(x)/dF_{\Gamma_E}(x)$. For this, we simplify (4), considering high interference scenario due to jammer as $\Gamma_E \simeq \gamma_{se}/\gamma_J$. We first obtain $F_{\Gamma_E}(x)$ using (13), apply $f_{\gamma_J}(x)$ along with the fact [13, Eq. 3.351.3], and then by differentiating the resultant, $f_{\Gamma_E}(x)$ can be evaluated as

$$
\begin{aligned}
f_{\Gamma_E}(x) = \alpha_e \sum_{r=0}^{m_e-1} \frac{\Xi_e(r)}{(\rho_e)^{r+1}} \sum_{s=0}^{r} \frac{r!}{s!} \left(\frac{m_j}{\eta_j}\right)^{m_j} \frac{\Gamma(s+m_j)}{\chi_e^{(r+1-s)}\Gamma(m_j)} \left(x\chi_e + \frac{m_j}{\eta_j}\right)^{-(s+m_j+1)} \\
\times \left(x^s(s+m_j)\chi_e - sx^{s-1}\left(x\chi_e + \frac{m_j}{\eta_j}\right)\right).
\end{aligned}
\tag{16}
$$

Finally, by inserting (12), since $\Gamma_D = \gamma_{su}$, and (16) into (15), performing the simplification using the identity [14, Eq. 22], and then solving the integration with the help of [13, Eq. 7.813.1], we obtain the expression for the SOP of the considered LMS system as

$$
\begin{aligned}
\mathcal{P}_{\text{sec}} = 1 - \alpha_u\alpha_e \sum_{\kappa=0}^{m_u-1} \frac{\Xi_u(\kappa)}{(\rho_u)^{\kappa+1}} \sum_{p=0}^{\kappa} \frac{\kappa!}{p!}\chi_u^{-(\kappa+1-p)} \sum_{r=0}^{m_e-1} \frac{\Xi_e(r)}{(\rho_e)^{r+1}} \sum_{s=0}^{r} \frac{r!}{s!}\chi_e^{-(r+1-s)} \frac{\Gamma(s+m_j)}{\Gamma(m_j)} \\
\times \left(\frac{(s+m_j)\chi_e\eta_j}{m_j\Gamma(m_j+s+1)(\gamma_s\chi_u)^{q+s+1}} G_{2,1}^{1,2}\left[\frac{\chi_e\eta_j}{\chi_u\gamma_s m_j}\middle| \begin{matrix} -(q+s), -(s+m_j) \\ 0 \end{matrix}\right] \right. \\
\left. - \frac{s(\gamma_s\chi_u)^{-(q+s)}}{\Gamma(m_j+s)} G_{2,1}^{1,2}\left[\frac{\chi_e\eta_j}{\chi_u\gamma_s m_j}\middle| \begin{matrix} -(q+s-1), 1-(s+m_j) \\ 0 \end{matrix}\right] \right),
\end{aligned}
\tag{17}
$$

where $G_{2,1}^{1,2}[\cdot]$ denotes the Meijer's G-function [13, Eq. 8.2.1.1].

3.2 Probability of SPSC

The SPSC is also a prime measure to analyze the secrecy performance which is said to occur when the secrecy capacity of the system becomes positive. Hence, the probability of SPSC can be expressed mathematically as

$$\mathcal{P}_{\text{SPSC}}^{\text{sec}} = 1 - \Pr[C_{\text{sec}} < 0]. \tag{18}$$

One can observe from (18) that the probability of SPSC can be evaluated readily using the expression of SOP by setting $R_s = 0$ in (14).

4 Numerical and Simulation Results

In this section, we present the numerical and Monte-Carlo simulation results to investigate the impact of terrestrial jamming and different shadowing scenarios of the satellite channels on the SOP and the probability of SPSC performance of the considered LMS system. For this, we assume that the satellite channels undergo heavy shadowing (HS) with parameters $(m_t, b_t, \Omega_t) = (1, 0.063, 0.0007)$ and average shadowing (AS) with parameters $(m_t, b_t, \Omega_t) = (5, 0.251, 0.279)$ [15]. Herein, we set $P_J = 5$ dB, $m_j = 1$, $\Omega_j = 1$, and $\rho_e = 0$ dB (unless stated otherwise).

In Fig. 2a, we plot the SOP curves versus ρ_u considering four different sets of shadowing scenarios (with AS and HS) for satellite channels. From this figure, it can be observed that the system achieves better SOP performance when the main channel experiences AS and wiretap channel undergoes HS scenario of shadowed-Rician fading. This is owing to the fact that, with the HS scenario, the wiretap channel condition becomes worse than that of the main channel. In contrast, when the main channel experiences HS and wiretap channel undergoes AS scenario, the system SOP performance deteriorates.

Figure 2b illustrates the impact of jamming power P_J and fading severity parameter m_j of $J \to E$ channel on the SOP performance. We consider here that the $S \to U$ experiences AS and $S \to E$ experiences HS scenarios of the shadowed-Rician fading. As expected, the SOP performance of the considered LMS system improves when

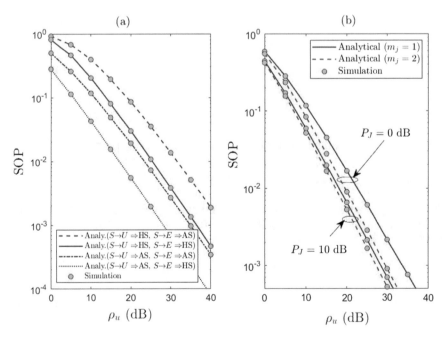

Fig. 2 a SOP for different shadowing scenarios with $R_s = 0.2$. **b** Impact of jamming power P_J and fading severity parameter m_j on the SOP with $R_s = 0.5$

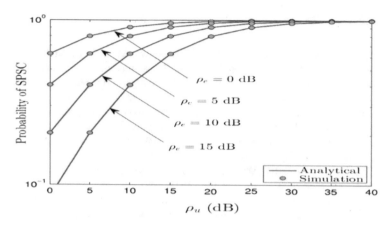

Fig. 3 Probability of SPSC for different values of ρ_e

either the jamming power or fading severity parameter m_j increases. This can be easily realized by comparing the SOP curves in Fig. 2b. With this, one can infer that better security can be ensured by employing the jammer with a high transmit power.

In Fig. 3, we plot the probability of SPSC curves for different values of ρ_e. For this, we assume that both main and wiretap channels undergo HS scenario of shadowed-Rician fading. We can see that the probability of SPSC performance degrades with an increase in ρ_e. This is due to the fact that the ability of eavesdropper to intercept the information improves with a high value of ρ_e and eavesdropper becomes more hazardous to the system.

5   Conclusion

We studied the secrecy performance of a downlink LMS system using a friendly jammer in the presence of an eavesdropper on the ground. We derived novel expressions of SOP and probability of SPSC for the considered system by adopting shadowed-Rician fading for the satellite channels and Nakagami-m fading for jamming channels. Our results demonstrated that while the satellite channel conditions have a severe impact on the LMS system performance, the jammer can notably improve the secrecy performance. In our future work, we investigate the PLS performance of a multi-user LMS system configuration by deploying multiple terrestrial jammers.

Acknowledgements. This publication is an outcome of the R&D work undertaken in the project under the Visvesvaraya Ph.D. Scheme of Ministry of Electronics & Information Technology (MeitY), Government of India, being implemented by Digital India Corporation (Formerly Media Lab Asia).

References

1. Abdi A, Lau W, Alouini M-S, Kaveh M (2003) A new simple model for land mobile satellite channels: First and second order statistics. IEEE Trans Wireless Commun 2(3):519–528
2. Sklavos N, Zhang X (2007) Wireless security and cryptography: Specifications and implementations, 1st edn. CRC Press, Baco Raton, FL, USA
3. Cruickshank H, Howarth M, Iyengar S, Sun Z, Claverotte L (2005) Securing multicast in DVB-RCS satellite systems. IEEE Wireless Commun 12(5):38–45
4. Zheng G, Arapoglou PD, Ottersten B (2012) Physical layer security in multibeam satellite systems. IEEE Trans Wireless Commun 11(2):852–863
5. Poor HV, Schaefer RF (2017) Wireless physical layer security. Nat Acad Sci 114(1):19–26
6. Bankey V, Upadhyay PK, Da Costa DB (2018) Physical layer security of interference-limited land mobile satellite communication systems. In: International conference on advanced communication technologies networking (CommNet), Marrakech, Morocco
7. An K et al (2016) Average secrecy capacity of land mobile satellite wiretap channels. In: Wireless communication and signal processing (WCSP), Yangzhou, China
8. Dong L, Han Z, Petropulu AP, Poor HV (2010) Improving wireless physical layer security via cooperating relays. IEEE Trans Signal Process 58(3):1875–1888
9. Li B et al (2019) Secrecy outage probability analysis of friendly jammer selection aided multiuser scheduling for wireless networks. IEEE Trans Commun 67(5):3482–3495
10. Xu S et al (2015) Improving secrecy for correlated main and wiretap channels using cooperative jamming. IEEE Access 7(1):23788–23797
11. Zheng G et al (2013) Improving physical layer secrecy using full-duplex jamming receivers. IEEE Trans Signal Process 61(20):4962–4974
12. Bankey V, Upadhyay PK (2017) Secrecy outage analysis of hybrid satellite-terrestrial relay networks with opportunistic relaying schemes. In: IEEE 85th vehicular technology conference (VTC), Sydney, Australia
13. Gradshteyn IS, Ryzhik IM (2000) Tables of integrals, series and products, 6th edn. Academic Press, New York
14. Bankey V, Upadhyay PK (2018) Ergodic capacity of multiuser hybrid satellite-terrestrial fixed-gain AF relay networks with CCI and outdated CSI. IEEE Trans Veh Technol 67(5):4666–4671
15. Miridakis NI, Vergados DD, Michalas A (2015) Dual-hop communication over a satellite relay and Shadowed-Rician channels. IEEE Trans Veh Technol 64(9):4031–4040

Experimental Validation of PVSYST Simulation for Fix Oriented and Azimuth Tracking Solar PV System

Fahad Faraz Ahmad[1]([⊠]), Mohamed Abdelsalam[1],
Abdul Kadir Hamid[2], Chaouki Ghenai[1,3], Walid Obaid[1],
and Maamar Bettayeb[2]

[1] Research Institute for Science and Engineering, University of Sharjah,
Sharjah, UAE
{ffahmad, cghenai, u00032590}@sharjah.ac.ae,
mabdalsalam13@gmail.com
[2] Electrical and Computer Engineering Department, College of Engineering,
University of Sharjah, Sharjah, UAE
{akhamid, maamar}@sharjah.ac.ae
[3] Sustainable and Renewable Energy Engineering Department,
College of Engineering, University of Sharjah, Sharjah, UAE

Abstract. The accurate prediction of power generation by the PV system is crucial during the designing stage and subsequently in the operation and maintenance phase. It provides a reference to evaluate the performance of the PV system. PVSYST is widely accepted simulation software for the PV system in the industry. In this study, a comparison of PVSYST simulation results and experimentally collected data for fix oriented and azimuth tracking solar system is analyzed. Five clear sunny days are selected for each case. The hourly average data has been used for comparison. The deviation of predicted values in case of fix oriented and azimuth tracking solar systems is 2.14% and 2.74%, respectively. This variation is primarily due to the mismatch of predicted weather data with real conditions. The results have concluded that PVSYST is reliable software to use for the prediction of PV energy generation with an acceptable margin. Further, the adaptation of azimuth tracking for the solar system is feasible and improves the average power production by 17.28% as compared to the fix-oriented solar system in the hot and humid environment of the UAE.

Keywords: PVSYST · PV system · Tracking system · Azimuth tracking

1 Introduction

The UAE has blessed with a substantial amount of solar light exposure, enabling it a great opportunity and huge potential for renewable and sustainable energy development [1]. The country is popular for its massive domestic oil and gas reserves. Since 2006, it is focused on renewable energy development and has been participating in several paralleled projects [2]. The UAE majorly depended on conventional energy resources for its energy requirements. However, a high step to achieve financial diversification, to

© Springer Nature Singapore Pte Ltd. 2020
N. Goel et al. (eds.), *Modelling, Simulation and Intelligent Computing*,
Lecture Notes in Electrical Engineering 659,
https://doi.org/10.1007/978-981-15-4775-1_25

prevent significant product downturns accompanied by the oil-relied economy, has resulted in new incentives and the appearance of the latest renewable energy technologies. Moreover, the population growth, quick industrialization and rapid increase in water need from desalination plants have led to high energy demands. Consequently, in the MENA, the nation is leading the renewable energy adaption and has resulted in establishing the Masdar, an Abu Dhabi Future Energy Company and the first carbon-neutral and zero-waste city in the region. According to Q1, 2019 report of the Business Monitor International, the solar energy turns into the main source of renewable energy in the UAE, with total solar generation improving from 1.25 terawatt-hour (TWh) in 2018 to 4.76 TWh in 2019 [3]. It is expected to grow up to 13.66 TWh in 2028. Therefore, it is predicted to improve the renewable part of the total power generation from 0.9% in 2018 to 3.4% in 2019 with a target of 6.9% in 2028. A number of under-construction projects in Dubai and Abu Dhabi illustrates the increasing trajectory. The UAE has also announced its aggressive "energy strategy 2050", which is the first energy policy to become law, for achieving the balance between supply and demand in the energy market [3]. This strategy will expand the country's energy production include 12% from clean coal, 38% from gas, 6% from nuclear energy and 44% from renewable energy (solar, wind and biofuels). The policy is willing to improve the part of renewable electricity generation capacity to 44% by 2050. This goes in partnership with the UAE's Vision 2021 to fulfil 27% of energy needs from clean sources, including nuclear power [4].

In this paper, a 2.88 kWp, the photovoltaic system is designed and installed on the rooftop of the University of Sharjah, UAE. The main objective of this study is to validate the simulation results and energy prediction of the residential-scale PV system by PVSYST. PVSYST is a widely accepted software in the solar industry for the simulation of an on-grid PV system [5–8]. Besides, the azimuth tracking system under the hot and humid environment of Sharjah, UAE has also been studied. Experimental work and data acquisition are conducted utilizing PV analyzer—Profitest. Finally, a comparison of fixed and azimuth tracking systems in terms of irradiance collection and thus power production has been conducted.

2   Photovoltaics and PV System

Photovoltaics, it is the production of electricity from light. Photovoltaic (PV) panels have been functioning from more than 50 years with diverse applications [9]. Three main classes of silicon-based materials were identified in use, monocrystalline, polycrystalline and amorphous cells. Besides the silicon, some other crystalline substances are also used for PV solar cells known as, third-generation solar cells, a novel compound semiconductor material, Gallium arsenide (GaAs). The organic PV cells are manufactured from polymers. The transparency property has made polymer solar cells useful in building integrated applications. On the other hand, one downside with third-generation solar cells is that they generate a relatively lower efficiency compared to silicon materials [10].

This experimental study compares two system configurations, fix oriented and azimuth tracking. Due to the movement of the Earth around the sun and the resulting impact on solar radiation, some tracking systems are designed to track the sun and consequently maximize the solar incident light that exposed on the modules by maintaining an optimum orientation between the sun and the solar panels. Respecting the tracking PV configuration, it usually employs a sophisticated control system. The tracking system can be categorized as a single-axis and dual-axis tracking system. On the other hand, in the case of the fix oriented PV system, the solar modules or arrays are permanently fixed at a specific angle and orientation towards the sun, with the aim of getting high solar exposure rate. In this study, the single-axis azimuth tracking system has been considered.

2.1   System Configuration

The experiment was conducted on a clear sky at 25.34° N, 55.42° E, University of Sharjah campus in Sharjah, United Arab Emirates. The system was installed on the rooftop of the W-12 building as shown in Fig. 1. Nine polycrystalline silicon solar panels of 320 Wp each, with an overall capacity of 2.88 kWp are installed. The total surface area of 17.5 m^2 is covered. The tilt angle of 20° is considered during the whole experiment. The solar irradiance sensor, ambient temperature sensor, and PV module temperature sensor are installed on the PV system to collect the data. The electrical power production by the PV system is recorded through the Profitest PV analyzer. The data is collected from 6:00 am to 5:00 pm with a resolution of 5 min.

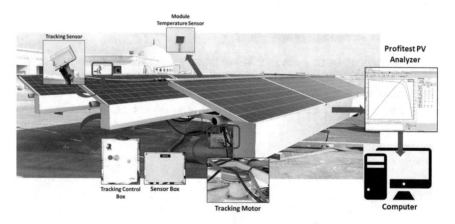

Fig. 1 Experimental system setup

3   Result and Discussion

3.1   PVSYST Simulation

For simulation, weather data is taken from the Meteonorm 7.1 software by adding the user defined location [5]. The monthly solar irradiance and the ambient temperature profile are presented in Fig. 2. The grid-connected solar system with a capacity of 2.88 kWp is simulated with PVSYST V6.4.3. Two separate cases fix orientation and azimuth-tracking are considered. The hourly data is extracted from the simulation results [11]. For each case, five clear sunny days are selected to compare the simulation results by PVSYST and real power production by the solar PV system. The average hourly data has been analyzed to compare the system.

3.2   Fixed Oriented System

The installed solar system has been fixed towards the south at the azimuth angle of $0°$. The tilt angle is fixed at $20°$. Six clear sunny days, 11th Sep., 27th Sep., 28th Sep., 2nd Oct., 4th Oct., and 5th Oct. of 2019 has been selected. The same day's data has also been extracted from PVSYST simulation to compare with real power production by the solar system. In Fig. 3, a comparison of solar irradiance, ambient temperature, module temperature and power generation by PVSYST and real measured values are presented. The percentage error in power generation estimation is 2.14%. The difference in irradiance value is 4.39% while in the case of ambient and module temperature is 13.06% and 10.67%, respectively, Table 1. The annual energy production estimation from PVSYST is 4.7 MWh/year with a specific yield of 1637 kWh/kWp/year. The predicted performance ratio of fix oriented PV system is 76.1%.

3.3   Azimuth Tracking System

An azimuth tracking system was installed to track the sun. For the tracking system, the six days, 9th Sep., 20th Sep., to 24th Sep., of 2019 has been selected to do the simulation and experimental work. The average values of hourly data of PVSYST and experimentally recorded are presented in Fig. 4. The error in the prediction of power

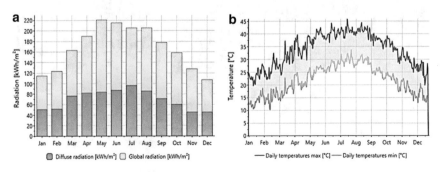

Fig. 2 Monthly profile **a** Global and diffuse radiation, **b** Daily max. and min. Ambient temperature

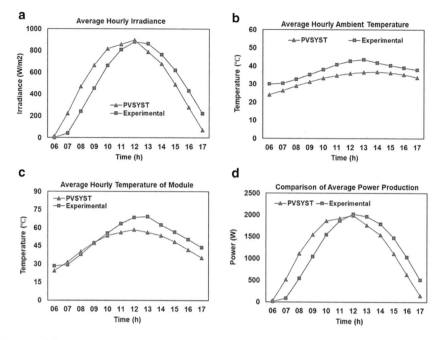

Fig. 3 PVSYST and experimental average hourly data for the fixed system: **a** Irradiance, **b** Ambient temperature, **c** Temperature of the module, **d** Power production

Table 1 Comparison of PVSYST and experimental data for the fixed system

Description	T ambient (°C)	Irradiance (W/m²)	T module (°C)	Power at MPP (W)
PV SYST	32.82	522.20	45.61	1182.09
Experimental	37.75	500.23	51.05	1157.38
Error (%)	13.06	4.39	10.67	2.14

production is 2.74%. The variation is due to the difference in the weather data. The error in the irradiance value is 0.65%. While the error in ambient and module temperatures is 9.58% and 6.84%, respectively, Table 2. The estimation of energy production by azimuth tracking system is 5.46 MWh/year with a specific production of 1894 kWh/kWp/year. The estimated performance ratio of azimuth tracking PV system is 76.6%.

The comparison has shown that the simulation through PVSYST is reliable with the difference of 2.14 and 2.74% in the case of fixed oriented and azimuth tracking system. It can be reliably used to estimate the power generation for the designed solar PV system on the desired location. The estimation model can be improved by providing more accurate weather data. In both cases, irradiance values considered by PVSYST are more close to the real value. On the other hand, ambient temperature and module

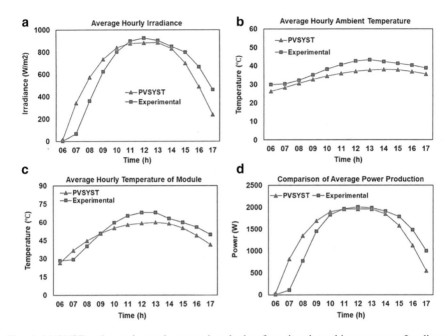

Fig. 4 PVSYST and experimental average hourly data for azimuth tracking system: **a** Irradiance, **b** Ambient temperature, **c** Temperature of the module, **d** Power production

Table 2 Comparison between PVSYST and experimental data for azimuth tracking system

Description	T ambient (°C)	Irradiance (W/m²)	T module (°C)	Power at MPP (W)
PV SYST	34.31	617.56	49.54	1394.61
Experimental	37.95	613.58	53.18	1357.40
Error (%)	9.58	0.65	6.84	2.74

temperature have a considerable deviation from the real values. It causes a significant effect on the estimation of power production by the PV system especially in hot weather areas like UAE.

3.4 Comparison of the Fix Oriented and the Azimuth Tracking PV System

In a fixed system, the solar panels are faced towards the south with the azimuth of 0°. However, in the case of a tracking system, the solar system will always be oriented directly towards the sun and will follow the sun from morning till evening. Hence, the average solar irradiance collected by the tracking system is more than a fix-oriented system and thus generates more power.

The comparison of power production by the fix oriented and tracking solar system is presented in Fig. 5. The increment in power production with hourly data is also

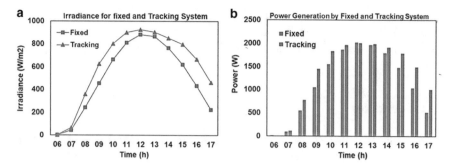

Fig. 5 Comparison of fixed and tracking system **a** Irradiance on the plane of solar PV panel, **b** Power generation

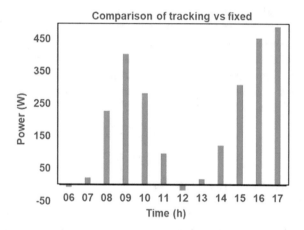

Fig. 6 Increment in power production by azimuth tracking system

illustrated in Fig. 6. It can be observed that the gain of power generation by azimuth tracking system oscillates from the peak in the morning, reaches the lowest value at noon and again increases in the evening time. The difference is negligible at noontime due to the orientation of the sun in the south. The peaks appear during sunrise and the sunset due to the perfect orientation of tracking solar PV systems towards the sun.

In Table 3, 22.6% more solar irradiance is collected by the solar PV system with azimuth tracking as compared to the fix oriented system. The collection of more solar

Table 3 Comparison of fix oriented and azimuth tracking system

Description	Irradiance (W/m²)	T module (°C)	Power at MPP (W)
Fixed	500.23	51.05	1157.38
Tracking	613.58	53.18	1357.40
Difference (%)	+22.66	+4.16	+17.28

irradiance causes an increment of 4.16% of the average module temperature. On average 17.28% more power is generated by the azimuth tracking system over a day. The power gain provides the basis to adopt the azimuth-tracking system for the solar system.

4   Conclusion

The experimental study is conducted to validate the simulation results from PVSYST software in case of fix oriented and azimuth tracking PV system. The variation in the prediction of power generation by PVSYST from the measured data in case of fix oriented and the azimuth-tracking system is 2.14% and 2.74%, respectively. The deviation is observed due to the variation of weather data taken from Meteonorm 7.1. It is also noted that the variance of considered solar irradiance value is 4.39% in the case of fix orientation and 0.65% in the case of the tracking system. On the other hand, variation in module operating temperature is considerably high which is the major cause of error in power production estimation. More accurate weather data can improve the prediction ability of PVSYST. It is also observed that the azimuth tracking system, improves 17.28% power production by PV system as compared to fix oriented solar system. Thus, it is feasible and promising to adopt azimuth tracking for PV System under UAE weather conditions.

References

1. Mokri A, Aal Ali M, Emziane M (2016) Solar energy in the United Arab Emirates: a review. Renew Sustain Energy Rev 28:340–375
2. Kazim M (2007) Assessments of primary energy consumption and its environmental consequences in the United Arab Emirates. Renew Sustain Energy Rev 11(3):426–446
3. UAE Energy strategy 2050. https://government.ae/en/about-the-uae/strategies-initiatives-and-awards/federal-governments-strategies-and-plans/uae-energy-strategy-2050. Last accessed 27 Oct 2019
4. Ministry of Climate Change and Environment (2017) National Climate Change Plan of the United Arab Emirates 2017–2050. Minist Clim Chang Environ 22–48
5. Kumar NM, Kumar MR, Rejoice PR, Mathew M (2017) Performance analysis of 100 kWp grid connected Si-poly photovoltaic system using PVsyst simulation tool. Energy Proc 117:180–189
6. Steiner M et al (2019) CPVIndia–Energy yield forecasting with PVsyst. In: AIP conference, vol proceedings, 2149(1), p 060005. AIP Publishing
7. Ma Z, Zhang Y, Li H (2019) Energy efficiency analysis of inland ship photovoltaic system based on PVsyst. In: IOP conference series: earth and environmental science, vol 242(2), pp 022057. IOP Publishing
8. Selmi T, Dhouibi H, Ghabi J (2019) Matlab/Simulink and PVSyst based modeling and validation of photovoltaic cells. Euro J Eng Res Sci 4(11):11–16
9. Spanggaard H, Krebs FC (2004) A brief history of the development of organic and polymeric photovoltaics. Sol Energy Mater Sol Cells 83:125–146

10. Ikedi C (2019) Experimental study of current-voltage characteristics for fixed and solar tracking photovoltaics systems. Recent developments in photovoltaic materials and devices (IntechOpen)
11. Sauer KJ, Roessler T, Hansen CW (2015) Modeling the irradiance and temperature dependence of photovoltaic modules in PVsyst. IEEE J Photovoltaics 5(1):152–158

Bequest of RETE Algorithm for Rule Assessment in Context Database

C. Shivakumar[1]([⊠]) and Siddhaling Urolagin[2]

[1] Computer Science Department, Jain University, Kanakapura, Ramanagaram
District, Karnataka 562117, India
shivusdmit@gmail.com
[2] Department of Computer Science, Birla Institute of Technology & Science
Pilani, Dubai Campus, Dubai, UAE
siddhaling@dubai.bits-pilani.ac.in

Abstract. In the current era, since software automation is buzzing everywhere
and it is creating huge opportunities to pay heed on generating rules and reasoning. In the same direction, our research work has been carried out. If any
system to be context-aware its database should be in a position to provide few
facilities to its system. So, this research work elucidates different contexts and its
dimensions through the context dimension tree. The novelty of this research
work is, it has used context data for generating the rules and reasoning using
RETE rule-based algorithm, Rule assessment against facts/data and ordering of
statement is been simplified using a particular algorithm.

Keywords: ADLs · RETE · Context · Context dimension tree

1 Introduction

When the user's context is observed, used same for monitoring purpose to know his/her
status and to respond accordingly the system should be context-aware. This type of
system can be achieved by generating rules using the available context data or facts.
When rules are been judged against facts or data in application scenario the process of
evaluation and ordering the statements will be costly. Using the RETE algorithm in a
context-aware environment we can avoid these shortcomings. This inference approach
saves both "Context is any information that can be used to characterize the situation of
an entity. It is a person, place, or object that is considered relevant to the interaction
between a user and an Activity of Daily Living application, including the user and
applications themselves." [1].

The parameters to characterize the user's contexts are role, interest topic, situation,
time, interface and location. Parameters will differ according to the application [2]. For
our research work, we have used the following parameters that are depicted pictorially.
The context dimension tree is a context data model which used to represent different
instances of the particular context. Context-aware software adapts according to the
location of use, the collection of nearby people, hosts, and accessible devices, as well
as changes to such things over time [3]. The context which we have used, its
dimensions and descriptions are as shown in Tables 1 and 2. If any system has the

N. Goel et al. (eds.), *Modelling, Simulation and Intelligent Computing*,
Lecture Notes in Electrical Engineering 659,
https://doi.org/10.1007/978-981-15-4775-1_26

Table 1 ADLs activity labels for user

UserID	Start_time	End_time	Activity
User1	2011-11-28 02:27:59	2011-11-28 10:18:11	Sleeping
User1	2011-11-28 10:21:24	2011-11-28 10:23:36	Toileting
User2	2011-11-28 10:34:23	2011-11-28 10:34:41	Breakfast
User2	2011-11-28 10:34:44	2011-11-28 10:37:17	Breakfast

Table 2 ADLs sensor events for user

UserID	Start time	End_time	Location	Place	Type
User1	2011-11-28 02:27:59	2011-11-28 10:18:11	Bed	Bedroom	Pressure
User1	2011-11-28 10:21:24	2011-11-28 10:23:36	Cabinet	Bathroom	Magnetic
User2	2011-11-28 10:34:23	2011-11-28 10:34:41	Fridge	Magnetic	Kitchen
User2	2011-11-28 10:34:44	2011-11-28 10:37:17	Cupboard	Magnetic	Kitchen

capability to interact with the user then such a system can be categorized as context-aware system [4]. If any system doesn't have this facility then such a system cannot deal with dynamic data but they will be confined to static data. When the system is context-aware it will be having a reasoning capability where it generates the rules by evaluating different patterns. In such scenarios where reasoning plays a vital role, where rule generation is required rule-based algorithms are of great help. So, in our research work, we have used RETE rule-based algorithm for rule generation. RETE algorithm is an efficient forward chaining algorithm.

In some organizations, someone writes the rules and reasoning will be done by someone else. Indeed, expert systems will be having separate knowledge part and reasoning part [4]. RETE algorithm is a pattern matching algorithm for implementing rules. It offers a logical portrayal accountable for matching facts. This rule system entails one or more conditions and diverse actions that may be commenced for each complete set of facts that match the condition [5]. Against a certain dataset, the RETE algorithm compares the rules [6] in the presented research work rules are defined to predict the context of the users. Its aim is to match a set of facts against a set of inference rules (productions) [7]. The rete algorithm is a well-known algorithm for efficiently perorating the many patterns/many objects match problem [8]. The viable anodyne of the Rete-evaluation would be the automatic runtime optimization of programs [9]. All data in the RETE network are firstly processed on root node of the Rete

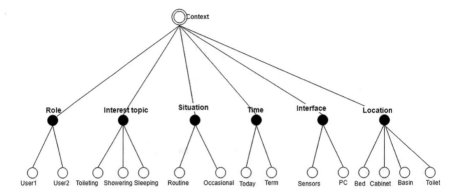

Fig. 1 Context dimension tree

network, and the data rhymed with conditions of each node are recurring to be processed on the lower nodes [10]. The Rete algorithm primarily figures a Rete network, and then it offers an efficient implementation for the expert system by the network for pattern matching [11]. The activity of Daily using dataset (ADLs) is used in this research work.

In Sect. 1 context data is used as an input for RETE algorithm where it generates the rules and context dimensions are been discussed. In Sect. 2 deals with methodology. Section 3 explains RETE algorithm framework and Sect. 4 briefs about Experimental results.

1.1   Context Dimension

Context Dimension Tree (CDT) [4, 12] as shown below is a common context data model as shown in Fig. 1. It is used to embody different contexts of the users. It embraces two types of graphical nodes one is black nodes and another is white nodes. Black nodes personify context dimensions and white nodes typify context values. CDT has one double circled node which is called a root node of the tree. Each leaf of the tree is a value node and it surfaces many parameters. Parameters are epitomized by white squares. So, CDT supports to characterize the different instances of users under various circumstances.

2   Methodology

The methodology comprises of three stages. The first stage is sensing the user/objects, second stage storing the sensed data and third stage generating rules and reasoning it. When actors/users are been observed by the sensor network it collects the context

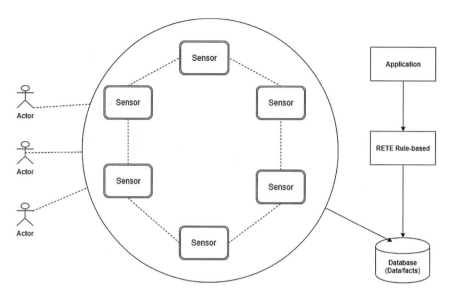

Fig. 2 Context rule-based architecture

information about them and stores in the database. The data/facts available in the database will be utilized by the RETE rule-based to generate rules, if conditions satisfied otherwise rules will be discarded. The architecture is depicted as shown in Fig. 2.

2.1 Dataset

3 RETE Algorithm Framework

Facts are an element of working memory and rules are of long-term memory. Match phase receipts data or facts as input and yields a set of rules which are matching with the data and these are called conflict sets. This will be fed to resolve the phase where it picks rules from the conflict set which is given to execute. Execute phase yields positive or negative token these are fed back into the network (Fig. 3).

3.1 Creating the RETE Network

RETE network is the heart of the RETE algorithm. It has nodes that consist of many objects which satisfy the specific or associated conditions. This algorithm works on facts. The first phase of the RETE network is a discrimination tree where it starts with

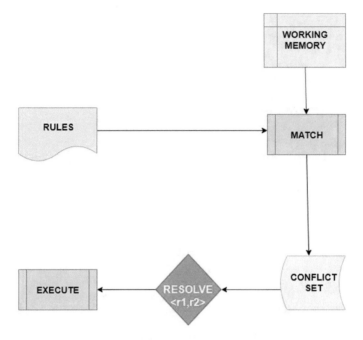

Fig. 3 RETE algorithm framework

Fig. 4 Alpha node creation

alpha nodes connected to classes [4, 13]. RETE network is a representation of rules. All instances of a given class will be listed in the alpha node. The network can be constructed as below.

a. First, alpha nodes are created for each class as shown in Fig. 4.
b. Conditions are then appended as shown in Fig. 5.
c. Finally, the nodes are connected across classes.
d. The path eventually ends with the action part of the rules (Fig. 6).

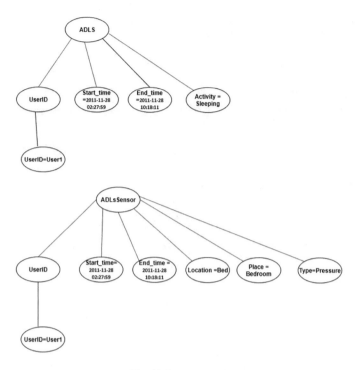

Fig. 5 Rete network

4 Experimental Results

After creating the root node of the discrimination tree and appending the conditions to it, different classes will be joined through the joint nodes and then finally evaluated. The data used in the evaluation phase is ADLs context data [12]. The evaluation phase consists of running the data through the RETE network to identify the applicable rules. If conditions are satisfied with some rules then they are active on the agenda. The agenda consists of a list of rules and objects which will be executed together that are responsible for the conditions to be true. As shown in Fig. 7 When some instances of classes match and if conditions are satisfied then the rules are valid otherwise, they will be invalid [14]. Our example follows a RETE network where it has alpha nodes, joint nodes through which two classes will be combined and the final stage represents the conditions as shown in Fig. 7.

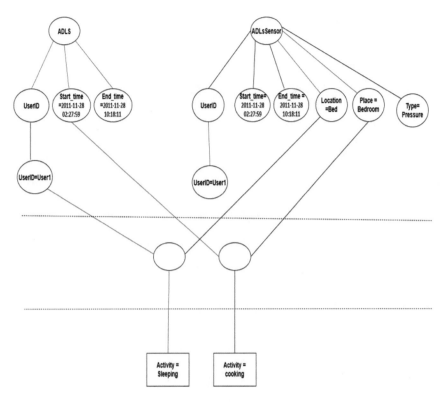

Fig. 6 Integration of classes

Table 3 shows evaluation results which is very clear that when the condition is satisfied with some rules then they are valid and will be active on the agenda. If the conditions are not satisfied with the rules then they are invalid and will not be active on the agenda.

Evaluation results are graphically represented below. Figure 8a–c depict the valid and invalid rule. When Location is bed it is clear that activity will be sleeping. In this context conditions are apt. So it is valid and rules are generated. In the second context, the data, conditions are not apt, place is bedroom and activity is cooking. So, it is invalid and no rules will be generated.

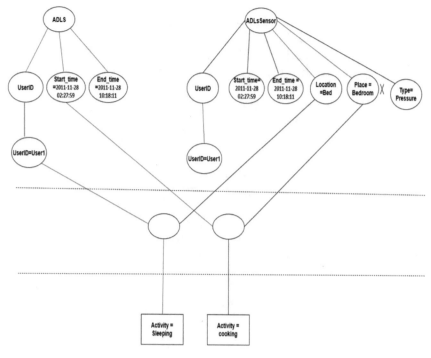

Fig. 7 Pattern matching

Table 3 Evaluation results

Rule	DATA		Condition	Remarks
Rule1	User1	Location='Bed'	Activity=Sleeping	Valid
Rule2	Start_time	Place=bedroom'	Activity=cooking	Invalid

5   Conclusion

In a context-aware system user's dynamicity can be captured and stored in a database. If the database has to be context-aware then it should interact with the user. For this data/fact stored in the database should be converted into rules and reasoned. In this research work ADLs context data is used and context dimensions descriptions are portrayed through context dimension tree. We have designed a context rule-based architecture that infuses sensor data into rule-base. In this research work, we have made an attempt to showcase how the discrimination tree is constructed and evaluated. This research work also explains the feasibility of using RETE rule-based algorithm and the way it simplifies the assessment of rules against data/facts and ordering of statements.

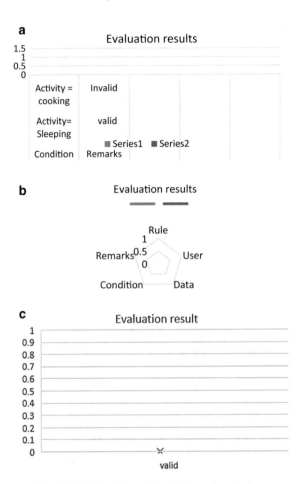

Fig. 8 Valid and invalid activity representation

References

1. Abowd GD, Dey AK, Brown PJ, Davies N, Smith M, Steggles P (1999) Towards a better understanding of context and context-awareness. In: International symposium on handheld and ubiquitous computing. Springer, Berlin, pp 304–307
2. Bolchini C, Schreiber FA, Tanca L (2004) A context-aware methodology for very small data base design. ACM SIGMOD Record 33(1):71–76
3. Schilit B, Adams N, Want R (1994) Context-aware computing applications. In: 1994 first workshop on mobile computing systems and applications. IEEE, New York, pp 85–90
4. Bolchini C, Salice F, Schreiber FA, Tanca L (2003) Logical and physical design issues for smart card databases. ACM Trans Inf Syst (TOIS) 21(3):254–285
5. Poblete R, De La Fuente D, Alonso M (2011) Using cloud computing with RETE algorithms in a platform as a service (PaaS) for business systems development. In: Proceedings on the international conference on artificial intelligence (ICAI), p 1. The Steering Committee of the

World Congress in Computer Science, Computer Engineering and Applied Computing (WorldComp)

6. Yay E, Madrid NM, Ramírez JAO (2014) Using an improved rule match algorithm in an expert system to detect broken driving rules for an energy-efficiency and safety relevant driving system. Proc Comput Sci 35, 127–136

7. Walzer K, Breddin T, Groch M (2008) Relative temporal constraints in the Rete algorithm for complex event detection. In: Proceedings of the second international conference on Distributed event-based systems. ACM, New York, pp 147–155

8. Cirstea H, Kirchner C, Moossen M, Moreau P-E (2004) Production systems and rete algorithm formalisation

9. Schmedding F, Sawas N, Lausen G (2007) Adapting the rete-algorithm to evaluate F-logic rules. In: International workshop on rules and rule markup languages for the semantic web. Springer, Berlin, pp 166–173

10. Kawakami T, Fujita N, Yoshihisa T, Tsukamoto M (2014) An evaluation and implementation of rule-based home energy management system using the Rete algorithm. Sci World J

11. Amailef K, Jie L (2013) Ontology-supported case-based reasoning approach for intelligent m-Government emergency response services. Decis Support Syst 55(1):79–97

12. Ordóñez F, de Toledo P, Sanchis A (2013) Activity recognition using hybrid generative/discriminative models on home environments using binary sensors. Sensors 13 (5), 5460–5477

13. Shivakumar C, Urolagin S (2017) Context data representation using context-aware model in ubiquitous computing scenario. Int J Innov Adv Comput Sci 6(5), ISSN 2347-8616

14. Kilic B (2019) Evaluation of a semantic IoT platform for reasoning and bigdata analytics. Master's thesis, Universitat Politècnica de Catalunya

LASF—A Lightweight Authentication Scheme for Fog-Assisted IoT Network

Ayan Kumar Das[1]([✉]) [iD], Sidra Kalam[1], Nausheen Sahar[1],
and Ditipriya Sinha[2]([✉]) [iD]

[1] Birla Institute of Technology, Mesra, Patna Campus, Patna 800014, India
das.ayan@bitmesra.ac.in, {sidrakalam1,nausheen.jiet}
@gmail.com
[2] National Institute of Technology Patna, Patna 800005, India
ditipriya.cse@nitp.ac.in

Abstract. Internet of things widens the scope of communication by connecting the physical objects to the Internet. These physical objects are vulnerable to various malicious activities, thus strong security features are required in IoT devices. Low power resource constrained-IoT devices limit the use of computational complex algorithm. In this paper, a lightweight authentication scheme has been proposed for fog-assisted IoT network to authenticate IoT devices at low computation cost. It uses three-way handshake with challenge response mechanism to verify the authenticity of the participating device. The performance is evaluated by using IFogSim tool kit and MATLAB, which shows that the proposed scheme is authenticating the user devices at low computational cost and storage utilization. It takes less handshake duration and average response time between the authenticating devices and the fog devices to improve the quality of service.

Keywords: Authentication · IoT · Fog · Handshake · User

1 Introduction

Internet of things (IoT) is a platform where different embedded devices are connected via Internet. These devices can gather and exchange data with each other as per the use of without any human interference. Wearable devices are in trends everywhere and it become very easy to monitor heart rate, fit bits, and other parameters that are required in medical field as it provide the accurate real-time data. Another application of IoT is smart farming in which sensors are used to measure the moisture needed to the soil and amount of fertilizer required in the soil. Along with lots of advantages, there are security breaches as the devices used in IoT system are vulnerable to the attackers. The aggregated data from various devices is uploaded in the cloud. The increase in number of devices causes increase in network latency. It also increases the chance for the adversary to attack at the cloud layer. It becomes very important to maintain the confidentiality of the sensitive data aggregated in the cloud layer. Security of the data from outsource cannot be guaranteed. Considering the drawbacks, fog layer is introduced in the IoT system. This layer is basically an extension of the cloud layer to

© Springer Nature Singapore Pte Ltd. 2020
N. Goel et al. (eds.), *Modelling, Simulation and Intelligent Computing*,
Lecture Notes in Electrical Engineering 659,
https://doi.org/10.1007/978-981-15-4775-1_27

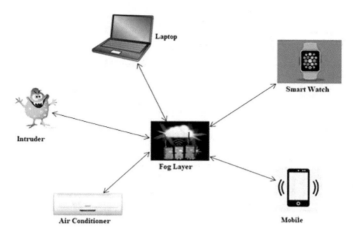

Fig. 1 Fog-assisted IoT environment

minimize the latency, collect and secure wide range of data, and better analysis of local data. Devises used in fog layer is called fog nodes that can be deployed anywhere with a network connectivity. In this paper, an authentication scheme named LASF—A Lightweight Authentication scheme for fog-assisted IoT network is proposed to authenticate the IoT user devices at low computation cost. In LASF, a fog-assisted IoT (FIoT) environment is designed as depicted in Fig. 1 to authenticate users by using two-way handshake between user and fog layer. The two-way authentication of this scheme prevents from many malicious activities, like replay attack, DoS, man-in-the-middle attack, and false data injection attack.

The rest of the paper is organized as follows: Sect. 2 deals with the state-of-the-art study in the related field Sect. 3 describes the proposed work, the performance is evaluated in Sect. 4 followed by conclusions in Sect. 5, and references.

2 Related Works

The state-of-the-art study reveals that in fog-enhanced IoT system, there are several issues that need to be solved for the secure and fast communication among the IoT devices. The research related to the existing security solutions in FIoT environment is discussed in this section.

Sicari and Rizzardi [1] have done a survey on authentication schemes and other security issues in IoT. The stored data in the fog layer is more open to the IoT attacks. Attribute-based encryption [2] is used to secure the data stored in fog layer and helps to prevent from chosen cypher-text attack (CCA). Sania and Yuan [3] have designed a cyber-security framework which uses an identity-based security mechanism (I-ICAAAN). This mechanism efficiently manages the energy requirement. They also propose an Intelligent Security System for Energy Management (ISSEM) to ensure the security of the system. No doubts that comparative to use of cloud computing, and fog

Table 1 Security comparison between existing schemes

Name of attacks	Ahmad et al. [6]	Farash et al. [5]	Wang et al. [4]	Sicari et al. [10]	Sani et al. [3]
Replay attack	✗	✗	✓	✓	✓
Man-in-the-middle attack	✓	✗	✗	✓	✓
DoS attack	✗	✗	✗	✗	✓
False data injection attack	✓	✓	✗	✓	✓
Sybil attack	✓	✓	✗	✓	✓
Repudiation attack	✓	✗	✗	✓	✓
Sinkhole attack	✗	✓	✓	✓	✗

computing is much better when it comes to low latency and security issues. There are still some security threats in the fog layer that need to be dealt in an efficient manner. In [4], the authors have designed an anonymous and secure aggregation scheme in fog-based IoT. In this approach, fog node helps the terminal device to collect and store the data in public cloud server. They have proposed a way to secure the data by using homomorphic encryption technique, in which pseudonyms is used to hide the identity of terminal device. Application of any security protocols in the IoT devices is difficult because of their constrained resources. Farash and Turkanovic [5] have proposed a user authentication and key aggregation scheme (UAKAS). They presented that user can authenticate directly to the particular sensor node in a heterogeneous WSN. No need to communicate with the gateway node. The data gathered by the sensor node can be accessed by the user. Jana and Khan [6] have proposed payload-based mutual authentication scheme in which four-way handshake process is used for verifying the authenticity of the participants. Implementation of CoAP [7] relies on DTLS [8] for the exchange of resources among the participating objects. This scheme uses payload technique in place of DTLS-enabled CoAP stack.

There are several papers for the detection and prevention of data from different attacks [9, 10]. However, Table 1 shows a comparative analysis of security issues addressed by various schemes.

3 Proposed Work

The resource-constrained IoT network is vulnerable to various types of attacks like replay attack, man-in-the-middle attack, DoS attack, etc. A huge amount of data is at risk which can interrupt the functioning of the IoT network. An adversary might intercept the data and replay the data to various IoT devices or gateway nodes. If any adversary will replay the message repeatedly to any fog, it will increase the number of requests for the fog. Thus results in DoS attack at the fog level, resulting which legitimate users will not get the service on time. To reduce various types of attacks, a lightweight secure protocol must be designed considering the resource constraint, limited power nature of the IoT device. Most of the authentication schemes are based on DTLS protocol which uses complex and resource consuming cipher for providing the security to the IoT device. It is expensive in terms of memory consumption, storage cost, and computation cost. The proposed scheme LASF is designed to address the above-mentioned issues. It is a lightweight authentication technique, which uses a three-way handshake scheme. It uses simple XOR operation along with challenge response pairs for authentication which results in low computation cost. The proposed scheme requires less storage space as the devices only need to store the pre-shared keys permanently, and challenge is stored for the short duration of time till the time response is received from the other party. This makes the proposed scheme suitable for low power resource-constrained IoT device.

In the proposed model, there are three steps for the two-way authentication of user and the fog layer. In two-way authentication, both the devices are authenticated to each other. A lightweight authentication scheme is designed that provides authentication using computationally inexpensive operation such as XOR. With the help of three handshake messages, authentication process is completed. Three-way handshake is a three-step method that authenticates the user and fog layer. The three steps are as follows: User challenge, fog response-challenge, and user response as depicted in Fig. 2.

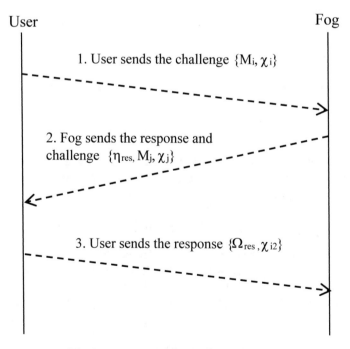

Fig. 2 Three-way authentication handshake

In these three steps, challenge and response are shared between the user and the fog layer. Once they are authenticated, a connection is created between the user and fog layer for the transmission of the message for a particular session.

An offline provisioning phase is introduced before the user challenge phase, in which user shares the secret key K_G with the fog layer. In the proposed model, it is assumed that secret keys are embedded in the physical object during the manufacturing phase. If any adversary will try to interfere with the secret key, then a notification will be generated. The generated notification will help the user to deal with the security breach.

In the user challenge phase, user sends a message, M_u to the fog layer, so that it can check the authenticity of the fog layer. Along with the message, it sends the random generated nonce χ_u. Nonce is a random-generated number that is used only once in the authentication process. It prepares and sends the cipher text encrypted using the shared session key to the fog layer, the packet format of which is as follows:

$$\text{Cipher1}[E|| \{ K_G||(M_u, \chi_u)\}]$$

In the next phase, i.e., fog response and challenge phase fog sends the response to the user in reply to the challenge. It also challenges the user, so that it can check whether user is authentic or not. When the fog receives the text, it decrypts the ciphered text and generates the response. To generate the response, fog computes a message by performing XOR operation on M_u and K_G as explained in Eq. 1.

$$\lambda_{\text{res}} = M_u \oplus K_G \tag{1}$$

Another XOR is performed on the resultant message λ_{res} of Eq. 1 and χ_u to generate the response for the user as in Eq. 2.

$$\eta_{\text{res}} = \lambda_{\text{res}} \oplus \chi_u \tag{2}$$

This resultant message η_{res} is transmitted as the fog response to the user along with the challenge M_F and the random nonce χ_F generated by the fog. All these details are transmitted in the ciphered form to the user. The packet format is

$$\text{Cipher2}[E||\{K_G||(\eta_{\text{res}}, M_F, \chi_F)\}]$$

Once the user receives the text, the user has to decipher the ciphered text. When the user deciphers the text, it gets the response and challenge of the fog layer. First, the user needs to extract λ_{res} from the response η_{res} of the fog by using its χ_u. The XOR operation is performed on η_{res} and χ_u by the user to get λ_{res}. To obtain the message, it again needs to perform the XOR operation on λ_{res} using the shared secret key K_G. Once user will extract the message sent by the fog, it will check whether the extracted message match with the message sent by the user or not. If both are a match then the fog is authenticated, otherwise it is unauthenticated.

In the user response phase, user will generate the response according to the challenge M_F sent by the fog layer. It computes the message by performing the XOR operation on the challenge message M_F and the secret key K_G shared between the user and fog layer as in Eq. 3.

$$\psi_{\text{res}} = M_F \oplus K_G \qquad (3)$$

To generate the response for the fog layer, user needs to perform the XOR operation on the computed ψ_{res} and χ_F as described in Eq. 4.

$$\Omega_{\text{res}} = \psi_{\text{res}} \oplus \chi_F \qquad (4)$$

Once all task are done, the user will send the response to the fog in ciphered text with the following packet format

$$\text{Cipher3}[E||\{K_G||(\Omega_{\text{res}}, \chi_{u2})\}]$$

Fog needs to decrypt the text after receiving the message. It extracts the value of the intermediate result ψ_{res} from the response Ω_{res} of the user by using its nonce χ_F. To obtain the message M_F sent by the user, fog again needs to perform the XOR operation on ψ_{res} and K_G. Once it extracts the message M_F sent by the user, it checks whether the extracted message matches with the message sent by the fog layer. If both match, then the user is authenticated otherwise user is unauthenticated.

4 Performance Analysis

In this section, the performance of the proposed scheme is evaluated. IfogSim toolkit is used for calculating the computation time and storage consumption, while MATLAB is used for calculating the handshake duration and response time. The comparison based on different parameters with other existing schemes is discussed in next subsections.

4.1 Handshake Duration

Handshake duration is the total time taken by the round trip message, i.e., user challenge phase, acknowledged by fog layer response and challenge, and again acknowledged back by the user response. Handshake duration is calculated at the user's end as defined in Eq. 5.

$$\sigma_{\text{hs}} = H_{\text{challenge}} + H_{\text{response}} + H_{\text{proc}} \qquad (5)$$

where $H_{\text{challenge}}$ is the time taken by the round trip of user challenge, H_{response} is the response time of the user to the fog's challenge, and H_{proc} is the time taken by the client to process the request. Handshake duration is calculated by executing 20 random handshakes. The standard deviation is calculated as in Eq. 6.

$$\sigma = \sqrt{1/H} \sum_{i=1}^{H} (n_i - X)^2 \qquad (6)$$

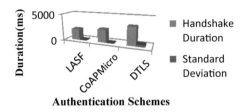

Fig. 3 Handshake duration of various schemes

where σ is the standard deviation, n is the duration of the handshake, X is the mean value, and H is the total number of actual handshake.

Figure 3 compares the proposed scheme LASF with the CoAP [6]-based payload authentication scheme and DTLS [8]. DTLS shows the handshake between client and server, where the execution of complex encryption and decryption added with exchange of complex ciphered text and certificates consumes more handshake duration. In CoAP-based payload authentication scheme AES encryption algorithm is used for encryption of the data for authentication purpose, whose execution takes longer time and hence increases the handshake duration. Thus, LASF performs better compared to DTLS and CoAP scheme.

4.2   Memory Consumption

Memory consumption is the amount of memory utilized in storing the details regarding the IoT devices and communicating messages. In LASF, only three handshake messages are used for two-way authentication. XOR operation has been used for computing the messages which is computationally cheap. The performance is compared with payload authentication scheme [5], CoAPBlip [7] and HTTP for a message of 500 bytes as depicted in Fig. 4, which shows proposed scheme LASF outperforms the other schemes.

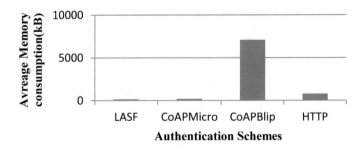

Fig. 4 Average memory consumption for various schemes

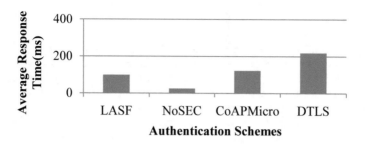

Fig. 5 Average response time of various schemes

Table 2 Computational cost of various schemes

Scheme	User	Fog layer
Farash et al.'s scheme [5]	$11T_h$	$14T_h$
Yuxiang Feng et al.'s [9]	$4T_h$	$6T_h$
Proposed scheme LASF	$2T_h$	$4T_h$

4.3 Average Response Time

Response time usually depends on the number of request handled by the fog layer at a given time. Figure 5 shows that the average response time of LASF is less compared to CoAPMicro [7] and DTLS [8]. However, it is higher than NoSEC that is sacrificing its security.

4.4 Computation Cost

Computation cost is the time required for the processing of information such as generating time stamp for time synchronization technique, computation of complex cipher, and encryption. Table 2 shows that the computational cost of LASF is less than other schemes as it is not using any complex cryptographic algorithm or time synchronization technique.

5 Conclusion

In this paper, a lightweight authentication scheme for fog-assisted IoT has been proposed to verify the authenticity of the user and fog layer, where three-way handshake with challenge response mechanism is used. The pre-shared symmetric key and random generated nonce are used to mask the messages. The experimental result and evaluation show that our scheme is computationally cheap, consumes less memory, and its handshake duration is also less as compared to other existing schemes.

References

1. Li T, Gao C, Jiang L, Pedrycz W, Shen J (2019) Publicly verifiable privacy-preserving aggregation and its application in IoT. J Network Comput Appl 126:39–44
2. Lee K, Kim D, Ha D, Rajput U, Oh H (2015) On security and privacy issues of fog computing supported Internet of Things environment. In: 6th international conference on the network of the future (NOF). IEEE, Canada, pp 1–3
3. Sani AS, Yuan D, Jin J, Gao L, Yu S, Dong ZY (2019) Cyber security framework for Internet of Things-based energy internet. Future Generation Comput Syst 93:849–859
4. Ying B, Naya A (2017) Anonymous and lightweight authentication for secure vehicular networks. IEEE Trans Veh Technol 66(12):10626–10636
5. Farash MS, Turkanović M, Kumari S, Hölbl M (2016) An efficient user authentication and key agreement scheme for heterogeneous wireless sensor network tailored for the Internet of Things environment. Ad Hoc Networks 36:152–176
6. Jan MA, Khan F, Alam M, Usman M (2019) A payload-based mutual authentication scheme for Internet of Things. Future Generation Comput Syst 92:1028–1039
7. Kuladinithi K, Bergmann O, Pötsch T, Becker M, Görg C (2011) Implementation of CoAP and its application in transport logistics. In: Proceedings of IP + SN, Chicago, IL, USA
8. Granjal J, Monteiro E, Silva JS (2012) On the effectiveness of end-to-end security for internet-integrated sensing applications. In: IEEE international conference on green computing and communications. IEEE, Besancon, pp 87–93
9. Wang H, Wang Z, Domingo-Ferrer J (2018) Anonymous and secure aggregation scheme in fog-based public cloud computing. Future Generation Comput Syst 78:712–719
10. Feng Y, Wang W, Weng Y, Zhang H (2017) A replay-attack resistant authentication scheme for the internet of things. In: IEEE international conference on computational science and engineering (CSE) and IEEE international conference on embedded and ubiquitous computing (EUC), vol 1. IEEE, New York, pp 541–547

Dimensionality Reduction for Water Quality Prediction from a Data Mining Perspective

J. Alamelu Mangai[1] and Bharat B. Gulyani[2(✉)]

[1] Department of Computer Science, Presidency University, Bengaluru, India
[2] Department of Chemical Engineering, Birla Institute of Technology and
Science Pilani, Dubai Campus, Dubai, UAE
gulyani@dubai.bits-pilani.ac.in

Abstract. Biochemical oxygen demand (BOD) is the measurement of the amount of dissolved oxygen used by aerobic microbes for oxidizing organic matter in water bodies and used for analyzing the water quality. The actual BOD prediction method is cumbersome. Instead an automatic prediction model is required that is accurate, faster and less expensive. This paper presents a data-driven model for predicting BOD, in a lower-dimensional space obtained using dimensionality reduction techniques that help remove irrelevant properties of high-dimensional data. Machine learning algorithms, namely decision stump, SVM, MLP, linear regression (LR), and instance-based learner (IBK), were trained with the full dataset with 11 parameters. The training set was later transformed into a lower-dimensional space using principal component analysis (PCA) and correlation-based feature selection (CFS). The performance of the learners on the full training set and transformed dataset was analyzed using correlation coefficient, RMSE, and MAE. The algorithms are able to preserve their predictive accuracy on the lower-dimensional space.

Keywords: BOD · Machine learning · SVM · MLP · LR · IBK · PCA · CFS

1 Introduction

Real-time monitoring of water quality for water bodies is required for efficient treatment plan and design because it provides information on the precise loading of pollutant to wastewater treatment facilities. There is a wide variety of organic pollutants (e.g., pesticides, furans, PAHs, bacteria, viruses, protozoa, etc.) usually found in surface water. Biochemical oxygen demand (BOD), an indirect indicator of organic matters, is a representative parameter for water quality [1]. However, it is very difficult to obtain continuous water quality data because of the scarcity of accessible space within the sewer systems and the necessity of separate laboratory experiments. Moreover, at least five days are required to acquire BOD data from the experiment and BOD itself may be biased by the presence of toxic substances that might cause the inhibition of the oxidizing bacteria.

Data mining is a powerful tool of applying a computer-assisted approach to obtain hidden meaning and information from the enormous datasets in order to make valid predictions on future data trends in areas like biological data analysis and other

© Springer Nature Singapore Pte Ltd. 2020
N. Goel et al. (eds.), *Modelling, Simulation and Intelligent Computing*,
Lecture Notes in Electrical Engineering 659,
https://doi.org/10.1007/978-981-15-4775-1_28

scientific approaches [2]. The data mining model can be built using supervised or unsupervised learning approaches.

Many data mining techniques are unable to accurately predict the future trends in high-dimensional space due to the high dimensionality. Dimensionality reduction (DR) allows the representation of observed properties of data in terms of minimum number of dimensions from the original dataset. DR helps in better visualization, noise removal, and compression of high-dimensional data [3]. The two techniques for DR are *feature projection* which maps the high dimensions from original dataset to lower dimensions by the formulation of linear combinations of the parameters in the original dataset (e.g., PCA), and *feature selection* which selects a subset of features that are most relevant to the feature to be predicted based on some heuristic measure (e.g., CFS).

In this paper, a data-driven model for predicting BOD is proposed. The machine learning algorithms used in modeling are regression models. The predictive accuracy of the models is analyzed both on the full training set and on the PCA and CFS transformed lower-dimensional datasets.

2 Related Work

Machine learning techniques find their applications in water quality prediction. The most common data mining-based models include artificial neural network (ANN), multivariate statistical analysis, regression models, and support vector machines (SVM). Data mining techniques have been applied to predict BOD by [4–6]. Multivariate statistical techniques using feature projection and feature selection algorithms for DR also find a wide range of applications for water quality prediction and management. Feature selection algorithm canonical correlation analysis (CCA) was used on data with 28 parameters collected over 5 different sites of Perak River basin followed by neural network predictive model, to determine the effects of BOD and COD on the river water quality [7]. *More description is given in the Appendix (available with authors).*

3 Methodology

The following sections give a description of the various machine learning algorithms that were used to induce a BOD prediction model in this paper.

3.1 Decision Stump

The decision stump was devised by Wayne Iba and Pat Langley in 1992 [8]. In machine learning, decision stump is a decision tree model consisting of a single layer, i.e., having just one internal node, the root node, connected to its terminal nodes. A stump, unlike a multilayer decision tree, ends the tree after its first split and thus predicts a decision based on single attribute input and is often used as base learners for ensemble methods. *More description is given in the Appendix.*

3.2 Support Vector Machine (SVM)

SVM in machine learning, developed by Cortes and Vapnik in 1995 [9], is models developed by supervised learning approach. This classification technique is based on statistical learning theory that is used in many practical applications of pattern recognition. SVM works well with high-dimensional data. The SVM determines an intuitive model indicating a technique for nonlinear classification, regression and outlier detection [10]. A recent efficient technique that uses unsupervised learning for training data using SVM [11] shows the improved ability to learn from large imbalanced dataset.

3.3 Multilayer Perceptron (MLP)

Multilayer perceptron is a feedforward neural network model consisting of more than one hidden layer in the network that uses supervised network learning [12]. MLP consists of more than one (multiple layers) of input nodes, each layer connected to the next one. It takes in the input data with a desired output for training the model. The learning of the MLP neural networks uses backpropagation algorithm. (*More description is given in the Appendix.*)

3.4 Multiple Linear Regression (MLR)

Linear regression is a statistical modeling approach and the most extensively used approach of regression analysis [13]. The model developed by linear regression is based on a relationship between one or more explanatory independent variables X with a dependent variable Y [14].

3.5 Instance-Based Learners (IBK)

Instance-based learning (IBK) is an extension of k-nearest neighbor (k-NN) supervised algorithms. IBK uses supervised learning algorithms where input is a series of instances each described by n feature-value pairs for training the model [15]. The IBK framework mainly includes a similarity function, classification function, and concept description updater. Regression function is used for predicting the target value of a new instance depending on the results of similarity function, concept description instances, and their performances [16].

3.6 Principal Component Analysis (PCA)

PCA is statistical technique to determine characteristics of similarities and differences among different attributes of huge dataset that cannot easily be represented graphically. PCA takes in largely correlated dataset as input and translates it to uncorrelated principal components easier for prediction and analysis [17].

PCA maximizes the linear combination of the factors or variables of the dataset. This is a technique focused on single sample of data with certain p variables that consist of observations without any groupings among them and no subset partitions of variables. The principal components obtained after analysis represent different dimensions. The first principal component is the linear combination of attributes having maximal variance among observations. The second principal component is also a linear combination with maximal variance; but it is in the orthogonal direction to the first principal component, and this goes on for rest of the components [18].

PCA uses statistical techniques such as finding mean and variance, calculating the covariance matrix or correlation matrix based on the type of dataset and evaluation of eigenvectors and eigenvalues [19].

The following steps include a description of the PCA algorithm:

Step 1. Calculate the mean of the given data dimension

$$\text{Mean}(\overline{X}) = \frac{1}{N} \sum_{i=1}^{N} X_i \tag{1}$$

where $X[i]$ is each data value and N is the total number of observations.

Step 2. Calculate the deviations from the mean; subtract the mean from each data value.

$$\text{Deviation } D = X_i - \overline{X} \tag{2}$$

where D is the new dataset containing calculated deviations.

Step 3. Calculate the covariance matrix of D (correlation matrix for invariant data)

$$S = \text{Cov}(x, y) = \frac{\sum_{i=1}^{n} (X_i - \overline{X})(Y_i - \overline{Y})}{n - 1}. \tag{3}$$

(where the formula for S is applicable to two-dimensional data x and y)

Step 4. Calculate eigenvectors and eigenvalues from covariance matrix.

This can be calculated geometrically or by algebraic methods.

Step 5. Rearrange the eigenvectors based on eigenvalues.

Sort the eigenvectors in the column of matrix (called FeatureVector matrix) in decreasing order of eigenvalues.

$$\text{FeatureVector} = (\text{eig}1, \text{eig}2, \ldots, \text{eig}n) \tag{4}$$

These eigenvectors are represented in the order of their significance. These are principal components.

Step 6. Choose eigenvectors depicting maximal representation of data from FeatureVector.

Eigenvectors are retained based on different strategies namely (1) based on decided threshold of variance %., for example, components that explain 95%, (2) retaining components having covariance greater than average covariance value of dataset, and (3) based on scree plot representations.

Step 7. Derive new dataset

$$FinalData = (FeatureData)T \times (D)T \tag{5}$$

where FinalData is data in terms of chosen eigenvectors, (FeatureData) T represents the eigenvectors in the rows of matrix $(D)T$ that represents each new dimension in each row.

Step 8. Get original data back in terms of selected principal components.

$$(D)T = (FeatureVector) \times (FinalData) + OriginalMean(\overline{X}) \tag{6}$$

Once principal components are found, dimensionality reduction of data can be done by removing certain less significant data without much loss of information from the original dataset.

3.7   Correlation-Based Feature Selection (CFS)

Feature selection algorithms help us to reduce the number of parameters and obtain a subset of original attributes that are most correlated with the class to be predicted. The correlation-based feature selection (CFS) approach uses correlation measure with a hypothesis that the features to be selected in the subset should be highly correlated to the parameter which is to be predicted by the data-driven model and should be less correlated with other features present in the original feature dataset [20]. For choosing the best feature subset among the many predicted, CFS uses search method to rank the feature subsets based on the heuristic correlation function for evaluation called the 'merit' of the subset.

The CFS algorithm consists of various steps (*see Appendix for description*).

4   Experimental Analysis

4.1   Dataset Description

Data for the experimental analysis was taken from the Web site of the Department of Environment, Food and Rural Affairs, UK Government [21]. The annual statistics include average concentrations of parameters of river water quality by its location. Parameters include temperature (°C), pH, conductivity (μS/cm), suspended solids (mg/L), DO (mg/L), ammoniacal nitrogen (mg/L), nitrate (mg/L), nitrite (mg/L), chloride (mg/L), total alkalinity (mg/L), orthophosphate (mg/L), and BOD (mg/L). *The dataset is given in the Appendix* as Table 1.

4.2   Performance Evaluation Metrics

For analyzing the performance of the models with the given dataset as well as with transformed dataset, root-mean-squared error (RMSE), mean absolute error (MAE), and correlation coefficient were used as the evaluation metrics. *Detailed description is given in the Appendix.*

The RMSE can be calculated using the following formula:

$$RMSE = \sqrt{\frac{\sum_{i=1}^{n}(Y_i - X_i)^2}{n}} \tag{7}$$

MAE is calculated using the following formula:

$$MAE = \frac{1}{N}\left(\sum_{i=1}^{N}|pi - ai|\right) \tag{8}$$

The coefficient correlation can be calculated using the following formula:

$$Correlation\ coefficient = \frac{1}{n-1}\sum_{i=1}^{n}\left(\frac{Pi - \bar{p}}{Sp}\right)\left(\frac{Xi - \bar{x}}{Sx}\right) \tag{9}$$

where $\bar{p}$ is the sample mean of predicted values, $\bar{x}$ is the sample mean of observed values while Sp and Sx are the sample variances of predicted values p and observed values x, respectively [6].

5   Results and Discussion

The performance of the various machine learning algorithms discussed in the previous section on the full dataset with 11 features was evaluated using the WEKA machine learning framework. The predictive models were trained using tenfold cross validation. Table 2 (*in the Appendix*) shows the parameter settings used for running these models.

The dataset was then transformed into a lower-dimensional space using PCA in Weka machine learning framework. It returns a set of transformed attributes namely PCA directions that can be ranked according to the amount of variance each accounts for. In this work, 'ranker' is the search method used for evaluating the attribute subsets. This method returns a list of principal components sorted according to their individual attribute scores called the cumulative variance.

Number of principal components is chosen such that it accounts for 95% of the cumulative variance. The dataset was also transformed to lower-dimensional space using CFS for evaluating the feature subset set.

The predictive accuracy of the various regression models on the full dataset with 11 features is given in Table 1. Predictive accuracy of these regressors is measured in terms of correlation coefficient, RMSE, and MAE.

The instance-based regression model (IBK) and the support vector machine-based regression model (SMO) have a high correlation coefficient. However, SMO also has a lower RMSE and MAE compared to all the other learners. Using PCA, the features are then transformed to a lower-dimensional space which has 7 attributes. The performance of the regression models on these transformed features is given in Table 2.

Any feature selection method should result in less number of features and either has to improve or maintain the predictive accuracy when compared with the original set of

Table 1 Performance analysis of various prediction models on full dataset

Prediction Models	Correlation Coefficient	RMSE	MAE
Decision stump	0.8492	0.8247	0.6699
SMO	0.9086	0.6588	0.4604
MLP	0.887	0.7514	0.545
MLR	0.898	0.6917	0.5109
IBK	0.9	0.7074	0.5485

Table 2 Performance analysis of various prediction models on PCA transformed dataset

Prediction models	Correlation coefficient	RMSE	MAE
Decision stump	0.8074	0.9225	0.7079
SMO	0.8811	0.7422	0.5677
MLP	0.8715	0.7826	0.5795
MLR	0.8856	0.727	0.5585
IBK	0.9049	0.6837	0.5289

features. The results in Table 2 show that the predictive accuracy of all the regression models after dimensionality reduction using PCA is similar to the higher-dimensional space with all the features. The performance of the IBK learner has improved in this lower-dimensional space in terms of all three performance measures. The CFS algorithm for feature selection has identified four features to be more predictive, namely pH, SS, DO, and nitrite. Table 3 shows the performance of the regression models trained using these features.

The results in Table 3 show that the performance of the SMO and MLP regression models with features selected by CFS is similar to that on the full dataset with all features. However, the performance of IBK and decision stump has degraded in terms of all three evaluation measures. Comparing the performance of the models on all three datasets namely dataset with full features, with features transformed using PCA and features selected using CFS, the following could be inferred that the performance of the IBK learners has improved on the features transformed using PCA. Since the principal components identified by PCA captures 95% of the variance in the data, this has helped IBK to identify the right set of nearest neighbors of a given test data.

Table 3 Performance analysis of various prediction models on CFS selected features

Prediction models	Correlation coefficient	RMSE	MAE
Decision stump	0.7947	0.9514	0.7155
SMO	0.8926	0.709	0.5188
MLP	0.865	0.7937	0.6471
MLR	0.8988	0.6855	0.5094
IBK	0.8545	0.8337	0.6232

6   Conclusions

This study has explored a data-driven model for predicting BOD in river water using data mining techniques. Two dimensionality reduction strategies, namely feature projection (PCA) and feature subset selection (CFS), have been used to reduce the size of the training data before training the prediction models. The performance of the regression models has been explored on the full dataset with all features and with the dataset after dimensionality reduction. Experimental results on a real-life dataset have shown that there is no compromise on the performance of the regression models on the reduced dataset. Dimensionality reduction methods have helped to remove the irrelevant and redundant features in the training phase.

References

1. Hach Robert CC, Klein L, Gibbs CR Jr (1997) Introduction to BIOCHEMICAL OXYGEN DEMAND. Technical information series, Booklet No. 7, USA: Hach Company
2. Kantardzic M, Data mining: concepts, models, methods, and algorithms, 2nd ed. University of Louisville. IEEE Press. Available: http://cecs.louisville.edu/datamining/PDF/0471228524.pdf
3. Sorzano COS, Vargas J, Pascual-Montano A, A survey of dimensionality reductiontechniques. Madrid, Spain
4. Emamgholizadeh S, Kashi H, Marofpoor I, Zalaghi E (2013) Prediction of water quality parameters of Karoon River (Iran) by artificial intelligence-based models. Int J Environ Sci Technol 11:645–656
5. Malek S, Mosleh M, Syed SM (2014) Dissolved oxygen prediction using support vector machine. Int J Comput Inf Syst Control Eng 8(1):46–50
6. Fathima A, Alamelu Mangai J, Gulyani BB (2014) An ensemble method for predicting biochemical oxygen demand in river water using data mining techniques. Int J River Basin Manag. https://doi.org/10.1080/15715124.2014.936442
7. Rahim NA, Ahmad Z (2013) Features selection in water quality prediction in neural network using canonical correspondence analysis (CCA). In: 6th international conference on process systems engineering (PSE ASIA), pp 25–27, June 2013
8. Iba W, Langley P (1992) Induction of one-level decision trees, in ML92. In: Proceedings of the ninth international conference on machine learning, Aberdeen, Scotland, 1–3 July 1992, San Francisco, CA. Morgan Kaufmann, pp 233–240
9. Cortes C, Vapnik V (1995) Support-vector network. Mach Learn 20:1–25
10. Meyer D (2014) Support vector machines the interface to libsvm in package e1071. FH Technikum, Wien, Austria, September 1, 2014
11. Nguyen G, Hoang SP, Bouzerdoum A (2010) Efficient SVM training with reduced weighted samples. In: IEEE World Congress on computational intelligence
12. Haykin S (1998) Neural networks—a comprehensive foundation, 2nd ed. Prentice-Hall, Englewood Cliffs
13. Hillenmeyer M (2005) Machine learning. Stanford University, pp 7–12, July 2005. Available at http://web.stanford.edu/ ~ maureenh/quals/pdf/ml.pdf. Year of access 2015
14. Carvalho CM, Section 2: multiple linear regression. The University of Texas McCombs School of Business. Available at http://faculty.mccombs.utexas.edu/carlos.carvalho/teaching/Section2.pdf

15. Aha D, Kibler D (1995) Instance-based learning algorithms. Mach Learn 6:37–66
16. Bhuvaneswari E, Sarma Dhulipala VR (2013) The study and analysis of classification algorithm for animal kingdom dataset. Inf Eng 2(1):6–13
17. Anh T, Magi S (2009, June) Principal component analysis. Final paper in Financial pricing, National Cheng Kung University
18. Rencher AC (2002) Methods of multivariate analysis, 2nd ed. Wiley, Canada
19. Smith LI (2002, February) A tutorial on principal components analysis, pp 2–8
20. Tuysuzoglu G, Birant D, Pala A (2018) Ensemble methods in environmental data mining. IntechOpen
21. Solanki A, Agrawal H, Khare K (2015) Predictive analysis of water quality parameters using deep learning. Int J Comput Appl 125(9):29–34

Automated Grading of Diabetic Macular Edema Using Deep Learning Techniques

Tanzeeha Sulaiman, J. Angel Arul Jothi$^{(\boxtimes)}$ (iD), and Shaleen Bengani

Department of Computer Science, Birla Institute of Technology and Science
Pilani, Dubai Campus, Dubai, UAE
{f20160007, angeljothi}@dubai.bits-pilani.ac.in,
shaleenbengani@gmail.com

Abstract. Diabetic macular edema (DME) is one of the major causes for visual impairment and can even lead to permanent blindness if not treated early. Manual screening by ophthalmologists is time-consuming and error-prone which necessitates the need for automated detection and grading of DME. In this paper, a deep learning-based DME-grading model is proposed for automatic DME grading of retinal fundus images. The model consists of an autoencoder network and a DME-grading network. The autoencoder network learns features of retinal fundus images. The DME-grading network uses the learned features to detect and grade the risk of DME. The proposed method is evaluated using the IDRiD dataset. The class imbalance of IDRiD dataset is overcome by using image augmentation and class weights. The highest accuracy, precision, recall, and $F1$-score achieved by the proposed method are 68%, 66%, 68%, 65%, respectively.

Keywords: Diabetic macular edema · Deep learning · Autoencoder · Semi-supervised learning · Transfer learning

1 Introduction

Diabetic macular edema (DME), the leading cause of permanent blindness, is a complication that may arise due to diabetic retinopathy (DR). Macular edema is the swelling of the macula, caused by the accumulation of fluids leaking from the vitreous humor or retinal blood vessels [1]. Fundus photography (FP), fluorescein angiography, and optical coherence tomography are the imaging methods usually used to diagnose DME out of which FP is most widely used because it is non-invasive and cost-effective.

According to [2], DME is diagnosed and is graded into three classes, namely Grade 0, Grade 1, and Grade 2, from fundus images by ophthalmologists. Examining fundus images by a specialist in order to detect and grade macular edema can be very time-consuming and is prone to human error. This arises the need for automated detection methods that are less time-consuming and can tackle these problems. These automated systems can accurately detect patients having DME, thereby reducing the workload of the ophthalmologists and helps in early diagnosis of DME which can reduce the risk of permanent visual impairment [3].

© Springer Nature Singapore Pte Ltd. 2020
N. Goel et al. (eds.), *Modelling, Simulation and Intelligent Computing*,
Lecture Notes in Electrical Engineering 659,
https://doi.org/10.1007/978-981-15-4775-1_29

The use of deep learning in medical image processing is a rapidly growing field. The key feature of deep learning algorithms is the ability to automatically learn and extract information or features [4]. In this paper, we propose a deep learning model that automatically grades retinal fundus images into three classes or grades as mentioned earlier. The proposed approach uses a combination of semi-supervised learning and transfers learning with fine-tuning for multi-class DME classification [5, 6].

The reminder of the paper is organized as follows: Sect. 2 provides the literature survey; Sect. 3 mentions the dataset used in this paper; Sect. 4 explains the method; Sect. 5 describes the various experiments conducted; Sect. 6 provides the results and discussions; and Sect. 7 concludes the paper.

2 Literature Survey

Literature reveals that a lot of works had used traditional machine learning techniques for DME grading [7–10]. The presence of exudates in the fundus images is used as a marker to identify and grade DME. One approach to grade DME is to segment exudates and classify the images. Prentašić and Loncaric [11] uses a deep learning convolutional neural network model to classify each pixel into exudate or non-exudate. Chudzik et al. [12] proposes a model using fully convolutional neural network (FCNN) to segment exudates with Inception modules along with transfer learning.

Perdomo et al. [13] proposes a two-stage classification that combines localization and segmentation of exudates with DME classification. In the first stage, an eight-layer CNN architecture is used to localize exudate. AlexNet architecture is used in the second stage for DME classification. The model proposed in [14] consists of three CNN architectures that are trained on one of three varying sizes of images for DME grading. Computation time is reduced by using the weights of the architecture trained using the smaller size images on the subsequent architecture. Juan et al. [15] proposes a two-stage model consisting of exudates segmentation followed by DME classification. In the first stage, fully connected residual networks are used to segment the exudates. Pixels having the maximum probability map are passed as input to another residual network in the second stage which performs DME classification.

3 Dataset Used

This work uses unlabeled images from Kaggle's diabetic retinopathy detection challenge dataset [16] to train the autoencoder network. Out of the 88,702 images in the dataset, 88,000 images are used for training and 702 images for validation.

Indian Diabetic Retinopathy Image Dataset (IDRiD) is used for training and testing the DME-grading network [2]. This dataset consists of 413 and 103 retinal fundus images for training and testing, respectively. Table 1 shows the different grades of DME and the distribution of the images in the IDRiD dataset into test and train set for each grade. The size of all images in the dataset is 4288×2848 pixels.

Table 1 IDRiD dataset

Grade	Train images	Test images
Grade 0	177	45
Grade 1	41	10
Grade 2	195	48

4   Method

The proposed method shown in Fig. 1 consists of a convolutional autoencoder network (model) and a DME-grading network. The convolutional autoencoder network is trained on unlabeled fundus images from Kaggle dataset so that it learns various features of the retinal fundus images. DME-grading network consists of the encoder layers from the autoencoder followed by additionally added fully connected layers. The weights of the pre-trained autoencoder network are used to initialize some layers of the DME-grading model which is then trained and tested on the Indian Diabetic Retinopathy Image Dataset (IDRiD). The following sections detail the proposed network.

4.1   Preprocessing

All of the input images from both the dataset are resized to 512×512 using bilinear interpolation. Real-time data augmentation is performed on the images so that there is an increase in the number of input images. The various augmentations performed are horizontal shift, vertical shift, and horizontal flip. All input images also are normalized so that each pixel is in the range [0, 1]. Contrast limited adaptive histogram equalization (CLAHE) is also applied on the low contrast images from the IDRiD to enhance the contrast of the images.

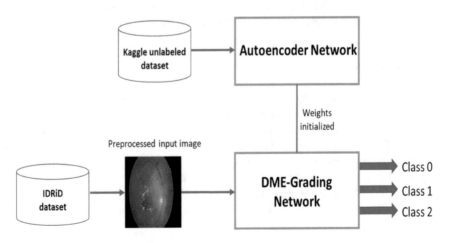

Fig. 1 Overview of the proposed method

4.2   Class Imbalance

It is evident from Table 1 that the total number of images in each grade (class) in the IDRiD is highly imbalanced. Only around 9% of the dataset belongs to Grade 1 and over 40% belongs to Grade 0 and Grade 2. In order to overcome the problem of class imbalance, we used either of the two methods, namely (1) augmenting training data in each grade up to a predefined number and (2) class weights.

Augmentation of training dataset. In this method, the total number of training images is increased to 1000 images per class. This is done by applying various augmentation on training set like flipping, addition of noise, Gaussian blur, and translation of images to left and right before runtime.

Class Weights. In this method, class weights are computed for each of the class so that each output class gets a weight or bias in order to balance the classes. The majority class gets a lower weight, whereas the minority class gets a higher weight.

4.3   The Autoencoder Network

Autoencoder is a type of unsupervised deep learning model that aims to copy the input as the output. The autoencoder consists of two parts, namely the encoder and the decoder. The encoder reduces the input image into its latent view representation (bottleneck) where it is compressed to the maximum, preserving important information. The decoder reconstructs the original image from the latent view representation.

In this work, we use a convolutional autoencoder model where the encoder consists of four convolutional layers of 64, 64, 32, and 32 filters of size 5 × 5. Each convolutional layer is followed by a max-pooling layer of size 2 × 2. Each convolutional layer extracts features of the input image and creates a feature map. This feature map is sensitive to the location of features and can cause overfitting.

Pooling layers downsample the feature maps to reduce the number of parameters at the same time preserving important information, thereby controlling overfitting as well as reducing the computational load. Similarly, in this work, the decoder consists of five convolutional layers of 32, 32, 64, 64, and 3 filters of size 5 × 5. The first four layers of the decoder are followed by a 2 × 2 upsampling layer.

Rectified linear unit (ReLu) is used as the activation function of each convolutional layer which adds nonlinearity. The autoencoder model consists of skip connections to connect different layers. Skip connections help to pass information from initial layers that might have been lost to latter layers. This helps in increasing the accuracy as more information will be retained in the latter layers.

4.4   DME-Grading Network

The DME-grading network consists of the encoder from the autoencoder network followed by a fully connected layer of 512 neurons and a classification layer of 3 neurons for classification. Figure 2 shows the structure of the DME-grading network. The fully connected layer consists of one dense layer of 512 neurons. The classification or the output layer consists of a dense layer of 3 neurons.

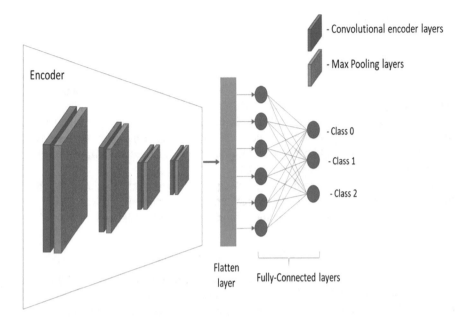

Fig. 2 Overall architecture of the DME-grading network

Since ReLu activation function results in dead neurons and bias shift, the DME-grading network uses the leaky rectified linear unit (Leaky ReLu) as the activation function for the fully connected layer. Softmax is used as the activation function for the output layer. The weights of the pre-trained encoder are used for training the classification model as the encoder has already learned the features of the retinal fundus images.

5   Experiments Conducted

5.1   Training of the Autoencoder

The autoencoder is trained on the Kaggle dataset for 20 epochs, 3000 steps per epoch with a batch size of 64. The optimizer used is Adam with a learning rate of 0.001.

5.2   Training of the DME-Grading Network

In order to find the best network configuration, the DME-grading network is trained on the IDRiD train dataset by varying parameters like epochs, learning rate, and fine-tuning (allowing few encoder layers to be trainable). All the experiments use a batch size of 64 and Adam optimizer. Categorical cross-entropy is used as loss function, and 5% of the training images are used for validation. Training is done either using class weights or augmentation of training dataset to tackle class imbalance.

Table 2 Training various DME-grading networks using class weights on the IDRiD train dataset

Model	Learning rate	Trainable	Epochs
M1	0.001	Last 1 covnet	131
M2	0.0001	Last 1 covnet	54
M3	0.0001	None	63
M4	0.001	Last 1 covnet	60
M5	0.001	None	35

None trainable indicates all the encoder layers are frozen. Last 1 covnet trainable indicates the last convolutional layer of the encoder is unfrozen

Training using class weights. Table 2 details the various experiments that are conducted using the preprocessed training dataset for different values of learning rates, fine-tuning details, and epochs using class weights. As shown in Table 2 while fine-tuning the DME-grading network, the encoder part is frozen completely (uses the pre-trained weights) or some of the layers can be unfrozen. Early stopping is used to prevent overfitting in these models.

Training using augmentation of dataset. Table 3 details the various experiments that are conducted using the preprocessed training dataset with 1000 images in each class for different values of learning rates, fine-tuning details, and epochs.

5.3 Testing of the DME-Grading Network

All the models specified in Tables 2 and 3 are evaluated on IDRiD test set which consists of 103 retinal fundus images. Preprocessing techniques used on the training set are applied on the testing set. The proposed method is evaluated based on accuracy, precision, recall, and $F1$-score.

Table 3 Training various DME-grading networks using augmentation on the IDRiD train dataset

Model	Learning rate	Trainable	Epochs
M6	0.001	Last 1 covnet	200
M7	0.0001	None	200
M8	0.001	None	200
M9	0.001	Last 1 covnet	150
M10	0.001	Last 1 covnet	100
M11	0.001	Last 2 covnet	150

6 Results and Discussion

Tables 4 and 5 provide the results of the DME-grading models mentioned in Tables 2 and 3, respectively. From Tables 4 and 5, it can be noted that the highest accuracy achieved is 56.31% (M5) and 68% (M9) for class weights and augmentation, respectively. Thus, the results indicate that augmenting the training dataset to have 1000 images per class achieves overall better results. Some of the possible reasons the proposed model achieved less accuracy is due to the quality of the images in IDRiD dataset and insufficient representative images for each grade. It can be noted from the literature that there are two other works [17, 18] that used IDRiD dataset for automatic DME grading using deep learning techniques and achieved better accuracy of 95.45% and 96.12%, respectively. These works augmented IDRiD dataset with other datasets and also used ensemble of CNNs to classify the images into different DME grades. In contrast to these works, the proposed method has not resorted to other data sources but aimed at solving the DME-grading problem with the available data. Also, the proposed approach used semi-supervised learning and transfer learning techniques.

Table 4 Results of various DME-grading networks mentioned in Table 2 on IDRiD test set

Model	Accuracy	Precision	Recall	$F1$-score
M1	46.6	59	47	51
M2	53.33	49	53	50
M3	54.36	49	54	51
M4	548	45	50	46
M5	**56.31**	51	56	53

Table 5 Results of various DME-grading networks mentioned in Table 3 on IDRiD test set

Model	Accuracy	Precision	Recall	$F1$-score
M6	64.07	65	64	64
M7	35.9	62	36	44
M8	46.6	59	47	51
M9	68	66	68	65
M10	62.13	59	62	60
M11	40.3	41	41	41

7 Conclusion

In this paper, we proposed a model for DME grading with the help of pre-trained convolutional autoencoder. The main problem of the dataset was the class imbalance. In order to resolve this, several deep learning models were trained using class weights or augmentation. Among these methods, the highest accuracy of 68% was achieved by the model which is trained on augmented dataset having 1000 images per class. The proposed method can be further enhanced using ensemble methods.

References

1. National Institute of Health, Macular edema. https://www.nei.nih.gov/learn-about-eye-health/eye-conditions-and-diseases/macular-edema. Last accessed 20 Dec 2019
2. Porwal P, Pachade S, Kamble R, Kokare M, Deshmukh G, Sahasrabuddhe V, Meriaudeau F (2018) Indian diabetic retinopathy image dataset (IDRiD). IEEE Dataport. http://dx.doi.org/10.21227/H25W98
3. Helmchen LA, Lehmann HP, Abràmoff, MD (2014) Automated detection of retinal disease. Am J Managed Care 20(11 Spec No. 17):eSP48–eSP522014)
4. Lee JG, Jun S, Cho YW, Lee H, Kim GB, Seo JB, Kim N (2017) Deep learning in medical imaging: general overview. Korean J Radiol 18(4):570–584
5. van Engelen JE, Hoos HH (2019) A survey on semi-supervised learning. Mach Learn
6. Tan C, Sun F, Kong T, Zhang W, Yang C, Liu C (2018) A survey on deep transfer learning. In: International conference on artificial neural networks. Springer, Berlin, pp 270–279
7. Kunwar A, Magotra S, Sarathi P (2015) Detection of high-risk macular edema using texture features and classification using SVM classifier. In: International conference on advances in computing, communications and informatics. IEEE, New York, pp 2285–2289
8. Wang Y, Yaonan Zhang Y, Zhaomin Y, Ruixue Z (2016) Machine learning based detection of age-related macular degeneration (AMD) and diabetic macular edema (DME) from optical coherence tomography (OCT) images. Biomed Opt Express 7(12):4928–4940
9. Punnolil A (2013) A novel approach for diagnosis and severity grading of diabetic maculopathy. In: International conference on advances in computing, communications and informatics. IEEE, New York, pp 1230–1235
10. Sukanesh R, Murugeswari S (2014) Detection of diabetic maculopathy using KNN algorithm. Appl Mech Mater 573:791–796
11. Prentašić P, Loncaric S (2016) Detection of exudates in fundus photographs using deep neural networks and anatomical landmark detection fusion. Comput Methods Programs Biomed 137:281–292
12. Chudzik P, Majumdar S, Calvia F, Al-Diri B, Hunter A (2018) Exudate segmentation using fully convolutional neural networks and inception modules. In: Proceedings of the SPIE 10574, Medical Imaging 2018: Image Processing, 1057430
13. Perdomo O, Otálora S, Juan F, Arevalo J, González F (2016) A novel machine learning model based on exudate localization to detect diabetic macular edema. In: Proceedings of the ophthalmic medical image analysis third international workshop, pp 137–144
14. Al-Bander B, Al-Nuaimy W, Al-Taee MA, Williams BM, Zheng Y (2016) Diabetic macular edema grading based on deep neural networks. In: Proceedings of the ophthalmic medical image analysis third international workshop, pp 121–128

15. Juan M, Lei Z, Yangqin F (2018) Exudate-based diabetic macular edema recognition in retinal images using cascaded deep residual networks. Neurocomputing 290:1601-171
16. Kaggle, Diabetic retinopathy detection. https://www.kaggle.com/c/diabetic-retinopathy-detection. Last accessed 20 Dec 2019
17. Kori A, Chennamsetty SS, Alex V (2018 Ensemble of convolutional neural networks for automatic grading of diabetic retinopathy and macular edema. arXiv preprint arXiv:1809.04228
18. Singh RK, Gorantla R (2019) DMENet: diabetic macular edema diagnosis using hierarchical ensemble of CNNs. BioRxiv, 712240

Modeling and Analysis of Stator Inter-turn Faults in a BLDC Motor Using Hybrid Analytical-Numerical Approach

Adil Usman$^{(\boxtimes)}$ ⓘ and Bharat Singh Rajpurohit

Indian Institute of Technology Mandi, Kamand, Mandi, Himachal Pradesh 175001, India
adilusman@ieee.org

Abstract. This paper shall model the Stator Inter-turn Faults (SITF) in a Brushless DC (BLDC) motor using a novel fault modeling approach. In order to comprehensively analyze the effect of SITF on the machine performance, the proposed hybrid analytical-numerical approach is adopted for modeling the BLDC motor under SITF conditions. The hybrid modeling techniques take less computation time and are more accurate than the existing analytical methods. The behavior of the motor in terms of phase currents, back-EMF (E_B), electromagnetic torque, and mechanical speed is studied to investigate the change in the characteristic performance of the machine during fault conditions. The significant change encountered in motor back-EMF is more realistic since the actual magnetic flux density (B_M) profile obtained through numerical analysis is emulated in the analytically developed model of a motor. In addition, the outcomes obtained through hybrid modeling techniques, are further validated completely through Numerical Methods (NMs) like Finite Element Analysis (FEA) to validate the authenticity of the proposed methodology. The significant changes investigated in motor electromagnetic quantities draws an inference to the SITF in the BLDC motor.

Keywords: Brushless DC (BLDC) motor · Electrical equivalent circuit (EEC) · Finite element analysis (FEA) · Numerical methods (Nms) · Stator Inter-Turn faults (Sitfs)

1 Introduction

1.1 Fault Modeling Methods in Brushless DC Motors

BRUSHLESS DIRECT CURRENT (BLDC) MOTORS are the special types of Permanent Magnet (PM) synchronous motors with high dynamic performance, high torque density, and better efficiency [1, 2]. Due to high deployment of BLDC motors in industrial applications, they are widely been operated for longer durations. During this continuous operation of these machines, they are subjected to unfavorable environmental conditions which include thermal and physical stresses leading to the emergence of fault. Faults in Brushless PM motors can be either on the stator or on the rotor of the machine [3]. It has been reported in [4] about the detailed classification of faults

© Springer Nature Singapore Pte Ltd. 2020
N. Goel et al. (eds.), *Modelling, Simulation and Intelligent Computing*,
Lecture Notes in Electrical Engineering 659,
https://doi.org/10.1007/978-981-15-4775-1_30

related to the PM motors and further an inference on the preferable methods for different types of fault conditions has been illustrated.

The performance of a BLDC motor during fault conditions can be better analyzed through mathematical representation of the machine which is obtained through modeling of a machine. It has been apprehended from vast literature [5–9] that for the modeling of demagnetization faults Magnetic Equivalent (MEC) based methods are used [6] while for the SITF, the use of Electrical Equivalent Circuit (EEC) based methods is mostly preferred [7]. Since both these methods compromise accuracy therefore, the fault diagnosis is obtained numerically using Numerical Methods (NMs) like Finite Element Analysis (FEA) for better accuracy [7–9]. However, NMs require more computational time.

The authors have therefore recently proposed a novel fault modeling technique using Hybrid Analytical-Numerical approach for modeling the faults in the BLDC motor. This fault modeling methods take less computational time and is more accurate than EEC and MEC based methods. Earlier in [10, 11] the demagnetization faults have been modeled using this technique. However, this paper shall model and analyze the SITF in BLDC motor using hybrid analytical-numerical approach, thereby investigating the motor performance in terms of phase currents, back-EMF, electromagnetic torque, and mechanical speed.

2   Modeling of BLDC Motors

2.1   Analytical Modeling of Ideal BLDC Motor

For the modeling of an ideal BLDC motor, the mathematical modeling equations can be referred from [1–4, 10, 11]. However, the modeling equations related to the voltage and ideal back-EMF shaping function of a motor are given from (1) to (6).

$$v_a = R_s i_{s,a} + (L - M)\left(\frac{di_a}{dt}\right) + \left(\frac{d\lambda_{PM,a}}{dt}\right) \tag{1}$$

$$v_a = R_s i_a + L_s\left(\frac{di_a}{dt}\right) + e_a \tag{2}$$

$$v_b = R_s i_b + L_s\left(\frac{di_b}{dt}\right) + e_b \tag{3}$$

$$v_c = R_s i_c + L_s\left(\frac{di_c}{dt}\right) + e_c \tag{4}$$

where
   v_a, v_b, v_c are the phase voltages,
   i_a, i_b, i_c are the phase current,
   λ_{PM} is the flux linkage of a PM for respective phases

Table 1 Analytical parameters

Symbol	Parameters	Value
P_o	Output power	850 W
V	Supply voltage	48 VDC
P_W	Windage loss	20 W
N	Rated speed	2650 rpm
I	Rated	21 A
T_e	Current rated torque	3.06 N m
φ	No. of phases	3
η	Efficiency	85%

$Ls = L - M$, where, L is a self-inductance while M is the mutual inductance between two windings.

e_a, e_b, e_c are the back-EMFs due to PMs given by

$$e_a = \frac{K_e}{2}\omega_m F(\theta_e), \quad \text{for phase } A \tag{5}$$

where θ_e is an electrical angle of rotor, ω_m is the mechanical speed and $F(\theta_e)$ is a back-EMF reference function as given in (6).

$$F(\theta_e) = \begin{cases} l1 & 0 \le \theta_e < \frac{2\pi}{3} \\ 1 - \frac{6}{\pi}\left(\theta_e - \frac{2\pi}{3}\right) & \frac{2\pi}{3} \le \theta_e < \pi \\ -1 & \pi \le \theta_e < \frac{5\pi}{3} \\ -1 + \frac{6}{\pi}\left(\theta_e - \frac{2\pi}{3}\right) & \frac{5\pi}{3} \le \theta_e < 2\pi \end{cases} \tag{6}$$

The closed-loop BLDC motor drive is developed using the parameters as given in Tables 1 and 2 and an equivalent block diagram is given in Fig. 1 while the ideal trapezoidal back-EMF obtained from (6) is given in Fig. 2.

Table 2 Numerical parameters of BLDC motor

Symbol	Material parameter	Material value
PM	Permanent magnet (8-poles)	NdFeB 40MGOe (N35SH)
	Stator material	Steel
H_c	Magnetic field strength	979,000 A/m
μ	Permeability	$4\pi \times 10^{-7}$ H/m
μ_{rm}	Relative permeability of the material	7400
$H_{c\ faulty}$	Faulty coercivity	489,500 A/m

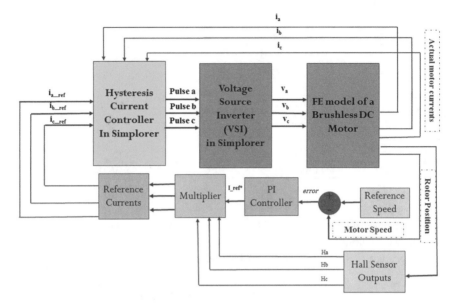

Fig. 1 Equivalent representation for a closed loop BLDC motor drive

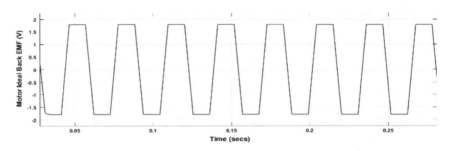

Fig. 2 Ideal back-EMF (*trapezoidal*) profile obtained through a shaping function given in (6) for a BLDC motor

2.2   Hybrid Analytical-Numerical Modeling Of A Healthy BLDC Motor

The hybrid analytical-numerical model of a BLDC motor can be developed through the replacement of an ideal back-EMF profile in the analytical model developed in the preceding subsection, with the numerically obtained B_M plots (*excluding slotting effect*). The Finite Element Method Magnetics (FEMM) tool is preferred over the other Numerical Methods (NMs) in this work, because of two significant advantages. First, it is an open-source software tool that is readily available and second in contrary to the transient analysis, the magnetostatics solver been used for analysis, analyzes the steady-state performance of the machine. This significantly reduces the computational time of the simulation. The numerical (*FEMM*) model of a BLDC motor is developed using the parameters listed in Tables 1 and 2. The magnetic flux density (B_M) profile, obtained

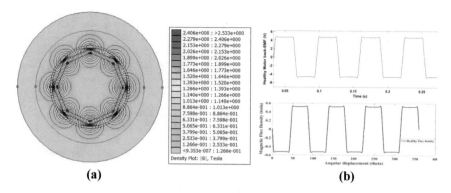

Fig. 3 **a** Meshed FE model of a healthy BLDC motor, **b** Magnetic density B_M profile of the PMs

from the FEMM simulation is extracted and replaces the back-EMF (E_B) shaping function block of the analytically modeled BLDC motor drive developed in Sect. 2. For the calculation of E_B, the motor back-EMF constant (K_E) obtained from the machine rating, is multiplied with the shaping function and mechanical speed of the motor. As given in (5), K_E is multiplied with the obtained B_M and the motor's speed to give the required back-EMF.

The FEMM model of a BLDC motor with PMs is shown in Fig. 3a. The model has eight poles with the corresponding airgap magnetic flux density of $B_M = 0.53$ T as shown in Fig. 3b. By adjusting the value of K_E in the analytical-numerical model of a motor, the actual back-EMF of the machine is obtained. For the $K_E = 0.0900$ V/rad/s (*as per the given machine rating*) and speed $\omega = 50$ rad/s, the corresponding back-EMF obtained is $E_B = 4.5$ V given in Fig. 4a. The respective phase currents I_{ph} = 3.2 A and expected electromagnetic torque $T_E = 0.6$ N m obtained is given from Fig. 4b, c. The steady-state speed is given in Fig. 4d.

2.3 Modeling of SITF in BLDC Motor

During Stator Inter-turn fault (SITF) in the machine, the insulation breakdown undergoes the change in resistance from infinite to a finite value which leads to the heavy inrush of current in the windings resulting in damage. In our study, the SITF is emulated in the hybrid model through the short-circuit of the winding coils through a finite resistance. The fault resistance in this study is taken as $R_f = 0.1$ Ω. The resistance and inductance have been split into two sub coils. The machine winding of phase a is shorted at 33.3% winding and the inductance for healthy and faulty parts are referred to as L_{a1} and L_{a2} respectively. The mutual inductance is represented as M_{a1a2} and given by:

$$L_{a1} = (1 - \mu)^2 L_a \tag{7}$$

$$M_{a1a2} = \mu(1 - \mu)L_a \tag{8}$$

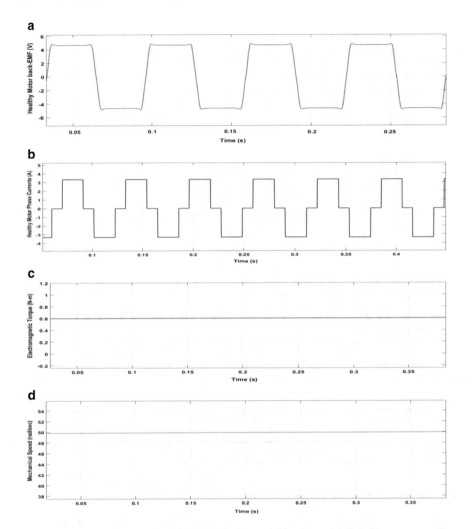

Fig. 4 **a** Back-EMF for a healthy Hybrid BLDC motor drive E_B = 4.5 V. **b** Phase currents for a healthy Hybrid BLDC motor drive I_{ph} = 3.2 A. **c** Healthy motor electromagnetic torque T_E = 0.6 N m. **d** Healthy motor speed ω = 50 rad/s

$$L_{a2} = (\mu)^2 L_a \qquad (9)$$

where μ represents the fraction per unit of faulty turns.

The analytical equations in support of the SITF in a BLDC motor are given from (7) to (10).

$$V_s = R_s[I_s] + [L_{ss}]\frac{d[I_s]}{dt} + [E_s] - [R_0]i_f - [L_0]\frac{di_f}{dt} \qquad (10)$$

where $[V_s]$, $[I_s]$ and $[E_s]$ are the stator voltage, current and back-EMF vectors given from (11) to (13)

$$[V_s] = [v_{as} \quad v_{bs} \quad v_{cs}]^T \tag{11}$$

$$[I_s] = [i_{as} \quad i_{bs} \quad i_{cs}]^T \tag{12}$$

$$[E_s] = [e_{as} \quad e_{bs} \quad e_{cs}]^T \tag{13}$$

R_s is the phase resistance and $[L_{ss}]$ is the inductance matrix of the healthy BLDC motor respectively. $[R_0]$ and $[L_0]$ are the fault resistance and inductance matrix in the faulty part of the winding given from (14) to (15). Inductance matrix is given in (16).

$$[R_0] = [R_{a2} \quad 0 \quad 0]^T \tag{14}$$

$$[L_0] = [L_{a2} + M_{a1a2} \quad M_{a2b} \quad M_{a2c}] \tag{15}$$

$$[L_{ss}] = \begin{bmatrix} L_s & M & M \\ M & L_s & M \\ M & M & L_s \end{bmatrix} \tag{16}$$

L_s is the phase self-inductance and M being the mutual inductance between phase windings of the healthy BLDC motor. R_{a2} and L_{a2} are the resistance and self-inductance of the faulty windings. M_{a1a2}, M_{a2b} and M_{a2c} are mutual inductances between the faulty winding and the healthy windings of phase a, b and c. The fault current between the insulation fault resistance r_f is referred to as i_f. Fault current can be evaluated from (17) as:

$$R_{a2}i_a + (L_{a2} + M_{a1a2} - M_{a2b})\frac{di_a}{dt} + M_{a2b}\frac{di_b}{dt} + M_{a2c}\frac{di_c}{dt}$$
$$+ e_{a2} = (R_{a2} + r_f)i_f + L_{a2}\frac{di_f}{dt} \tag{17}$$

The faulty electromagnetic torque is given by (18) and (19):

$$T_e = \frac{[E_s]^T[I_s] - e_{a2}i_f}{\omega} \tag{18}$$

$$T_e - T_l = J\frac{d\omega}{dt} \tag{19}$$

J is the moment of inertia and T_l is the load torque, ω is an angular mechanical speed.

During the SITF in the hybrid BLDC motor, the machine quantities undergo a realistic change which is investigated through the results shown in Fig. 5. The expected change in electromagnetic quantities, viz. phase currents (Fig. 5a), motor back-EMF

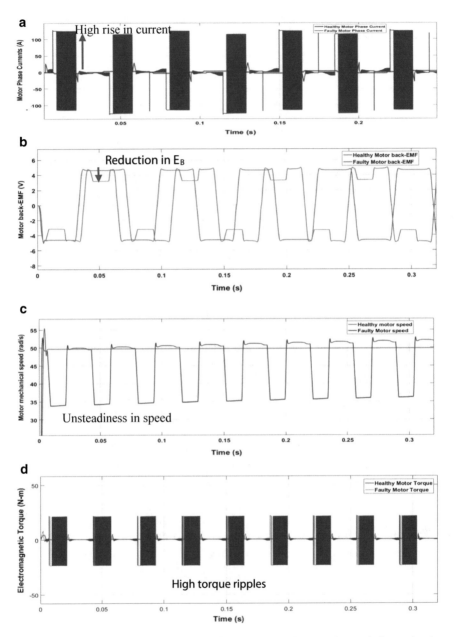

Fig. 5 a High rise in phase currents under SITF conditions. **b** Reduction and change in phase sequence in back-EMF of a machine under SITF conditions. **c** Unsteadiness in mechanical speed for SITF conditions. **d** High ripples in motor electromagnetic torque for SITF conditions

(Fig. 5b), mechanical speed (Fig. 5c) and electromagnetic torque (Fig. 5d) deduced from (19) validates with the analytical equations discussed above. As reported in the literature [7–9] that during the SITF conditions, there is a high rise in the motor phase current while the back-EMF decreases significantly. This is well validated through Fig. 5a where the current increases to approximately ten times the rated current value during the SITF conditions. The motor back-EMF has a significant drop in magnitude as can be observed from Fig. 5b. This is due to an increase in current in order to maintain the constant power operation. Since the mechanical speed is directly proportional to the motor back-EMF, therefore any reduction in back-EMF will have a direct impact on the motor speed. This can be validated through Fig. 5c where the motor speed decreases consequently. Cumulatively, the change in speed will adversely affect the motor back-EMF as its frequency will decrease gradually.

The expected change in electromagnetic torque in terms of high ripples is observed through Fig. 5d. The detrimental changes observed under SITF conditions in the BLDC motor, modeled through hybrid analytical-numerical methods, thus authenticate the validity of the proposed approach. Moreover, the results obtained under SITF conditions for a hybrid model are further compared and validated with the outcomes obtained through numerical analysis, as will be discussed in the subsequent section.

3   Numerical Analysis of SITF in BLDC Motor

The FE model of a BLDC motor is developed using the parameters given in Tables 1 and 2. SITF in the BLDC motor is induced through the short-circuit of the winding coils through a finite resistance, $R_f = 0.1 \, \Omega$. The turns per coil $n_T = 6$. The coil_1 and coil_1re have been split into two sub coils as shown in Fig. 6. Each disintegrated coil is assigned two turns while the parental coil is given four turns each (coil_1 and coil_1re) respectively. The machine winding of phase a is shorted at 33.3% (two out of six turns per coil are short-circuited) winding. The model is validated using (7)–(19) as discusses before. The developed SITF model is co-simulated at rated speed of $N = 2650$ rpm

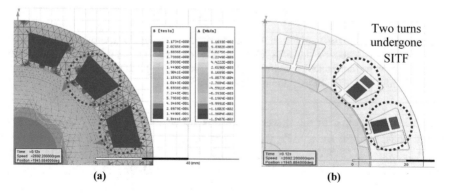

(a)   (b)

Fig. 6 a The meshed co-simulated model of a BLDC motor (*quadrant view*) showing two turns undergone SITF. **b** Splitting of a healthy coil into sub coil to emulate SITF

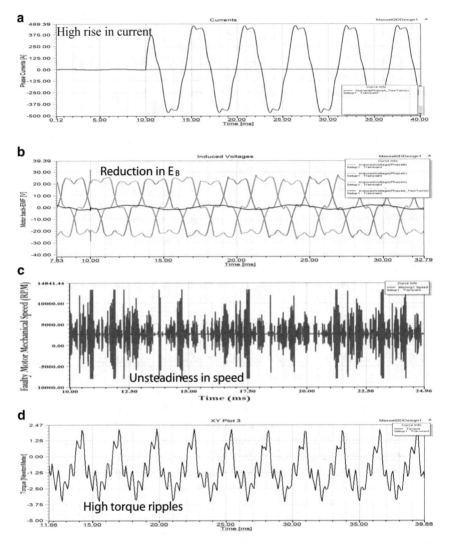

Fig. 7 **a** Sudden high rise in phase current for the shorted windings. The current increases to 480 from 2.7 A. **b** Reduction in back-EMF for the shorted turns during SITF conditions. E_{B_faulty} = 5 V. **c** Mechanical speed of a BLDC motor during SITF conditions. **d** High rise accompanied with torque ripples during SITF conditions

(ω = 277.50 rad/s), with the driver circuit developed in Simplorer. The sudden high rise in phase current for a SITF winding can be observed from Fig. 7a. At the rated speed and rated back-EMF constant (*same as in Hybrid model*), K_E= 0.0900 rad/s, the expected E_B is found to be 25 V which is given in Fig. 7b. The other changes in electromagnetic quantities are given in Fig. 7c, d.

4   Results and Discussions

The detrimental high rise in phase currents to more than ten times the rated value was observed in the winding under SITF conditions as shown in Fig. 7a. The rise in current is the same as had been investigated in the hybrid model of SITF in BLDC motor given in Fig. 5a. The decrease in motor back-EMF under SITF condition can be observed through Fig. 7b where the winding (under SITF condition) reduces its back-EMF to one-fifth of the healthy value along with changing the phase sequence. This has been earlier observed in the case of hybrid analytical-numerical approach also, where the back-EMF reduced significantly, and the phase sequence changed significantly as given in Fig. 5b. Moreover, the changes observed in mechanical speed and electromagnetic torque validates with the changing behavior as has been earlier observed through hybrid modeling approach. The unsteadiness in mechanical speed shown in Fig. 7c (and also, in Fig. 5c), and the existence of high ripples in the torque as can be observed from Fig. 7d (and also, in Fig. 5d) thereby validates the authenticity of the proposed approach in modeling the SITF in the BLDC motor. It is to be noted that the computational time taken for the numerical analysis was very high in comparison to the simulation time taken by the hybrid analytical-numerical approach. Thus, the proposed hybrid modeling approach has the advantage of consuming less computational time for modeling the motor under fault conditions.

5   Conclusions

The proposed hybrid analytical-numerical method replaces the default ideal back-EMF of the BLDC motor, with the more realistic numerically obtained back-EMF profile. The distinguished performance of the machine is obtained in terms of phase currents, motor back-EMF, electromagnetic torque, and mechanical speed. In contrary to other existing fault modeling methods, the proposed hybrid model approach is found to have better accuracy, consuming less computation time. Moreover, through this approach, the electromagnetic quantities inclusive of machine geometrical effect like shape of the magnet (which alters the ideal trapezoidal shape of the back-EMF to the more realistic characteristic) can be investigated. The other significant changes in machine quantities like high rise in phase currents and the reduction in machine back-EMF confirms the existence of stator inter-turn fault in the motor. The ongoing experimental validation is in progress and shall be the extension of this research work.

References

1. Hendershot JR, Miller TJE (1994) Design of Brushless permanent magnet motors. Oxford University Press, New York
2. Toliyat H, Nandi S, Choi S, Meshgin-Kelk H (2016) Electric machines: modeling, condition monitoring, and fault diagnosis. CRC Press, Boca Raton
3. Krause PC, Wasynczuk O, Sudhoff SD (2002) Analysis of electric machinery and drive systems. IEEE Press, New York

4. Usman A, Joshi BM, Rajpurohit BS (2017) Review of fault modeling methods for permanent magnet synchronous motors and their comparison. In: 2017 IEEE 11th international symposium on diagnostics for electrical machines, power electronics and drives (SDEMPED), Tinos
5. Nandi S, Toliyat HA (1999) Condition monitoring and fault diagnosis of electrical machines—a review. In: 34th IEEE industry applications society annual meeting conference, pp 197–204
6. Faiz J, Nejadi-Koti H (2016) Demagnetization fault indexes in permanent magnet synchronous motors—an overview. IEEE Trans Magn 52(4):1–11
7. Moon S, Jeong H, Lee H, Kim SW (2017) Detection and classification of demagnetization and interturn short faults of IPMSMs. IEEE Trans Industr Electron 64(12):9433–9441
8. Mohammed OA, Liu Z, Liu S, Abed NY (2007) Internal short circuit fault diagnosis for PM machines using FE-based phase variable model and wavelets analysis. IEEE Trans Magn 43(4):1729–1732
9. Qi Y, Zafarani M, Gurusamy V, Akin B (2019) Advanced severity monitoring of interturn short circuit faults in PMSMs. IEEE Trans Transp Electr 5(2):395–404
10. Usman A, Rajpurohit BS (2019) Comprehensive analysis of demagnetization faults in BLDC motors using novel hybrid electrical equivalent circuit and numerical based approach. IEEE Access 7:147542–147552
11. Usman A, Joshi BM, Rajpurohit BS (2019) Modeling and analysis of demagnetization faults in BLDC motor using hybrid analytical-numerical approach. In: IEEE 45th annual conference of the IEEE industrial electronics society, IECON'19, Lisbon, Portugal, Oct. (2019), pp 1034–1039

A Novel Approach to Design Single-Phase Cycloconverter Using SiC MOSFET and Its Performance Analysis Over IGBT

Maithili Shetty, Karthik K. Bhat, Anoop Narayana, and Melisa Miranda[✉]

Department of Electronics and Communication Engineering, PES University, Bengaluru, India
melisamiranda@pes.edu

Abstract. Silicon Carbide (SiC) MOSFET devices exhibiting several advantages, including high blocking voltage, lower conduction losses, and lower switching losses, when compared to silicon-based devices have become commercially available, enabling their adoption into power supply products. This paper presents a novel approach to designing a cycloconverter using SiC MOSFETs as opposed to the conventional usage of IGBT. A comparative study is attempted between the two with respect to distortion and system efficiency. MATLAB/Simulink models and simulations are used to analyze the results for the above.

Keywords: Cycloconverter · IGBT · Silicon carbide MOSFET · System efficiency · Switching frequency

1 Introduction

Wide-bandgap (WBG) based semiconductors such as Silicon Carbide (SiC) or Gallium Nitride (GaN) are ready to carve out a niche in applications that demand the ability to work at high voltages and temperatures while demonstrating high efficiency and relatively smaller dimensions owing to their intrinsic properties. These WBG-based semiconductors offer several advantages over the equivalent silicon devices available in the market today, few of which include, lower leakage current, significantly higher operating temperatures, better conduction, and switching properties. For these reasons, the WBG devices have been identified to have a promising future in the power semiconductor industry.

Here the focus is only on the Silicon Carbide based power devices and its applications. There has been a tremendous amount of research effort on developing power semiconductor devices with Silicon Carbide (SiC) in the pursuit of higher efficiency and smaller dimensions [1, 2]. The availability of SiC wafers on a commercial basis has led to the demonstration of many types of metal-oxide-semiconductor (MOS)-gated devices that exploit its unique properties. These emerging Silicon Carbide (SiC) MOSFET power devices promise to displace silicon IGBTs from the majority of challenging power electronics applications by enabling superior efficiency and power

© Springer Nature Singapore Pte Ltd. 2020
N. Goel et al. (eds.), *Modelling, Simulation and Intelligent Computing*,
Lecture Notes in Electrical Engineering 659,
https://doi.org/10.1007/978-981-15-4775-1_31

density, as well as the capability to operate at higher temperatures [3]. Research-based on the comparison of a SiC-based DC/DC converter and an IGBT based DC/DC converter concluded that the efficiency of a SiC converter is greater than that of the IGBT converter over an output power range [4]. An electro-thermal analysis of an automotive traction inverter platform based on SiC MOSFET and SiC IGBT technology is discussed in [5] and the results show that there is a higher total loss reduction in the SiC MOSFET model compared to the IGBT model. For all these reasons, in this paper, we are designing a cycloconverter using SiC MOSFET as opposed to the usage of IGBT in doing the same.

In a cycloconverter, a constant voltage and frequency AC waveform is converted into another AC waveform of lower or higher frequency without using any DC link in the conversion process thus making it highly efficient [6]. A single-phase to single-phase cycloconverter consists of two full-wave converters that are linked back to back. There has been extensive research carried out to explore the several possibilities for realizing an AC variable speed drive with cycloconverter. Various different solutions and the performance analysis of the cycloconverter in rolling mill drive applications is presented in [7]. A cycloconverter can be programmed to generate variable-frequency variable-voltage to drive an induction motor [8].

The objective of this paper is to design an efficient cycloconverter using SiC MOSFET and compare the performance of that with a cycloconverter designed using IGBT. The forthcoming sections give a better understanding of the above. SiC MOSFET is modeled using MATLAB/Simulink and a novel approach to design a cycloconverter using the same is presented. An analysis of all the simulation results and comparison of the performance of SiC MOSFET with IGBT pertaining to various characteristics such as system efficiency and distortion is dealt with upon in the later sections. Thereafter all the main results are concluded.

2   Simulink Model Analysis

2.1   Silicon Carbide MOSFET Model

An accurate SiC MOSFET Model is built using MATLAB/Simulink. Extensive research on the SiC device has demonstrated it to be a superior material to silicon in many properties for the construction of power switching devices [9]. The SiC MOSFET as a majority carrier switch eliminates the minority carrier current tail experienced with silicon IGBTs, resulting in much lower switching losses. An added benefit of using the SiC MOSFET in place of the conventional IGBT is their overall system efficiency improvement, the capability of higher frequency operation, and the reduction in size. Figure 1 shows the Simulink model of the SiC MOSFET.

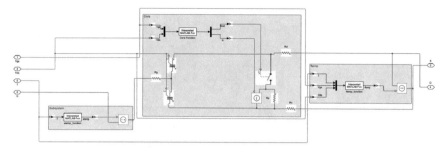

Fig. 1 Silicon carbide MOSFET simulink model

The above SiC MOSFET Model uses three main MATLAB functions, namely, the Core Design Function, ETemp Function, and the FTemp Function.

Core Function Design The core model which uses the Core Design MATLAB function is shown in Fig. 2.

The core function includes the SiC MOSFET characteristics with respect to the drain current and the drain-source voltage ($I_d - V_{ds}$). The relationship between the two is described by the following equations [10]:

$$If\ V_{gs} < V_{th}$$

$$I_d = 0 \tag{1}$$

$$If\ 0 < V_{ds} < \frac{(V_{gs} - V_{th})}{(1 + a)}$$

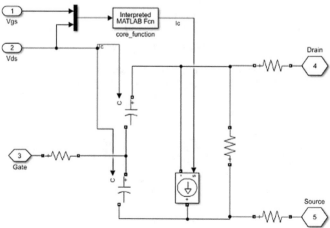

Fig. 2 Core simulink model

$$I_d = k \left[\left(V_{gs} - V_{th} \right) V_{ds} - \frac{(1+a)V_{ds}^2}{2} \right] (1 + \lambda V_{ds}) \tag{2}$$

$$If \ 0 < \frac{\left(V_{gs} - V_{th} \right)}{(1+a)} < V_{ds}$$

$$I_d = \frac{K}{2(1+a)} (1 + \lambda V_{ds}) \left(V_{gs} - V_{th} \right)^2 \tag{3}$$

In the above equations, the parameter 'a' represents the growth of the depletion layer and depends on the intrinsic structure of the SiC material and its properties. λ is the channel width modulation. K is the transistor gain which is related to the electron mobility by:

$$K = \mu C_{ox} \frac{W}{2L} \tag{4}$$

where L is the channel length, W is the channel width and C_{ox} is the oxide capacitance.

The mobility μ [11] is directly proportional to the drain current I_d and the transconductance g_m. The threshold voltage characteristic equation for the SiC MOSFET can be written as [12]:

$$V_{th} = V_{fb} + V_{it} + \Psi_{sin v} - \frac{Q_b}{C_{ox}} \tag{5}$$

$$V_{fb} = \Phi_m - \Phi_{Sic} - \frac{Q_f + Q_{it}}{C_{ox}} \tag{6}$$

Here, V_{fb} is the flat band voltage and V_{it} is the interface trap voltage. V_{fb} is related to the work function of the metal contact before the gate-oxide (Φ_m), the work function of the SiC (Φ_{Sic}), the Fermi potential in the bulk and the thermal voltage. The threshold voltage V_{th} contains a linear temperature dependence.

ETemp Function The linear temperature dependence of the threshold voltage V_{th} is expressed by the ideal voltage generator Etemp whose Simulink model is shown in Fig. 3.

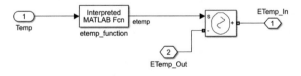

Fig. 3 ETemp simulation block

The voltage generator ETemp, which is present at the gate terminal adds its contribution in opposition to the gate voltage. The ETemp function can be described as follows: where α is the slope of the temperature variation of the threshold voltage V_{th} which can be represented as:

$$ETemp = [(T - T_{std})\alpha] \tag{7}$$

$$\alpha = \left(\frac{V_{th2} - V_{th1}}{T_2 - T_1}\right) = \left(\frac{1.8\,V - 2.5\,V}{125\,°C - 25\,°C}\right) = -0.007\,\frac{V}{°C} \tag{8}$$

where V_{th2} and V_{th1} are threshold voltages evaluated at T_2 and T_1, respectively. The standard temperature, T_{std}, is 25 °C.

FTemp Function FTemp is a current generator and it adds its contribution in the same direction to that of the drain-source current. Its Simulink model is depicted in Fig. 4. The carrier mobility increases in the working temperature range of [300–500]K for each of the operating regions of the device such as subthreshold, linear, and saturation region [13]. This behavior is due to the decrease in the occupied trap charge density with the rising temperature. The consequence of this is that more electrons in the channel are available at a given gate voltage, hence, when the temperature increases, a movement of Fermi level towards the bandgap can be observed. At temperatures higher than 500 K [14], the mobility decreases since the lattice scattering dominates and begins to release the interface trap charges.

According to the above considerations, the main mechanisms affecting the carrier mobility of the SiC MOSFET inversion layer are the phonon and interface traps scattering. The mobility of the MOSFET strongly depends on temperature and it can be expressed as follows [15],

$$\mu(T, E_{eff})\alpha\left(\frac{1}{\mu_{ph}(T, E_{eff})} + \frac{1}{\mu_{it}(T, E_{eff})}\right) \tag{9}$$

Here, the phonon scattering mobility μ_{ph} depends on the temperature through the following expression,

$$\mu_{ph}\alpha T^{-1.5} \tag{10}$$

For the interface traps, it is possible to consider a quite similar behavior, i.e.,

$$\mu_{it}\alpha T^{\beta} \tag{11}$$

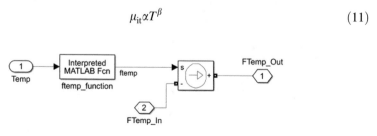

Fig. 4 FTemp simulation block

According to Eqs. (9), (10), and (11), an ideal current generator FTemp can be introduced and its value can be expressed as,

$$\text{FTemp} = I_{d_{std}}\left[\left(\frac{T}{T_{std}}\right)^{\beta} - \left(\frac{T}{T_{std}}\right)^{-1.5}\right] \tag{12}$$

Here, the two parameters 'a' (in Eqs. (2) and (3)) and 'β' [16] (in Eq. (11)), are determined with a least-square fit procedure, making a regression using the device curves given by the manufacturer.

2.2 Cycloconverter Design Using IGBT

A cycloconverter for converting AC power at supply frequency to AC power at a higher frequency is modeled in Simulink using IGBT as the switching element. This is shown in Fig. 5. To control the cycloconverter circuit, we need to produce trigger pulses in a particular sequence and feed them to the two separate converters called the P-converter and the N-converter; each behaving like an H-bridge inverter. These signals are produced using the signal builder and the transport delay block which helps in the switching operation of the IGBTs by carefully triggering the gate terminals. For the positive cycle of the input voltage, the IGBTs P1 and P2 conduct for a particular time period. Next, for the same time period, P1 and P2 are switched off which results in the conduction of IGBTs N1 and N2. The same operation continues for the negative cycle of the input voltage where IGBTs P3, P4, N3, and N4 take part in the conduction. The below model is used as the standard of comparison for the SiC model which is designed in this paper.

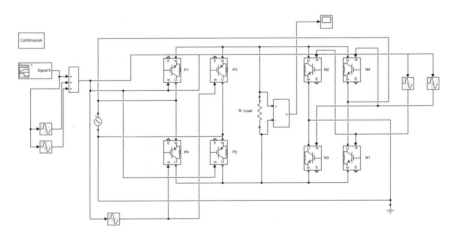

Fig. 5 Cycloconverter model using IGBT

The input voltage to the cycloconverter is 100 V at a frequency of 50 Hz shown in Fig. 6. The output frequency observed across the resistive load of 1 ohm is six times that of the supplied input frequency, i.e., the output is alternating at a frequency of 300 Hz at 33 V as seen in Fig. 7.

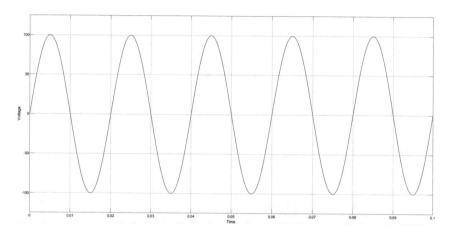

Fig. 6 IGBT cycloconverter

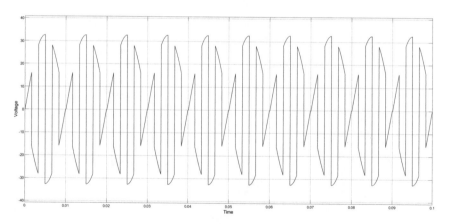

Fig. 7 IGBT cycloconverter output waveform

3 Simulation Analysis

3.1 Silicon Carbide MOSFET Simulation

The final subsystem model which will be used in the design of the cycloconverter is depicted in Fig. 8. The proposed Simulink model is validated using the standard SiC MOSFET datasheet.

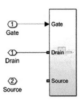

Fig. 8 Silicon carbide MOSFET simulink subsystem

The parameters used for the design of the SiC MOSFET are obtained from the manufacturer's datasheet. The model in Fig. 1 is simulated with all the key formulae and values plugged into the interpreted MATLAB functions which are the core function, ETemp function, and the FTemp function. The output characteristics, i.e., the drain current vs the drain-source voltage ($I_d - V_{ds}$) plot is shown in Fig. 9.

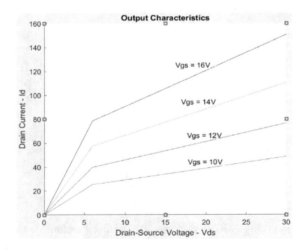

Fig. 9 Output characteristics of silicon carbide MOSFET from simulation

The $I_d - V_{ds}$ plots are obtained for four different values of V_{gs}, i.e., 10, 12, 14, and 16 V. Furthermore, the transfer characteristics for the simulated model at $V_{ds} = 20$ V is shown in Fig. 10. The cutoff voltage (V_{th}) for the MOSFET is taken as 2.1 V as obtained from the datasheet.

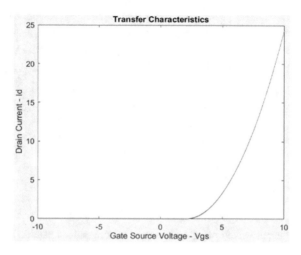

Fig. 10 Transfer characteristics of silicon carbide MOSFET from simulation

3.2 Cycloconverter Using SiC MOSFET

The cycloconverter modeled in Fig. 5 using IGBT is replaced by the Silicon Carbide MOSFET subsystem as seen in Fig. 11.

The model uses the same configuration as the IGBT cycloconverter with eight SiC MOSFETs, in place of IGBT, with the back-to-back connections to form two full-wave rectifiers. The input voltage and frequency given to this model are the same as that of the IGBT cycloconverter in Fig. 5, i.e., 100 V and 50 Hz respectively. The output voltage across the resistive load of 1 O is measured and depicted in Fig. 12. The output frequency is 6 times the input frequency, which results in a 300 Hz signal same as obtained in the IGBT model. The model is simulated for 0.1 s.

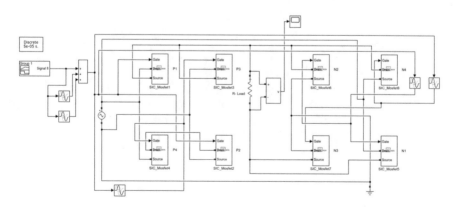

Fig. 11 Cycloconverter simulation model using silicon carbide MOSFET

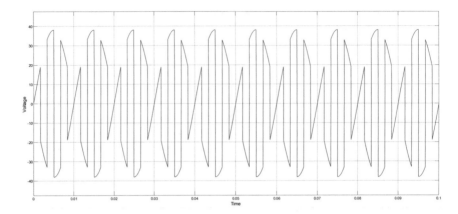

Fig. 12 Output waveform of cycloconverter using silicon carbide MOSFET

3.3   Performance Comparison

The output graphs in Figs. 7 and 12 for IGBT and SiC Cycloconverter respectively, are observed and it is seen that a slight distortion exists at the zero crossing in the IGBT cycloconverter while the switching effect is taking place. The distortion can be clearly observed in Fig. 13. This distortion is absent in the SiC Cycloconverter as seen evidently in Fig. 14. Thus, the cycloconverter using SiC can change from one cycle to another more smoothly with no crossover distortion which can be particularly useful for higher frequency operation. Therefore, the better switching application of the SiC MOSFET can be concluded from above.

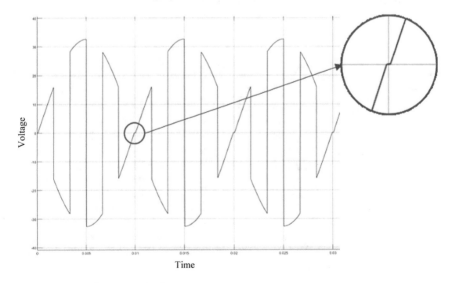

Fig. 13 Zero-crossing observed in IGBT cycloconverter

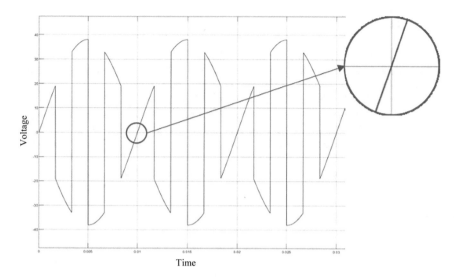

Fig. 14 Zero-crossing observed in SiC cycloconverter

Furthermore, a comparatively greater voltage drop is also observed in the IGBT model for the same input voltage of 100 V. The peak voltage of the IGBT cyclo-converter is 33 V as opposed to the peak voltage of the SiC MOSFET cycloconverter which is 38 V. Thus, the efficiency of the SiC model is 1.15 times higher than that of the IGBT model.

4 Conclusion

A novel approach to design a cycloconverter using SiC MOSFET in MATLAB/Simulink has been proposed in this paper. The SiC MOSFET used in the cycloconverter is designed in Sect. 2 and the simulation results of this MOSFET involving the output characteristics, transfer characteristics, and threshold voltage are matched and validated by comparing it with the characteristics of the actual SiC MOSFET pertaining to its datasheet.

The performance analysis of the IGBT and the SiC Cycloconverter model concluded that the peak voltage obtained at the output in the SiC Cycloconverter is less attenuated resulting in a higher efficiency by a factor of 1.15. Furthermore, the zero crossing distortion is absent in the SiC Cycloconverter model which makes it useful for higher frequency operation.

References

1. Biela J, Schweizer M, Waffler S, Kolar JW (2011) SiC versus Si—evaluation of potentials for performance improvement of inverter and DC-DC converter systems by SiC power semiconductors. IEEE Trans Industr Electron 58(7):2872–2882
2. Stevanovic LD, Matocha KS, Losee PA, Glaser JS, Nasadoski JJ, Arthur SD (2010) Recent advances in silicon carbide MOSFET power devices. In: Twenty-fifth annual IEEE applied power electronics conference and exposition (APEC), Palm Springs, CA, 2010
3. McBryde J, Kadavelugu A, Compton B, Bhattacharya S, Das M, Agarwal A (2010) Performance comparison of 1200 V Silicon and SiC devices for UPS application. In: IECON 2010—36th annual conference on IEEE industrial electronics society, Glendale, 2010
4. Nielsen RØ, Török L, Munk-Nielsen S, Blaabjerg F (2013) Efficiency and cost comparison of Si IGBT and SiC JFET isolated DC/DC converters. In: IECON 2013—39th annual conference of the IEEE industrial electronics society, Vienna, 2013
5. Kempitiya A, Chou W (2018) An electro-thermal performance analysis of SiC MOSFET vs Si IGBT and diode automotive traction inverters under various drive cycles. In: 34th thermal measurement, modeling & management symposium (SEMI-THERM), San Jose, 2018
6. Biabani MAKA, Pasha MA (2016) Performance analysis of step up and step down cyclo converter. In: 2016 international conference on electrical, electronics, and optimization techniques (ICEEOT), Chennai, 2016
7. Hagmann R (1991) AC-cycloconverter drives for cold and hot rolling mill applications. In: Conference record of the 1991 IEEE industry applications society annual meeting, Dearborn, Michigan
8. Brindha B, Porselvi T, Ilayaraja R (2018) Speed control of single and three phase induction motor using full bridge cycloconverter. In: 2018 international conference on power, energy, control and transmission systems (ICPECTS), Chennai, 2018
9. Shillington R, Gaynor P, Harrison M, Heffernan B (2010) Applications of silicon carbide JFETs in power converters. In: 2010 20th Australasian universities power engineering conference, Christchurch, 2010, pp 1–6
10. Pratap R, Singh RK, Agarwal V (2012) SPICE model development for SiC power MOSFET. In: 2012 IEEE international conference on power electronics, drives and energy systems (PEDES), Bengaluru, 2012, pp 1–5
11. Kaushik N, Haldar S, Gupta M, Gupta RS (2006) Interface traps distribution and temperature distribution and temperature-dependent 6H-SiC MOSFET analysis. Semiconductor Sci Technol 21(6):6–12 (2006)
12. Potbhare S, Goldsman N, Lelis A, McGarrity JM, McLean FB, Habersat D (2008) A physical model of high-temperature 4H-SiC MOSFETs. IEEE Trans Electron Dev 55 (8):2029–2040
13. Chain K, Huang JH, Duster J (1997) A MOSFET electron mobility model of wide temperature range for (77–400 K) IC simulation. In: Semiconductor science and technology, vol 12, issue 4, pp 355–358 (1997)
14. Sze SM (1981) Physics of semiconductor devices. John Wiley and Sons
15. Ryu SH (1997) Development of CMOS technology for smart power applications in silicon carbide, PhD dissertation, Dept Elect and Comp Eng Purdue Univ West Lafayette, US (1997)
16. Wang J et al (2008) Characterization, modeling, and application of 10-kV SiC MOSFET. IEEE Trans Electron Dev 55(8):1798–1806 (2008)

Demagnetization Fault Diagnosis in BLDC Motor Using Low-Cost Hall Effect Sensors

Vivek Kumar Sharma$^{(\boxtimes)}$ (iD), Adil Usman,
and Bharat Singh Rajpurohit

Indian Institute of Technology Mandi, Kamand, Mandi, HP 175001, India
vivek.sharma@ieee.org

Abstract. This paper proposes a demagnetization fault diagnostic method for a brushless direct current (BLDC) motor using Hall effect (HE) sensor sequence. The low-cost Hall effect sensors positioned 120° apart are used on the stator of a BLDC motor in order to monitor the change in the sequence during demagnetization fault conditions. Demagnetization effect in the permanent magnets (PMs) can be due to various causes; however, the broken PM defects are taken in our study. In contrary to the healthy operation of the motor, the significant change observed in the Hall sequence pattern during the demagnetization fault conditions is used as the fault signatures in this study. The change in machine quantities in terms of phase currents and magnetic flux density (B_M) characteristics is investigated numerically through a developed co-simulation model of a BLDC motor with the Simplorer drive circuit in Maxwell 2D tool. The numerically obtained results are validated experimentally, and the change in Hall sequence signals is investigated in order to detect, diagnose and identify the demagnetization fault in the BLDC motor. The diagnosis of the broken PM demagnetization fault using the low-cost Hall sensor's sequence is the novel contribution to this work.

Keywords: Brushless direct current (BLDC) motors · Demagnetization · Hall effect (HE) sensors · Permanent magnet (PM)

1 Introduction

Brushless direct current (BLDC) motors are the special type of synchronous motors having permanent magnets (PMs) on the rotor and three-phase concentrated windings on the stator [1, 2]. The applications of BLDC motors have increased significantly in the past few years due to its several effective impacts like higher torque density, higher efficiency and better dynamic performance [3]. Due to the continuous operation of the motor and the adverse ambient conditions, they are subjected to faults. Fault in the machine can degrade its output performance characteristics which can finally lead to the complete failure of the system [4]. Thus, the diagnosis of fault in its incipient stage is vital to prevent the complete system from breakdown.

In a BLDC, motor fault can be either on the stator or rotor of the machine. The stator-related faults mainly include the stator winding faults such as stator inter-turn faults and inter-phase faults [5, 6], while the rotor-related faults comprise of

© Springer Nature Singapore Pte Ltd. 2020
N. Goel et al. (eds.), *Modelling, Simulation and Intelligent Computing*,
Lecture Notes in Electrical Engineering 659,
https://doi.org/10.1007/978-981-15-4775-1_32

demagnetization faults and rotor eccentricity faults, which have been widely reported in the literature [7, 8]. However, there is very less work been done on the rotor magnet defects, which is also a type of demagnetization faults accounted widely in BLDC motors [9]. Broken PMs demagnetization fault is more vulnerable in the BLDC motor running at varying loads, bearing high operating stresses and ageing of magnets with time. Broken PM fault detection has been reported in some literature through phase currents and back-EMF signatures [8–10]. Advance methods of diagnosing the demagnetization faults using magnetic flux signatures have also been recently used in [9, 10].

Since it has been observed that for the detection and diagnosis of faults in the machine motor current signature, back-EMF and flux density or flux linkage signatures are only used. However, in this study the authors have given a novel contribution of diagnosing the broken PM demagnetization fault through the output of a Hall effect sensor sequence. The change in Hall sensor sequence observed during the demagnetization fault conditions validates the demagnetization fault conditions.

It is to be noted that since the Hall sensor sequence is monitored directly through the PM magnetic flux density (B_M) in the machine, therefore the change in Hall sensor sequence is caused due to the changing B_M during the demagnetization fault conditions. Thus, in this study the numerical model of a BLDC motor is developed under healthy and broken PM fault conditions. The change in magnetic flux density is observed which is ultimately the cause of changing Hall sensor sequence during broken PM fault conditions. Other electromagnetic signatures like phase currents and the motor back-EMF are also investigated, and the detrimental changes are further reported in this work.

In this paper, Sect. 2 elaborates on the modelling of demagnetization effect in BLDC motor using electrical equivalent circuit (EEC) and numerical methods (NMs)-based approaches. Section 3 illustrates the experimental investigation of the broken PM defects in a BLDC motor using author's proposed low-cost Hall sensor's sequence. The outcomes obtained are validated and further explained in detail for the authenticity of the proposed diagnostic technique.

2 Modelling of Demagnetization Effect in BLDC Motor

The broken PM defects are modelled as a demagnetization fault in BLDC motor. The demagnetization fault effects can be modelled mathematically through electrical equivalent circuit (EEC)-based equations and numerically through finite element analysis (FEA). The subsequent sections will elaborate on both the methods.

2.1 Electrical Equivalent Circuit (EEC)-Based Analysis of a BLDC Motor

The EEC-based modelling of a BLDC motor can be referred from the earlier work in [1–3]. However, the significant equations required for modelling the BLDC motor and the PM demagnetization effect through back-EMF (E_B) and magnetic flux density (B_M) signatures can be given from (1) to (4).

$$v_a = R_s i_{s,a} + (L - M)\left(\frac{di_a}{dt}dt\right) + \left(\frac{d\lambda_{PM,a}}{dt}\right) \tag{1}$$

$$v_a = R_s i_a + L_s \left(\frac{di_a}{dt}\right) + e_a \tag{2}$$

$$e_a = \frac{K_e}{2} w_m F(\theta_e) \tag{3}$$

The comprehensive mathematical modelling of the broken PM demagnetization effect can be studied in detail in [5]. However, the investigation through the change in E_B and B_M can be illustrated from (4).

$$e_a = \frac{d\varphi_a}{dt}d_t = \frac{d\theta_r}{dt}d_t\frac{d\varphi_a}{d\theta_r}d\theta_r = \omega_m \frac{d\varphi_a}{d\theta_r}d\theta_r \tag{4}$$

Similarly, the equations for v_b & v_c and e_b & e_c can be written.where

$$L_s = L - M$$

v_a, v_b, v_c	are the phase voltages,
i_a, i_b, i_c	are the phase currents,
e_a, e_b, e_c	are the back-EMFs due to PMs,
θ_m	is a mechanical angle of rotor,
θ_e	is an electrical angle of rotor.

$$\theta_e = \frac{p}{2}\theta_m$$

p	is number of poles,
$F(\theta_e)$	is a back-EMF reference function as can be referred in detail from [10].

2.2   Numerical Methods (Nms)-Based Analysis of a BLDC Motor

The finite element method (FEM) is used for the numerical analysis of a BLDC motor and the effect of demagnetization faults. The FE model of a 16-poles, 18-slots outer rotor BLDC motor is developed on a Maxwell 2D tool using the parameters as given in Table 1. The Simplorer-based electric drive circuit operating through hysteresis current controller (HCC) is co-simulated under healthy and faulty conditions. The co-simulation is run for time $t = 0.12$ s with a step size of 0.001 s, as shown in shown in Fig. 1. The findings from the co-simulation-based mathematical modelling of the BLDC motor under healthy and broken magnets fault conditions have been illustrated through Fig. 2a–e. The no-load phase currents for a healthy motor can be observed from Fig. 2a. The peak value is found to be 1.5 A, while during the broken PM fault, this phase current as shown in Fig. 2b increases significantly. The detrimental

Table 1 BLDC motor parameters

Symbol	Parameters	Value
V	Rated voltage	24 VDC
P	No. of poles	16
θ	Rotor position	Outer
P_o	Output power	600 W
N	Reference speed	3000 rpm
OD_S	Outer stator	8.12 cm
ID_S	Inner stator	3.4 cm
l_s	Length of stator	3.87 cm
N_T	No. of slots	18
OD_R	Outer rotor	9.26 cm
ID_R	Inner rotor	8.681 cm
l_R	Length of rotor	3.62 cm
$NdFeB$	Magnet type	Surface mounted
b_T	Magnet thickness	0.25 cm

reduction and distortion in back-EMF can be observed from Fig. 2c. The magnetic flux density (B_M) for a healthy BLDC motor as shown in Fig. 2d is found to be 0.3T. During the broken magnet fault conditions, the B_M gets distorted, and the peak magnitude decreases to 0.22T with respect to 0.3T of other healthy poles as shown in Fig. 2e. Therefore, the significant changes in phase currents, back-EMF and B_M signify the existence of broken PM demagnetization fault in the machine.

3   Experimental Investigations of Demagnetization Fault

3.1   Development of a Test Bed for a Broken PM Demagnetization Fault in a BLDC Motor

An 18-slots, 16 poles outer rotor motor BLDC motor as shown in Fig. 3a is used in our study. The stator is embedded with three Hall sensors positioned at 120° apart from each other as shown in Fig. 3b, c. To imitate the broken PM demagnetization fault in a BLDC motor under study, one of the poles (PM) is broken and paced in the rotor of the machine as shown in Fig. 3c.

A complete experimental test bed along with the Hall sequence measuring arrangement for the investigation of rotor demagnetization faults (*broken PM*) has been set-up as shown in Fig. 4a. A set of three-BLDC motor which comprises of a demagnetized motor and two healthy motors is used for simple switching between the healthy and faulty operations. A three-phase voltage source inverter (VSI) and a DAQ is used for the operation and recording of data. The Hall sensor sequence arrangement is exclusively shown in Fig. 4b.

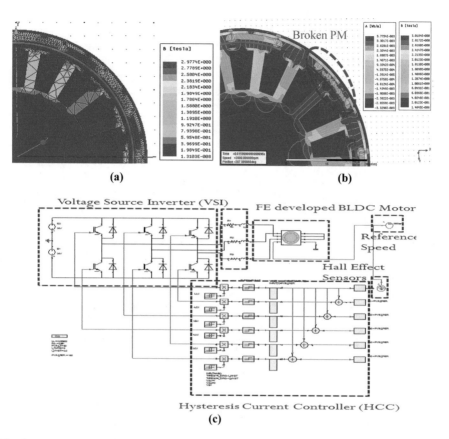

Fig. 1 **a** FE model of healthy BLDC motor. **b** FE model of a broken magnet in BLDC motor. 6th Pole undergoes demagnetization effect. **c** Co-simulation of FE built BLDC motor and Simplorer built drive system

The three Hall effect sensors positioned at 120° apart from each other in the stator of a machine are used for fault diagnosis in our study. These sensors used for acquiring the rotor position and also act as an electronic commutator. The change in Hall sensor sequence has been considered as the fault diagnosis signature for the investigation of demagnetization fault in the BLDC motor. It is to be noted that the Hall sensor sequence is changed due to the change in flux caused due to demagnetization effect; therefore, the relation of Hall voltage (sequence) and the magnetic flux can be derived as follows:

$$V_{\mathrm{H}} = \frac{-\mathrm{IB}}{ned} = R_{\mathrm{H}} \frac{\mathrm{IB}}{d} \mathrm{d} \tag{5}$$

where n = carrier density, d = conductor length and R_{H} is a Hall coefficient.

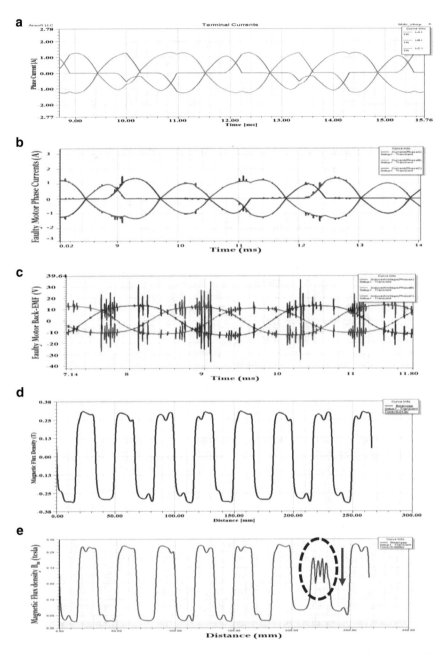

Fig. 2 a Phase currents of healthy BLDC motor. No-load current of 1.5 A flows through the stator winding of a motor. **b** Phase currents of broken PM BLDC motor. The current peak value increases and has harmonics due to the occurrence of faults. **c** Change in motor back-EMF during broken PM demagnetization fault. **d** Magnetic flux density of $B_M = 0.3T$ of a healthy BLDC motor. **e** Distorted B_M for broken 6th pole

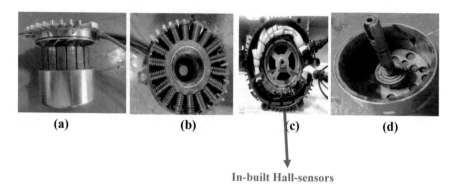

In-built Hall-sensors

Fig. 3 Internal structure of a BLDC motor used for the experimental study. **a** Outer rotor motor and the Stator getting detached. **b** 18-slots stator with three-phase winding. **c** 16-poles outer rotor with a broken PM

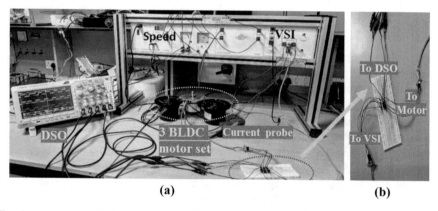

Fig. 4 a Test bed for broken magnet BLDC motor with current and hall sequence measurement. **b** Zoomed view of input, output and measurement junction of Hall sensor

Thus, through the significant change in Hall sensor sequence due to the change in flux density, the effect of demagnetization can be validated. Moreover, the change in magnetic flux density (B_M) as has been earlier investigated through analytical and numerical analysis for demagnetization fault diagnosis is now experimentally validated through the change in Hall sensor sequence ($V_H \propto B$) which will be discussed in the subsequent section. This is an additional information to the existing fault detection signatories used in BLDC motors.

3.2 Results and Discussions

The phase currents of BLDC motor during healthy and broken PM demagnetization are shown in Fig. 5a, b. For a healthy BLDC motor, the expected no-load phase current (*phase A*) of 1.24 A can be observed from Fig. 5a. During the demagnetization fault,

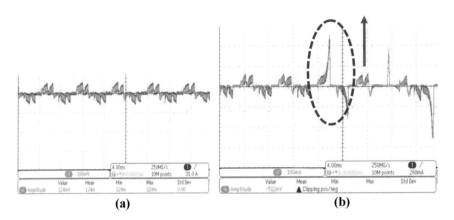

Fig. 5 **a** Experimentally obtained phase current (phase A) of healthy BLDC motor. **b** Exprementally obtained phase current (phase A) of broken magnet demagnetized BLDC motor [*Y-axis: 1 div = 2 A, X-axis:1 div = 4 ms*]

the significant rise in phase current is observed which is shown in Fig. 5b. The change in current is observed when phase A comes in vicinity of a broken PM and is found to be almost five times the healthy current (as encircled in Fig. 5b).

The three Hall sensor sequences, i.e. output of inbuilt Hall effect sensors for both healthy and broken PM faulty condition, are shown in Fig. 6a, b. The three Hall sequences of a healthy BLDC motor display a constant magnitude of 4 V having 120° displacement with respect to each other as shown in Fig. 6a. Moreover, the constant duty cycle is also observed throughout the complete rotation of a BLDC motor. However, for a broken PM demagnetization fault, the duty cycle observed from Fig. 6b is not constant (as encircled in Fig. 6b) throughout the complete rotation of the motor even though the phase displacement and magnitude still remain the same as that of a healthy motor. In addition, the motor back-EMF also reduces (as encircled in Fig. 6c) for the corresponding broken PM, illustrated through Fig. 6c.

Therefore, the significant observable changes reflected in the Hall sensor sequence during the broken PM defects justify the existence of demagnetization faults in the machine. The changes in Hall sequence are due to the change in magnetic flux density (B_M) due to broken PM, which directly affects the Hall voltage (v_H) and hence the Hall sequence. In contrary to the healthy operation of the motor, the Hall sequence does not remain constant and varies consequently during the demagnetization fault conditions, justifying the broken PM fault existence. The changes in other electromagnetic signatures (*like phase currents, back-EMF and magnetic flux density*), as investigated experimentally and through numerical analysis, validate the proposed broken PM demagnetization fault diagnostic technique using inbuilt Hall sensors. Thus, it can be inferred that the change in Hall sequence characteristics and manifests the existence of broken PM fault in a BLDC motor. This Hall sequence signature can be used as an auxiliary information for demagnetization fault diagnosis along with the conventionally adopted currents and back-EMF signatures.

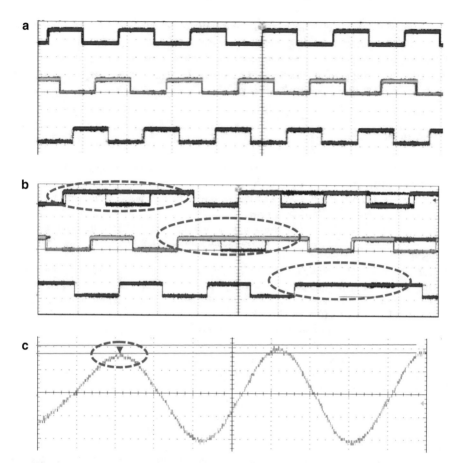

Fig. 6 a Exprementally obtained output (*Hall sequence*) of in-built Hall sensors of healthy BLDC motor [*Y-axis: 1 div = 4 V, X-axis:1 div = 4 ms*]. **b** Exprementally obtained output (*Hall sequence*) of in-built Hall sensors of BLDC motor during broken magnet demagnetization fault [*Y-axis: 1 div = 4 V, X-axis:1 div = 4 ms*]. **c** Exprementally obtained back-EMF of BLDC motor during broken PM demagnetization fault depicting the significant change in magnitude [*Y-axis: 1 div = 10 V, X-axis:1 div = 4 ms*]

4   Conclusion

In this paper, author has detected demagnetization fault on the basis of Hall sequence, i.e. output of inbuilt Hall sensor of BLDC motor used as electronic commutator and also used for acquiring rotor position. This Hall sequence can be used as an additional signature along with conventionally used back-EMF and current signatures for detecting abnormality of BLDC motor. Although the output of Hall sensor, i.e. Hall sequence during fault, was 120° phase displaced like healthy motor and having same magnitude, duty cycle was not constant throughout the rotation as it was in healthy motor and the variation in duty cycle was found to follow a particular pattern. The inconstant nature of duty cycle manifests the broken magnet demagnetization fault. Numerical modelling of the same BLDC motor is also shown along with suitable results to validate the experimental results.

References

1. Krause PC, Wasynczuk O, Sudhoff SD (2002) Analysis of electric machinery and drive systems. IEEE Press, New York, USA
2. Hendershot JR, Miller TJE (1994) Design of brushless permanent magnet motors. Oxford University Press, New York, NY, USA
3. Nandi S, Toliyat HA (1999) Condition monitoring and fault diagnosis of electrical machines-a review. In: 34th IEEE industry applications society annual meeting conference, pp 197–204
4. Usman A, Joshi BM, Rajpurohit BS (2017) Review of fault modeling methods for permanent magnet synchronous motors and their comparison. In: 2017 IEEE 11th international symposium on diagnostics for electrical machines, power electronics and drives (SDEMPED), Tinos, 2017, pp 141–146
5. Qi Y, Zafarani M, Gurusamy V, Akin B (2019) Advanced severity monitoring of interturn short circuit faults in PMSMs. IEEE Trans Transp Electrif 5(2):395–404
6. Moon S, Jeong H, Lee H, Kim SW (2017) Detection and classification of demagnetization and interturn short faults of IPMSMs. IEEE Trans Industr Electron 64(12):9433–9441
7. Faiz J, Nejadi-Koti H (2016) Demagnetization fault indexes in permanent magnet synchronous motors—an overview. IEEE Trans Magn 52(4):1–11
8. Park Y et al (2019) Online detection of rotor eccentricity and demagnetization faults in PMSMs based on hall-effect field sensor measurements. IEEE Trans Ind Appl 55(3):2499–2509
9. Goktas T, Zafarani M, Lee KW, Akin B, Sculley T (2017) Comprehensive analysis of magnet defect fault monitoring through leakage flux. IEEE Trans Magn 53(4):1–10
10. Usman A, Rajpurohit BS (2019) Comprehensive analysis of demagnetization faults in BLDC motors using novel hybrid electrical equivalent circuit and numerical based approach. IEEE Access 7:147542–147552

Design of Electrical Power Systems for Satellites

Aashna Kapoor$^{(\boxtimes)}$ ⓘ and A. R. Abdul Rajak ⓘ

Birla Institute of Technology and Science Pilani, Dubai Campus, Dubai, UAE
{f20160171, abdulrazak}@dubai.bits-pilani.ac.in

Abstract. An integral subsystem of a satellite is its Electrical Power System (EPS). Spacecraft power systems have undergone significant new developments in the last decade and will continue to do so even at a faster rate in the current decade. The EPS functions to supply continuous power during the satellite mission life, control and distribute power, support power requirements for peak and average electrical load, and protect payload operations against failures within the EPS. The design of the solar panels and batteries depends on the payload/s power demand and the mission lifetime. This paper studies the design, management and characteristics of the power subsystem of small satellites. The EPS for small satellites is required to have high efficiencies and low masses because of volume and weight constraints. Based on the power utilized by onboard equipment and devices, this paper proposes an efficient and durable power system.

Keywords: Electrical power systems · Design · Satellite · Solar array · Battery

1 Introduction

A satellite is an artificial-specialized transmitter/receiver that is intentionally placed in an orbit around the earth or any other body in space. Satellites are indispensable for performing vital functions such as weather forecasting, navigation, broadcasting, search and rescue operations, international communication, research and experimentation in space, and more [1]. However, their design and construction involve adherence to stringent requirements, such as size, cost, technical efficiency, reliability, and capability to resist significant changes in temperature and exposure to severe radiation.

A typical satellite system is segregated into numerous subsystems. This paper will mainly focus on the Electrical Power Subsystem. The EPS is responsible for generation, storage, conditioning, and supply of power to the satellite bus and payload. For a large 3-axis body stabilized satellite, the EPS contributes to approximately 30% of the spacecraft's dry mass [3]. Regardless of specific design prerequisites, the basic building blocks for any satellite power generation and distribution system are all essentially the same. Because satellites are non-serviceable autonomous remote machines, it is crucial to have a secure and reliable power system with a long lifespan [2]. Without power availability, we can neither locate the spacecraft nor monitor its system performance. Without power, a satellite is simply debris in space.

© Springer Nature Singapore Pte Ltd. 2020
N. Goel et al. (eds.), *Modelling, Simulation and Intelligent Computing*,
Lecture Notes in Electrical Engineering 659,
https://doi.org/10.1007/978-981-15-4775-1_33

2 Electrical Power System

The fundamental units of any satellite power system are the primary power source, backup batteries, bus voltage regulators, fuses, load switches, and the distribution harness (Fig. 1).

Solar radiation is the only available external source of energy in space. A satellite EPS not using solar energy must be fitted with its own onboard energy source such as a primary battery, fuel cells, or even nuclear and chemical fuels [3]. The most widely used sources of power for satellites that do consume solar energy, are solar photovoltaic (PV) cells arranged into arrays that provide power during sun-on periods and also serve to recharge the battery packs during sun-off periods. The battery functions as a backup power source in the event of a solar eclipse. The main power line connects to the solar cells and battery packs and also feeds a number of DC–DC converters that are responsible for regulating and supplying intermediate voltages to the various satellite electronic devices. These include onboard computers, cameras, transceivers, amplifiers, magnetometers, magnetorquers, etc. Unlike earth-bound systems, satellites cannot plug into power outlets to satisfy their power demands. Therefore, mechanical and electronic switches in a power switching network are needed to manage satellite power consumption and distribution. A battery monitor is also incorporated in the system design to assess and facilitate sustainable power management [4].

In order to reduce solar array size, battery size, and overall costs, the EPS design is configured to minimize the required load power in standby mode. This is accomplished by providing power only to the components that are ON constantly. For instance, the CDHS receives power supply continuously whereas the communication security network and radio frequency (RF) sections receive power only when the satellite is communicating with the ground station [5]. Furthermore, the satellite EPS can supply conditioned power to the other systems with the help of the DC–DC converters as mentioned earlier. The converters themselves are synchronized to be switched on and off with the payload. This ON/OFF duty cycle conserves energy and decreases converter contribution to the satellite standby power mode.

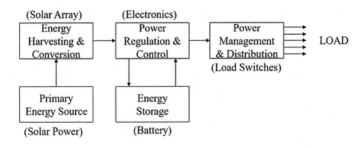

Fig. 1 Satellite EPS block diagram

3 Solar Arrays

3.1 Array Power Generation

A photovoltaic cell harvests energy by converting solar radiation into electrical energy. Silicon (Si) and Gallium Arsenide (GaAs) with efficiencies ranging from 14–19% to 18.4–28.8%, respectively, are the semiconductors typically used for satellite solar cells. Solar panels need to have a lot of surface area that can be pointed toward the sun as the spacecraft moves. More exposed surface area means more DC voltage can be created from the incident solar energy. However, since spacecrafts must be small, compact, and economical, this limits the amount of power that can be produced.

Any steady-state satellites EPS module assumes that the power collected during the daylight periods provides all the power required by the load during that period and to recharge the batteries from their use during the eclipse period. Based on this assumption, the figure below represents a simplified equivalent circuit for the power system.

The equations in the following discussion have been developed with the help of [6]. Based on our assumption, the solar array power (P_{sa}) should be sufficient to simultaneously charge the batteries (P_{bc}) and also supply the load during charging (P_{Lc}). P_d is the average power required by the load during daylight. Now, consider the EPS model to be predicated on a single orbit energy balance equation. This signifies that the amount of energy consumed from the batteries in a single orbit must be restored during the satellite's non-eclipse period on that same orbit. This can be mathematically represented by:

$$P_d \times T_d \leq P_{bc} \times T_e \tag{1}$$

where T_d and T_e refer the eclipse and daylight durations, respectively. The P_{sa} based on the EPS model is hence given by:

$$P_{sa} = \frac{\frac{P_d T_d}{\eta_2 \eta_c \eta_d \eta_3} + \frac{P_{Lc} T_e}{\eta_1}}{T_e} \tag{2}$$

The efficiencies in Eq. (2) depend on the structure of the EPS equivalent circuit such as the one shown in Fig. 2. The efficiencies $\eta_2, \eta_c, \eta_d, \eta_3$ can be clubbed together as the variable Xe representing the efficiency of the satellite EPS to supply the power from the array to the batteries for charging during daylight and from the batteries to the loads during eclipse. Similarly, η_1 and any other efficiencies depending on the satellite equivalent circuit can be collectively termed as X_d to represent the efficiency of the satellite EPS to supply the power from the array to the loads in daylight. The term PL_c, the supply to the load during charging, is essentially the average power required by the load during eclipse and can hence be termed as P_e for simplicity. Therefore, the solar array power capability is given by:

$$P_{sa} = \frac{\frac{P_d T_d}{X_d} + \frac{P_e T_e}{X_e}}{T_d} \tag{3}$$

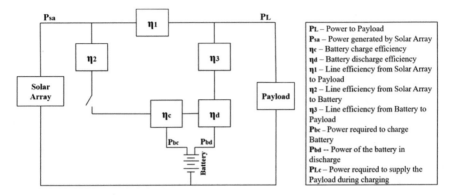

Fig. 2 A general simplified EPS equivalent circuit [6]

3.2 Array Performance Characteristics

A vital measure of the photovoltaic cell performance is its photoelectric energy conversion efficiency and its cost per watt capacity. These two parameters determine the competitive advantage of the PV cell over alternative power generation technologies [3].

Because a satellite revolves around the earth, its solar panels generate power intermittently. That is, the sun and eclipse phases alternate, and each panel can generate power only during their respective sun phase.

With increase in temperature, the PV cell open-circuit voltage (V_{oc}) decreases, whereas the short-circuit current (I_{sc}) increases, with an overall decrease in power. In fact, for Silicon cells, the power output decreases by 0.5% for every degree centigrade rise in operating temperature [3].

Array performance is also sensitive to the input solar irradiation. Its I–V characteristic shifts down at lower sun intensity with a slight reduction in voltage. However, the efficiency of the cell is indifferent to the solar radiation in the practical working range [3]. Another characteristic is the efficiency degradation of the satellite solar array. Due to ultraviolet degradation, fatigue (thermal cycling), material outgassing, and micrometeoroid loss, the efficiency of the solar arrays decreases over time [1].

Overall, the amount of power generated by all solar panels (Sup(t)) in a satellite at time 't' is a function of solar panel configuration at design time such as the number, the arrangement, and the type of solar panels (e.g., mono crystalline or polycrystalline panels) represented by n_S, the satellite's position and attitude $s(t)$, solar panel efficiency $e_S(t)$, and temperature $\theta(t)$. The function is as follows:

$$\text{Sup}(t) = F_{\text{sup}}(n_S, s(t), e_S(t), \theta(t)) \tag{4}$$

4 Battery System

Batteries are the most widely used energy storage technology. Internally, a battery converts chemical energy into electrical energy during discharge and the opposite during charge.

The internal construction of a typical cell includes positive and negative electrode plates with insulating separators submersed in a chemical electrolyte in between. Each cell stores electrochemical energy at a low electrical potential. The cell voltage depends solely on the electrochemistry, whereas the cell capacity (C) measured by Ampere-hours depends on physical size. The battery voltage rating is stated in terms of the average voltage during discharge. The higher the required voltage, higher is the number of cells required in series [3].

A battery is made up of multiple of electrochemical cells connected in series–parallel circuit combinations to obtain the necessary current and voltage value. Two battery assembly configurations can be proposed depending on the mission type and requirements: The S–P topology is characterized by serial strings of cells assembled in parallel. The P–S topology constitutes cells being connected in parallel as a module or cell package that defines the battery capacity and these modules are then serially connected to comply with the satellite voltage range [7] (Fig. 3).

4.1 Battery Performance and Properties

Rechargeable battery cells are known to possess unique, non-linear characteristics such as discharge efficiency. The more the discharge rate the lesser is its efficiency [1]. Like solar arrays, batteries too are affected by temperature and efficiency. As per the calendar fade effect, battery performance or efficiency at time 't', represented by $e_B(t)$, deteriorates over time regardless of whether it is used or not [1].

In both charge and discharge modes, a small percentage of energy is dissipated as heat. The battery charge status is represented using the term state of charge (SoC). The SoC at a given time is directly proportional to the available charge at that instant. The battery SoC affects the cell voltage, electrolyte freezing point and specific gravity [3].

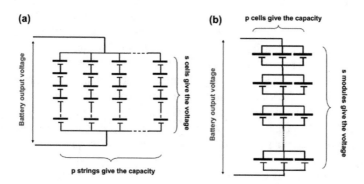

Fig. 3 Two satellite battery assembly configurations. **a** S–P topology; **b** P–S topology [7]

The SoC is usually expressed as a percentage and can be mathematically represented by:

$$SoC = \frac{\text{Ah capacity remaining in the battery}}{\text{Rated Ah capacity}} \quad (5)$$

The battery depth of discharge (DoD) indicating the percentage of the battery that has been discharged relative to the overall capacity of the battery is simply given by:

$$DoD = 1 - SoC \quad (6)$$

The three most commonly used batteries for spacecraft missions are Nickel–Cadmium (NiCad), Nickel–Hydrogen (NiH_2), and Lithium-ion (Li-ion) batteries. NiH_2 batteries can provide twice the specific energy than that of NiCad. Li-ion cells notably provide higher energy levels, lower thermal dissipation, and a longer cycle life at a lower weight and in smaller volumes than any NiCad or NiH_2 batteries [7]. Today, 98% of newly manufactured satellites are powered by Li-ion batteries.

Some of the parameters that affect battery selection are maximum load current, operating temperature range, charge cycles, shelf-life, specific energy, cost, time-averaged DoD, resistance to shock and vibration, and power management schemes [5].

4.2   Battery Overcharge

Controlling battery overcharge is of immense importance in order to maintain battery temperature and overall battery life. When a battery is completely charged, it reaches the maximum voltage beyond which all input power is converted into heat. For every electrochemistry, there is a well-defined relationship between voltage and temperature (V–T) at the end of charging [3]. This relationship can be used to determine when the battery is fully charged and then cut back to the trickle charge rate equal to the self-discharge rate. For Li-ion batteries, generally, failure to accommodate the limitations of the electrochemistry of a cell, with regard to trickle charging after reaching a fully charged state, can lead to overheating, release of gasses and, possibly to a fire or explosion [12].

To diminish any damage done to the battery as a result of overcharge, the EPS should be able to detect when the reconstitution of the active chemicals is complete and to stop the charging process before any damage is done while at all times maintaining the cell temperature within its safe limits [13]. Identifying this cut off point and terminating the charge are critical in preserving battery life. This can be done by setting a predetermined upper voltage limit, often called the termination voltage. As soon as termination voltage is reached, charging is discontinued.

A risk of overcharging the battery can still exist, either from errors in determining the cutoff point or from any battery damage. A resettable fuse can be used to disconnect the charger when danger signs appear. This additional safety precaution is especially important for high power batteries used in large satellites where the consequences of failure can be both serious and expensive [13]. Many Lithium-ion cells have built-in current interruption devices (CIDs) that help protect them from overcharging. The CID is activated when the cells build up excessive pressure that typically occurs when the cells are overcharged to voltages close to or above 5 V. Large cells consist of shutdown separators, vents, and a fusible link to the electrode as supplementary levels of protection [14].

5 Power System Design Process

5.1 Power Budget

In order to begin with the design process, we need to compile a detailed load power profile of each satellite component that must be powered. This is known as the power budget or power allocation. The power for the ON and OFF durations of all equipment, peak power during ON time, quiescent power during OFF time, duty ratio, and thermal dissipation are all listed to create a load power profile. Any parameters that may vary during sunlight, eclipse, during BOL, or EOL are accounted for.

Duty ratio D is expressed as a percentage of the ON time. The average power consumed by any equipment is the product of this duty ratio to its peak power. The average power is used to calculate the solar array and battery ratings.

The power budget for a generic small satellite has been created using Microsoft Excel in Fig. 4. Details regarding the mission have been provided in Table 1.

Considering a 30% design margin, the day power and eclipse power comes out to 12.26 W and 4.86 W respectively. These are calculated keeping in mind all individual

Spacecraft Power Budget	Voltage (V)	Power (W)	Day/Eclipse/ Both	Day Power (W)	Day Duty Cycle	Day Avg Power (W)	Eclipse Power (W)	Eclipse Duty Cycle	Eclipse Avg Power (W)	Orbit Avg Power(W)	Peak Power(W)
Payload		8.00		8.00		2.57	1.00		1.00	1.93	8.00
Camera (Active)	12	6.00	Day	6.00	0.01	0.06	0.00	0.00	0.00	0.04	
Camera (Idle)	12	0.00	Both	0.00	0.99	0.00	0.00	0.00	0.00	0.00	
Heater (Active)	12	1.00	Both	1.00	1.00	1.00	1.00	1.00	1.00	1.00	
Heater (Idle)	12	0.00	Both	0.00	0.80	0.00	0.00	0.40	0.00	0.00	
Lamp (Active)	12	1.00	Day	1.00	0.01	0.01	0.00	0.00	0.00	0.01	
Lamp (Idle)	12	0.00	Both	0.00	0.99	0.00	0.00	0.00	0.00	0.00	
		1.50	Both	1.50	1.00	1.50	0.00	1.00	0.00	1.50	
Secondary Camera		6.00		6.00		0.06	0.00		0.00	0.04	6.00
Secondary camera (Active)	12	6.00	Day	6.00	0.01	0.06	0.00	0.00	0.00	0.04	
Secondary camera (Idle)	12	0.00	Both	0.00	0.99	0.00	0.00	0.00	0.00	0.00	
ADCS Subsystem		0.51		0.51		0.41	0.41		0.31	0.37	0.51
ADCS Controller	3.3	0.4	Both	0.40	1.00	0.40	0.3	1	0.3	0.359	
Magnetorquers	5	0.1	Both	0.10	0.02	0.002	0.1	0.02	0.002	0.002	
Sensors	3.3	0.01	Both	0.01	1.00	0.01	0.01	1	0.01	0.01	
Communication Subsystem		28.28		28.28		1.70	28.28		1.50	1.62	22.00
Transmitter (Active)	12	10.00	Both	10.00	0.02	0.20	10.00	0.00	0.00	0.12	
Transmitter (Idle)	12	5.14	Both	5.14	0.00	0.00	5.14	0.00	0.00	0.00	
Transmitter (off)	0	0.00	Both	0.00	0.98	0.00	0.00	1.00	0.00	0.00	
Receiver (Active)	12	12.00	Both	12.00	0.13	1.50	12.00	0.13	1.50	1.50	
Receiver(Idle)	12	1.14	Both	1.14	0.00	0.00	1.14	0.00	0.00	0.00	
Receiver(off)	0	0.00	Both	0.00	0.87	0.00	0.00	0.87	0.00	0.00	
Processor Subsystem		1.00		1.00		0.32	1.00		0.30	0.31	1.00
C&DH (Active)	5 & 3.3	0.70	Both	0.70	0.04	0.03	0.70	0.01	0.01	0.02	
Memory Module (Idle)	5 & 3.3	0.30	Both	0.30	0.96	0.29	0.30	0.99	0.30	0.29	
Thermal Control (none)											
Total Power				37.79		5.06	30.69		3.12	4.27	37.51
Power Design Margin		30%		11.34		1.52	9.21		0.93	1.28	11.25
Subtotal Power				49.13		6.58	39.90		4.05	5.55	48.76
Electrical Power Subsystem						5.68			0.81	3.71	17.94
Battery Charging (Efficiency 90%)		3.31				3.64			0.00	2.17	6.83
Power Conversion and Regulation (80%)						2.04			0.81	1.54	11.12

	DAY	ECLIPSE	PEAK POWER
Spacecraft Power Required (W)	12.26	4.86	55.45
Spacecraft Power Available (W)	12.28	0.00	
Spacecraft Power Balance (W)	0.02	-4.86	

Fig. 4 Satellite power budget

Table 1 Satellite mission information

Initial orbital period	90.39	Mins
Percent of time in sunlight	59.49%	Percent
Daylight duration (T_d)	53.77	Mins
Eclipse duration (T_e)	36.62	Mins
Mission lifetime	0.5	Years

power demands of the satellite subsystems, including the battery charging circuitry and the energy used during power conversion and regulation. Now we can design the solar arrays and batteries necessary to fulfill these power demands.

5.2 Solar Array Design

The solar array area A_{sa} is calculated by dividing Eq. (3) by the amount of power developed by a solar cell per unit area. The amount of output power per unit area (P_o) is the product of the production efficiency of solar cells (η_{cell}) and the solar intensity (approximately 1358 W/m^2). P_o is presented as:

$$P_o = 1358 \times \eta_{cell} \left(\frac{\text{Watts}}{\text{m}^2}\right) \tag{7}$$

Variation in solar intensity is ±4% from the average. Hence, the value of P_o varies depending on the season and due to solar cell efficiency degradation over time.

The Beginning of Life (BOL) power per unit area is given by the equation:

$$P_{BOL} = P_o I_d \cos \theta \tag{8}$$

where I_d is the inherent degradation referring to temperature related losses, design and assembly losses, and even shadowing due to appendages. The value of I_d typically varies from 0.49 to 0.88. The incidence angle θ constantly varies making Eq. (8) an approximation.

L_d, the life degradation factor throughout the satellite lifetime is calculated as:

$$L_d = (1 - \text{degradation per year})^{\text{satellite life}} \tag{9}$$

The End of Life (EOL) power per unit area is given by the equation:

$$P_{EOL} = L_d P_{BOL} \tag{10}$$

Now, solar array area (A_{sa}) required to meet the power requirement of the satellite is represented by:

$$A_{sa} = \frac{P_{sa}}{P_{EOL}} \qquad (11)$$

Finally, to estimate the solar array mass M_{sa}, take power density to be 45 W/kg. Usually power density varies from 14 to 47 W/kg. M_{sa} is calculated as:

$$M_{sa} = \left(\frac{1}{45}\right) P_{sa} \qquad (12)$$

To determine the total array string length, we divide P_{sa} by P_{cell}, the power produced by a single solar cell.

$$N = \frac{P_{sa}}{P_{cell}} \qquad (13)$$

125×125 mm monocrystalline Silicon cells (Mono Cell 125×125 Series) have been considered for the satellite solar array. These cells have a maximum power capacity of 2.30 Watts each with an efficiency of 15.5% [9]. Based on the calculations conducted in Fig. 5, it is found that five Silicon cells would be required to constitute the solar array. The solar array would hence be able to produce a P_{sa} of 12.28 W. The solar array area is estimated to be 0.083 m^2 and the solar array mass is estimated to be 0.27 kg, assuming a specific performance of 45 W/kg.

5.3 Battery Design

The first step to determine the size of the battery is to calculate the number of cells (N). This is calculated as:

$$N = \frac{V_{bus}}{V_{cell}} \qquad (14)$$

The amount of energy the battery must be able to store is called the capacity C_{req}. This is expressed as follows:

Spacecraft Solar Array Sizing		
Mission Life Time	0.5	years
Maximum Eclipse	36.62	mins
Power Input	1358	Watts/m^2
Efficiency	15.5%	
Power Out (Po)	210.49	Watts/m^2
Inheret Degradation (Id)	0.77	
Power Degredation (3.75%/yr) (Ld)	0.98	

Silicon Cells of Dimensions	Mono Cell 125 x 125 Series	
Thickness	200	μm
Width	12.500	cm
Length	12.500	cm
Width	0.125	m
Length	0.125	m
Area	0.0156	m^2

Parameters		
Pcell	2.30	W
Icell	4.58	A
Vcell	0.503	V
Number of cells in Solar Array (N)	5	cells

Power Generated	0°	30°	60°
Power Generated (BOL) (W)	10.13	8.77	5.06
Power Generated (EOL) (W)	9.94	8.61	4.97

Design Parameters		
Psa (Average)	12.28	W
Solar Array Area (Asa)	0.083	m^2
Solar Array Mass (Msa) [Sp. Perf = 45W/kg]	0.27	kg

Fig. 5 Solar array sizing design calculations

$$C_{\text{req}} = P_{\text{e}} \times T_{\text{e}} [\text{Wh}] \tag{15}$$

The total capacity at EOL (C_{EOL}) of the battery is:

$$C_{\text{EOL}} = \frac{C_{\text{req}}}{\text{DOD} \times \eta_{\text{battery}}} [\text{Wh}] \tag{16}$$

Over the satellite lifetime (N_{years}), the batteries will degrade. Due to this degradation, it is clear that the EOL capacity will be considerably lower than the BOL capacity. This is accounted for by using the fading factor F_{fading} which typically has a value of 0.92%/year for Li-ion batteries [8]. The battery cell storage capacity at BOL C_{BOL} hence be given by:

$$C_{\text{BOL}} = \frac{C_{\text{EOL}}}{F_{\text{fading}} \times N_{\text{years}}} [\text{Wh}] \tag{17}$$

The final step is to estimate the total mass and volume of the batteries. This is done using empirical relations, using specific mass (m_{sp}) and energy density (e_{density}) [8]. The m_{sp} for Lithium-ion batteries typically ranges from 100 to 265 Wh/kg and the e_{density} from 250 to 693 Wh/L. The mass and volume of the battery are calculated as follows:

$$M_{\text{battery}} = \frac{C_{\text{BOL}}}{m_{\text{sp}}} [\text{kg}] \tag{18}$$

$$V_{\text{battery}} = \frac{C_{\text{BOL}}}{e_{\text{density}}} [\text{l}] \tag{19}$$

As mentioned earlier, there are several battery choices for use in satellites. For comparison, calculations have been performed for both Nickel–Cadmium and Lithium-ion batteries. The VRE Cs 1600 Series NiCad batteries with 1.6 Ah capacity and the

Battery Sizing Parameters	NiCad	Li-Ion	
Vbus - Bus Voltage	14.4	14.4	Volts
Depth of Discharge (Maximum)	35%	35%	percent
Creq - Stored Energy Requirement	2.97	2.97	Watt hours
CEOL - Capacity at End of Life	9.42	9.42	Watt hours
CBOL - Capacity at Beginning of Life	15.70	20.47	Watt hours

Cell Properties	VRE Cs 1600	VL 18650	Series
Vcell - Cell Voltage	1.2	3.7	Volts
Ccell - Cell Capacity	1.6	2.2	Amp hours
msp - Specific Mass	50	210	Wh/kg
edensity - Energy Density	120	400	Wh/L
CB - Cell Energy Capacity	1.92	8.14	Watt hours
Cells Per Battery Pack (N)	12	4	Cells
Total Battery Pack Capacity	23.04	33	Watt hours
DOD Typical	13%	9%	percent

Battery Charging	NiCad	Li-Ion	
Charge Energy	2.97	2.97	Watt hours
Orbit Day Length	0.90	0.90	hours
Charge Power	3.31	3.31	Watts

Battery Sizing Parameters		
Orbit Period	1.51	hours
Mission Life	4380	hours
Charge/Discharge Cycles	2907	cycles
Eclipse Time (Te)	0.61	hours
Eclipse Power Consumption (Pe)	4.86	Watts

Required Capacity Calculation		
DOD = Limit on Battery Depth of Discharge	30%	percent
Xe = Transmission Efficiency between battery and Load	0.9	
C = Required Total Battery Capacity (A-hr)	0.92	A-hr
Bus Line (V)	12	Volts
CB = Required Battery Storage Capacity (W-hrs)	10.99	W-hr

Summary - Batteries		
NiCad Total Capacity	23	Watt hours
Mass of NiCad Battery	0.3139	kg
Volume of NiCad Battery	0.1308	Litre
Li-Ion Total Capacity	33	Watt hours
Mass of Li-Ion Battery	0.0975	kg
Volume of Li-Ion Battery	0.0512	Litre

Fig. 6 Battery design calculations

VL 18650 Series Li-ion batteries with 2.2 Ah capacity were selected for the design process [10]. Calculated results for NiCad show a requirement of 12 cells that will together contribute to a capacity of 23 Wh, a mass of 0.314 kg, and a volume of 0.13 L. Similarly, the calculated results for Li-ion show a requirement of four cells that will together contribute to a capacity of 33 Wh, a mass of 0.098 kg, and a volume of 0.051 L. As shown in the results, the selected Li ion batteries can fulfill the same storage power requirements using a lower mass, volume, and capacity constraints, as compared to the NiCad batteries [11] (Fig. 6).

6   Conclusion

In this paper, an optimum design of a satellite power system has been presented. The main idea behind the design procedure is to enhance the satellite power during daylight and eclipse while profitably reducing the area, mass, volume, and budget constraints. The construction features and performance of solar arrays and batteries are discussed to understand the available technology prior to design. A systematic power budget is assembled for a generic nanosatellite to determine the individual subsystem and the overall satellite power demands during daylight and eclipse periods. As per the data collected from the power budget, the solar arrays and backup batteries are designed accordingly. The design considers Silicon solar cells for energy harvesting and conversion and Lithium-ion batteries for energy storage. As per numerical analysis, the Lithium-ion batteries outperform the Nickel–Cadmium batteries by a significant margin in the required storage capacity, total mass, and total volume.

References

1. Lee J, Kim E, Shin KG (2013) Design and management of satellite power systems. In: IEEE 34th real-time systems symposium (2013)
2. Tan BL, Tseng KJ (2003) Intelligent and reliable power supply system for small satellites. In: The 25th international telecommunications energy conference, 2003, INTELEC '03
3. Patel MR (2005) Spacecraft power systems, 1st edn. CRC Press, 2000 N.W. Corporate Blvd., Boca Raton, Florida 33431 (2005)
4. Akagi JM (2003) Power generation and distribution system design for the Leonidas Cubesat network. In: Undergraduate fellowship reports, pp 1–6. Hawai'i Space Grant Consortium, Honolulu (2006)
5. Salim A, Dakermanji G Small satellite power system design. Fairchild Space Company, Germantown, Maryland 20874
6. Melon CW (2009) Preliminary design, simulation, and test of the electrical power subsystem of the Tinyscope Nanosatellite. Naval Postgraduate School
7. Borthomieu Y (2014) Lithium-ion batteries, 1st edn. Elsevier
8. How to (‖): Size a satellite battery, https://www.valispace.com/how-to-size-a-satellite-battery/. Last accessed 19 October 2019
9. https://www.enfsolar.com/pv/celldatasheet/1825?utm_source=ENF&utm_medium=cell_list&utm_campaign=enquiry_product_directory&utm_content=14452. Last accessed 15 September 2019

10. Saft rechargeable battery systems, http://www.powerpack.com.sg/theme_clean/static/src/img/document/Saft_RBS_Handbook.pdf. Last accessed 20 October 2019
11. Lotfy A, Anis WR, Atalla MA, Halim JVM, Abouelatta M (2017) Design an optimum PV system for the satellite technology using high efficiency solar cells. Int J Comput Appl (0975–8887) 168(3) (2017)
12. Battery protection methods, https://www.mpoweruk.com/protection.htm. Last accessed 27 September 2019
13. Battery chargers and charging methods, https://www.mpoweruk.com/chargers.htm. Last accessed 27 September 2019
14. McKissock B, Loyselle P, Vogel E (2008) Guidelines on lithium-ion battery use in space applications. In: NASA aerospace flight battery program, Part 1 2 (2008)
15. Battery Management Systems (BMS), https://www.mpoweruk.com/bms.htm#smartbats. Last accessed 27 September 2019
16. Mahdi MC, Jaafer JS, Shehab AAR (2014) Design and implementation of an effective electrical power system for nanosatellite. Int J Sci Eng Res 5(5)
17. Mohammed HI, Ahmed HN, Eliwa AE, Sabry W, Mostafa R Innovative design of a cube satellite power distribution and control subsystem. Online J Electron Electr Eng (OJEEE) 1(1)
18. Aoudeche A, Zhao X, Kerrouche KD (2018) Design of a high-performance electrical power system for an earth observation nanosatellite. Association of Computing Machinery (2018)
19. Brown CD (2002) Elements of spacecraft design. American Institute of Aeronautics and Astronautics. Inc, 1801 Alexander Bell Drive, Reston, VA 20191–4344 (2002)

Single OTRA-Based Implementation of Second-Order Band Reject Filter (Three Configurations)

Mourina Ghosh[1(✉)], Subhasish Banerjee[2], Shekhar Suman Borah[1], and Pulak Mondal[3]

[1] Department of ECE, IIIT Guwahati, Guwahati, India
mourina_06@rediffmail.com
[2] Department of ECE, MCKV Institute of Engineering, Howrah, India
[3] Institute of Radio Physics and Electronics, University of Calcutta, Kolkata, India

Abstract. This paper presents a way to realize three configurations of second-order band reject filters (BRFs). Only single operational transresistance amplifier (OTRA) and few passive components have been used in this work. CMOS realization of OTRA using 180 nm model has been implemented in PSPICE for simulation works. Simulated filter characteristics matches very well with the theoretical analysis for all the three configurations. Monte Carlo analysis and non-ideal analysis have been done. Analog design environment (ADE) tool of cadence virtuoso has been used to perform the layout of the proposed configurations.

Keywords: Band reject filter · CMOS · Layout · OTRA · PSPICE

1 Introduction

The band reject filters are used to remove a selected band of frequency components from the signal to obtain the desired output. Many works which are reported in literatures showing the realization of different filters using various building blocks discussed below. Implementation of MOS-C voltage-mode low-pass filter (LPF), high-pass filter (HPF), band pass filter (BPF), notch filter (NF) and all pass filter (APF) has been described in [1] with OTRA as an active block. A single OTRA-based voltage-mode APF and NF have been implemented in [2]. First-order APF operation has been demonstrated in [3] using OTRA. Another OTRA-based first-order APF operation has been realized in [4]. In [5], current differencing buffered amplifier (CDBA)-based APF and BPF operations have been demonstrated. A method for controlling the quality factor has also been described in this paper. Realization of voltage-mode LPF, HPF and BPF using OTRA has been described in [6]. Another CDBA-based LPF, BPF, HPF, BRF and APF operations have been established in [7]. So, the implementation of more than one configuration of band reject filter using OTRA in a single work is limited. This work describes a method to realize three configurations of OTRA-based second-order BRF operations. Subsequent paragraphs, however, are indented.

N. Goel et al. (eds.), *Modelling, Simulation and Intelligent Computing*,
Lecture Notes in Electrical Engineering 659,
https://doi.org/10.1007/978-981-15-4775-1_34

2   Circuit Description

To overcome the limitations of conventional op-amps (limited gain–bandwidth product and limited slew rate), the OTRA has drawn the attention of many researchers because of the advantages offered by the current mode circuits. The OTRA is a three-terminal analog building block represented in Fig. 1. The CMOS realization of OTRA is shown in Fig. 2 [8]. The port relationship is characterized by the following equation as,

$$V_p = V_n = 0, \ V_0 = I_p R_m - I_n R_m \tag{1}$$

where the voltage and current at input terminal p are represented by V_p and I_p, the voltage and current at input terminal n are denoted by V_n and I_n and V_0 is the output voltage at terminal z, respectively. R_m, the transresistance gain is infinity in ideal situation which forces the currents at the input terminals to be equal. So, OTRA has to be used in a negative feedback configuration. The input terminals of OTRA are virtually grounded and the difference of the two input currents multiplied by the transresistance gain produces the output voltage. OTRA also provides high bandwidth independent of closed-loop gain [9].

The proposed configuration of BRFs along with the component values has been shown in Fig. 3. Three different configurations of BRF can be realized by varying the status of the switches SW1, SW2 and SW3 which is given in Table 1.

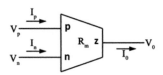

Fig. 1 Block diagram of OTRA

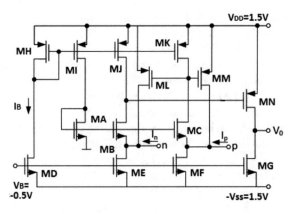

Fig. 2 CMOS implementation of OTRA

Table 1 Status of the switches

Conf.	Status of the switches		
	SW1	SW2	SW3
1st	Open	Close	Open
2nd	Close	Open	Open
3rd	Close	Open	Close

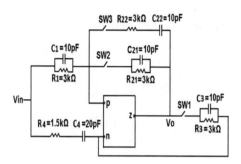

Fig. 3 BRFs: proposed configurations

For 1st configuration of BRF the transfer function can be derived as,

$$\left(\frac{V_0}{V_{in}}\right)_1 = \frac{R_{21}[s^2 C_1 C_4 R_1 R_4 + s(C_1 R_1 + C_4 R_4 - C_4 R_1) + 1]}{R_1[s^2 C_{21} C_4 R_{21} R_4 + s(C_{21} R_{21} + C_4 R_4) + 1]} \tag{2}$$

Centre frequency $(f_0)_{BRF1}$ and quality factor $(Q)_{BRF1}$ for 1st configuration can be derived from Eq. (2) as,

$$(f_0)_{BRF1} = \frac{1}{2\pi} \sqrt{\frac{1}{C_{21} C_4 R_{21} R_4}} \tag{3}$$

$$(Q)_{BRF1} = \frac{\sqrt{C_{21} C_4 R_{21} R_4}}{C_{21} R_{21} + C_4 R_4} \tag{4}$$

The transfer function of the BRF for 2nd configuration can be represented as,

$$\left(\frac{V_0}{V_{in}}\right)_2 = \frac{R_3[s^2 C_1 C_4 R_1 R_4 + s(C_1 R_1 + C_4 R_4 - C_4 R_1) + 1]}{R_1[s^2 C_3 C_4 R_3 R_4 + s(C_3 R_3 + C_4 R_4) + 1]} \tag{5}$$

From the above expression, the centre frequency $(f_0)_{BRF2}$ and quality factor $(Q)_{BRF2}$ for 2nd configuration is given by,

$$(f_0)_{BRF2} = \frac{1}{2\pi} \sqrt{\frac{1}{C_3 C_4 R_3 R_4}} \tag{6}$$

$$(Q)_{BRF2} = \frac{\sqrt{C_3 C_4 R_3 R_4}}{C_3 R_3 + R_4 C_4} \tag{7}$$

Similarly for 3rd configuration the transfer function of the BRF can be given by,

$$\left(\frac{V_0}{V_{in}}\right)_3 = \frac{s^2 C_1 C_4 R_1 R_4 + s(C_1 R_1 + C_4 R_4 - C_4 R_1) + 1}{s^2 C_{22} C_3 R_{22} R_3 + s(C_{22} R_{22} + C_3 R_3 - C_{22} R_3) + 1} \tag{8}$$

Again for 3rd configuration, from Eq. (8), the centre frequency $(f_0)_{BRF3}$ and quality factor $(Q)_{BRF3}$ is given by,

$$(f_0)_{BRF3} = \frac{1}{2\pi} \sqrt{\frac{1}{C_{22} C_3 R_{22} R_3}} \tag{9}$$

$$(Q)_{BRF3} = \frac{\sqrt{C_{22} C_3 R_{22} R_3}}{C_{22} R_{22} + C_3 R_3 - C_{22} R_3} \tag{10}$$

3   Simulation Outputs

The CMOS realization of OTRA as shown in Fig. 2 using 180 nm technology has been implemented for simulation work. The aspect ratio of the transistors as given in [9] has been represented in Table 2. From Eqs. (3), (6) and (9), the theoretical centre frequency of BRF for all configurations can be calculated as 5.31 MHz. Figure 4a–c, respectively, gives the theoretical and simulated frequency response characteristics for all three configurations along with the corresponding phase plots. From the simulated characteristics, the centre frequency of operation has been found as 5.42 MHz for all configurations which is in close accord with the calculated value. Also, the simulated result matches well with the theoretical characteristics. From Eqs. (4), (7) and (10), the quality factor can be calculated as 0.5 for 1st and 2nd configuration and 1 for the 3rd configuration. Monte Carlo analysis has been performed for 100 samples with 5% variation in the values of all passive components and the result is shown in Fig. 5.

4   Layout

To validate the compatibility of the work with integrated circuit applications, layout of the proposed circuits has been performed in ADE tool of cadence virtuoso. Figure 6a represents the layout of the 1st configuration which consumes an effective area of 7022.196 μm^2 (91.15 $\mu m \times$ 77.04 μm). The values of the passive components chosen are $R_1 = 3$ kΩ, $C_1 = 1$ pF, $R_{21} = 3$ kΩ, $C_{21} = 2$ pF, $R_4 = 1$ kΩ and $C_4 = 5$ pF.

Similarly for layout of the 2nd configuration, the chosen passive components values are $R_1 = 500$ Ω, $C_1 = 1$ pF, $R_3 = 500$ Ω, $C_3 = 3$ pF, $R_4 = 1$ kΩ and $C_4 = 3$ pF. Figure 6b gives the layout of the proposed BRF for 2nd configuration which consumes an

Table 2 Aspect ratios

Transistor No.	W/L (μm)
MA, MB, MC	36/0.9
MD	3.6/0.9
ME, MF	10.8/0.9
MG	3.6/0.9
MH, MI, MJ, MK	18/0.9
ML, MM	36/0.9
MN	18/0.18

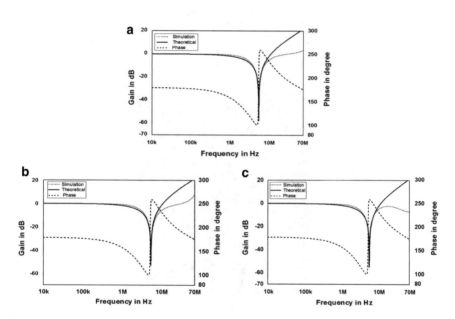

Fig. 4 Characteristics of BRF for; **a** 1st configuration for **b** 2nd configuration, **c** 3rd configuration

effective area of 7663.7214 μm² (92.49 μm × 82.86 μm). Again in Fig. 6c, the layout of the proposed BRF for 3rd configuration is achieved by choosing the passive components of $R_1 = 1$ kΩ, $C_1 = 3$ pF, $R_{22} = 1$ kΩ, $C_{22} = 3$ pF, $R_3 = 1$ kΩ, $C_3 = 3$ pF, $R_4 = 1$ kΩ and $C_4 = 1.2$ pF, which consumes an effective area of 9801.8277.196 μm² (109.31 μm × 89.67 μm). With the above value of passive components for layout, the centre frequency for 1st configuration is 29.05 MHz, for 2nd configuration is 75.02 MHz and for 3rd configuration is 53.05 MHz, respectively. Post-layout results are shown in Fig. 7a–c respectively. The centre frequency obtained from post-layout simulations is 22.18 MHz for 1st configuration, 67.76 MHz for 2nd configuration and 44.72 MHz for 3rd configuration respectively. It is examined that all results are in close matching.

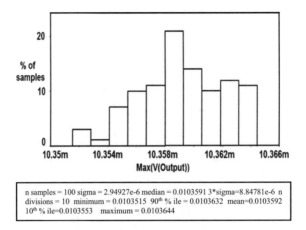

n samples = 100 sigma = 2.94927e-6 median = 0.0103591 3*sigma=8.84781e-6 n
divisions = 10 minimum = 0.0103515 90th % ile = 0.0103632 mean=0.0103592
10th % ile=0.0103553 maximum = 0.0103644

Fig. 5 Monte Carlo analysis of BRF configurations showing the variation of quality factor

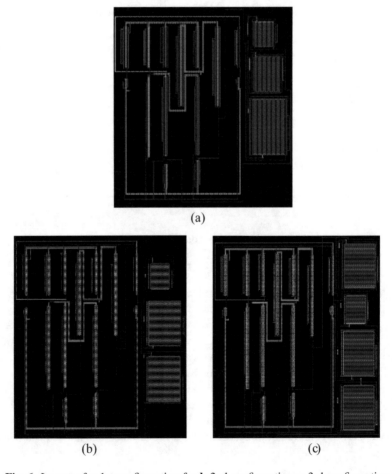

Fig. 6 Layout of **a** 1st configuration for **b** 2nd configuration, **c** 3rd configuration

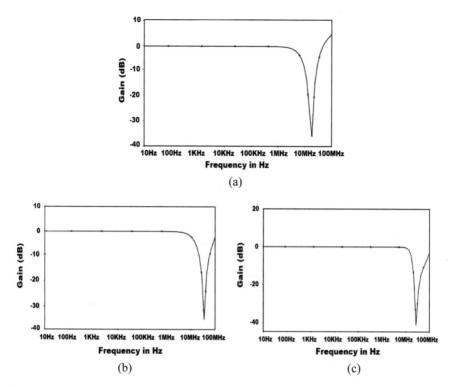

Fig. 7 Post-layout output for **a** 1st configuration for **b** 2nd configuration, **c** 3rd configuration

5   Non-ideality Analysis

The transresistance gain of OTRA actually has a finite value, although considered as infinite quantity in ideal case [9]. For a single-pole model of R_m we get,

$$R_m(s) = \frac{R_0}{1 + \frac{s}{w_0}} = \frac{R_0 w_0}{s + w_0} = \frac{1}{\frac{s}{R_0 w_0} + \frac{1}{R_0}} \tag{11}$$

In Eq. (11), w_0 and R_0, respectively, give the pole angular frequency and dc transresistance gain of OTRA. For filter operations, considering large frequency values and by putting $C_p = 1/(R_0 w_0)$ Eq. (11) reduces to,

$$R_m(s) \approx \frac{1}{sC_p} \tag{12}$$

Taking this into account, for 1st configuration the non-ideal transfer function is given by,

$$\left(\frac{V_0}{V_{in}}\right)_1 = \frac{\frac{sC_4}{1+sC_4R_4} - \frac{1}{R_1} - sC_1}{\frac{1}{R_{21}} + sC_{21} - sC_p} \tag{13}$$

By similar analysis, the non-ideal transfer function for 2nd and 3rd configuration can be derived as,

$$\left(\frac{V_0}{V_{in}}\right)_2 = \frac{\frac{1}{R_1} + sC_1 - \frac{sC_4}{1+sC_4R_4}}{\frac{1}{R_3} + sC_3 + sC_p} \tag{14}$$

$$\left(\frac{V_0}{V_{in}}\right)_3 = \frac{\frac{1}{R_1} + sC_1 - \frac{sC_4}{1+sC_4R_4}}{\frac{1}{R_3} + sC_3 - \frac{sC_{22}}{1+sC_{22}R_{22}} + sC_p} \tag{15}$$

Inspection of Eq. (13) reveals that the effect of C_p can be compensated by increasing the capacitance value C_{21} by an amount of C_p. Similarly from Eqs. (14) and (15), it can be seen that the effect of nonlinearities can be absorbed within the circuit itself by decreasing the value of C_3 by an amount of C_p. Thus the circuit provides complete self-compensation and there is no need to connect external capacitors for compensation.

6   Comparative Study

A comparative study of the proposed configuration with the previously published works [1–7] is shown Table 3.

It is observed that although topologies [1, 2] and [7] have implemented BRF, but in all cases only one configuration has been reported, whereas in the proposed work, three configurations of BRF are shown. Moreover, topology [1], topology [5] and topology [7] use five active devices, three active devices and two active devices, respectively, but the proposed configuration uses only one active device. Moreover, the proposed work uses comparable no of passive components as in others.

Table 3 Comparison table

Ref. No.	Active block and No.	R count	C count	Type of filter	Order
1	OTRA-5	12	2	LPF, BPF, HPF, BRF, NF	2nd
2	OTRA-1	2–3	2–3	APF, NF	1st 2nd
3	OTRA-1	3–4	1–2	APF	1st 2nd
4	OTRA-1	2	1	APF	1st
5	CDBA-3	5	3	APF, BPF	1st
6	OTRA-1	4	2	LPF, HPF, BPF	2nd
7	CDBA-2	4	2	LPF, HPF, BPF, BRF, APF	2nd
Proposed	OTRA-1	3–4	3–4	BRF(3 configurations)	2nd

7 Conclusion

This work proposes three novel realization of second-order BRF using a single OTRA and few passive components (resistors and capacitors). Some significant features of this work are pointed out in this section. (i) Without modifying the entire circuit, three BRF functions can be realized by altering the condition of the switches, (ii) Post-layout outputs confirm the viability of the work, (iii) frequency response characteristics (both simulated and theoretical) are in close agreement, (iv) centre frequency of operation obtained from simulated characteristics matches very well with the calculated values, (v) hardware outputs authenticate the feasibility of the proposed configuration.

References

1. Pandey R, Pandey N, Paul SK, Singh A, Sriram B, Trivedi K (2012) Voltage mode OTRA MOS-C single input multi output biquadratic universal filter. Theor Appl Electr Eng 10 (5):337–344
2. Kilnic S, Cam U (2005) Cascadable allpass and notch filters employing single operational transresistance amplifier. Comput Electr Eng 31(6):391–401
3. Cakir C, Cam U, Cicekoglu O (2005) Novel allpass filter configuration employing single OTRA. IEEE Trans Circ Syst II Exp Briefs 52(3):122–125
4. Cam U, Cakir C, Cicekoglu O (2004) Novel transimpedance type first-order all-pass filter using single OTRA. Int J Electron Commun 58(4):296–298
5. Toker A, Ozoguz S, Cicekoglu O, Acar C (2000) Current-mode all-pass filters using current differencing buffered amplifier and a new high-bandpass filter configuration. IEEE Trans Circuits Syst II Analog Digital Signal Process 47(9):949–954
6. Pandey R, Pandey N, Paul SK, Singh M, Jain M (2012) Voltage mode single OTRA based biquadratic filters. In: Proceedings of third international conference on computer and communication technology, pp 63–66
7. Tangsrirat W, Pukkalaun T, Surakampontorn W (2008) CDBA-based universal biquad filter and quadrature oscillator. Act Passive Electron Compon 2008(1):1–6
8. Singh AK, Gupta A, Senani R (2017) OTRA-based multi-function inverse filter configuration. Theor Appl Electr Eng 15(5):846–856
9. Komanapalli G, Pandey R, Pandey N (2019) New sinusoidal oscillator configurations using operational transresistance amplifier. Int J Circuit Theory Appl 47(5):1–20

Novel Distance-Based Subcarrier Number Estimation Method for OFDM System

J. Tarun Kumar and V. S. Kumar[(✉)]

Dept of ECE, SR Engineering College, Warangal 506371, India
{tarunjuluru, kumar.s.vngl}@gmail.com

Abstract. Aiming at the problem of orthogonal frequency division multiplexing (OFDM) signal subcarrier number estimation, a subcarrier number estimation method based on Novel Test (NT) distance is proposed using the Gaussian nature of OFDM signal. The NT distance output at detection-end DFT module is smallest when DFT points match the transmitter. Theoretical analysis and simulation results show that this method can distinguish Gaussian distribution from non-Gaussian distribution and correctly estimate the number of subcarriers of OFDM signal.

Keywords: Novel test (NT) · Empirical distribution function (EDF) · Subcarrier number estimation · Blind estimation · Orthogonal frequency division multiplexing (OFDM)

1 Introduction

Orthogonal frequency division multiplexing (OFDM) allocates high-speed data streams to orthogonal subcarrier for transmission, effectively reducing the symbol rate of each path. In actual communication, OFDM can select a subchannel with better conditions according to the channel situation, instead of using [1, 9] per-path carrier. In OFDM blind parameter estimation, modulation recognition is carried out first, and it is confirmed that parameters are estimated only after OFDM modulation, such as carrier frequency, symbol width, cyclic prefix length, subcarrier number, and so on. Among them, estimation of the number of subcarriers is an important item. At present, it is mostly the length estimation and cyclic prefix estimation of IFFT when OFDM modulation is produced, instead of the number of subcarrier actually used by OFDM. If the number of subcarrier is further estimated, subcarrier interval can be estimated.

Small research has been done on subcarrier estimation. In the literature [2], the inverted spectrum is introduced into the estimation of OFDM subcarrier number, but this method does not perform well under the condition of low signal-to-noise ratio. In the literature [3], a method is proposed to estimate using high-order cyclic cumulates. These methods are essentially high-order statistics, resulting in a large amount of computation, which is not conducive to practical use. In the literature [4], OFDM signal subcarrier number blind recognition algorithm based on AWGN channel is proposed. The algorithm utilizes the characteristics of strong normality of baseband OFDM modulation signal when FFT points mismatch and uses Gaussian detection algorithm to estimate the number of OFDM signal subcarrier. The Gaussian test of samples in

© Springer Nature Singapore Pte Ltd. 2020
N. Goel et al. (eds.), *Modelling, Simulation and Intelligent Computing*,
Lecture Notes in Electrical Engineering 659,
https://doi.org/10.1007/978-981-15-4775-1_35

statistics has been widely studied. In this paper, based on the practical application, using the asymptotic Gaussianity of OFDM signals, the Novel Test (NT) method in mathematical statistics is introduced to realize the fast identification of a number of OFDM signal subcarriers.

2   Empirical Distribution Function and Novel Test

2.1   Empirical Distribution Function

Let $(x_1, x_2, \ldots, x_n)$ be a set of sample observations of the population X, and arrange them in order of magnitude $x_{(1)} \leq x_{(2)} \leq \ldots \leq x_{(n)}, x$ is any real number and its weighing function is:

$$F_{(n)}(x) = \begin{cases} 0 & x < x_{(1)} \\ \frac{k}{n} & x_{(k)} \leq x < x_{(k+1)} \\ 1 & x \geq x_{(n)} \end{cases} \tag{1}$$

It is an empirical distribution function (EDF). The graph of the empirical distribution function is a step curve. If the observed value is not repeated, each hop of the step is $1/n$; if there is a repetition, it is stepped up by a multiple of $1/n$. For any real number x, the value of $F_{(n)}(x)$ is equal to the frequency of the sample observation $(x_1, x_2, \ldots, x_n)$ that does not exceed x. From the relationship between frequency and probability, $F_{(n)}(x)$ can be used as an approximation of the distribution function $F(x)$ of the population X. As n increases, the degree of approximation is better. For the empirical distribution function $F_{(n)}(x)$, the following result is proved: for any real number x, when $n \to \infty$, then $F_{(n)}(x)$ converges to the distribution function $F(x)$ with probability 1, i.e.,

$$P\left\{ \lim_{n \to \infty} \underbrace{\sup}_{\text{all} x} (|F_n(x) - F(x)|) = 0 \right\} = 1 \tag{2}$$

The empirical distribution function is defined by Eq. (1). Figure 1 depicts an EDF with a sample size of 100 for the signal sample sequence to be tested. The purpose is to test whether the sample is from a standard normal population, and the distribution function (CDF) of $N(0, 1)$ is also depicted in Fig. 1. In short, the object of the EDF test is the maximum distance between the two curves in Fig. 1. If the maximum distance of the signal sampling sequence EDF from the ideal normal distribution is small, it can be considered as a Gaussian distribution.

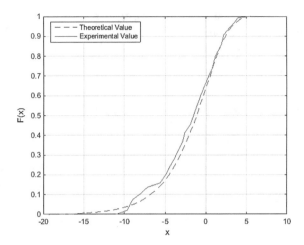

Fig. 1 Empirical distribution function and theoretical distribution function

There are several options for $F_{(n)}(x)$, such as $F_{1(n)}(x)$, $F_{2(n)}(x)$ [5, 6], and this article uses the following $F_{(n)}(x)$:

$$F_n(x) = \begin{cases} \frac{1}{n+2} & x < x_{(1)} \\ \frac{k+1}{n+2} & x_{(k)} \le x \le x_{(k+1)} \\ \frac{n+1}{n+2} & x \ge x_{(n)} \end{cases} \qquad (3)$$

The NT is based on EDF and is used to determine if a sample is from a population of a particular distribution [7]. Let $(x_1, x_2, \ldots, x_n)$ be a sample from the total X sample size n, sort them in ascending order, and form order statistics $x_{(1)} \le x_{(2)} \le \ldots \le x_{(n)}$. Then, construct the NT statistic (NT distance) as:

$$D_n = \underbrace{\max}_{\text{all}x} (|F_n(x_i) - F(x_i, \theta)|) \qquad (4)$$

where $F_n(x_i)$, the empirical distribution, is a function of the signal sample; $F(x_i, \theta)$ is theoretical distribution of the parameter vector θ estimated by the signal sample. The normal distribution is taken as an example $\widehat{\theta} = (\widehat{u}, \widehat{\sigma})$, which is the sample mean and standard deviation. The null hypothesis is accepted or rejected at significance level α by calculating the maximum distance between $F_n(x_i)$ and $F(x_i, \theta)$ and then comparing it with normal distribution threshold. If the NT statistic is smaller than the critical value, accept the null hypothesis that the sample is from a normal distribution population [8–14].

3   OFDM Subcarrier Number Estimation Model

3.1   Analysis of OFDM Received Signals with Unknown DFT Points

Regardless of Gaussian white noise, it is assumed that the IDFT conversion point of the transmitting end is N, the DFT point of the DFT module of the detecting end is MN, and M is a positive number. The OFDM transmit signal expression is:

$$y_n^m = \frac{1}{\sqrt{N}} \sum_{i=0}^{N-1} d_i^m \exp\left(j\frac{2\pi in}{N}\right), \quad i = 0, 1, \ldots, N-1 \tag{5}$$

where y_n^m is the nth IDFT output of the mth OFDM symbol; d_i^m is the ith bit of mth baseband symbol. The signal processed by the data stream through DFT module of the detection end is:

$$
\begin{aligned}
Y_i &= \frac{1}{\sqrt{MN}} \sum_{m=0}^{M-1} \sum_{i=0}^{N-1} y_n^m \exp\left(-j\frac{2\pi i}{MN}(mN+n)\right) \\
&= \frac{1}{\sqrt{MN}} \sum_{m=0}^{M-1} \sum_{l=0}^{N-1} d_l^m \exp\left(-j\frac{2\pi im}{M}\right) \cdot \sum_{n=0}^{N-1} \exp\left(j\frac{2\pi n}{N}\left(l - \frac{i}{M}\right)\right)
\end{aligned}
\tag{6}
$$

The description of the derivation (6) is that if the transformation point of the receiver is assumed to be greater than the transformation points N of the transmitter, the receiving end of the DFT transformation will have more than one OFDM symbol, as shown in Fig. 2. If $\frac{i}{M}$ is a positive integer and $l = \frac{i}{M}$, Eq. (6) is further simplified:

$$Y_i = \frac{1}{\sqrt{M}} \sum_{m=0}^{M-1} \sum_{i=0}^{N-1} d_l^m \exp\left(-j\frac{2\pi im}{M}\right)\delta\left(l - \frac{i}{M}\right) = \frac{1}{\sqrt{M}} \sum_{m=0}^{M-1} d_{\frac{i}{M}}^m \tag{7}$$

where $\delta(\cdot)$ is a unit sample function. Equation (7) can be understood as follows. If i is an integer multiple of M, then the DFT output of MN point is only the sum of M binary symbols. In general, the multiple M is not too large, and Y_i will exhibit non-Gaussian. If $\frac{i}{M}$ is not an integer, i.e.,

$$\sum_{n=0}^{N-1} \exp\left(j\frac{2\pi n}{N}\left(l - \frac{i}{M}\right)\right) = \frac{1 - \exp\left(j2\pi\left(l - \frac{i}{M}\right)\right)}{1 - \exp\left(j\frac{2\pi}{N}\left(l - \frac{i}{M}\right)\right)} \tag{8}$$

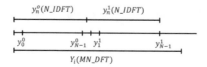

Fig. 2 Schematic diagram of DFT transformation at the receiving end

Substituting Eq. (8) into Eq. (6) gives:

$$Y_i = \frac{1}{N\sqrt{M}} \sum_{n=0}^{N-1} \frac{1 - \exp\left(j2\pi\left(l - \frac{i}{M}\right)\right) \sin\left(\pi\left(l - \frac{i}{M}\right)\right)}{1 - \exp\left(j\frac{2\pi}{N}\left(l - \frac{i}{M}\right)\right) \sin\left(\pi\left(l - \frac{i}{M}\right)\right)} \cdot \left[\sum_{i=0}^{M-1} d_l^m \exp\left(-j\frac{2\pi im}{M}\right)\right] \quad (9)$$

Therefore, if i is a non-integer multiple of M, then the DFT output of MN point is that each point includes the sum of code bits between code bits and OFDM symbols within each OFDM symbol, i.e., the sum of multiple random variables, and Y_i will show obvious Gaussian nature.

3.2   Subcarrier Estimation Process Based on NT

According to the above analysis, the OFDM signal can be subjected to DFT modules with different transform points to obtain DFT outputs with different transform points, and then NT check is performed to select the largest value, and the maximum number of transform points corresponding to the value is OFDM. The number of subcarriers with the rapid development of hardware technology, the speed of DFT computing is no longer a problem. Moreover, the number of existing OFDM signal subcarriers is basically a number of powers of two, then the subcarrier number set $(N = 2^K, K > 1, k \in Z)$ can be constructed, so that a few special points (such as 16, 32, 64, 128, 256, etc.) and several points in the vicinity are tested to further speed up the identification process. Taking the 512 subcarrier OFDM signal as an example, the hypothesis test model is established as follows:

H_0: The OFDM signal subcarrier number is 512.

H_1: The OFDM signal subcarrier number is not 512. Inductive rapid estimation of OFDM subcarrier number processing steps is:

Step 1 DFT is performed on the input OFDM signal y_n^m and the real part (or imaginary part) of the output (Y_i) is obtained, and N samples $(x_1, x_2, \ldots, x_n)$ are obtained.

Step 2 Estimate the parameter vector $\widehat{\theta} = (\widehat{u}, \widehat{\sigma})$ by means of maximum likelihood estimation (MLE) or moment estimation to obtain a normal distribution function $F(x_i, \theta)$.

Step 3 The real part (or imaginary part) of N samples from large to small, composing order statistics, and calculating the empirical distribution function $F_n(x_i)$.

Step 4 Traverse the real part (or imaginary part) of the sample, and select the maximum value of absolute value D_n of subtraction between the two in Step 2 and Step 3 as the NT distance under the N value.

Step 5 Change the value N and repeat Step 1–Step 4 to get the NT distance under different N. The largest one corresponds to N.

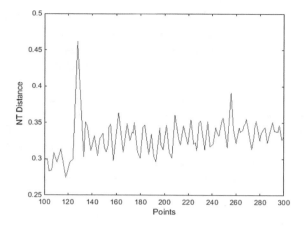

Fig. 3 NT distance under different points

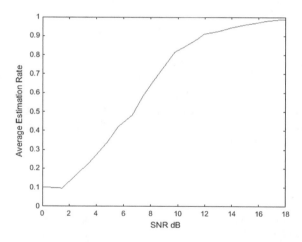

Fig. 4 Relationship between average correct estimation rate and SNR in dB

4 Numerical Analysis

4.1 NT Distance with Different DFT Points

Simulation conditions: The number of OFDM signal subcarrier is 512, and the signal-to-noise ratio of Gaussian white noise channel in the transmission channel is more than 10 dB. The DFT operation of the point 300–500 is obtained, and the output sequence of each transformation point is calculated, and then the maximum distance between the empirical distribution function and the theoretical distribution function of the corresponding sample points is calculated, i.e., NT distance. As shown in Fig. 3, NT distance has a maximum value of $N = 512$, which is consistent with the theoretical analysis.

The simulation conditions are unchanged and 10^{12} Monte Carlo simulations are performed under different SNR conditions. If the maximum value of the simulation is 512, then this simulation is correct to determine the average correct estimation rate for this method. It can be seen from Fig. 4 that the OFDM subcarrier number estimation method based on the maximum NT distance has a probability of correctly estimating 0.9 when the signal-to-noise ratio is 11.5 dB, which is effective.

5 Conclusion

Blind parameter estimation of OFDM signals is becoming increasingly important with the widespread use of OFDM techniques in the field of radio spectrum management, communication measures, and cognitive radio. The estimation of the number of subcarriers is an important part of parameter estimation. Many studies have focused on OFDM symbol width and cyclic prefix length estimates with less carrier estimates. In this paper, the method of statistics is introduced into the OFDM signal subcarrier estimation. Through theoretical derivation, the feasibility of estimating the carrier number based on the maximum NT distance is proved. Finally, the simulation results show that the proposed method can effectively estimate the number of OFDM subcarriers. Introducing other methods of normality testing into subcarrier estimation is a problem to be studied in the future research problem.

References

1. Chen Shaoping, Yao Tianren (2004) Intercarrier interference suppression and channel estimation for OFDM systems in time-varying frequency-selective fading channels. IEEE Trans Consum Electron 50(2):429–435
2. Liedtke F, Ulrike A (2008) Evaluation of features for the automatic recognition of OFDM signals in monitoring or cognitive receivers. J Telecommun Inf Technol: 30–36
3. Zheng G, Zhao W, and Ming L (2008) Blind estimation of OFDM Sub-carrier frequencies based on the high-order cyclic cumulants. J Electron Inf Technol 2
4. Shi M, Yeheskel BN, Wei S (2007) Blind OFDM systems parameters estimation for software defined radio. In: 2007 2nd IEEE international symposium on new frontiers in dynamic spectrum access networks. IEEE
5. Ghasemi A, Zahediasl S (2012) Normality tests for statistical analysis: a guide for non-statisticians. Int J Endocrinol Metab 10(2):486
6. Bain L (2017) Statistical analysis of reliability and life-testing models: theory and methods. Routledge
7. Nguyen HT, Wu B (2006) Fundamentals of statistics with fuzzy data, vol 198. Springer, New York

8. Laird N (1978) Nonparametric maximum likelihood estimation of a mixing distribution. J Am Stat Assoc 73(364):805–811
9. Navatha K, Tarun Kumar J, Ganguly P AES implementation on Virtex-6 FPGA for enhanced performance using pipelining and partial reconfiguration techniques
10. Kumar VS, Tarun Kumar J (2019) NC-OFDM/ OQAM based cognitive radio network. LAP Lambert Academic Publishing, 978-620-0-45455-3, 22 Oct 2019
11. Xue L et al (2019) An improved interference cancellation channel estimation method for OQAM/OFDM system. J Phys Conf Ser 1169(1)
12. Zhang J et al (2019) Real-valued orthogonal sequences for iterative channel estimation in MIMO-FBMC systems. IEEE Access
13. Pranitha B et al (2018) BER performance investigation of MIMO underwater acoustic communications. In: 2018 11th international symposium on communication systems, networks and digital signal processing (CSNDSP). IEEE
14. Tarun Kumar J (2018) Equalizer design to compensate impairments in OFDM system. J Adv Res Dyn Control Syst: 1819–1826, ISSN: 1943-023X, Sep 2018

A Novel Optimization Algorithm for Spectrum Sensing Parameters in Cognitive Radio System

J. Tarun Kumar and V. S. Kumar[✉]

Department of ECE, SR Engineering College, Warangal, India
{tarunjulru, kumar.s.vngl}@gmail.com

Abstract. In this paper, an optimization algorithm for cooperative spectrum sensing in cognitive radio (CR) is proposed, to maximize the spectrum sensing efficiency under the condition of limited interference, an optimization scheme of cooperative spectrum sensing mechanism for cognitive radio system. The system model is defined, and the cooperative spectrum sensing is used to jointly optimize the system targets, including sensing time, transmission time, and the number of sensing users participating in the collaboration. The simulation results show that the optimization scheme can maximize the spectrum sensing efficiency under the condition that the interference is limited.

Keywords: Cognitive radio (CR) · Cooperative spectrum sensing (CSS)

1 Introduction

In recent years, cognitive radio (CR) has received more and more attention because of its ability to obtain sufficient spectrum efficiency through spectrum sharing and dynamic access. It is considered to be an important way for next-generation wireless communication systems to solve the problem of spectrum scarcity. As an intelligent wireless communication system, cognitive radio first needs to perceive the surrounding radio environment. Therefore, spectrum perception is an important prerequisite for cognitive radio communication. Spectrum sensing must be able to quickly and accurately detect the presence of the primary user signal, so that the secondary user dynamically occupies the licensed frequency band with low utilization and at the same time avoids harmful interference to the primary user. Once the main user appears, the secondary user should be able to detect it immediately and quickly exit the frequency band it occupied [1, 2]. Commonly used spectrum sensing methods include matched filter detection, energy detection, and cyclo-stationary feature detection. Energy detection is simple to implement and performs well in the case of high signal-to-noise ratio. Under actual conditions, the secondary performance is seriously degraded due to factors such as channel fading and shadowing. Cooperative spectrum sensing is a method that can effectively improve the secondary user performance, and has become a research hotspot.

The main goal of spectrum sensing is to maximize the spectrum sensing efficiency of cognitive radio systems while meeting the constraints of interference. In the periodic spectrum sensing process, the spectrum sensing and data transmission of the sensing

© Springer Nature Singapore Pte Ltd. 2020
N. Goel et al. (eds.), *Modelling, Simulation and Intelligent Computing*,
Lecture Notes in Electrical Engineering 659,
https://doi.org/10.1007/978-981-15-4775-1_36

user cannot be performed simultaneously. Therefore, it is necessary to comprehensively consider the allocation of the sensing time and the transmission time. The longer the sensing time, the higher the accuracy, and the less interference it can cause to the primary user. However, when the sensing time becomes longer, the transmission time of the secondary user will be reduced, resulting in a decrease in the sensing efficiency of the secondary user. Therefore, sensing time and transmission time are two key parameters affecting spectrum sensing efficiency and interference problems. The choice of this parameter will have an important impact on the performance of cognitive radio networks. At the same time, in cooperative spectrum sensing, the number of users participating in cooperative network also affects spectrum sensing efficiency.

At present, there has been a lot of research on the parameter optimization problem of cooperative spectrum sensing. Literature [3, 4] optimizes the number of sensing users participating in cooperation to minimize the probability of error or maximize the probability of detection. In [5], a set of parameters including the perceived time and the number of recognized users participating in the cooperation are optimized under the condition, satisfying the system detection performance, so that the cognitive user's throughput is maximized. Literature [6–11] studied the problem of maximizing the perceived efficiency in the case of limited interference, taking into account the influence of two key parameters of sensing time and transmission time. In this paper, by jointly optimizing the sensed parameters including the sensing time, the transmission time, and the number of sensed users participating in the collaboration, the spectrum sensing efficiency is maximized under the condition of limited interference.

2 System Model

2.1 Problem Definition

CR Network: M secondary users, one base station, and one primary user. Each sensing user makes a decision whether the primary user signal exists by periodic spectrum sensing, and sends the decision result to the base station, and the base station makes a final decision according to the received information. The typical frame structure of periodic spectrum sensing (see Fig. 1) includes the sensing time (T_s) and the transmission time (T) [5].

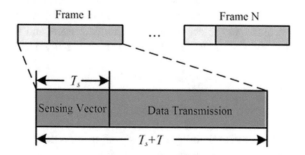

Fig. 1 Spectrum sensing frame structure

To solve the problem, the following questions are first introduced:

Step 1 Sensing efficiency refers to the ratio of the resources used to transmit data to the entire frame resource after spectrum sensing. The definition is as follows:

$$\eta = 1 - \frac{mT_s}{M.T + mT_s} \tag{1}$$

where M is the number of perceived users in the CR network; m is the number of sensed users participating in the collaboration.

Step 2 The ratio of interference time to the duration of data transmission is defined as follows:

$$\varepsilon = \frac{T_I}{T} \tag{2}$$

where T_I represents the duration of interference, i.e., the time when primary frequency band is busy, and the user is sensed to communicate.

The goal of spectrum sensing is to find a set of optimal sensing parameters that maximize the sensed efficiency when the interference is limited, i.e.,

$$\text{Find: } m^*, T_s^*, T^*$$

$$\max: \eta(m^*, T_s^*, T^*) = 1 - \frac{mT_s}{MT + mT_s} \tag{3}$$

$$\text{s.t. } \varepsilon \leq \Gamma, \quad 1 \leq m \leq M$$

where m^*, T_s^*, and T^*, respectively, represent the optimal number of sensed users participating in the cooperation, the optimal sensed time, and the optimal transmission time; Γ represents the maximum interference limit that the system can tolerate.

2.2 Energy Detection Based on Statistical Characteristics of Licensed Frequency Band Occupancy Status

The occupancy statistics of the licensed band is seen as a periodic change in the transition between busy and idle states. The busy and idle states, respectively, indicate that the band is occupied and unoccupied by authorized users, and the busy and idle durations are subject to exponential distribution. The probability density functions are:

$$f_B(t) = \alpha e^{-\alpha t} \tag{4}$$

$$f_I(t) = \beta e^{-\beta t} \tag{5}$$

where α and β represent the probability of transitioning from the busy state to the idle state and the probability of transitioning from the idle state to the busy state, respectively. These two values are estimated from the spectrum sensing of the licensed frequency band, so they are assumed to be known to the cognitive user. Then, for a certain frequency band, the probability of being in a busy state and an idle state is: $P_B = \frac{\beta}{\alpha+\beta}$ and $P_I = \frac{\alpha}{\alpha+\beta}$.

H_0 indicates absence of primary user signal, H_1 indicates presence of primary user signal, and the signal received at ith sensed user is:

$$x_i(k) = \begin{cases} n_i(k)H_0 \\ s(k) + n_i(k)H_1 \end{cases}, \quad i = 1, 2, \ldots, M \tag{6}$$

where $s(k)$ represents the primary user transmitting signal; $n_i(k)$ represents the additive white Gaussian noise with mean 0, and variance is δ_n^2; $s(k)$ and $n_i(k)$ are independent of each other. When the number of sampling points N is large enough, according to the central limit theorem, Y_i approximates a Gaussian distribution:

$$Y_i = \begin{cases} N\left(N\delta^2, 2N\delta^4\right)H_0 \\ N\left(N(1+\gamma_i)\delta^2, 2N\left(1+2\gamma_i^2\right)\delta^4\right)H_1 \end{cases} \tag{7}$$

According to literature [4], the statistical characteristics of the licensed band are taken into consideration, for a given licensed frequency band, assuming $N = 2T_sW$, W for the bandwidth, and then, the first user's false alarm probability (P_F) and detection probability (P_D) are:

$$P_{F,i} = \frac{\alpha}{\alpha+\beta}Q\left(\frac{\lambda_i - 2T_sW\delta^2}{2\delta\sqrt{T_sW}}\right) \tag{8}$$

$$P_{D,i} = \frac{\beta}{\alpha+\beta}Q\left(\frac{\lambda_i - 2T_sW(1+\gamma_i)\delta^2}{2\delta^2\sqrt{T_sW(1+2\gamma_i^2)}}\right) \tag{9}$$

where γ_i is the ith perceived user receiving signal-to-noise ratio: $Q(t) = \frac{1}{2\pi}\int_t^{+\infty} e^{\frac{-u^2}{2}}du$.

2.3 Cooperative Spectrum Sensing

It is assumed that there are m users participating in the spectrum sensing in the above cognitive radio network, i.e., $1 \leq m \leq M$. Using the logical "OR" criterion, the detection probability and false alarm probability of the cooperative spectrum sensing are:

$$Q_F = 1 - \prod_{i=1}^{m}(1 - P_{F,i}) \tag{10}$$

$$Q_D = 1 - \prod_{i=1}^{m}(1 - P_{D,i}) \tag{11}$$

Without loss of generality, it is assumed that all sensed users have the same sensing performance, so here, $P_{F,i}$ and $P_{D,i}$ are represented as P_F and P_D. The false alarm probability and detection probability of a single-sensed user are:

$$P_F = 1 - \sqrt{(1 - Q_F)^m} \tag{12}$$

$$P_D = 1 - \sqrt{(1 - Q_D)^m} \tag{13}$$

3 Optimization of Cooperative Sensing Parameters

3.1 Interference Analysis Model of Licensed Bands

The interference duration in the entire frame is first analyzed. In fact, the probability of licensed band state changes two or more times during each data transfer process which is small and is ignored. Therefore, the user communication can cause interference to the primary user in the following two cases:

1. During the data transmission process, the licensed frequency band is in a busy state and the sensing result is judged as an idle state;
2. The sensing result is idle, and the licensed frequency band is changed from the idle state to the busy state during the data transmission.

Assuming that $T_{I,1}$ and $T_{I,0}$ represent the average interference time in these two cases, respectively, the total statistical average interference time is:

$$T_I = T_{I,1}(1 - Q_D) + T_{I,0}(1 - Q_F) \tag{14}$$

When sensing error occurs, the user is sensed to be harmful to the primary user during data transmission. If the licensed band status changes from the time of data transmission to the idle state after the time τ, the time of occurrence of the interference is τ, then $T_{I,1}$ is:

$$T_{I,1} = \int_{T}^{\infty} f_B(\tau).T\mathrm{d}\tau + \int_{0}^{T} f_B(\tau).\tau\mathrm{d}\tau \tag{15}$$

If the licensed band changes from data transfer to a busy state after the time τ, the interference occurs at $t - \tau$, then $T_{I,0}$ is:

$$T_{I,0} = \int_{0}^{T} f_B(\tau).(T - \tau)\mathrm{d}\tau \tag{16}$$

In the periodic frame structure sensed, considering the strict interference limitation to the primary user, the average time for the licensed band to be in the busy and idle state is much larger than the data transmission time, i.e., $T = \frac{1}{\alpha}$, and according to the Taylor series, the formula (14)–(16) is substituted into Eq. (2), which gives:

$$\varepsilon = 1 - Q_D + \frac{\alpha T}{2}(Q_D - Q_F) \tag{17}$$

3.2   Optimization of Cooperative Sensing Parameters

Assume that the false alarm probability $\overline{Q}_F$ and the detection probability $\overline{Q}_D$ that a given system should satisfy.

1. Sensing time T_s combining Eqs. (8) and (11), respectively, to determine the detection thresholds and make them equal, and the sensing time is:

$$T_s = \frac{1}{W\gamma^2}\left[Q^{-1}\left(\frac{1 - \sqrt{(1-\overline{Q}_F)^m}}{P_I}\right) - \sqrt{1+2\gamma^2}.Q^{-1}\left(\frac{1-\sqrt{(1-\overline{Q}_D)^2}}{P_B}\right)\right]^2 \tag{18}$$

2. The transmission time T is obtained by Eqs. (3) and (17), and the transmission time must satisfy:

$$T \leq \frac{2}{\alpha}\frac{\Gamma - 1 + \overline{Q}_D}{\overline{Q}_D - \overline{Q}_F} \tag{19}$$

Using Eq. (3), T is derived by partially differentiating Eq. (3) with respect to T, i.e., $\frac{\partial \eta}{\partial T} > 0$ is seen. Therefore, the maximum value of T is taken.

3.3   Program Steps

The proposed scenario uses a one-dimensional search, and the detailed steps are as follows:

Step 1   Initialization $\alpha, \beta, \overline{Q}_D, \overline{Q}_F, \Gamma, \eta_{max} = 0$, calculation $P_I, P_B, T = \frac{2}{\alpha}\frac{\Gamma-1+\overline{Q}_D}{\overline{Q}_D-\overline{Q}_F}$;

Step 2   For m from   1   to M,   calculate T_s, η,   if $\eta > \eta_{max}$,   then $m^* = m, T_s^* = T_s, T^* = T$.

4   Simulation Results

The simulation conditions are as follows: $M = 70, \overline{Q}_F$ and $\overline{Q}_D$ are 0.02% and 99.98%, respectively, signal-to-noise ratio γ^{-4} dB, and bandwidth $W = 10^5$ Hz. Figure 2 shows the relationship between the sensed efficiency of cooperative spectrum sensing and the number of sensed users participating in cooperation in different licensed band occupancy.

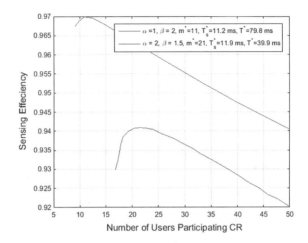

Fig. 2 Comparison of cooperative spectrum sensing efficiency under different licensed frequency band occupancy probabilities

It is observed that when the statistical characteristics of the licensed frequency band are certain, as the number of participating user's increases, the sensed efficiency η increases initially and then decreases, there is a maximum value, and there is a set of optimal cooperative sensing parameters (see Fig. 2 for labeling). Maximize the spectrum sensing efficiency of cognitive radio networks. When comparing the statistical characteristics of different licensed frequency bands, the frequency band with high availability needs to allocate more resources for sensing, so its sensing efficiency is lower than the frequency sensing efficiency with lower availability.

Figure 3 shows the relationship between sensed efficiency and the number of sensed users participating in cooperation under different interference constraints. It is observed that when the interference limit that the system can tolerate is certain, there is a set of optimal cooperative sensing parameters (labeled in Fig. 3) for the convex function of the number m of users participating in the collaboration when the same efficiency η is perceived. The spectrum sensing efficiency of the radio network is maximized. For interference limits that is tolerated by different systems, the greater the interference limit, the longer the transmission time, and the higher the perceived efficiency.

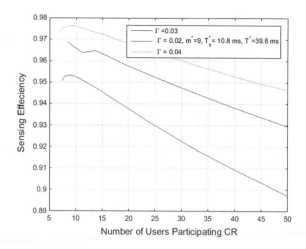

Fig. 3 Comparison of cooperative spectrum sensing efficiency under different interference constraints

5 Conclusion

In this paper an optimized algorithm for spectrum sensing for CR system is proposed. By jointly optimizing the perceptual parameters including sensing time, transmission time, and the number of sensing users participating in collaboration, the spectrum sensing is analyzed under the condition of limited interference. The problem of maximizing efficiency is verified by simulation. The current work is mainly for the case of a single primary user channel, and the next step will be to study the situation of multiple primary user channels.

References

1. Mishra SM, Sahai A, Broderson RW (2006) Cooperative sensing among cognitive radios. In: Proceedings of IEEE international conference on communications. Istanbul, Turkey
2. Ghasemi A, Sousa ES (2008) Spectrum sensing in cognitive radio networks: requirements, challenges and design trade-offs. IEEE Commun Mag 46(3):32–39
3. Peh ECY, Liang Y-C (2007) Optimization for cooperative sensing in cognitive radio networks. In: Proceedings of IEEE international wireless communications networking conference. Hong Kong, China
4. Zhang W, Mallik RK, Letaief KB (2008) Cooperative spectrum sensing optimization in cognitive radio networks. In: IEEE international conference on communications. Beijing, China
5. Liang Y, Zeng Y, Peh ECY et al (2008) Sensing throughput tradeoff for cognitive radio networks. IEEE Trans Wireless Commun 7(4):1326–1337
6. Lee W-Y, Akyildiz IF (2008) Optimal spectrum sensing framework for cognitive radio networks. IEEE Trans Wireless Commun 7(10):3845–3857

7. Navatha K, Tarun Kumar J, Ganguly P AES implementation on Virtex-6 FPGA for enhanced performance using pipelining and partial reconfiguration techniques
8. Kumar VS, Tarun Kumar J (2019) NC-OFDM/ OQAM based cognitive radio network. LAP Lambert Academic Publishing, 978-620-0-45455-3, 22 Oct 2019
9. Zhang J et al (2019) Real-valued orthogonal sequences for iterative channel estimation in MIMO-FBMC systems. IEEE Access
10. Pranitha B et al (2018) BER performance investigation of MIMO underwater acoustic communications. In: 2018 11th international symposium on communication systems, networks and digital signal processing (CSNDSP). IEEE
11. TarunKumar J (2018) Equalizer design to compensate impairments in OFDM System. J Adv Res Dyn Control Syst: 819–1826, ISSN: 1943-023X, Sep 2018

Multimodal Multilevel Fusion of Face Ear Iris with Multiple Classifications

Himanshu Purohit and Pawan K. Ajmera[⊠]

Birla Institute of Technology and Science, Pilani, Pilani Campus, Pilani, Rajasthan, India
{p2015502,pawan.ajmera}@pilani.bits-pilani.ac.in

Abstract. With the advancement in the computational efficiency, there is also simultaneous increase in many efficient and secure biometric systems that are capable for the use of multiple sources of access authorization. Single biometric systems are inefficient and less secure which give rise to the advancement of multimodal biometric systems. Also, fusion of many biometric modalities is high area of interest, and here, many methods are deployed for the fusion of biometric data. Multimodal biometric system provides many evidences for the same person. In this paper, the design of multimodal biometrics based on face, ear, and iris modalities with multilevel fusion-based approach is preferred. In the presented work with multilevel multimodal fusion, 95.09% accuracy has been obtained which is better than highest unimodal accuracy; in this case, it is iris 94.06%. The obtained results are better than similar multimodal fusion-based model with single classifiers such as RNN with 90.58% accuracy and KNN classifier with 91.22% accuracy. So, in this work multilevel fusion of (i) different unimodal methods with (ii) feature level fusion of multiple traits has been proposed for person identification.

Keywords: Multilevel fusion · Biometric modalities · Multimodal biometrics · Feature level · Score level

1 Introduction

As to fulfill the present higher user demand and drastic advancement in the technology, unimodal biometric system faces serious problems. So, the solution for this is multimodal biometric system as it possesses better noise sensitivity, accuracy, reliability, interclass variability, and interclass similarity. Here, in this presented research work, the main objective is multilevel fusion of feature level results with other level of unimodal results and to design efficient authentication system for face, ear, and iris. Here, in this present context of research, fusion at feature level is applied for the consolidation of the information presented by face, ear, and iris feature sets obtained from the same person. Also, this feature level fusion depicts valuable information as compared to match score level or decision level fusion. Here, multiple-level fusion strategy is proposed to enhance system accuracy. It is found that multilevel combination of fusion increases accuracy at the cost of complexity.

© Springer Nature Singapore Pte Ltd. 2020
N. Goel et al. (eds.), *Modelling, Simulation and Intelligent Computing,*
Lecture Notes in Electrical Engineering 659,
https://doi.org/10.1007/978-981-15-4775-1_37

Now here, for the classification, KNN classifier and SVM both are used. KNN is basically an instance-based learning algorithm in which the estimation of function is performed locally, and all other computations are performed during classification. Apart from KNN, SVM, i.e., support vector machine, is also preferred for classification of input vectors in this research work.

2 Related Work

Basically, in this section, we discuss the relevant literature present in the field of development for multimodal biometric system (MBS) which is based on multilevel fusion-based methodology. Since our proposed system works using face, iris, and ear as biometric traits so for multiclassification purpose, multiclass SVM and KNN has been deployed. Feature extraction is major step of image processing including identification and authentication systems. V. H. Gaidhane, et al. have presented a specific experimental analysis where flame and fire image analysis which is based on local binary patterns, double thresholding, and Levenberg–Marquardt optimization. In this work, the presented algorithm detects the sharp edge and removes the noise and irrelevant artifacts. This gives the explanation of specific feature-based analysis [1]. In fusion-based biometric model classification is crucial, and Levenberg–Marquardt algorithm-based classifier can be classical way to find out accurate results. V. H. Gaidhane et al. have presented emotion recognition using eigenvalues and Levenberg–Marquardt algorithm-based classifier. The robustness of this algorithm is tested on low-resolution images and recognition rate of 94.6 and above were reported with different databases [2]. Kang et al. stated rank-level fusion for vein and single finger geometry as biometric trait. This reduces the size of the recognition device which is involved in the mechanism. Now for the extraction of feature, they used Fourier descriptors, and for rank-level fusion, they applied max, min, and sum rule. They demonstrated the results by showing that there will be a decrease in the equal error rate of this proposed method by 1.089% as compared to finger vein recognition method and 1.627% for finger geometry recognition methods [3]. Poh et al. also discussed about the classification of fusion techniques. According to the authors, fusion before matching and fusion after matching are the two broad classifications of fusion techniques [4]. Vishi et al. experimented the overall performance of MBS by taking help of fusion at score level. Authors first extracted the feature from fingerprint and iris, and after comparing feature template with database, each score took into account and these scores are combined by min score, max score, simple sum, and user weighted sum rules. Then, final decision is made on the basis of comparison between the combined score and threshold [5]. N. T. Vetrekar, et al. suggested various classes of fusion approach named hyperspectral method which is quite different from conventional fusion procedure. They could obtain the recognition rate of around 96.92% at Rank-1 [6]. Paul et al. introduced a new method for decision fusion using social network analysis. For the feature extraction of multibiometric traits, they have taken Fisher linear discriminant analysis and measured the similarity and then the result is fused with the social network analysis for each trait [7]. Mondal et al. have designed a feature-level fusion of multibiometric recognition system in which a feature adaptive approach is defined to improve the multimodal recognition [8]. Somashekhar et al. proposed two types of fusion, viz. feature level and decision level for the purpose of face- and fingerprint-based MBS designing [9]. Prabhakar et al. suggested nonparametric approach for score fusion which is based on joint multivariate densities, and based on those, posterior probabilities are computed by the help of Bayes rule [10]. Hao et al. proposed an effective method for the improvement of K-nearest neighbor algorithm-based text classifier and conducted exhaustive experiments to illustrate that a significant improvement in performance of classification has been achieved by the help of adapted KNN [11]. U. Gawande, et al. proposed the development of a fingerprint and iris fusion system which uses a single hamming distance-based matcher to provide higher accuracy than the individual uni-modal system which results in accuracy of 94.07% with FAR and FRR of 1.46% and 6.87%, respectively [12]. Xiaona et al. have suggested an algorithm which is based on KCCA, i.e., kernel canonical correlation analysis, and they also tested it for face and ear biometrics. By this method, authors have suggested a nonlinear associated feature which was basically proposed for recognition and classification purpose. [13]. Kaur et al. provided a detailed review of rank-, feature-, and decision-level fusion for multimodal biometric system. Their proposed system works very well for 50 test images with EER of 0.45% [14]. Elmir et al. proposed feature- and score-level-based multilevel fusion model where they have used max of scores as multimodal fusion approach for face, voice, and online signature modalities [15]. Soltane suggested feature-level-based multilevel fusion model and as a part of multimodal fusion approach. For the experimental purpose, they have used eNTERFACE 2005 multi-modal biometric databases and obtained the test performance with EER of 5.04% for GEM (Greedy EM Algorithm) and EER of 3.91% for FJ (Figueiredo Jain Algorithm) [16].

3   Methodology

For the purpose of passive individual distinguishing proof, human ear is the good solution because ear pictures are easy to take and even their structure and shape does not change drastically as the time elapsed. Now, in our research work, as illustrated in Fig. 1, three biometric traits such as face, ear, and iris are taken for fusion-based multimodal biometric system. The ear has certain unique biometric features such as helix, anti-helix, tragus, anti-tragus and lobe. The geometric and shape-based analysis may generate unique feature set for identification. Skin color segmentation approach with standard feature extraction gives improved feature set. The universality and ease of data collection makes ear a good choice for biometric systems. The complete process is divided into three stages as segmentation stage, feature extraction stage, and recognition stage.

3.1   Segmentation Stage

Segmentation is basically a mechanism which is used to partition a digital image into multiple objects. The main goal of this segmentation is to change the image repre-sentation to make it more meaningful for analysis purpose. Now here, all the three

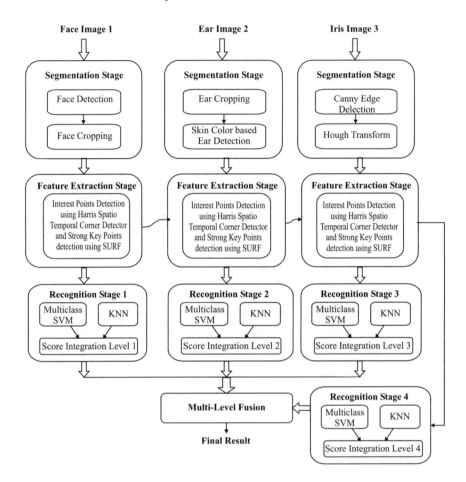

Fig. 1 Structure of proposed method

input test images of face, ear, and iris have undergone through segmentation mechanism. Firstly, for face detection, test images of face are acquired from the gallery and then computer vision face detection tool is used to detect the face present in the test image. After that, the detected face was cropped for the next process of feature extraction. Secondly for ear detection, ear was cropped by automatic cropping method and then it was detected by skin color-based ear detection mechanism. Thirdly for iris detection, Canny edge detection is used for the generation of edge map, linear Hough transform is used for localizing occluding eyelids, whereas circular Hough transform is used for localizing the iris and pupil regions.

3.2   Feature Extraction Stage

In this stage, for the respective output image from previous stage using Harris Spatio, interest points are detected. In our research work, we have used SURF, viz. speeded-up robust features for the temporal corner detector and strong key points detection. SURF is basically an in-plane rotation detector and descriptor in which the detector locates the key points in the image and the descriptor describes the features of the key points and then it constructs the feature vectors of the key points (Figs. 2 and 3).

Fig. 2 Face processing

SURF is a fast, reliable, and robust algorithm for local, similarity invariant representation and comparison of images which involves two steps: One is feature extraction and the other is feature description. Given a point $p = (x, y)$ in an image I, the Hessian matrix $H(p, \sigma)$ at point p and scale σ is:

$$S(x, y) = \sum_{i=0}^{x} \sum_{j=0}^{y} I(i, j) \tag{1}$$

$$H(\text{Þ}, \varrho) = \begin{Bmatrix} L_{xx}(\text{Þ}, \sigma), L_{xy}(\text{Þ}, \sigma) \\ L_{yx}(\text{Þ}, \sigma), L_{yy}(\text{Þ}, \sigma) \end{Bmatrix} \tag{2}$$

where $L_{xx}(\text{Þ}, \sigma)$ etc. is the convolution of the second-order derivative of gaussian with the image $I(i, j)$ at the point.

$$\{\backslash \text{p}\} \sigma_{\text{approximation}} = \text{current filter size (base filter scale/base filter size)} \tag{3}$$

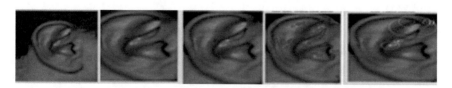

Fig. 3 Ear processing

3.3   Fusion

In first step, extracted features from different traits are fused together to create multi-modal feature vector, and this vector is compared with stored template for getting feature-level score. This score gives accuracy of feature-level fusion. In alternate process, individual traits are matched with respective stored template and a matching score is generated for each trait. Respective scores of traits generate individual accuracy for unimodal systems. These scores are also fused together to generate major score index for multimodal score level fusion.

At final stage, score generated from feature level fusion process is again fused with multimodal score and a final score is generated for entire system.

4   Results

The experiments were performed on the system with Intel core i5 processor using MATLAB 9.4, R2018a Software and its image processing toolbox as the simulation platform.

For face, ear, and iris, unimodal accuracy obtained is 92.4%, 93.0%, and 94.06%, respectively. For feature-level fusion, some improvement has been noted and result was 94.16%. In the proposed method, accuracy was improved significantly compared with unimodal accuracies, and the resultant accuracy was 95.09%. Figures 4, 5, 6, 7, 8 and 9 depict above-mentioned details, and Table 1 summaries all methods. The recall rate, precision, and specificity were 96.16%, 94.12%, and 93.03%, respectively. The performance of multilevel fusion has been reported better in many experiments with constrained databases. The work with multilevel fusion of multimodal biometrics is comparatively obtained with increase in complexity but with improved accuracy (Fig. 10 and Table 2).

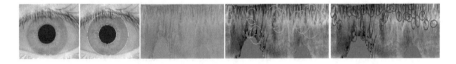

Fig. 4 Iris processing

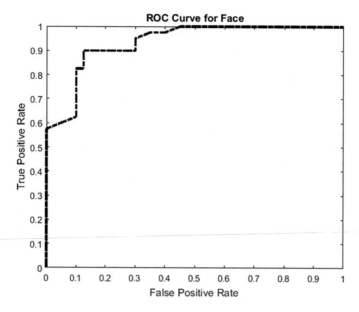

Fig. 5 ROC curve for unimodal face

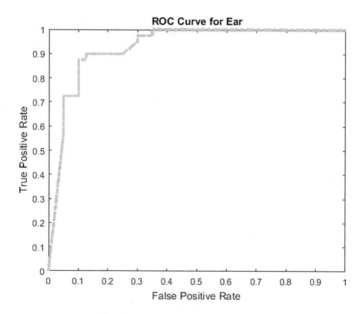

Fig. 6 ROC curve for unimodal ear

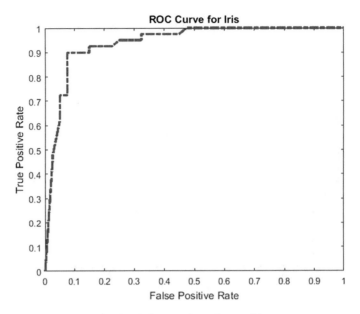

Fig. 7 ROC curve for unimodal iris

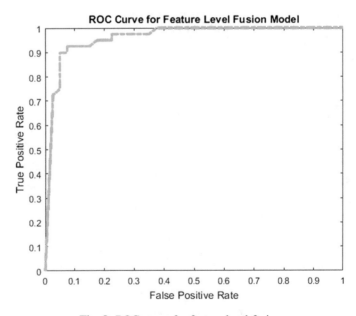

Fig. 8 ROC curve for feature level fusion

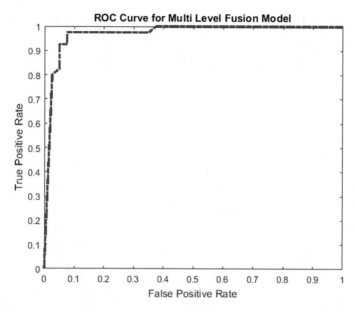

Fig. 9 ROC curve for proposed multilevel fusion

Table 1 Accuracy obtained for different experiments

Parameter	Face (%)	Ear (%)	Iris (%)	Feature fusion (%)	Proposed multilevel fusion method (%)
Accuracy	92.44	93.00	94.06	94.16	95.09

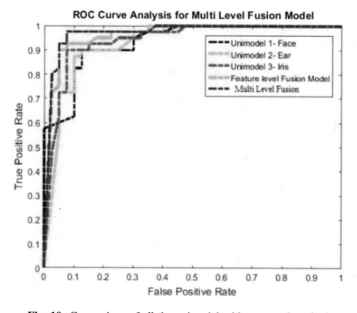

Fig. 10 Comparison of all the unimodal with proposed method

Table 2 Performance comparison with other methods

Performance parameters	MBS with RNN classification (%)	MBS with SIFT based feature extraction with KNN (%)	Proposed MBS with SURF based feature extraction with SVM and KNN (%)
Accuracy	90.58	91.22	95.09

5   Conclusion

Here in this work, MBS which is based on multilevel fusion of face, iris, and ear was presented. The feature-level fusion is applied as this technique generates feature set which contains excellent information regarding unprocessed biometric data as compared to other fusion methodologies. On next stage, score-level fusion is also applied, and score generated is mixed with the results of feature-level fusion results, obtained in the previous stage. The accuracy achieved from this proposed multilevel fusion is 95.09% which is far better than other competitive methods. Apart from this, the experimental values obtained by the proposed method for precision, specificity, and recall rate are 94.12%, 93.03%, and 96.16%, respectively. The work can be enhanced by adding more fusion schemes like decision and rank-level fusion with used methods. In future, the results can also be tested with other standard databases and compared with neural network-based results in similar conditions.

References

1. Gaidhane VH, Hote YV (2018) An efficient edge extraction approach for flame image analysis. Pattern Anal Appl 21(4):1139–1150
2. Gaidhane VH, Hote YV, Singh V (2016) Emotion recognition using eigenvalues and Levenberg–Marquardt algorithm-based classifier. Sādhanā 41(4):415–423
3. Kang BJ, Park KR (2010) Multimodal biometric method based on vein and geometry of a single finger. IET Comput Vision 4(3):209–217
4. Poh N, Kittler J (2010) Multimodal information fusion. In: Thiran J-P, Marqués F, Bourlard H (eds) Multimodal signal processing. Academic Press, pp 153–169
5. Vishi, K, Yayilgan SY (2013) Multimodal biometric authentication using fingerprint and iris recognition in identity management. In: 9th international conference on intelligent information hiding and multimedia signal processing. IEEE, China
6. Vetrekar NT, Gad RS (2018) Multi spectral imaging for robust ocular biometrics. In: International conference on biometrics, ICB. IEEE, Australia
7. Paul PP, Gavrilova ML, Alhajj R (2014) Decision fusion for multimodal biometrics using social network analysis. IEEE Trans Syst Man Cybern Syst 44(11):1522–1533
8. Mondal A, Kaur A (2016) Comparative study of feature level and decision level fusion in multimodal biometric recognition of face, ear and iris. IJCSMC 5(5):822–842
9. Somashekhar BM, Nijagunarya YS (2018) Face and fingerprint fusion system for identity authentication using fusion classifiers. IJCSES 9(1)

10. Prabhakar S, Jain AK (2002) Decision level fusion in fingerprint verification. Pattern Recogn 35(4):861–874
11. Hao X, Tao X, Zhang C, Hu Y (2007) An effective method to improve KNN Text classifier. In: 8th ACIS international conference on software engineering, AI, networking and parallel/Distributed computing. IEEE, pp 379–384
12. Gawande U, Sapre A, Jain A, Bhriegu S, Sharma S (2013) Fingerprint-iris fusion based multimodal biometric system using single hamming distance matcher. Int J Eng Inventions 2 (4):54–61
13. Xu X, Zhao Y, Li H (2012) The study of feature level fusion algorithm for multimodal recognition. In: 8th international conference on computing technology and information management. IEEE, South Korea
14. Kaur G, Bhushan S, Singh S (2017) Fusion in multimodal biometric system: a review. Indian J Sci Technol 10(28)
15. Elmir Y, Elberrichi Z, Adjoudj R (2014) Multimodal biometric using a hierarchical fusion of a person's face, voice and online signature. JIPS 10(4):555–567
16. Soltane M (2013) Soft decision level fusion approach to a combined behavioral speech signature biometrics verification. IJSIP 6(1):1–16

Bounded Rate of Control-Based Guidance for Targets Exhibiting Higher Accelerations

Anil Kumar Pal[(✉)], Ankit Sachan, Rahul Kumar Sharma,
Shyam Kamal, and Shyam Krishna Nagar

Department of Electrical Engineering, Indian Institute of Technology (BHU),
Varanasi, India
{anilkp.rs.eee15,ankits.rs.eee15,rahulks.rs.eee16,
shyamkamal.eee,sknagar.eee}@iitbhu.ac.in

Abstract. With the advent of technology, the modern warfare systems are becoming sophisticated. Such systems working as targets have very high acceleration capabilities. In future, it is expected that this acceleration capacity is going to increase many times. To chase such targets is a difficult task for the present-day missiles. Such an attempt would produce quite high demand of lateral acceleration. To address these issues, a novel guidance strategy is presented here which not only tracks such targets but also ensures that the required latex is contained.

Keywords: Bounded rate of control (BRC) · Sliding mode control (SMC) · Control rate · Missile guidance

1 Introduction

Modern war technologies are getting smarter and stronger day by day. Various self-defense measures used in modern ships are discussed in [1]. The close-in weapon system (CIWS) is an important attribute of this self-defense mechanism used in ships for providing protection against missiles. Sliding mode-based guidance considering impact time and impact angle constraints can be found in [2]. Evasive targets with significant acceleration capabilities are quite ubiquitous. In the days to come, we could expect even smarter targets with even higher acceleration and maneuverability. Tracking such targets with contained lateral acceleration would be a challenge. We should update our existing technologies for such a cause. Therefore, in this paper, we primarily consider such class of targets and provide a guidance structure that tracks such targets with acceptable miss-distance and keeps the required latex within bounds.

The case of stationary target and non-maneuvering target moving with small velocity is given in [2]. Sliding mode strategy has been extensively used in designing the guidance laws owing to its robustness character. But sliding mode introduces the chattering phenomenon. Some methods to reduce the required latex are discussed in [3]. There the authors have considered highly maneuverable target by considering weaving target scenario, but they have taken the maximum acceleration to be 80 m/s^2.

Proportional navigation (PN) is a widely used successful guidance strategy for stationary and non-maneuvering targets [4]. Investigation of the performance of

© Springer Nature Singapore Pte Ltd. 2020
N. Goel et al. (eds.), *Modelling, Simulation and Intelligent Computing*,
Lecture Notes in Electrical Engineering 659,
https://doi.org/10.1007/978-981-15-4775-1_38

proportional navigation against weaving targets has been carried on in [5, 6]. If some relaxation in miss-distance is allowed, then one can consider PN guidance law for chasing weaving targets.

Another important aspect that has been overlooked so far is the amount of extra burden that the actuator has to face if the boundedness of the rate of control is not considered. By rate of control, we mean the rate of change of latex. It is an important consideration since it decides the faithfulness of the actuator. A lot of published works that use SMC imply signum function (or its approximation) in the control structure and thus have inevitably introduced unbounded rate of change of control at the points of discontinuity [2, 3, 7, 8]. The objective of this paper is to formulate a guidance law form that gives a bounded rate of control action. Targets with very high acceleration capabilities have been considered here, and the proposed guidance law is used to intercept them with contained latex rate.

2   Problem Formulation

For the missile system, the control is actually the lateral acceleration given to the missile. For SMC-based guidance, this control comprises of signum function. For the simple scalar case, control is $u = k\mathrm{sgn}(x)$, where $u \in \mathbb{R}$ and $x \in \mathbb{R}$. It is evident that the control rate $\dot{u}$ contains impulse and thus becomes unbounded.

Thus, in sliding mode-based guidance strategies, the implied signum function could produce unbounded control rate. This can be considered to be one drawback of using sliding mode-based guidance. The usual practice is to use a sigmoidal function instead of signum function, but even this approximation leads to high control rate at the moment states tend to zero. This exerts enormous pressure on the actuator. So, the prime objective is to remove this burden from the missile actuator. The proposed technique tries to find out a solution to this problem, taking inspiration from the work of Jonathan Laporte et al. [8]. There, the authors have spoken about the boundedness of control and its successive derivatives. Moving along the same line, we implement this idea for missile guidance application.

3   Bounded Rate of Control-Based Guidance

3.1   Engagement Geometry

Consider the planar 2D engagement between missile and target (Fig. 1). The relative kinematics between missile and target is presented here. We have considered a 2D planar model. Missile and target are assumed to be point masses. Moreover, their velocities are assumed to be constant throughout the engagement. The autopilot and seeker dynamics are assumed to be perfect.

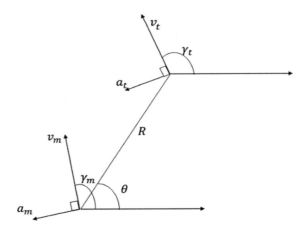

Fig. 1 Missile and target engagement

The dynamical equations can be written as:

$$\dot{R} = v_t \cos(\gamma_t - \theta) - v_m \cos(\gamma_m - \theta)$$

$$\dot{\theta} = \frac{1}{R}[v_t \sin(\gamma_t - \theta) - v_m \sin(\gamma_m - \theta)] \tag{1}$$

$$\dot{\gamma}_t = \frac{a_t}{v_t}, \dot{\gamma}_m = \frac{a_m}{v_t} = \frac{a_c}{v_t}$$

where $R, \dot{R}$ = relative range and range rate between missile and target; v_m, v_t = missile velocity and target velocity, respectively; a_m, a_t = missile lateral acceleration and target lateral acceleration, respectively; a_c = commanded acceleration of the missile; γ_m, γ_t = flight path angle of missile and target, respectively; $\theta, \dot{\theta}$ = LOS angle and LOS angle rate.

3.2   Bounded Rate of Control (BRC)-Based Guidance

From Eq. (1), we have,

$$\ddot{\theta} = \frac{1}{R}\left[-2\dot{R}\dot{\theta} + a_t \cos(\gamma_t - \theta) - a_m \cos(\gamma_m - \theta)\right] \tag{2}$$

This is a second-order system and a_m is the actual control that can be exerted upon the missile. Consider the guidance law form given by $a_c = N'V_c \dfrac{\dot{\theta}}{\sqrt{1+\dot{\theta}^2}}$, where a_c = commanded acceleration of the missile, N' = effective navigation constant, and V_c = closing velocity between missile and target. This law is actually derived from the PN guidance law by replacing $\dot{\theta}$ with $\dfrac{\dot{\theta}}{\sqrt{1+\dot{\theta}^2}}$. This type of transformation brings a saturating effect which bounds the value of a_c and its rate. It may be noted that we have used 2-norm; however, any other p-norm can also be used. It is to be observed that if the values of $\dot\theta$ are very small then this law is equivalent to the PN law. This law is effective for the situation where a higher LOS rate condition is encountered. For such a case, PN law produces high lateral acceleration demand, whereas this law keeps that demand within bounds.

For implementation, the proposed guidance law is as simple as PN law but is still very elegant and exhibits its own benefits. The sliding mode-based guidance strategy usually tends to produce complex guidance law forms, which are challenging to implement.

3.3 Comparison of Rate of Control for PN and BRC

For simplicity, we consider the form of PN law as:

$$a_c = N' v_m \dot\theta \tag{3}$$

So, the rate of control for PN law can be given as:

$$\dot a_c = N' v_m \ddot\theta \tag{4}$$

Similarly, for BRC we consider $a_c = N' v_m \dfrac{\dot\theta}{\sqrt{1+\dot\theta^2}}$ and the rate of control for BRC can be evaluated to be,

$$\dot a_c = N' v_m \frac{\ddot\theta}{\left(1+\dot\theta^2\right)^{3/2}} \tag{5}$$

From (4) and (5), it can be inferred that for similar variations in $\ddot\theta$ the rate of control is less in BRC as compared to PN law. However, it may be noted that the range of $\ddot\theta$ depends upon the guidance law form, and hence, it is possible to encounter situations where BRC might produce higher variations in $\ddot\theta$. For those cases, BRC may produce a comparable or even higher rate of control as compared to PN law.

4 Stability Analysis

For a non-maneuvering target, $a_t = 0$. Therefore, Eq. (2) can be rewritten as,

$$\ddot\theta = \frac{1}{R}\left[-2\dot R\dot\theta - a_c \cos(\gamma_t - \theta)\right] \tag{6}$$

Let us consider the Lyapunov function, $V = \frac{1}{2}\dot\theta^2$. Then, its time derivative is given by $\dot V = \ddot\theta\dot\theta = \frac{\dot\theta}{R}\left[-2\dot R\dot\theta - a_c \cos(\gamma_t - \theta)\right]$. Now, for PN law, we have $a_c = \frac{-N'\dot R\dot\theta}{\cos(\gamma_m - \theta)}$. Thus, $\dot V = \frac{\dot\theta}{R}\left[-2\dot R\dot\theta + N'\dot R\dot\theta\right] = [N' - 2]R\frac{\dot\theta^2}{R}$, since $\dot R < 0$ and $R > 0$ during the engagement. Therefore, for $N' \geq 2$, $\dot{V} \leq 0$. This assures the stability. If $N' > 2$, then $\dot{V} < 0$, asymptotic stability is achieved. Consider BRC, $a_c = \frac{-N'\dot{R}}{\cos(\gamma_m - \theta)} \frac{\dot{\theta}}{\sqrt{1 + \dot{\theta}^2}}$. Then, $\dot{V} = [N' - 2] \frac{R\dot{\theta}^2}{R\sqrt{1 + \dot{\theta}^2}} < 0$ if $N' > 2$. So, this assures asymptotic stability. Therefore, similar to the PN law, BRC works fine for the non-maneuvering targets for $N' > 2$.

5   Simulation and Results

Missile and target engagement that incurs a higher LOS rate can be effectively handled by this guidance law. Such a class of targets is weaving targets. Weaving targets have been considered to be difficult to track. We have considered weaving targets with high acceleration capabilities. The utility of this law can be seen easily in tracking such targets. The following assumptions are considered:

(i)   The point mass models for missile and target are considered.
(ii) The missile is capable of producing large lateral acceleration.
(iii) The angle between missile velocity vector and LOS remains acute so that the target remains in the field of view of the seeker throughout the engagement.

The acceleration profile is considered as $a_t = a\sin(\pi t)$ where, for case (i), $a = 500$, for case (ii), $a = 550$, and for case (iii), $a = 600$. For simulations, the initial geometry is taken as in [3]. For comparative analysis, we have considered classical PN law and SBPN guidance law [6]. The initial geometry and other data required [3] are as follows: $R(0) = 4500$ m, $\theta(0) = 200°$, $\gamma_t = 140°$, $\gamma_t = 60°$, $v_m = 500 \frac{m}{s}$, $v_t = 500 \frac{m}{s}$.

The classical PN guidance law form used for comparison is $a_c(\text{PN}) = N'V_c\dot{\theta}$ where closing velocity, $V_c = -\dot{R}$. The SBPN guidance law [6] is,

$$a_c(\text{SBPN}) = \frac{1}{\cos(\gamma_m - \theta)}\left[-N'\dot{R}\dot{\theta} + w\text{sgn}\left(\dot{\theta}\right)\right]$$

An approximation for the signum function is used so as to produce the following guidance law which has been used for the comparative study,

$$a_c(\text{SBPN}) = \frac{1}{\cos(\gamma_m - \theta)}\left[-N'\dot{R}\dot{\theta} + W\left(\frac{\dot{\theta}}{|\dot{\theta}| + \delta}\right)\right]$$

where W is the bound on target acceleration taken as 500 for the case (i), 550 for case (ii), and 600 for case (iii) for simulations under ideal conditions.

5.1   Simulations with Constant Missile Velocity

Here, we consider constant missile velocity. The target is also considered to be moving with a constant velocity. Only the kinematic model is used for this simulation. The initial conditions and target acceleration profiles are given as in the above section. The control appears in the form of the commanded latex. In this study, the prime focus is on bounded rate of control. Therefore, we are concerned with the rate of change of commanded latex. This factor is associated with the pressure that the actuator goes through during engagement. Table 1 gives a comparison of the three laws.

In the table, the maximum values of latex and rate of change of latex during the engagement are given. These maximas usually occur in the end phase. However, in the case of SBPN apart from the end-phase spikes, there are peaks in between as well (as evident in Fig. 2). These are the zero-crossing points of LOS rate. The higher number of peaks in the SBPN case presents the limitation of using sliding mode-based techniques. This also leads to an increased burden on the actuator. It can be observed that the maximum latex rate is minimum in the case of BRC. Even the maximum latex is minimum in case of BRC with the exception of case (i), where the slight increase in maximum latex is due to higher variations in $\dot{\theta}$ in comparison with other laws. Leaving aside the end phase, it can be observed from Fig. 2 that the variations in $\dot{a}_c$ are similar in case of PN and BRC. We have also found in simulations that the variations in a_c are similar for PN and BRC.

5.2 Simulations Under Realistic Conditions

Now, we consider the realistic scenario where we take into account the effects of thrust and drag. With slight alterations, most of the data for realistic simulation is similar as in [3]. Here, the missile velocity varies as $\dot{v} = \frac{T-D}{m}$ where T is thrust, D is the drag force acting on the missile, and m is its mass.

The mass m includes the mass of the propellant m_p (here it is taken to be 7 kg). When the propulsion system is ON, then the propellant is being utilized, and hence, the mass varies as $m(t) = m_i - \dot{m}_f(t)$ where m_i is the initial mass of missile (taken here as 167 kg), and $\dot{m}_f$ is the fuel mass flow rate given by $\dot{m}_f = \frac{m_p}{t_b}$ where t_b is the burn time

Table 1 End-phase performance comparison among PN, BRC, and SBPN

	PN	SBPN	BRC
Case (i)	$a_t = 500 \sin(\pi t)$		
Miss dist.	35	73.07	39.61
$\max|a_c|$	1064.10	1353.10	1083.20
$\max|\dot{a}_c|$	1.91×10^5	1.87×10^6	1.80×10^5
Case (ii)	$a_t = 550 \sin(\pi t)$		
Miss dist.	28.47	34.68	35.75
$\max|a_c|$	1051.80	3093.60	1045.60
$\max|\dot{a}_c|$	2.10×10^5	4.00×10^6	1.84×10^5
Case (iii)	$a_t = 600 \sin(\pi t)$		
Miss dist.	30.34	80.22	40.94
$\max|a_c|$	1003.40	1542.30	980.70
$\max|\dot{a}_c|$	1.93×10^5	2.30×10^6	1.58×10^5

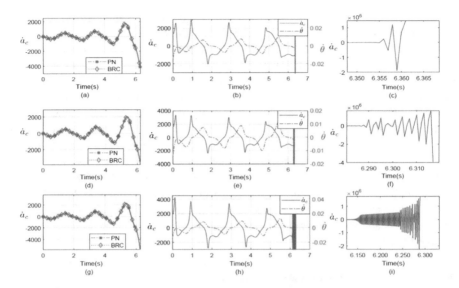

Fig. 2 Performance comparison between PN, BRC, and SBPN. **a** Case (i) latex rate for PN and BRC. **b** Case (i) latex rate and LOS rate for SBPN. **c** Case (i) end-phase latex rate for SBPN. **d** Case (ii) latex rate for PN and BRC. **e** Case (ii) latex rate and LOS rate for SBPN. **f** Case (ii) end-phase latex rate for SBPN. **g** Case (iii) latex rate for PN and BRC. **h** Case (iii) latex rate and LOS rate for SBPN. **i** Case (iii) end-phase latex rate for SBPN

(taken here as 3.5 s). For simplicity, it is assumed that this engagement occurs at constant altitude on a horizontal plane. The altitude here is assumed to be 3000 m, and following standard atmosphere, the corresponding density of air is $\rho = 0.909$ kg/m³. The drag force acting on the missile is given by $D\frac{1}{2}\rho v_{\mathrm{m}}^2 C_D A$ where ρ is the density of air, A is the reference area, and C_D is the drag force coefficient.

Let us assume the parabolic model for the drag coefficient given by $C_D = C_{D0} + kC_L^2$ where C_{D0} is the zero-lift drag coefficient, C_L is the lift force coefficient, and k is the induced drag parameter. As the lift force is given by $L = \frac{1}{2}\rho v_{\mathrm{m}}^2 C_L A$, the resulting lateral acceleration is given by $a_c = L/m$. The drag equation can be rewritten as $D = k_1 v_{\mathrm{m}}^2 + k_2 \frac{a_c^2}{v_{\mathrm{m}}^2}$ where $k_1 = \frac{1}{2}\rho C D_0 A$ and $k_2 = 2k_{\mathrm{m}}^2 \rho A$. For the purpose of simulation, the following data is considered [4]: $C_{D0} = 0.74$, $k = 0.03$, $A = 0.0324$, $k_1 = 0.0109$ kg/m, and $k_2 = 55,464$ kg-m. The thrust of 6200 N acts for a duration of 3.5 s. The initial velocity of the missile is taken as $v_{\mathrm{m}}(0) = 900$ m/s.

The maximas for the latex and rate of change of latex, given in Table 2, have occurred in the end phase (in the last few milliseconds). The overall performance of the BRC is found to be satisfactory. For the case (i), both maximum latex and maximum rate of change of latex are less in BRC as compared to PN and SBPN. For the case (ii), both maximum latex and maximum rate of change of latex are less in BRC as compared to PN, but higher as compared to SBPN. But, SBPN produces mid-phase maximas. Therefore, BRC performance is still better than SBPN. Similar performance can be seen for BRC in case (iii) but with an added advantage of lesser miss-distance.

Table 2 End-phase performance comparison among PN, BRC, and SBPN under unrealistic condition

	PN	SBPN	BRC
Case (i)	$a_t = 500\sin(\pi t)$		
Miss dist.	70.10	22.68	81.22
$\max\lvert a_c\rvert$	940.79	974.14	940.39
$\max\lvert \dot{a}_c\rvert$	9.31×10^4	1.86×10^5	8.20×10^4
Case (ii)	$a_t = 550\sin(\pi t)$		
Miss dist.	22.82	38.35	25.48
$\max\lvert a_c\rvert$	2189.40	1128.50	1246.10
$\max\lvert \dot{a}_c\rvert$	1.11×10^6	1.69×10^5	2.32×10^5
Case (iii)	$a_t = 600\sin(\pi t)$		
Miss dist.	97.01	406.40	75.69
$\max\lvert a_c\rvert$	6731.10	5488.80	2020.60
$\max\lvert \dot{a}_c\rvert$	3.74×10^6	1.56×10^6	3.09×10^4

For realistic case, the bound W as used for ideal case simulation for SBPN case yields huge miss-distance and hence has not been used. The formulations given in [6] for choosing gains are rather complex to be used and neither using W as bound on target acceleration as suggested in [3] yields realizable results. Using W as a bound on target acceleration works fine to suppress the variations in $\dot{\theta}$ but does not fulfill the objective of achieving acceptable miss-distance. This presents a generic issue with the sliding mode technique where we find the target is able to escape even though the variations in $\dot{\theta}$ are minimized. The value used for W for case (i) is 80, case (ii) is 20, and for case (iii) is 10. However, in Table 2, it can be seen that for the case (iii), still a large miss-distance is observed. Decreasing the value of W could produce lesser miss-distance but only at the cost of higher jerk produced in the end phase.

From Fig. 2, it is evident that in case of SBPN, the rate of control (i.e., rate of change of latex) has peaks at the points where $\dot{\theta}$ lies in the vicinity of zero. These peaks are obviously due to the use of signum approximation in SBPN.

6 Conclusion

A new dimension of performance analysis, i.e., latex rate, has been explored in this paper. With the prime focus on this performance parameter, a novel idea of the bounded rate of control (BRC)-based guidance strategy has been introduced to reduce the burden on the actuator. The performance of this strategy has been compared with the classical PN law and sliding mode-based SBPN law for the class of weaving targets with large accelerations. It is seen that BRC has an end-phase advantage over PN law. The BRC law is also superior as compared to SBPN since SBPN produces significant latex rates during the engagement.

References

1. Jeon I-S, Lee J-I, Tahk M-J (2006) Impact-time-control guidance law for anti-ship missiles. IEEE Trans Control Syst Technol 14(2):260–266
2. Harl N, Balakrishnan S (2012) Impact time and angle guidance with sliding mode control. IEEE Trans Control Syst Technol 20(6):1436–1449
3. Phadke S, Talole SE (2012) Sliding mode and inertial delay control based missile guidance. IEEE Trans Aerosp Electron Syst 48(4):3331–3346
4. Zarchan P (1994) Tactical and strategic missile guidance. American Institute of Aeronautics and Astronautics Inc, Washington, DC
5. Zarchan P (1995) Proportional navigation and weaving targets. J Guid Control Dyn 18 (5):969–974
6. Babu KR, Sarma I, Swamy K (1994) Switched bias proportional navigation for homing guidance against highly maneuvering targets. J Guid Control Dyn 17(6):1357–1363
7. Shtessel YB, Shkolnikov IA, Levant A (2007) Smooth second-order sliding modes: missile guidance application. Automatica 43(8):1470–1476
8. Laporte J, Chaillet A, Chitour Y (2017) Global stabilization of linear systems with bounds on the feedback and its successive derivatives. SIAM J Control Optim 55(5):2783–2810

PWM-Based Proxy Sliding Mode Controller for DC–DC Buck Converters

Kumar Abhishek Singh[✉], Sandeep Soni, Ankit Sachan,
and Kalpana Chaudhary

Department of Electrical Engineering, Indian Institute of Technology (BHU),
Varanasi, India
{kabhisheksingh.rs.eee17,sandeepkrsoni.rs.eee17,
ankits.rs.eee15,kchaudhary.eee}@iitbhu.ac.in

Abstract. In this paper, a proxy sliding mode control (PSMC) is designed for a DC–DC buck converter. The mathematical form of the controller combines the proportional–integral–derivative controller and a sliding mode controller in an algebraic way. The objective of the control is to regulate the output voltage of the DC–DC buck converter in the presence of line voltage and load uncertainty. Simulation and experimental results of the DC–DC buck converter are carried out to demonstrate the efficacy of the proposed controller.

Keywords: Sliding mode control (SMC) · Proportional–integral–derivative (PID) · Buck converter

1 Introduction

Sliding mode controllers (SMCs) have been widely used for DC–DC converters because of their inherently structured behavior [1–3]. In [4–6], SMCs are based on hysteresis or delta modulation. A demerit of this method is the variable switching frequency of operation. To overcome this issue, an additional mechanism required to ensure a constant switching frequency. An alternative way is the use of an equivalent control (derived using a sliding mode control approach) to modulate the pulse width modulator. In this way, the controller acts like a traditional duty cycle controller with a constant switching frequency [7–10].

In the proposed paper, we use proxy sliding mode control (PSMC) as an equivalent control to modulate the pulse width modulator. PSMC controller is a modified version of SMC control and PID controller. A physical interpretation of the above approach can be understood with the aid of Fig. 1.

In a PSMC, the virtual object, also known as a proxy, is connected to an actual controlled object (DC–DC buck converter) through a virtual coupling that acts as PID control. The other end of the proxy is connected to the SMC controller. In the theory of PSMC, the discontinuous signum-type function algebraically transferred to saturation function, which provides chattering-free phenomena.

The primary motivation to use PSMC is firstly to avoid the chattering effect, which is undesired behavior for practical systems [11, 12]; secondly, robustness against load and input voltage uncertainties. Authors Kikuuwe and Fujimoto developed the PSMC

© Springer Nature Singapore Pte Ltd. 2020
N. Goel et al. (eds.), *Modelling, Simulation and Intelligent Computing*,
Lecture Notes in Electrical Engineering 659,
https://doi.org/10.1007/978-981-15-4775-1_39

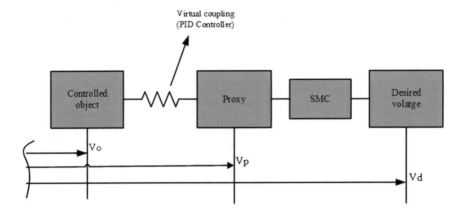

Fig. 1 Physical illustration of PSMC

control approach in [13–15]. Due to various industrial applications, this concept attracts attention from numerous control research group. To the best of the author's knowledge, this is the first attempt to use a PWM-based PSMC controller for DC–DC buck converter.

This paper is constructed as follows: Sect. 2 presents mathematical preliminaries and the system model description. Section 3 provides the PSMC design for the DC–DC buck converter. Section 4 demonstrates the simulation and experimental results. Later, concluding remarks are provided in Sect. 5.

2   Mathematical Preliminaries and Model Description of DC–DC Buck Converter

2.1   Mathematical Preliminaries

In this paper, $\mathbb{R}$ denotes the set of all real numbers, $\mathbb{N}$ denotes the set of all the natural numbers, $\| \cdot \|_p$ denotes the p-norm, whereas $\| \cdot \|_2$ denotes the 2-norm or induced matrix norm. The signum and saturation function is defined as follows:

$$\text{sign}(z) = \begin{cases} -1 & \text{if } z < 0 \\ [-1, 1] & \text{if } z = 0 \\ 1 & \text{if } z > 0. \end{cases} \quad \text{sat}(z) = \begin{cases} -1 & \text{if } z < -1 \\ z & \text{if } z \in [-1, 1] \\ 1 & \text{if } z > 1. \end{cases} \tag{1}$$

The analytical relation between signum function and saturation function is given here:

$$x = \text{sign}(z - x) \Rightarrow x = \text{sat}(z), \quad \forall x, z \in \mathbb{R}. \tag{2}$$

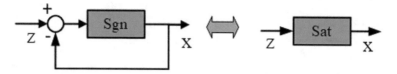

Fig. 2 Block diagram representation of (2)

The proof of the above relation is as follows:

$$x = \text{sign}(z - x) \qquad\qquad x = \text{sat}(z)$$

$$x = \begin{cases} -1 & \text{if } z - x < 0 \\ [-1, 1] & \text{if } z - x = 0 \\ 1 & \text{if } z - x > 0. \end{cases} \qquad x = \begin{cases} -1 & \text{if } z < -1 \\ z & \text{if } z \in [-1, 1] \\ 1 & \text{if } z > 1. \end{cases}$$

The graphical representation of relation (2) is shown in Fig. 2. Some other important relations which are used in this paper are defined as follows:

$$x = W\text{sign}(Y(z - x)) \Rightarrow x = W\text{sat}\left(\frac{z}{W}\right). \tag{3}$$

In a similar manner

$$x + Uy = Y\text{sign}(z - Wx) \Rightarrow x = -Uy + Y\text{sat}\left(\frac{\frac{z}{W} + Uy}{Y}\right). \tag{4}$$

where $U, Y, W > 0$ and $x, y, z \in \mathbb{R}$. This relation holds because of $\text{sign}(Yz) = \text{sign}(z), \quad \forall Y > 0, z \in \mathbb{R}$.

2.2 Model Description of DC–DC Buck Converter System

A basic PWM-based DC–DC buck converter is shown in Fig. 3, where i_L is the inductor current, v_{in} is the input voltage, v_o is the capacitor voltage, L is the inductor, C is the capacitor, and R is the load resistance. Under continuous conduction mode, the dynamics of the buck converter is as follows:

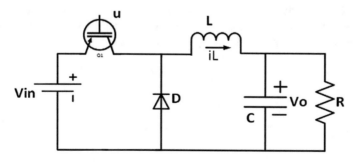

Fig. 3 Average model circuit of the buck converter

$$\begin{cases} L\dfrac{di_L}{dt} = uv_{in} - v_o \\ C\dfrac{dv_o}{dt} = i_L - \dfrac{v_o}{R}. \end{cases} \tag{5}$$

3 Proxy-Based Sliding Mode Control (PSMC)

The physical model of PSMC is illustrated in Fig. 1. In PSMC, a controlled object (DC–DC buck converter) is connected to a proxy through a virtual coupling that performs a PID control action to maintain its length to be zero.

Let $v_o \in \mathbb{R}$ is the output voltage of the controlled object (DC–DC buck converter), $v_p \in \mathbb{R}$ is the voltage of proxy object, and $v_d \in R$ is the desired output voltage for the controlled object. $F_{PID} \in \mathbb{R}$ is the force produced by the PID-type virtual coupling which can be defined as

$$F_{PID} := K_I \alpha + K_P \dot{\alpha} + K_D \ddot{\alpha} \tag{6}$$

where

$$\alpha := \int_0^t (v_p - v_o)\,dt \tag{7}$$

and K_I, K_P, K_D are positive real coefficient which represents the integral, proportional, and derivative gains, respectively. The selection of these parameters value is such that v_o is controlled to follow v_p. Force $F_{SMC} \in \mathbb{R}$ produced by the SMC controller, which is applied to the virtual object, as follows

$$F_{SMC} := K\,\mathrm{sign}\big(v_d - v_p + H(\dot{v}_d - \dot{v}_p)\big) \tag{8}$$

where $K > 0$. The force F_{PID} is directly connected to the controlled object (DC–DC buck converter), and its repulsive force is applied to the virtual object (proxy). Therefore, the dynamics of the proxy can be defined as

$$m\ddot{v}_p = F_{SMC} - F_{PID}. \tag{9}$$

Figure 4a shows the block diagram interpretation of the controller along with m. A proxy is a virtual object whose mass can be selected as zero. Then, (9) can be written as $F = F_{PID} = F_{SMC}$. Equations (6) and (8) can be rewritten as

$$\sigma := (v_d - v_o + H(\dot{v}_d - \dot{v}_o)) \tag{10a}$$

$$F = K_I \alpha + K_P \dot{\alpha} + K_D \ddot{\alpha} \tag{10b}$$

$$0 = F - K\,\mathrm{sign}(\sigma - \dot{\alpha} - H\ddot{\alpha}). \tag{10c}$$

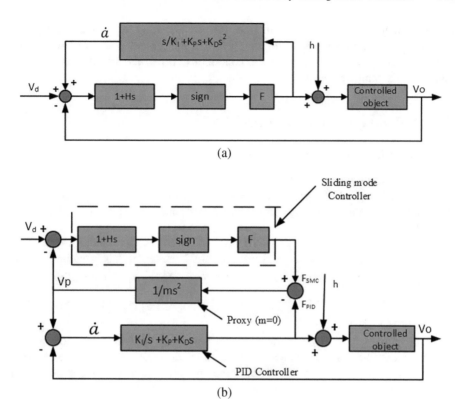

Fig. 4 Block diagram representation of PSMC

These equations are a continuous-time state-space representation of PSMC. Equation (10c) can be seen as an implicit state equation representing the state vector $[\alpha, \dot{\alpha}]$. Equation (10b) presents the actuator force F, which should be instructed to the actuator. Figure 4b is the block diagram of (10) and the block of F and a is interchanged. In Fig. 4, signum function is enclosed within a closed loop in the controller. Therefore, a signum function can be algebraically transferred to saturation (sat) function.

Using Eq. (4), we can rewrite (10) as follows:

$$\sigma = (v_d - v_o + H(\dot{v}_d - \dot{v}_o)) \tag{11a}$$

$$\ddot{\alpha} = -\frac{K_P\dot{a} + K_I\alpha}{K_D} + \frac{K}{K_D}\,\text{sat}\left(\frac{K_D}{K}\left\{\frac{\sigma - \dot{\alpha}}{H} + \frac{K_P\dot{a} + K_I\alpha}{K_D}\right\}\right) \tag{11b}$$

$$F = K\,\text{sat}\left(\frac{K_D}{K}\left\{\frac{\sigma - \dot{\alpha}}{H} + \frac{K_P\dot{a} + K_I\alpha}{K_D}\right\}\right). \tag{11c}$$

From the above discussion, the signum function can be rolled out as a sat function. Equations (10) and (11) are algebraically similar, which implies that (10) is a saturated controller that is genuinely different from directly changing the signum function by saturation function via approximations. Therefore, the presented control law does not cause chattering. Afterward, apply the controller (11c) in the PWM block to get a constant switching frequency.

4 Simulation and Experimental Results

4.1 Numerical Simulations

First, the average model (5) of the DC–DC buck converter is numerically simulated. To show the advantage of the proposed method over conventional sliding mode control, we will use simulation results to compare the performance among them for the average model (5) (Table 1).

The parameter values for PSMC controller is $K = 5, K_P = 300, K_D = 50, K_I = 25, H = 0.1$. Assume that the conventional SMC controller has a sliding surface $s := K_1 e_1 + e_2$, where $e_1 := v_o - v_d$ and $e_2 := {}^{i_L}/_C - {}^{v_o}/_{(RC)}$. At the sliding surface, one can calculate $u := {}^{(LCv + v_d)}/_{v_{in}}$, where $v := -e_2 + \frac{e_1}{(LC)} + \frac{e_2}{(LC)} - K_2\text{sign}(s)$,, with $K_1 = 1, K_2 = 10$.

With these values of the parameter, we have simulated both the controller in MATLAB/Simulink, as we can see from the simulation result of Fig. 5 the proposed controller provides better results over conventional SMC Controller in the presence of sudden input voltage changes. Therefore, the simlation results confirm that the proposed controller is robust against sudden input changes.

4.2 Experimental Results

To verify the efficacy of the proposed controller, experimental results are presented. The experimental setup is shown in Fig. 6. The experimental hardware is performed with a DSP processor with MATLAB/Simulink, and parameter values are the same as the simulation part. Buck converter performance under varying input voltage (v_{in}) is shown in Fig. 7.

Table 1 Component values of DC–DC buck converter

Description	Parameter	Nominal value
Input voltage	v_{in}	20 V
Desired output voltage	v_d	13 V
Inductance	L	1.5 mH
Capacitance	C	100 pF
Load resistance	R	100 Ω

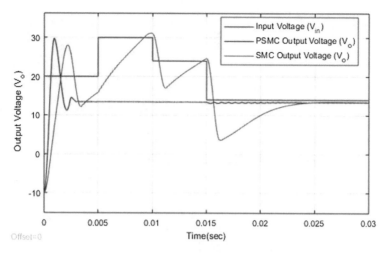

Fig. 5 Output voltage of the DC–DC buck converter using both PSMC and conventional SMC

Figure 7 shows that despite the change in input voltage (v_{in}), the proposed controller maintains output voltage (v_o) as desired and inductor current (i_L) constant. Figure 8 shows that despite the variation in load resistance (one can see via varying inductor current), the proposed controller maintains output voltage as desired. Therefore, the experimental results of the DC–DC buck converter show that the proposed converter is robust against sudden change in input voltage and load variations.

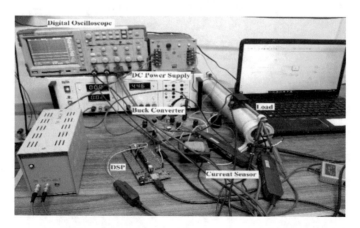

Fig. 6 Experimental setup of the DC–DC buck converter

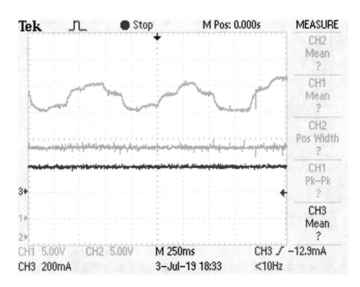

Fig. 7 Performance of the converter with the input voltage varying v_{in} periodically stepwise between 20 and 30 V. (Top) Input voltage v_{in} (5 V/div). (Middle) Output voltage v_o (5 V/div). (Bottom) Inductor current i_L (0.2A/div)

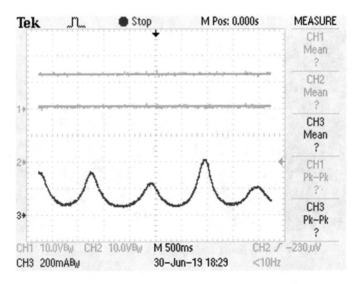

Fig. 8 Performance of the converter with the load resistor varying periodically stepwise between 30 and 300 Ω. (Top) Output voltage v_o (10 V/div). (Middle) Input voltage v_{in} (10 V/div). (Bottom) Inductor current i_L (0.2 A/div)

5 Concluding Remarks

This paper is proposed a PSMC controller for the DC–DC buck converter. With the help of experimental and simulation results, it has been shown that the controller is robust in the presence of sudden input voltage changes and load variations. The advantage of PSMC is chattering-free action, which is desirable condition for the industrial applications. The future extension of this work can be used in the design of another DC–DC converter such as buck, buck–boost, and Cuk converter.

References

1. Vidal-Idiarte E, Carrejo CE, Calvente J, Martinez-Salamero L (2010) Two-loop digital sliding mode control of dc-dc power converters based on predictive interpolation. IEEE Trans Industr Electron 58(6):2491–2501
2. Wai R-J, Shih L-C (2010) Design of voltage tracking control for DC-DC boost converter via total sliding-mode technique. IEEE Trans Industr Electron 58(6):2502–2511
3. Cardim R, Teixeira MC, Assuncao E, Covacic MR (2009) Variable- structure control design of switched systems with an application to a dc-dc power converter. IEEE Trans Industr Electron 56(9):3505–3513
4. Cardoso B, Moreira A, Menezes B, Cortizo P (1992) Analysis of switching frequency reduction methods applied to sliding mode con-trolled dc-dc converters. In: Proceedings of APEC'92 seventh annual applied power electronics conference and exposition. IEEE, pp 403–410
5. Malesani L, Spiazzi R, Tenti P (1995) Performance optimization of cuk converters by sliding-mode control. IEEE Trans Power Electron 10(3):302–309
6. Mattavelli P, Rossetto L, Spiazzi G (1997) Small-signal analysis of DC–DC converters with sliding mode control. IEEE Trans Power Electron 12(1):96–102
7. Mahdavi J, Emadi A, Toliyat H (1997) Application of state space averaging method to sliding mode control of PWM DC/DC converters. In: IAS'97 conference record of the 1997 IEEE industry applications conference thirty-second IAS annual meeting, vol 2. IEEE, pp 820–827
8. Mahdavi J, Nasiri MR, Agah A, Emadi A (2005) Application of neural networks and state-space averaging to dc/dc PWM converters in sliding -mode operation. IEEE/ASME Trans Mechatron 10(1):60–67
9. Tan S-C, Lai Y-M, Tse CM (2006) A unified approach to the design of PWM-based sliding-mode voltage controllers for basic DC–DC converters in continuous conduction mode. In: IEEE transactions on circuits and systems. I regular papers
10. Tan S-C, Lai Y-M, Chi KT (2008) General design issues of sliding- mode controllers in dc-dc converters. IEEE Trans Industr Electron 55(3):1160–1174
11. Utkin V, Guldner J, Shi J (2009) Sliding mode control in electro-mechanical systems. CRC press
12. Edwards C, Spurgeon S (1998) Sliding mode control: theory and applications. CRC Press
13. Kikuuwe R, Fujimoto H (2006) Proxy-based sliding mode control for accurate and safe position control. In: Proceedings 2006 IEEE international conference on robotics and automation. ICRA 2006. IEEE, pp 25–30

14. Kikuuwe R, Yasukouchi S, Fujimoto H, Yamamoto M (2010) Proxy-based sliding mode control: a safer extension of PID position control. IEEE Trans Rob 26(4):670–683
15. Kikuuwe R (2017) Some stability proofs on proxy-based sliding mode control. IMA J Math Control Inf 35(4):1319–1341

Real-Time Air Quality Estimation from Station Data Using Extended Fractional Kalman Filter

Bijoy Krishna Mukherjee[1](✉)(iD) and Santanu Metia[2,3,4](iD)

[1] Department of Electrical and Electronics Engineering, Birla Institute of Technology and Science Pilani, Pilani Campus, Pilani, Rajasthan, India
bijoy.mukherjee@pilani.bits-pilani.ac.in
[2] Environmental Quality, Atmospheric Science & Climate Change Research Group, Ton Duc Thang University, Ho Chi Minh City 700000, Vietnam
metia.santanu@tdtu.edu.vn
[3] Faculty of Environment & Labour Safety, Ton Duc Thang University, Ho Chi Minh City 700000, Vietnam
[4] Faculty of Engineering & IT University of Technology, University of Technology Sydney, Broadway NSW 2007, Sydney, Australia

Abstract. Air, soil and water pollutions have the greatest risk factors for human health. There are different types of air pollutants which are emitted from human activities. One of these pollutants is nitrogen dioxide (NO_2) which is produced from fossil fuel-based energy and use of motor vehicles. Since India is facing deteriorated air quality due to economic development, air quality management is becoming a real challenge. In 2015, an emission inventory (EI) was developed for India with 2015 as the base year. This EI is developed on an engineering model approach which is based on a technology-linked energy emission modeling approach. Accurate EI is important for future air quality modeling and air quality management. Since EI has uncertainties in data, some kind of estimation is essential. Estimation through extended fractional Kalman filter (EFKF) is considered in the present paper, and its performance is found to be superior as compared to a standard extended Kalman filter (EKF).

Keywords: Extended fractional Kalman filter · Emission inventory · Nitrogen dioxide

1 Introduction

Nitrogen oxides (NO_X) are pollutants which are playing an important role in both stratospheric and tropospheric chemistry. NO_X are defined as the sum of nitrogen dioxide (NO_2) and nitrogen oxide (NO). Formation of ozone (O_3) in the troposphere is due to the catalytic reaction of NO_X which contributes to the formation of secondary aerosols in atmosphere [1, 2]. NO_X also play an important role in the production of greenhouse gases like ground level O_3 in local, regional and global scales. Power plants and manufacturing industries are the main sources of NO_X. Road transports have also contributed NO_X level increase in India. The potential for increase in NO_X emitted outdoors to exacerbate climate change by increasing the amount of NO and NO_2

© Springer Nature Singapore Pte Ltd. 2020
N. Goel et al. (eds.), *Modelling, Simulation and Intelligent Computing*,
Lecture Notes in Electrical Engineering 659,
https://doi.org/10.1007/978-981-15-4775-1_40

released to the atmosphere is receiving less attention in India. Climate change study will require accurate emission inventories for both NO and NO_2.

There are two types of emission inventory (EI) estimations which are based on the top-down approach and the bottom-up approach. In [3], authors used the top-down approach to develop EI which was more accurate. The top-down approach is based on real data, whereas the bottom-up approach is based on detailed provincial economic and energy data [4]. However, in the real data having high process uncertainties and measured only at some specific location, some kind of estimation is unavoidable for a more reliable monitoring and management of air quality across a vast area. To this end, Matern function-based extended Kalman filter (EKF) and extended fractional Kalman filter (EFKF) have been used in the literature for air quality estimation in Sydney, Australia [5]. In the present paper, the same is extended for the state of West Bengal, India, for the base year data of 2015 [6]. Moreover, the fractional orders in EFKF are tuned using the Grünwald–Letnikov method (unlike genetic algorithm as done previously in [5]), and a higher-order Matern function is considered for modeling (fourth order instead of three). The results show definite improvement in estimation of NO_2 level as compared to an EKF.

2   Method

Methodology for this study is an extension of that presented in [3]. The information from [3] is briefly summarized here, and then the extended data analysis is described in detail.

2.1   Study Area

Near-road measurements of air pollutants were obtained at a study area in the state of West Bengal, India (Fig. 1). NO_2 data were recorded at 24 monitoring stations across West Bengal; out of them one was in a metropolitan city Kolkata and one in a major industrial city Durgapur. Data were recorded from the near-road site between January 1, 2015, and December 31, 2015. Other 22 locations are suitably chosen covering the whole state.

2.2   Emission Inventories and Uncertainties

Figure 2 shows the top-down approach to develop EI. Monitoring data have uncertainties due to measurement errors, process uncertainties and other factors. The extended fractional Kalman filter has the potential of providing an improved estimate of the actual pollutant levels.

2.3   Extended Fractional Kalman Filter

The EFKF is particularly suitable for accurate and effective state estimation of highly nonlinear systems, where additive noise, initial deviation, process disturbance and inevitably missing measurements do not affect the prediction performance appreciably

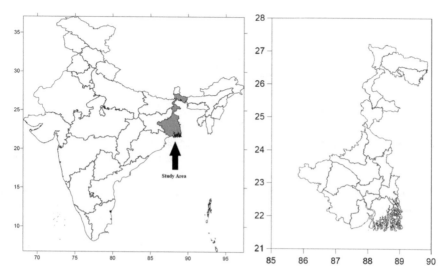

Fig. 1 Study area, West Bengal in India

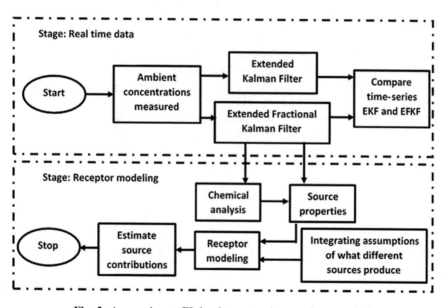

Fig. 2 Approaches to EI development using top-down method

[7]. In outdoor air quality modeling, an EFKF with Matèrn function-based covariances has been applied for pollutant prediction [5] to improve the accuracy of inventories to complement missing data taking into account the spatial distribution of the indoor air quality profiles. Here, by adopting a Matèrn correlation function for a length scale $l = \sqrt{5}/\lambda$, the model for EKF is proposed as

$$\frac{df(t_n)}{dt} = \begin{bmatrix} 0 & 1 & 0 & 0 \\ 0 & 0 & 1 & 0 \\ 0 & 0 & 0 & 1 \\ -\lambda^4 & -4\lambda^3 & -6\lambda^2 & -4\lambda \end{bmatrix} f(t_n) + \begin{bmatrix} 0 \\ 0 \\ 0 \\ 1 \end{bmatrix} w(t_n)$$

$$y(t_n) = \begin{bmatrix} 1 & 0 0 & 0 \end{bmatrix} f(t_n) + d(t_n) \tag{1}$$

where λ is a positive constant for the system quadruple pole (at—λ) depending on the correlation length l of the Gaussian process involved [5], $f(t_n)$ represents monitoring station data assumed to have initial zero mean and covariance matrix $diag\{0.1\}$ with measurement variance 0.5^2 and spectral density of process noise 10^{-6}. In the present case, the length scale is taken as the average distance between the monitoring stations which turns out to be about 85 km.

2.4   Fractional-Order Identification

Fractional-order systems are considered as a generalization of integer-order systems to improve the mathematical representation of the actual dynamic system. In this work, the fractional-order derivative is modeled using the Grünwald–Letnikov method and implemented using the fractional-order modeling and control (FOMCON) toolbox in MATLAB [8] with data collected in the time domain from monitoring station data. Air pollutant concentrations, after conversion, are to be processed for prediction of abnormalities using the EFKF where the fractional-order transfer function is identified. Here, the black box modeling [9, 10] is applied to infer a dynamic system model based upon experimentally collected data. This filter model represents a relationship between system inputs and outputs under external stimuli in order to determine and predict the system behavior. Let y_r denote the experimental pollutant profile using Eq. (1) as the plant output and y_m the identified model output. We consider the single-input and single-output (SISO) case where both y_r and y_m are $N \times 1$ vectors with the model output error:

$$\varepsilon = y_r - y_m, \tag{2}$$

where estimation performance can be evaluated via the mean squared error:

$$\varepsilon_{MSE} = \frac{1}{N} \sum_{i=0}^{N} \varepsilon_i^2 = \frac{\|\varepsilon\|_2^2}{N} \tag{3}$$

From conventional system identification methods, a corrected indoor air quality profile can be obtained from the corresponding rational transfer function as follows:

$$F(s) = \frac{1}{a_4 s^4 + a_3 s^3 + a_2 s^2 + a_1 s + a_0} \tag{4}$$

Table 1 Fractional-order system estimated by using FOMCON

Station	Fractional-order system	ε_{MSE}
Durgapur	$1/\left(3.2s^{3.5} + 10.63s^{2.8} + 16.8s^{1.5} + 0.5s^{0.67} + 6.8s^{0.01}\right)$	0.21
Haldia	$1/\left(1.2s^{4.5} + 18.63s^{2.7} - 1.5s^{1.05} + 9.8s^{0.91} + 0.01s^{0.61}\right)$	0.58
Kolkata	$1/\left(6.7s^{3.1} + 18.1s^{2.6} + 0.8s^{1.08} + 7.5s^{0.86} + 0.8s^{0.05}\right)$	0.33
Siliguri	$1/\left(0.2s^{4.1} + 81.64s^{2.7} + 1.66s^{1.8} + 0.01s^{0.91} + 0.12s^{0.72}\right)$	0.41
Malda	$1/\left(0.02s^{4.1} + 11.3s^{2.8} + 0.7s^{1.6} + 1.5s^{0.99} + 0.17s^{0.55}\right)$	0.53

where $a_4 = 1$, $a_3 = 1.058 \times 10^{-1}$, $a_2 = 4.2 \times 10^{-3}$, $a_1 = 7.408 \times 10^{-5}$, and $a_0 = 4.9 \times 10^{-7}$.

In fractional-order modeling, right-hand side of Eq. (1) is considered to be a fractional-order derivative, and therefore, Eq. (4) generalizes to a fractional-order transfer function given by

$$F^{\alpha}(s) = \frac{1}{a_4 s^{\alpha_{a_4}} + a_3 s^{\alpha_{a_3}} + a_2 s^{\alpha_{a_2}} + a_1 s^{\alpha_1} + a_0 s^{\alpha_{a_0}}}. \tag{5}$$

Table 1 shows the values of fractional orders obtained by using the FOMCON toolbox with the initial transfer function from Eq. 5 for five monitoring stations.

3   Results

In order to evaluate the performance of prediction, we introduce several model performance measures including mean absolute percentage error (MAPE), root mean square error (RMSE) and R^2 (the coefficient of determination), defined, respectively, as follows:

$$\text{MAPE} = \frac{100}{n} \sum_{j=1}^{n} \left(\frac{|a_j - b_j|}{|a_j|} \right), \tag{6}$$

$$\text{RMSE} = \sqrt{\frac{1}{n} \sum_{j=1}^{n} (a_j - b_j)^2}, \tag{7}$$

$$R^2 = 1 - \left(\frac{\sum_{j=1}^{n} (b_j - a_j)^2}{\sum_{j=1}^{n} (b_j)^2} \right), \tag{8}$$

where a_j and b_j are the forecast and observed values, and n is the number of samples. MAPE and RMSE are applied as performance criteria of the prediction model to quantify the errors of forecasting values. The coefficient of determination R^2 is used to

Table 2 Performance statistics of EFK and EFKF

Station	Latitude	Longitude	EKF			EFKF		
			MAPE (%)	RMSE	R^2	MAPE (%)	RMSE	R^2
Durgapur	23.5204°	87.3119°	4.10	2.9707	0.931	3.10	2.3142	0.957
Haldia	22.0627°	88.0833°	7.60	1.8831	0.839	6.24	1.5195	0.898
Kolkata	22.5726°	88.3639°	7.79	5.6743	0.954	6.60	4.7103	0.968
Siliguri	26.7271°	88.3953°	7.78	2.4279	0.857	6.15	1.9317	0.911
Malda	25.0108°	88.1411°	5.31	1.4439	0.845	4.30	1.2136	0.893

assess the strength of the relationship of the estimation to the accurate observation. Table 2 provides comparative statistics of EKF and EFKF prediction data at the five key monitoring stations considered.

Figure 3a, b shows the time series and estimation error plots for NO_2 emission at the Kolkata monitoring station. The EKF shows 4.1% as MAPE, and the EFKF shows 3.1% MAPE at the Durgapur monitoring station. From Table 2, it can be easily

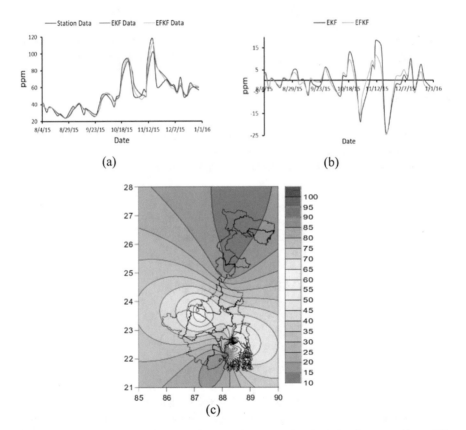

(a)   (b)

(c)

Fig. 3 **a** Time series plot of NO_2 at Kolkata monitoring station; **b** Estimation error plot; **c** NO_2 distribution on November 16, 2015, based on station data

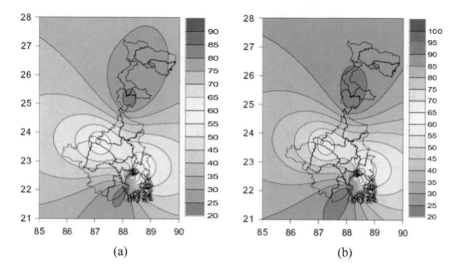

Fig. 4 NO$_2$ distribution on November 16, 2015, based on filter outputs: **a** EKF; **b** EFKF

concluded that the EFKF is more accurate than the EKF. The EFKF can handle uncertainties better than the EKF in terms of the pollutant profile estimation. Figure 3c shows the spatial plot of NO$_2$ level across the whole state and the surrounding areas obtained from extrapolating the 24 station data using Kriging method. From the climatic point of view, it can also be interpreted from Fig. 3c that NO$_2$ emissions produced by motor vehicle exhaust and industrial plants in Kolkata and surrounding areas have high concentration levels.

Durgapur, being another industrial town having power plants, steel manufacturing plants and other industries, similar trend is observed therein. Similar trend is observed from the extrapolation of the two filter outputs in Fig. 4a, b. The EKF estimated data and the EFKF estimated data are compared in Fig. 4. Figure 5 shows the difference between monitoring station data and the outputs of the two filters, namely EKF and EFKF. It is readily observed that EFKF has more accurate estimation than the EKF. Therefore, to develop EI, the EFKF is more reliable than the EKF.

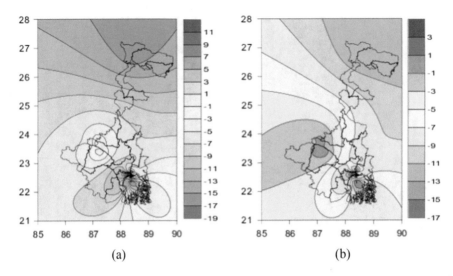

Fig. 5 NO$_2$ distribution difference between station data and filter outputs on November 16, 2015: **a** EKF; **b** EFKF

4   Conclusion

NO$_2$ concentration level is an important factor of air quality of an area. Hence, monitoring this level in key NO$_2$ producing cities such as Kolkata and Durgapur in West Bengal, India, forms a part of the policy of the state government. For accurate air quality monitoring across the state, either the underlying mathematical model of air quality dynamics should be accurately known or a large number of monitoring stations should be established. In the present paper, it is shown that instead a Matèrn covariance function-based EFKF can very faithfully estimate both in time and spatial domains the NO$_2$ level from a few key station data. The proposed filter is shown to outperform a standard EKF. Accurate estimation of NO$_2$ pollutant in achieved by using the EFKF with Matèrn function-based covariance taking into account the correlation length and smoothness of the spatiotemporal pollutant profile, which is not possible with conventional EKF. As further research, other filters such as the unscented Kalman filter or H$_\alpha$ filter may be extended to fractional versions, and it may be investigated if they can provide further accuracy in air quality estimation.

References

1. Wang J, Yang Y, Zhang Y, Niu T, Jiang X, Wang Y, Che H (2019) Influence of meteorological conditions on explosive increase in O$_3$ concentration in troposphere. Sci Total Environ 652:1228–1241
2. Wang R, Tie X, Li G, Zhao S, Long X, Johansson L, An Z (2019) Effect of ship emissions on O$_3$ in the Yangtze River Delta region of China: analysis of WRF-Chem modeling. Sci Total Environ 683:360–370

3. Saide P, Zah R, Osses M, Ossés de Eicker M (2009) Spatial disaggregation of traffic emission inventories in large cities using simplified top–down methods. Atmos Environ 43 (32):4914–4923

4. Zhao Y, Nielsen CP, McElroy MB (2012) China's CO_2 emissions estimated from the bottom up: recent trends, spatial distributions, and quantification of uncertainties. Atmos Environ 59:214–223

5. Metia S, Oduro SD, Duc HN, Ha Q (2016) Inverse air-pollutant emission and prediction using extended fractional Kalman filtering. IEEE J Sel Topics Appl Earth Obs Remote Sens 9(5):2051–2063

6. http://www.data.gov.in

7. Hu X, Yuan H, Zou C, Li Z, Zhang L (2018) Co-estimation of state of charge and state of health for lithium-ion batteries based on fractional order calculus. IEEE Trans Veh Technol 6 (11):10319–10329

8. Tepljakov A (2017) Fractional-order modeling and control of dynamic systems (Springer Theses). Springer

9. Ljung L (1999) System identification: theory for the user, 2nd edn. Prentice Hall PTR, Upper Saddle River, NJ, USA

10. Metia S, Oduro SD, Sinha AP (2020) Pollutant profile estimation using unscented Kalman filter. In: Basu T, Goswami S, Sanyal N (eds) Advances in Control, Signal Processing and Energy Systems. Lecture Notes in Electrical Engineering vol 591. Springer, Singapore

Negotiating Deals Using Artificial Intelligence Models

Sujith Sizon, Somil Mathur, and Nilesh Goel[✉]

Birla Institute of Technology and Science Pilani, Dubai Campus, Academic City,
Dubai, United Arab Emirates
f20160277@dubai.bits-pilani.ac.in, goel.
nilesh@gmail.com

Abstract. Recent years have seen an increased demand in negotiation technologies, seen as a key coordination mechanism for the interaction of providers
and consumers that optimize the selling of different kind of goods in industries
like real estate and used car marketplace. Suggested applications range from
modeling interactions between customers and merchants in retail electronic
commerce, to the online sale of information goods, or reducing operational
procurement costs of large companies. A new tenant could use an AI agent to
negotiate the final lease for his or her apartment. Usually tenants, landlords, and
their respective brokers typically shed in a lot of time discussing and negotiating
details of the lease terms, to achieve fair pricing. AI can help the negotiation
process by grounding it in hard data and clear analysis. During this project,
several laboratory-based focus group studies will be held to generate initial
dataset and train a neural network-based AI model to negotiate the best deals.

Keywords: Artificial intelligence · Negotiation and bargaining

1 Introduction

Negotiation comes into play in almost all kinds of social, political, and economic
setting, but still people avoid it because of lack of skill or domain expertise in that
particular sector like the real estate market, etc. This ends up creating a high-income
inequality and results in unfair trade. An attempt to fix these issues has led to an
increasing focus on development of artificial intelligence-based negotiation agents
which automatically negotiate with others using its vast domain knowledge and
negotiation experience.

Negotiation is simultaneously both a language-based and reasoning-based problem
in which an idea/deal must be formulized and the projected using dialogues. Such
dialogues will contain both cooperative and non-cooperative aspects which an autonomous agent must understand plan and generate to win the negotiation battle. We have
used a mechanical Turk [1], to collect information from human users, negotiating a
better deal on a few items where the mode of payment is items kind of like a barter
system. The goal is to show that end to end neural models can be trained to negotiate
by allowing them to make human like dialogues. Providing an expandable and topic
independent does not incorporate the skills required to negotiate perfectly. The goal of

© Springer Nature Singapore Pte Ltd. 2020
N. Goel et al. (eds.), *Modelling, Simulation and Intelligent Computing*,
Lecture Notes in Electrical Engineering 659,
https://doi.org/10.1007/978-981-15-4775-1_41

Fig. 1 A sample user experience of our negotiation interface

the autonomous agents during training is to maximize reward instead of just closing a deal; this is implemented by using self-play reinforcement learning. In total, we generated and used up to 5000 dialogue pieces between two humans to land multiple negotiation deals [2]. On the mechanical Turk interface, each user has a set of items in his possession with a value for each; they are asked to negotiate and exchange those items with another person who has different items but undisclosed cost function (Fig. 1).

Artificial intelligence-based negotiation agents can offer a lot of perks like better winning deal ratio, reduction in negotiation time, cost, stress, and cognitive and intellectual effort required by the negotiator. To accommodate the wide spectrum of negotiation, the AI agent has to have different kinds of features including self-reliance and freedom to perform its actions and consequently remain dependent on the user. Design and implementation of a fully self-sufficient, self-directed, and interdependent AI-based negotiation agent is what we are trying to achieve in this paper. We examine and address the main challenges associated with autonomous negotiation and how we attempt to tackle them based on various frameworks and neural networks [3].

After analyzing the results given by our autonomous agents, we find complex never before seen negotiation strategies. For example, we find our agents fake value a valueless case scenario or item so that it can later win a better deal by reversing the situation. Lying is a complex skill that requires understanding the other agents' beliefs and usually found in newborn children also. Even our autonomous agents learnt the art to lie to steer the deal in their best interests.

The applications of autonomous negotiating agents can be found in almost all industries nowadays. Salesperson or customer service representative could asses a particular situation and offer a better more accurate proposal or discount to solve the situation. The AI can process a lot of data across multiple real-life representatives and try to understand wherever there are learning opportunities and better negotiation opportunities. The AI-based negotiation agent can be applied as a chatbot, which are really good at negotiating (Fig. 2).

Autonomous negotiation largely consists of three main features. First one is self-sufficiency; it is the ability to take care of itself. Secondly, it should have self-

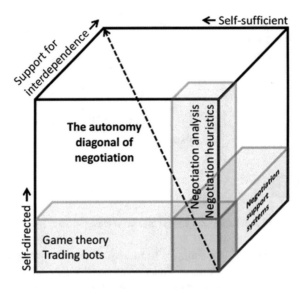

Fig. 2 Three perpendicular axes of autonomous negotiation research

directedness, ability to act under certain restrictions, or directions preset; thirdly, the support for interdependence so that the agent can be influenced and trained to behave based on certain dialogues by the user.

2 Accumulating Training Dataset

2.1 Overview

To initially train our AI negotiation agent, we initially build up a novel negotiation task and furthermore collected a dataset of human–human dialogues for this training. This undertaking and dataset pursue our proposed general system for examining semi-cooperative discourse. At first, every agent is provided information indicating the nature of the situation or the deal and a reward function which will score the result of the negotiation. These agents alternatively send each other messages until a final decision has reached. At the point when one agent chooses that an understanding has been made, the two agents freely yield what they think the concurred choice was. In the event that clashing choices are made, the two agents are given zero reward.

2.2 Task for Negotiation

Two agents are both indicated a similar gathering of things and trained to partition them with the goal that everything doled out to one operator. Every agent is given an alternate arbitrarily produced worth capacity, which gives a non-negative an incentive for every item. The value functions of each item are set to certain restrictions such as:

(1) Every product has non-negative value
(2) The accumulative total value of each person's total assets should be 10.

These requirements uphold that it is not workable for the two agents to get the highest score and that no item is useless to the two agents, so the negotiation will be always aggressive. After 10 turns, we permit agents the alternative to finish the negotiation with no understanding, which is worth 0 points to the two clients. We utilize 3 thing types (books, caps, balls), and somewhere in the range of 5 and 7 absolute things in the pool. Figure 1 shows our interface.

2.3   Data Collection

We gathered a lot of human–human discoursed utilizing Amazon Mechanical Turk. Laborers were paid $0.15 per exchange, with a $0.05 reward for maximal scores. We just utilized specialists situated in the USA with a 95% endorsement rating and at any rate 5000 past HITs. Our information gathering interface was adjusted based on Policy Network with two stage training for dialogue systems [4]. We gathered a sum of 5000 discoursed, in view of 2200 interesting situations (where a situation is the accessible things and qualities for the two clients). We held out a test set of 250 situations (500 exchanges). Holding out test situations implies that models must sum up to new case scenarios.

3   Building AI Models

We propose a straightforward however viable standard model for our conversational agent, in which a grouping to arrangement model is prepared to create the total exchange, adapted on an agent's info.

3.1   Data Encoding

Every conversation is changed over into two training datasets, indicating the total discussion from the point of view of every agent. The models vary on their information objectives, yield decision, and whether dialogues were read or composed. Training models contain a negotiation task objective g, indicating the available things and their qualities, a chat message x, and a yield choice o determining which things every autonomous agent will get. In particular, we speak to g as a rundown of six whole numbers relating to the check and estimation of every one of the three thing types. Chat message x is a rundown of tokens x_{0-t} (where t represents the time during the running period of our RNN model) [5], containing the turns of every operator interleaved with images checking whether a turn was composed by the agent or their accomplice, ending in an exceptional token showing one specialist has denoted that an understanding has been made. End result o is six whole numbers portraying what number of every one of the three thing types are allocated to every specialist. See Fig. 3.

Crowd Sourced Dialogue

Perspective Agent 1

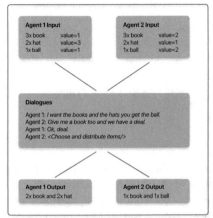

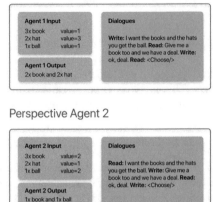

Fig. 3 Publicly collected chat messages (left) encoded into two training models (right)

3.2   Training the Agents

We trained an end-to-end network to automize the negotiation agent's dialogues based on input chat messages and required deal output requirements. Our training network included 4 recurrent neural networks (RNN). It is trained and modeled as gated recurrent unit (GRU) [6], which is basically a gating mechanism for the normal RNN models that were initially used. We use the input variables defined in Sect. 3.1 to define the states for our AI model. The negotiation scenario's objectives g is represented as GRU_g. For the training of the model, we represent each token by x_t. Each step or state derives values from its previous state. The time during running period of the training experiment is given by t. The final state represented as h^g is given by the following equation:

$$h_t = GRU_w(h_{t-1}, [Ex_{t-1}, h^g])$$ (1)

The above equation predicts the tokens for each step using an embedding matrix E using the previous hidden state h_{t-1} and previous token x_{t-1}. This model derives alternating agent's dialogues and allowing to be forwarded it to the consequent models as shown in Fig. 4. Toward the finishing point of the deal, the autonomous agents give a progression of tokens or let the other party know about its choice of items. We produce each yield restrictively freely, utilizing a different classifier for each negotiation event occurrence.

Supervised learning [7] plans to mimic the activities of human clients; however, it does not expressively attempt to boost an agent's objectives. Rather, we investigate pretraining with supervised learning, and afterward calibrating against the assessment metric utilizing reinforcement learning [8]. During reinforcement learning, an autonomous agent endeavors to improve its parameters from discussions with another agent B. While the other agent B could be a human, in our tests, we utilized our fixed

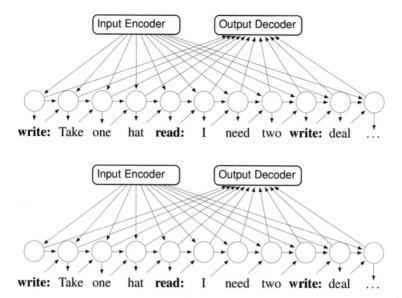

Fig. 4 The first shows AI model in supervised training stage and the second during RL stage

supervised model that was prepared to emulate humans. The subsequent model is fixed as we found that dynamically updating the parameters of both the agents prompted disparity from the human language. As a result, agent A figures out how to improve by reenacting discussions with the assistance of a surrogate forward model.

4 Experimental Observations

All the testings for our autonomous negotiation agents were carried out using PyTorch. All the variables were selected from the test dataset which we mined using Amazon mechanical Turk. During our learning state, we compared different approaches. One was supervised learning; another one was reinforcement learning (deal output-based self-play) and last a mixture of both. Our results show that normal supervised learning model produces the best humanly dialogues and the alternative one causes a negative effect on the dialogues making it more abstract. Nevertheless, majority output of the dialogues was human-readable responses.

4.1 Model Comparisons

We then compared the end-to-end output results of both supervised learning model and human-to-human dialogues on mechanical Turk. We analyzed a few stats for our model like:

(1) Negotiation Score: A ranking out of 10 for the autonomous agents which could be real life or our AI.

Table 1 Some stats of our mechanical Turk dataset

Metric	Training result
Number of dialogues	5000
Average duration of deal	6.6 turns
Average length of dialogue	7.6 words
Agreement score	80.1%
Avg. negotiation score	6.0/10.0
Pareto optimality	76.9%

(2) Agreement Level: Cases where equality of opinions was established that is, they both wanted the same set of items.

(3) Pareto Optimality [9]: Pareto optimality refers to a situation where both the agents scores are balanced like a see-saw if one has to increase the other has to decrease and, in our case, decrease in score means insufficient negotiation.

Be that as it may, we likewise find fundamentally more instances of the objective-based models neglecting to concur an arrangement with people—to a great extent a result of their progressively forceful negotiation strategies. Another major conclusion derived from our analysis is that "Goal-based models negotiate harder". This model has more conversation duration than supervised training model, suggesting that the autonomous agents are hard at work on their negotiation tactics and not quickly giving in on the deal and trying to achieve a max possible score for themselves. But side effect of this is that people could walk away from the deal and that would end up in a negative scoring for the agent. Figure 5 shows one of the negotiation situations where our model is very persistent and has a long conversation.

Input		
Human	3xbook *value*=2 3xhat *value*=1 1xball *value*=1	
RL+ROLLOUTS	3xbook *value*=1 3xhat *value*=2 1xball *value*=1	
Human	I'd like the books and the hats.	
RL+ROLLOUTS	I will take the books and hats	
Human	I really need the books and at least one other object.	
RL+ROLLOUTS	Ok, you can have one book and one ball	
Human	If I can have all the books, I can leave you the rest.	
RL+ROLLOUTS	Deal	
Output		**Reward**
RL+ROLLOUTS	3xhat 1xball	7/10
Human	3xbook	6/10

Fig. 5 Example of a very persistent negotiation agent holding out for higher score

During our training sessions, we also found out that "Models learn to deceive" during negotiation. One of the methods used by our autonomous agents is to negotiate a high scoring deal. Another common behavior we noticed was that the agents started out with a low-ball offer to later propose a counter offer in their favor.

5 Conclusion

We have used end-to-end learning of natural language negotiations as an assignment for AI, contending that it challenges both semantic and thinking abilities while having vigorous assessment measurements. We assembled a huge dataset of human–human negotiations, which contain an assortment of fascinating strategies. We have demonstrated that it is conceivable to prepare chatbots from start to finish and, however, that their capacity can be significantly better by training and decoding to amplify their objectives, as opposed to increasing their probability. The formation of autonomous negotiating agents that can bolster people with their negotiations is a multidisciplinary challenge. The characteristic semantic issue and the emotional matters involved implied that negotiation cannot be taken care of by AI alone, and a human–agent collaboration is required at multiple levels. By training several neural network-based GRU models on the negotiation datasets, we realized the potential advantages and shortcomings of the autonomous agents, key zones of agent innovation, man-made tactics, and AI-made strategies: Autonomous negotiating agents can beat individuals in terms of deal optimality, with influencing figuring for taking care of feelings, and inclination to realize what is significant in the negotiation.

References

1. Kimball A (2014) The statistical power of mechanical Turk. In: Proceedings of the purdue linguistics society. vol 9, pp 113–119
2. Tim B, Katsuhide F, Enrico HG, Koen H, Ito Takayuki, Nicholas RJ, Jonker C, Kraus Sarit, Lin Raz, Valentin R et al (2013) Evaluating practical negotiating agents: results and analysis of the 2011 international competition. Artif Intell 198:73–103
3. Bengio Y (2013) Deep learning of representations: looking forward. In: NIPS 2009, vol 7978, pp. 113–120
4. Fatemi M, Asri LE, Schulz H, He J, Suleman K (2016) Policy networks with two-stage training for dialogue systems. In: Conference: proceedings of the 17th annual meeting of the special interest group on discourse and dialogue, vol 17 ICML, pp 1873–1881
5. Sherstinsky A (2016) fundamentals of recurrent neural network (RNN) and long short-term memory (LSTM) network, vol arXiv:1808.03314. pp 192–245
6. Chung J, Gulcehre C, Cho KH, Bengio Y (2014) Empirical evaluation of gated recurrent neural networks on sequence modeling. In: NIPS 2014 workshop on deep learning
7. Osisanwo FY, Akinsola JET, Awodele Hinmikaiye J, Olakanmi, Akinjobi (2017) Supervised machine learning algorithms: classification and comparison. Int J Comput Trends Technol (IJCTT) 48(3):128–138

8. Gosavi Abhijit (2009) Learning Reinforcement: a tutorial survey and recent advances. Informs J Comput 21(2):178–192
9. Desai N, Critch A, Russell S (2018) Negotiable reinforcement learning for pareto optimal sequential decision-making. In: 32nd conference on neural information processing systems, (NeurIPS 2018), Montreal, Canada, pp 4717–4725

Level-Dependent Changes in Concurrent Vowel Scores Using the Multi-layer Perceptron

Akshay Joshi and Anantha Krishna Chintanpalli[✉]

Department of Electrical and Electronics Engineering, Birla Institute
of Technology and Science Pilani, Pilani Campus, VidyaVihar,
Rajasthan 333031, India
anantha.krishna@pilani.bits-pilani.ac.in

Abstract. An alternative computational model was developed to predict the level-dependent changes in the identification scores of both vowels for same and different fundamental frequency (F0) conditions. In this current study, the temporal-responses of the auditory-nerve model were the input layer to a multi-layer perceptron for predicting the identification scores of both vowels. The perceptron was trained to obtain the similar identification score (as observed in normal-hearing listeners) for different-F0 condition at 50 dB SPL. The training was done using the gradient descent with momentum and adaptive learning rate backpropagation algorithm, Finally, the perceptron was tested for same-and different-F0 conditions across various range of vowel levels. The model was successful qualitatively in predicting the level-dependent changes in concurrent vowel scores for same-and different-F0 conditions.

Keywords: Concurrent vowel identification · Neural network · F0 difference

1 Introduction

To study the effect of fundamental frequency (F0) difference, the concurrent vowel identification experiment is often studied (e.g., [1, 2, 5–7, 15]). In this experiment, two synthetic vowels, with equal level and duration, are presented simultaneously to listeners' one ear (i.e., monaural presentation) and they responded by identifying both the vowels, that were present. The general observation from these behavioral studies is that the percent identification score of both vowels improves with increasing F0 difference and then eventually asymptotes at ~ 3-Hz F0 difference or higher. Additionally, these studies are conducted at a specific vowel level. To understand the level-dependent changes in the ability to utilize F0 difference cue for identification, Chintanpalli et al. [4] collected the concurrent vowel data as a function of vowel level for same (0-Hz) and different (26-Hz) F0 conditions. Figure 1 shows the percent correct identification of both vowels for same- and different-F0 conditions, as a function of vowel level (modified from [4]). Their younger normal-hearing listeners (aged between 20 and 26 years) showed that there was an increase in concurrent vowel score as the level was increased from 25 to 50 dB SPL and then the score was decreased as the level was changed from 65 to 85 dB SPL. The F0 benefit is a quantitative metric used in the

© Springer Nature Singapore Pte Ltd. 2020
N. Goel et al. (eds.), *Modelling, Simulation and Intelligent Computing*,
Lecture Notes in Electrical Engineering 659,
https://doi.org/10.1007/978-981-15-4775-1_42

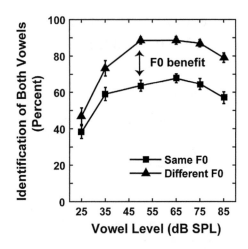

Fig. 1 Percent identification scores of both the vowels for same F0 (squares) and different F0 (triangles) as a function of the vowel level. Note that the figure shown is the percent identification scores rather than the rationalized arcsine transformed scores, modified from Chintanpalli et al. [4]. Error bar denotes ± 1SEM. The F0 benefit is indicated by an arrow

concurrent vowel literature and is a difference in the identification scores between the 26-Hz and 0-Hz F0 difference conditions. The F0 benefit was increased from 25 to 50 dB SPL and then fairly remained constant from 65 to 85dB SPL. Settibhaktini and Chintanpalli [14] developed a computational model, involving the auditory-nerve responses [17] with a modified version of Meddis and Hewitt's [11] F0 segregation algorithm, to successfully capture these level-dependent changes in scores and F0 benefit, at least qualitatively.

The current study here attempts to provide an alternative computational model for predicting these level-dependent changes in identifications scores of concurrent vowels for same-and different-F0 conditions. The proposed model utilized the same auditory-nerve responses but a multi-layer perceptron was used for identification instead of the F0-based segregation algorithm. This type of neural network-based modeling is limited in concurrent vowel literature and only Culling and Darwin [8] had used this framework to predict the identification scores of both vowels with increasing F0 difference.

2   Methods

2.1   Stimuli

The current study utilized the same procedure for generating the concurrent vowels, which was used in Chintanpalli et al. [4] and Settibhaktini and Chintanpalli [14]. Five synthetic English vowels (/i/,/ɑ/,/u/,/æ/,/ɹ/) were generated using a cascade formant synthesizer [9]. Each vowel is characterized by a fundamental frequency (F0) and formant frequencies (F1–F5). The formant frequencies and its bandwidth associated

Table 1 Formants in Hz for five different vowels. Values in parenthesis of first column correspond to bandwidth around each formant (in Hz)

Vowel	/i/	/ɑ/	/u/	/æ/	/ɝ/
F1 (90)	250	750	250	750	450
F2 (110)	2250	1050	850	1450	1150
F3 (170)	3050	2950	2250	2450	1250
F4 (250)	3350	3350	3350	3350	3350
F5 (300)	3850	3850	3850	3850	3850

with each of these vowels are shown in Table 1 and were similar to those used in previous studies on concurrent vowel identification (e.g., [1, 4, 14, 15]). The duration of individual vowel was 400-ms, including 15-ms raised cosine rise and fall ramps.

A concurrent vowel pair was obtained by adding any two individual vowels. For 26-Hz (or different) F0 condition, one vowel had F0 = 100 Hz and the other vowel had F0 = 126 Hz. There were 25 vowel pairs for this condition. To maintain the equal number of vowel pairs, the 0-Hz (or same) F0 condition had five identical vowel pairs and ten different vowel pairs but repeated twice. A total of 50 concurrent vowel pairs (25 vowel pairs × 2 F0 difference conditions) were used at each level. The individual vowels were ranged from 25 to 85 dB SPL. Overall, there were 300 vowel pairs (50 vowel pairs × 6 levels). These vowel pairs were presented as an input to the computational model for predicting the concurrent vowel scores across F0 difference conditions and levels. There were two stages to the computational model: (1) population responses of the auditory–nerve fibers, and (2) multi-layer perceptron for concurrent vowel identification.

2.2 Computational Model Stage 1: Auditory-Nerve Responses

Zilany et al. [17] auditory model was used to predict the auditory-nerve responses to concurrent vowels. This phenomenological model is an extension of previous models (e.g., [3, 16, 18]), that had been successfully tested against neurophysiological data obtained from cats to simple and complex stimuli including vowels. The model also captures the level-dependent changes in phase locking of auditory-nerve fibers (as reflected due to changes in cochlear nonlinearities) to concurrent vowel pairs. This feature is relevant for studying the effect of level to avail F0 difference cue on concurrent vowel identification.

The input to the model was the vowel pair and the output was the time-varying discharge rate of a single auditory-nerve fiber (in spikes/sec) from a specific characteristic frequency (CF). The 100 CFs were selected that were ranged between 125 Hz and 4000 Hz (logarithmically spaced). The overall discharge rate at each CF was the weighted sum of the discharge rates as per the distributions of the spontaneous rate (high SR = 0.61, medium SR = 0.23 and low SR = 0.16; Liberman [10]). The sampling frequency was 100 kHz, as this is the minimum value required to run Zilany et al. [17] model in MATLAB. These were the same discharge rates that were collected and used in Settibhaktini and Chintanpalli [14].

2.3   Computational Model Stage 2: Supervised Multi-Layer Perceptron

The discharge rates from auditory-nerve fibers were the input layer to a multi-layer perceptron (MLP) for concurrent vowel identification. The auditory-nerve responses were obtained for 100 CFs to each 400-ms concurrent vowel. Thus, each vowel pair had a matrix dimensionality of 100 (i.e., the number of CFs used) × 40000 (i.e., 400 ms × 100 kHz sampling frequency). Overall, for all 25 concurrent vowel pairs, the matrix dimensionality of the input layer was 2500 (100 CFs × 25 vowel pairs) × 40000 for each level. For the output layer, one hot encoding matrix with 2500 × 25 (i.e., the number of vowel pairs) dimension was used as the target patterns. In each column, only the CF rows corresponding to a specific vowel pair were set to 1; otherwise 0. For example, /i, i/ was allotted to the first column and only the first 100 rows were set to 1 and the rest of the 2400 rows were set to 0.

As the first formant (F1) and the second formant (F2) are generally important for vowel identification [12], this allowed us to reduce the dimensionalities of the input and output layers. Only discharge rates those CFs were around ± 0.5 octaves of F1 and F2 of each vowel of the pair were selected. All five vowels (except /æ/) had 20 CFs around F1 and 20 CFs around F2. However, /æ/ had 20 CFs around F1 and 19 CFs around F2. Hence, the vowel pair /æ, æ/ had 78 CFs, four different vowel pairs /æ, x/ had 79 CFs, four different vowel pairs /x, æ/ had 79 CFs and 80 CFs for other 16 vowel pairs, where /x, x/, where x is some other vowel except /æ/. Thus, the number of CFs now reduced to 1990 (78 + 8 × 79 + 16 × 80) instead of 2500 in both the input and output layers. Only corresponding 1990 CFs were selected, and thus, this reduction in the input features from 2500 × 40000 to 1990 × 40000 was beneficial for training the neural network.

A nine-layer (i.e., with eight hidden layers) perceptron was used. The number of neurons in each of the hidden layers was 40. The log-sigmoidal activation function was used for each neuron in this network architecture, indicating that the output ranged between 0 and 1. Each row of the input features (i.e., 1990 × 40000) was passed to the input layer and its corresponding output at the output layer from the perceptron was computed. This procedure was repeated to all the rows (i.e., CFs associated with F1 and F2 of the vowel pair) of the input features. Finally, at the output layer, the softmax activation function was applied to the neuron's actual response to generate the probabilistic value for each CFs of the concurrent vowel pair. The weights for each of the layers in the nine-layer perceptron were trained using the gradient descent with momentum and adaptive learning rate backpropagation algorithm [13]. The weight updation at a current instant was equal to the sum of the current weight updation using the gradient descent algorithm and the previous instant's weight updation multiplied by the momentum value. The inclusion of the momentum with the previous instant's weight updation has been shown to increase the convergence rate than using the traditional gradient descent algorithm. In MATLAB, "traingdx" command was used for the gradient descent with momentum and adaptive learning rate backpropagation algorithm. The momentum value and the initial learning rate (η) were 0.9 and 0.01, respectively. For each epoch, if the performance decreased toward the error goal (default value = 0), then η was increased by the factor of 1.05. If the performance was greater than the maximum performance (default = 1.04), then η was decreased by the factor of 0.7 and there was no change in weight updation. The network training was stopped when validation error increased more than 6 times since it was last decreased (early stopping). All these are the default values, used in the MATLAB. The cross entropy cost function was used to measure the performance of the neural network with a small regularization value of 0.1024. This regularization value was used to improve the accuracy of model's scores and to prevent overfitting and underfitting of the data.

3   Methods

The nine-layer perceptron with the input dimensionality of 1990×40000 and the output dimensionality of 1990×25 was used to predict the level-dependent changes in concurrent vowel scores for same-and different-F0 conditions. The MLP model was trained for 50 dB SPL at different F0 difference because the F0 benefit was maximum at this level in the concurrent vowel data (see Fig. 1). For a given vowel pair, the outputs of the softmax activation function were averaged across the corresponding CF range of that particular vowel pair. For example, /ɑ, ɑ/ was allotted to the second column of the output dimensionality and only the rows from 81 to 160 (i.e., corresponding to its CFs) of the 1990 were averaged. If the individual output was greater than the average value, then the output was set to 1; otherwise 0. The MLP model identified the correct vowel pair if more than 60% (user-defined value) of the outputs produced 1. More specifically, if more than 60% of CFs are responding correctly, then the MLP model identified the correct vowel pair. This procedure was repeated for all 25 vowel pairs at 50 dB SPL for different F0 condition to obtain an overall accuracy rate = 76%. In order to achieve this accuracy rate (similar to the score in the concurrent vowel data) at the training stage, the number of hidden layers had to be 8 and number of neurons in each hidden layer had to be 40. Finally, the MLP model was then tested for 25 vowel pairs for same- and different-F0 conditions across six vowel levels, ranged between 25 and 85 dB SPL.

Figure 2A shows the MLP model scores for both vowels as a function of vowel level for same (squares) and different (triangles) F0 conditions. The identification score improved as vowel level increased from low-to-mid levels and declined at higher levels for both F0 conditions. These patterns of identification scores are qualitatively similar to that of listeners' identification scores (see Fig. 1) and model scores (see Fig. 2B) from Settibhaktini and Chintanpalli [14].

Figure 3 shows the actual (dashed line) and predicted F0 benefit (squared symbol) at each vowel level. Even though the MLP model's F0 benefit was lower across levels, the model was successful in capturing the pattern of variation in actual F0 benefit with vowel level, qualitatively. Additionally, the predicted F0 benefit from this current study matched perfectly with Settibhaktini and Chintanpalli [14], except at 35 dB SPL, where the MLP model matched closely with concurrent vowel data (compare circle and squared symbols of the solid lines in Fig. 3).

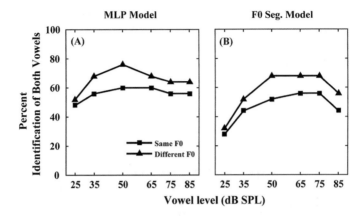

Fig. 2 Percent identification of both vowels for same and different F0 conditions as a function of vowel level. (A) MLP model, and (B) F0-based segregation model [14]. Both the models used the same peripheral model [17]. Note that the model scores from Settibhaktini and Chintanpalli [14] are used in panel (B) for comparison purposes only

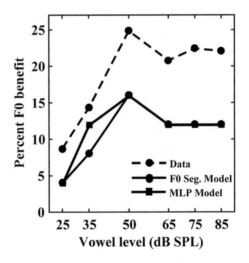

Fig. 3 Effect of vowel level on percent F0 benefit on concurrent vowel identification. Legend indicates different studies. Note that the predicted F0 benefit from Settibhaktini and Chintanpalli [14] is shown only for purpose comparisons

4   General Discussions

The current neural network model provides an alternative approach that is qualitatively successful in predicting the concurrent vowel scores for same-and different-F0 conditions across vowel levels. In both F0 difference conditions, the MLP model score increased from low-to-mid levels and then decreased from mid-to-high levels (see Fig. 2A). The predicted F0 benefits across vowel levels from this study were qualitatively similar to concurrent vowel data (see Fig. 3, compare dashed line with squared symbol). Additionally, the predicted F0 benefit matched very closely with Settibhaktini and Chintanpalli [14] as a function of vowel level (see Fig. 3, compare circle and squared symbols of the solid lines).

For all 25 concurrent vowel pairs, the matrix dimensionality of the input layer was 2500×40000 for each level. With these inputs, the MLP model had very slow convergence rate and also yielded poor scores across levels. However, when the CFs were extracted based on F1 and F2, the MLP model produced the scores that were qualitatively (see Fig. 2A) similar to the concurrent vowel data for same-and different-F0 conditions. This suggests that F1 and F2 of each vowel of the pair contribute to the correct identification [4].

The concurrent vowel data [4] as well as the previous modeling study [14] had shown that the correct identification of one vowel of the pair was always 100% regardless of the F0 difference and vowel level. However, the current MLP model had failed to predict this effect of one vowel correct identification. One possible explanation could be that the MLP model was designed to train only for identification of two vowels. Perhaps, two stages of training may be required: (1) Identification of one vowel of the pair, and (2) Identification of both vowels of the pair. One possible approach could be that two neural networks can be cascaded together such that the first network can be used to predict the identification of one vowel and the second network can be used to predict the identification of both vowels. Nevertheless, the current study had open up new approach based on neural network to predict the concurrent vowel data for normal-hearing and hearing-impaired listeners that are published in the literature.

Acknowledgements. This work was supported by second author's Outstanding Potential for Excellence in Research and Academics (OPERA) Grant and Research Initiation Grant (RIG), awarded by BITS Pilani, Pilani campus, Rajasthan, India. The level-dependent concurrent vowel data was used from Chintanpalli et al. [4], which was collected at Medical University of South Carolina, USA, under the supervision of Dr. Judy R. Dubno (NIH/NIDCD: R01DC000184 and P50 DC000422, and NIH/NCRR: UL1RR029882), Professor in Department of Otolaryngology-Head and Neck Surgery. Many thanks to Mr. Harshavardhan Settibhaktini for generating the figures in a publishable format.

References

1. Arehart KH, Rossi-Katz J, Swensson-Prutsman J (2005) Double-vowel perception in listeners with cochlear hearing loss: Differences in fundamental frequency, ear of presentation, and relative amplitude. J Speech Lang Hear Res 48(1):236–252. https://doi.org/10.1044/1092-4388(2005/017)
2. Assmann PF, Summerfield Q (1990) Modeling the perception of concurrent vowels: vowels with different fundamental frequencies. J Acoust Soc Am 88(2):680–697. https://doi.org/10.1121/1.399772
3. Carney LH (1993) A model for the responses of low-frequency auditory-nerve fibers in cat. J Acoust Soc Am 93(1):401–417. https://doi.org/10.1121/1.405620

4. Chintanpalli A, Ahlstrom JB, Dubno JR (2014) Computational model predictions of cues for concurrent vowel identification. J Assoc Res Otolaryngol 15(5):823–837. https://doi.org/10. 1007/s10162-014-0475-7

5. Chintanpalli A, Ahlstrom JB, Dubno JR (2016) Effects of age and hearing loss on concurrent vowel identification. J Acoust Soc Am 140(6):4142–4153. https://doi.org/10.1121/1. 4968781

6. Chintanpalli A, Heinz MG (2013) The use of confusion patterns to evaluate the neural basis for concurrent vowel identification. J Acoust Soc Am 134(4):2988–3000. https://doi.org/10. 1121/1.4820888

7. Culling JF, Darwin CJ (1993) Perceptual separation of simultaneous vowels: within and across-formant grouping by F0. J Acoust Soc Am 93(6):3454–3467. https://doi.org/10.1121/ 1.405675

8. Culling JF, Darwin CJ (1994) Perceptual and computational separation of simultaneous vowels: cues arising from low-frequency beating. J Acoust Soc Am 95(3):1559–1569. https://doi.org/10.1121/1.408543

9. Klatt DH (1980) Software for a cascade/parallel formant synthesizer. J Acoust Soc Am 67 (3):971–995. https://doi.org/10.1121/1.383940

10. Liberman MC (1978) Auditory-nerve response from cats raised in a low-noise chamber. J Acoust Soc Am 63(2):442–455. https://doi.org/10.1121/1.381736

11. Meddis R, Hewitt MJ (1992) Modeling the identification of concurrent vowels with different fundamental frequencies. J Acoust Soc Am 91(1):233–245. https://doi.org/10.1121/1. 402767

12. Peterson GE, Barney HL (1952) Control methods used in the study of vowels. J Acoust Soc Am 24:175–184. https://doi.org/10.1121/1.4991346

13. Rumelhart DE, Hinton GE, Williams RJ (1986) Learning internal representations by error propagation. In: Rumelhart DE, McClelland JL (eds) Parallel distributed processing, vol 1. MIT Press, Cambridge, MA, pp 318–362

14. Settibhaktini H, Chintanpalli A (2018) Modeling the level-dependent changes of concurrent vowel scores. J Acoust Soc Am 143(1):440–449. https://doi.org/10.1121/1.5021330

15. Summers V, Leek M (1998) F0 processing and the separation of competing speech signals by listeners with normal hearing and with hearing loss. J Speech Lang Hear Res 41(6):1294–1306

16. Zhang X, Heinz MG, Bruce IC, Carney LH (2001) A phenomenological model for the responses of auditory-nerve fibers: I. Non-linear tuning with compression and suppression. J Acoust Soc Am 109(2):648–670. https://doi.org/10.1121/1.1336503

17. Zilany MSA, Bruce IC, Carney LH (2014) Updated parameters and expanded simulation options for a model of the auditory periphery. J Acoust Soc Am 135(1):283–286. https://doi. org/10.1121/1.4837815

18. Zilany MSA, Bruce IC, Nelson PC, Carney LH (2009) A phenomenological model of the synapse between the inner hair cell and auditory nerve: long-term adaptation with power-law dynamics. J Acoust Soc Am 126(5):2390–2412. https://doi.org/10.1121/1.3238250

A Dual-Band Modified Quadrilateral Square Slotted Rectenna for RF Energy Harvesting

Geriki Polaiah$^{(\boxtimes)}$ (iD), K. Krishnamoorthy, and Muralidhar Kulkarni

Department of Electronics and Communication Engineering, National Institute
of Technology Karnataka Surathkal, Mangalore 575025, India
{polaiahgeriki,mkuldce}@gmail.com,
krishnak_ece@yahoo.com

Abstract. A dual-band planar rectenna, consisting of a modified quadrilateral square slot antenna with rectangular microstrip patch connect to a 50 Ω feed line to improve the impedance matching and a single-series diode configuration-based half-wave rectifying circuit for high conversion efficiency, operate in frequency bands of universal mobile telecommunication service UMTS (2.1 GHz) and higher WLAN/Wi-Fi (5 GHz), is proposed for RF energy harvesting and wireless power transmission. The inverted L-section transmission line is introduced between the diode and dc pass filter to eliminate the harmonics within the operating frequencies. The antenna is connected to the rectifying circuit by using a pair of 50 Ω SMA coaxial connectors. The peak measured conversion efficiency of proposed rectenna is 59.4% achieved at the input power of −9.8 dBm and optimized load resistance of 560 Ω, respectively.

Keywords: Dual-band antenna · Feedline patch · Rectenna · Series diode rectifier · UMTS and Wi-Fi bands

1 Introduction

Energy harvesting and wireless power transmission have received more attraction recently in RF and microwave regime for the implementation of autonomous battery-less wireless sensor nodes, radio frequency identification devices (RFID), and electronic devices powered by low input dc power. A decoupled dual-dipole antenna integrated with the Villard voltage doubler rectifier for 2.4 GHz ISM band is described in [1]. A compact folded dipole-based slot-loaded dual-band rectenna operating in frequencies of 915 MHz and 2.45 GHz has been presented in [2] with the measured efficiency of 37% at the input power of −9 dBm and optimized load resistance of 2.2 K Ω. In [3], the combination of solar cell and dual-band rectenna in a compact structure for ambient electromagnetic energy harvesting is proposed. This rectenna obtain RF-dc conversion efficiency of 15% at the input power of −20 dBm and at the frequencies of 850 and 1850 MHz. The implantable rectenna working in the medical communication service band for lithium-ion rechargeable batteries is presented in [4]. A dual-band rectenna at 915 MHz and 2.45 GHz and on-body wireless energy harvester at 460 MHz are implemented [5]. A dual ISM band (915 MHz and 2.45 GHz) Yagi-Uda, narrowband 1.96 GHz, and broadband 2–18 GHz rectenna arrays have been presented

© Springer Nature Singapore Pte Ltd. 2020
N. Goel et al. (eds.), *Modelling, Simulation and Intelligent Computing*,
Lecture Notes in Electrical Engineering 659,
https://doi.org/10.1007/978-981-15-4775-1_43

in [6]. A broadband rectenna over the frequency range from 1.8 to 2.5 GHz is reported in [7] with the measured efficiency of 55% at the input power of −10 dBm. A dual-band rectifier with an extended input power range from 0 to 15 dBm using a single-series diode-transistor configuration is discussed in [8] with 30% power conversion efficiency with the input power range from −15 to 20 dBm. Loop antenna over the AMC surface is implemented for dual-band rectenna operating at digital TV and cellular band frequencies [9].

In this paper, a modified quadrilateral square slotted antenna with a rectangular microstrip patch connected to a 50 Ω feedline is proposed to enhance the performance of rectenna for RF power harvest from UMTS and Wi-Fi bands. The rectifier is designed by integrating a T-shape matching network with a half-wave rectifying circuit. Simulation and measurements are performed to validate the proposed rectenna. The proposed design is useful for harvesting RF energy from both cellular and Wi-Fi bands simultaneously.

2   Antenna Design Details and Result Analysis

2.1   Design Procedure

The configuration of proposed dual-band receiving antenna, designed and fabricated on a low-cost FR4 substrate with the characteristics of $\tan\delta = 0.025$, $\varepsilon_r = 4.3$, and $t = 1.6$ mm is displayed in Fig. 1. The substrate dimension is 84×74 mm^2 ($L_s \times W_s$). The antenna consists of a square slot with the quadrilateral side length of 43 mm, and two symmetrical V-shaped and U-shaped slots are extended with this slot to enhance the bandwidth of operating frequency bands, and rectangular feed line patch (RFLP) connects to a feed line for improving the impedance matching. This slot is excited by a 50 Ω feed line along with the RFLP. The side length of symmetrical V-shaped slot is 13 mm, vertical and inclined side lengths of symmetrical U-shaped slot are 9 mm and 13 mm, respectively. The dimensions of rectangular patch and feedline are 39×20 mm^2 ($L_p \times W_p$) and 32×3.1 mm^2 ($L_f \times W_f$). The proposed modified quadrilateral square slot dimensions are keeping constant and only varying the feed length and dimension of rectangular patch to obtain the required resonant frequencies. The performance of antenna in terms of reflection coefficient magnitude $|S_{11}|$ (dB) with the parametric analysis of feed line length and a rectangular patch dimensions, and 2D radiation patterns are discussed in the Sects. 2.2 and 2.3.

2.2   Simulation Results and Parametric Analysis

The parametric analysis of proposed antenna has been carried out by varying the feed line length and dimensions of a rectangular feed line patch (RFLP). The variation of simulated result of $|S_{11}|$ (dB) versus frequency of the antenna without and with RFLP at different feed lengths (L_f) and with RFLP at fixed feed length are labeled in Fig. 2. The poor impedance matching at lower and higher feed lengths and good impedance matching at medium feed lengths are observed in Fig. 2a for without RFLP. This configuration possesses resonance at 5 and 9 GHz frequencies. The antenna size is

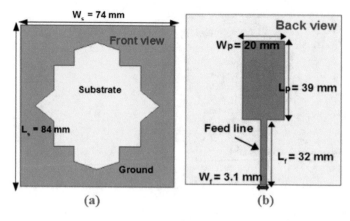

Fig. 1 Configuration of the proposed dual-band antenna **a** front view, **b** back view

increased by connecting the rectangular patch to the feed line in order to obtain the required resonance at lower frequencies. It is observed that the impedance matching at required frequencies of 2.1 and 5 GHz is absolutely controlled by dimensions of RFLP is shown in Fig. 2b, c. The proposed dual-bands are achieved at optimized feed length of $L_f = 32$ mm, and rectangular patch dimensions of $L_p = 39$ mm, $W_f = 20$ mm, respectively and without altering the modified quadrilateral square slot dimensions. This proposed slot is similar to the "Modern Aztec Quatrefoil Shape" which is popular in textile designs. These fixed slot dimensions are randomly chosen according to the general compact structure of planar antennas. The realized gain values of 2.5 and 8.2 dBi at the operating frequencies of 2.1 GHz and 5 GHz, respectively, are obtained. The omnidirectional radiation pattern is accomplished at a lower band frequency of 2.1 GHz.

2.3 Antenna Measurement Results

The proposed configuration of antenna design and simulation of various parameters is carried out by using Computer Simulation Technologies (CST) Microwave Studio, fabricated using the S103 Proto Mat LPKF PCB machine and measured using the Agilent Technologies E8363C PNA network analyzer. The simulated and measured | S_{11}| (dB) versus frequency and 2D radiation patterns of the antenna are shown in Figs. 3 and 4, respectively. The dual-band frequencies of measured results exactly match with the simulated result, but less impedance matching noticed because of the variation of dielectric constant, thickness of the substrate, and fabrication tolerances.

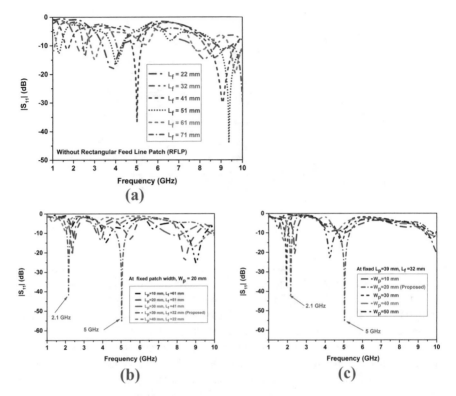

Fig. 2 Variation of simulated result of $|S_{11}|$ versus frequency of dual-band antenna **a** without RFLP and at different feed lengths (L_f), **b** with RFLP-fixed patch width (W_p), and at different patch lengths (L_p) and feed lengths (L_f), **c** with RFLP-fixed patch length (L_p) and feed length (L_f), and at different patch widths (W_p)

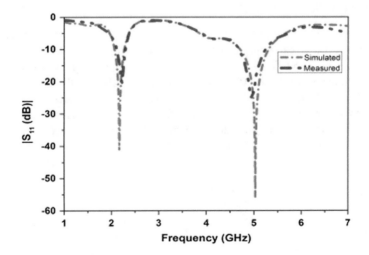

Fig. 3 $|S_{11}|$ versus frequency of the proposed dual-band antenna

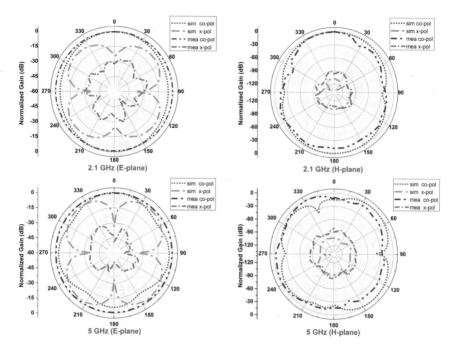

Fig. 4 Simulated and measured 2D radiation patterns of the proposed dual-band antenna

3 Rectifier Design and Result Analysis

3.1 Design Details

The proposed dual-band rectifier operating at the same frequencies (2.1 and 5 GHz) of the antenna is designed and fabricated using FR4 substrate with the characteristics of $\tan\delta = 0.025$, $\varepsilon_r = 4.3$, and $t = 1.6$ mm. The dimension of rectifier circuit is 104×52 mm^2. The rectifier design and simulations are carried out by using Keysight Technologies Advanced Design System (ADS) high-frequency RF simulator. The schematic of proposed dual-band rectifier along with impedance matching network, inverted L-section transmission line for harmonics elimination, and dc pass filter is displayed in Fig. 5.

The Schottky diode (SMS7630-079LF) from Skyworks is used for converting the input RF signals to output dc voltage. The single T-shape dual-band matching network is designed for matching the input impedance of dual-band (2.1 and 5 GHz) to 50 Ω source impedance, and a dc pass filter is designed with the help of short stub transmission line along with 100 pF chip capacitor from Murata for blocking the RF signals and not reaches to the load. The ceramic resistor with an optimized load resistance (R_L) of 560 Ω (using simulation) is connected between the top strip line and bottom ground to measure the output dc voltage. In addition to this, an inverted L-section transmission line for harmonics elimination between two operating bands is introduced between the diode and dc pass filter without disturbing the resonant frequencies. The Schottky diode

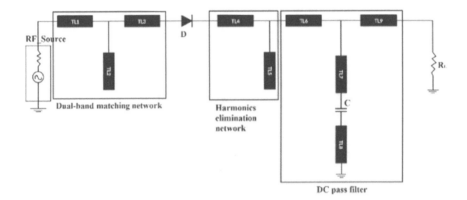

Fig. 5 Schematic of the proposed dual-band rectifier

Table 1 Parameters and dimensions of the rectifier with the dual-band matching network

Parameter	Length/width (mm)	Parameter	Length/width (mm)
TL1	18.5/1.2	TL6	23.8/1.6
TL2	15.4/1.4	TL7	14/1.6
TL3	21.2/1.2	TL8	15.8/1.6
TL4	14.4/1.7	TL9	21/1.6
TL5	19/1.8		

has been chosen with a series resistance (R_s) of 20 Ω, breakdown voltage (V_{br}) of 2 V, a threshold voltage (V_{th}) of 147 mV, and zero-bias junction capacitance (C_{j0}) of 0.14 pF [10], respectively. The design parameters of the rectifier circuit with the dual-band matching network are given in Table 1.

3.2   Result Analysis

The simulated and measured reflection coefficient magnitude $|S_{11}|$ versus frequency of the proposed dual-band rectifier is shown in Fig. 6. The measured result matches with the simulated one with small deviation due to tolerances of the substrate characteristics and fabrication. The variation of output voltage and efficiency at different load resistances and input powers are shown in Figs. 7 and 8. The simulated and measured maximum RF-dc conversion efficiency of 73.6 and 65.7% have been achieved at the input power of 0 dBm and load resistance of 560 Ω. It is observed that the efficiency is increased and maintains almost stability within the load resistance range from 0 to 5 K Ω when the input power is less than 0 dBm. Similarly, the efficiency is decreased gradually when the input power and load resistance are increased.

The proposed single-series diode configuration-based half-wave rectifier circuit is more suitable for achieving the high conversion efficiency at low input power. The simulated and measured maximum output dc voltage of 2.8 and 2.1 V have been

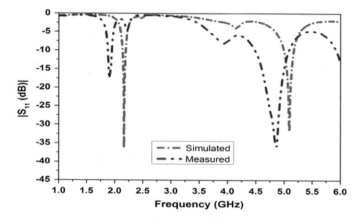

Fig. 6 $|S_{11}|$ versus frequency of the proposed dual-band rectifier

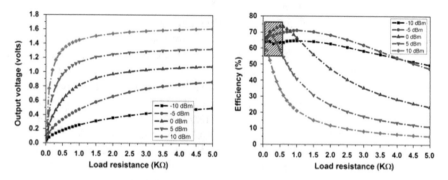

Fig. 7 Simulated output voltage and efficiency versus load resistance (R_L) of the proposed dual-band rectifier at different input powers (P_{in})

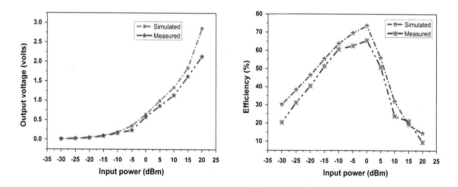

Fig. 8 Simulated and measured output voltage and efficiency versus input power (P_{in}) of the proposed dual-band rectifier at optimized load resistance (R_L) of 560 Ω

obtained at the input power of 20 dBm and optimized load resistance of 560 Ω. It is observed from Fig. 8, the diode goes through an electrical break down when the input power is beyond the value of 0 dBm, and then the conversion efficiency is also decreased and reaches a minimum value. The rectifier design and simulations of S-parameter and harmonic balance are carried out for reflection coefficient $|S_{11}|$ and output voltage by using Keysight Technologies Advanced Design System (ADS) high-frequency RF simulator.

4   Simulated and Measured Results of Rectenna

4.1   Simulation Results

The co-simulation of antenna and rectifier circuit is important before analyzing the measurement results of the rectenna. The Z-parameter file of receiving antenna from the CST simulator is exported into ADS simulator and integrated to the rectifier circuit instead of the position of RF power source. This power source could be considered as antenna while executing all the rectenna simulations. The simulated result of $|S_{11}|$ versus frequency and output voltage and efficiency of proposed dual-band rectenna are shown in Fig. 9. The results of $|S_{11}|$ versus frequency of antenna and rectifier and rectenna are analyzed. The good agreement has been achieved among each other. From Figs. 6 and 9, an intermediate resonance distinguished between the desired operating frequencies and it is recognized closely at 4 GHz, which is controlled by the harmonics elimination network. So, the proposed rectifying circuit not only resonates at essential frequencies but rejection of harmonics in the middle of two operating frequency bands and at higher frequency bands also. The rectenna maximum simulated efficiency of 61.8% has been achieved at the output voltage of 0.84 V and input power of –5 dBm, respectively. The antenna is connected to the rectifying circuit by using a pair of 50 Ω SMA coaxial connectors in order to transfer the maximum RF power from antenna to rectifier and for measurement of complete rectenna prototype.

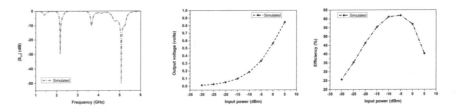

Fig. 9 Simulated result of $|S_{11}|$ versus frequency and output voltage and efficiency versus input power of the proposed dual-band rectenna

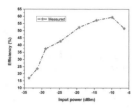 

Fig. 10 Measured efficiency and fabricated prototype of the proposed dual-band rectenna

4.2 Measurement Results

A standard double ridged broadband transmitting horn antenna has varying gain from 3 to 16 dB in 0.8 to 18 GHz range of frequency is used for the measurement of proposed rectenna. The distance between transmitting antenna and rectenna has been chosen such that the far-field condition is valid for two operating frequencies. The power received by the receiving antenna is measured using the spectrum analyzer. The distance increased gradually by 0.2 m greater than the far-field distance and measures the received power at different distances. The rectifier is connected to antenna in the position of a spectrum analyzer. The output voltage is measured using the multimeter at similar distances at which the received power was measured. The RF-dc conversion efficiency of rectifier and rectenna using corresponding output voltages and load resistance of 560 Ω can be calculated using the following Eq. (1).

$$\text{Efficiency}(\%) = \frac{V_0^2}{R_L} \times \frac{1}{P_{in}} \times 100 \tag{1}$$

where V_0 is output dc voltage at the load resistor (R_L) and P_{in} is input power received by the antenna.

The peak measured conversion efficiency of rectenna is 59.4% achieved at the input power of −9.8 dBm is shown in Fig. 10. The fabrication prototype of antenna and rectifier connected by a pair of 50 Ω SMA coaxial connectors is displayed in Fig. 10. The input impedance (Z_{in}) of rectifier is a function of input power (P_{in}), operating frequency (f), and load resistance (R_L). The input RF power in an environment is continuously varying and available at low level. The complex impedance compression networks (CICN) are use to compress the variable input impedance (Z_{in}) in the medium range of input power in order to achieve the constant output voltage and conversion efficiency in addition to the frequently used general conventional rectifier topologies of half-wave rectifier circuit, single-stage voltage doubler, Greinacher charge pump, and Dickson charge pump circuits in the RF energy harvesting and wireless power transmission applications.

5 Conclusion

In this paper, a modified quadrilateral square slotted rectenna working in UMTS (2.1 GHz), and Wi-Fi/WLAN (5 GHz) frequencies have been proposed. The designed antenna has been realized by etching one square, two V-shaped, and two U-shaped slots coupled to each other in the ground plane. The measured gains of antenna at 2.1 and 5 GHz of frequencies are 1.9 dB and 7.2 dB, respectively. The dual-band rectifier, along with the single T-shape matching circuit and inverted L-section harmonic network has been separately designed, fabricated, and measured to achieve the desired performance. Finally, the quadrilateral modified square slot antenna has been connected with the rectifier circuit by a pair of 50 Ω SMA coaxial connectors to realize the proposed dual-band rectenna. The maximum measured conversion efficiencies of rectifier and rectenna are 65.7 and 59.4% achieved. The proposed dual-band rectenna is suitable for RF energy harvesting and wireless power transmission applications at low input RF power levels.

References

1. Yeoh WS, Rowe WST, Wong KL (2012) Decoupled dual-dipole rectennas on a conducting surface at 2.4 GHz for wireless battery charging. IET Microw Antennas Propag 49(7):238–244
2. Niotaki K, Kim S, Jeong S, Collado A, Georgiadis A, Tentzeris MM (2013) A compact dual-band rectenna using slot-loaded dual-band folded dipole antenna. IEEE Antennas Wirel Propag Lett 6:1634–1637
3. Collado A, Georgiadis A (2013) Conformal hybrid solar and electromagnetic (EM) energy harvesting rectenna. IEEE Trans Circuits Syst-I Regular Papers 60(8):2225–2234
4. Cheng HW, Yu TC, Luo CH (2013) Direct current driving impedance matching method for rectenna using medical implant communication service band for wireless battery charging. IET Microw Antennas Propag 7(4):277–282
5. Kim BS, Vyas R, Bito J, Niotaki K, Collado A, Georgiadis A, Tentzeris MM (2014) Ambient RF energy-harvesting technologies for self-sustainable standalone wireless sensor platforms. In: Proceedings of the IEEE, vol 102, no 11, pp 1649–1666
6. Popovic Z, Sean K, Dunbar S, Robert S, Dolgov A, Zane R, Erez F, Hagerty J (2014) Scalable RF energy harvesting. IEEE Trans Microw Theory Tech 62(4):1046–1056
7. Song C, Yi H, Zhou J, Zhang J, Yuan S, Carter P (2015) A high-efficiency broadband rectenna for ambient wireless energy harvesting. IEEE Trans Antennas Propag 63(8):3486–3495
8. Liu Z, Zhong Z, Guo YX (2015) Enhanced dual-band ambient RF energy harvesting with ultra-wide power range. IEEE Microw Wirel Propag Lett 29(9):630–632
9. Kamoda H, Kitazawa S, Kukutsu N, Kobayashi K (2015) Loop antenna over artificial magnetic conductor surface and its application to dual-band RF energy harvesting. IEEE Trans Antennas Propag 63(10):4408–4417
10. SMS7630-079LF Schottky Diode Datasheet (2018) Skyworks Solutions

Modeling, Simulation, and Comparison of Different Ferrite Layer Geometries for Inductive Wireless Electric Vehicle Chargers

Eiman A. Elghanam, Mohamed S. Hassan$^{(\boxtimes)}$, and Ahmed Osman

American University of Sharjah, Sharjah, UAE
{eelghanam, mshassan, aosmanahmed}@aus.edu

Abstract. In order to maximize the inductive link efficiency in wireless electric vehicle (EV) chargers, a ferrite layer is added to focus the magnetic field lines, improve the coupling performance, and reduce the leakage of flux to the surrounding ferrous materials. The geometry of ferrite directly affects the self and mutual inductances of the primary and secondary coils and accordingly their coupling factor. Three ferrite geometries are investigated in this work, and their coupling behavior is studied and compared to that of a ferrite sheet. Due to the inherent misalignment variations in wireless EV chargers, the simulation is conducted over a range of air gaps as well as lateral and longitudinal misalignments. Based on the simulation results, the geometry with long ferrite bars is recommended for dynamic EV charging scenarios in which large lateral misalignments are expected, whereas shorter bars are more recommended for static charging scenarios due to their smaller volume and cost-effectiveness despite their lesser tolerance for lateral misalignment in comparison to long ferrite bars.

Keywords: Electric vehicle · Wireless charging · Rectangular coil · Ferrite geometry · Wireless power transfer · Inductive link efficiency

1 Introduction

With the increasing interest in adopting electric vehicles (EVs) as part of establishing an environment-friendly transportation system, extensive research is globally conducted to develop efficient EV charging solutions. Although plug-in EV chargers were initially developed, the associated hazards and limitations of wired charging have motivated the development of wireless EV chargers. In particular, resonant inductive power transfer (RIPT) systems are being utilized due to their high efficiencies and high safety levels [1, 2]. RIPT systems utilize concepts of Ampere's and Faraday's laws to wirelessly couple AC power from a primary coil buried under the ground to a secondary coil fitted at the bottom of the EV. This inductive charging process can take place while the vehicle is statically parked at a charging station or in a dynamic mode while the vehicle is in motion. Dynamic EV charging aims to relieve the range anxiety experienced by EV users due to fear of battery depletion during their journeys [3] by enabling the EV to recharge its battery from wireless charging pad installed along its route.

© Springer Nature Singapore Pte Ltd. 2020
N. Goel et al. (eds.), *Modelling, Simulation and Intelligent Computing*,
Lecture Notes in Electrical Engineering 659,
https://doi.org/10.1007/978-981-15-4775-1_44

The key objective for designing a RIPT system, for both stationary and dynamic charging modes, is to maximize the power transfer efficiency from the mains grid to the EV battery. Several circuits constitute this power transfer system, each of which needs to be optimized to achieve this objective. Nevertheless, one of the most important design stages is the design of the inductive link comprising the primary and secondary coils, as it directly impacts the wireless power transfer process by controlling the flow of magnetic fields from the primary coil to the secondary one. Several coil structures have been used in the literature including circular, rectangular, and bipolar coils [3–6]. For each coil structure, a ferrite layer is typically included on the primary and secondary sides, in order to improve the self and mutual inductances of the coils, their coupling performance and their power transfer efficiencies. This work focuses on studying the impact of the geometry of the ferrite layer on the different parameters of the IPT coils in order to provide recommendations on the selection of the ferrite geometry for both stationary and dynamic wireless charging systems.

The rest of this paper is organized as follows: Sect. 2 provides preliminary mathematical analysis on the inductive link efficiency and highlights the need for enhancements in the coils' inductive parameters to maximize the power transfer efficiency. Section 3 then provides the details of the design specifications and the simulation setup required to study the effect of different ferrite geometries on the coil parameters. Results of the conducted simulations are then reported and discussed in Sect. 4 before the paper is concluded in Sect. 5.

2   System Model

The block diagram of a typical RIPT EV charging system is shown in Fig. 1.

In order to evaluate the grid-to-vehicle power transfer efficiency of this wireless EV charging system, the efficiency of each block in Fig. 1 needs to be studied. The inductive link efficiency is particularly investigated in this work, as it comprises the primary and secondary coils and their corresponding equivalent series resistances (ESRs), and hence can be used to evaluate the coupling performance of these coils with different ferrite geometries, independent of the loading conditions. The expression for the inductive link efficiency, denoted by η_{link}, is given by Zargham and Gulak [7]:

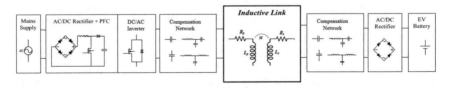

Fig. 1 Block diagram of RIPT EV charging system

$$\eta_{\text{link}} = \frac{k^2 Q_p Q_s}{\left(1 + \sqrt{1 + k^2 Q_p Q_s}\right)^2}, \tag{1}$$

where k is the coupling factor and Q_p and Q_s are the quality factors of the primary and secondary coils, respectively. The coupling factor, k, can be calculated as follows:

$$k = \frac{M}{\sqrt{L_p L_s}}, \tag{2}$$

where M is the mutual inductance between the primary and secondary coils, and L_p and L_s are the corresponding self inductances of the two coils, respectively. The coil quality factors, on the other hand, can be evaluated using the following expressions:

$$Q_p = \frac{\omega_0 L_p}{R_p}, \quad Q_s = \frac{\omega_0 L_s}{R_s}, \tag{3}$$

where ω_0 denotes the operating frequency of the power transfer system, and R_p and R_s represent the ESRs of the primary and secondary coils, respectively.

In order to increase the inductive link efficiency in (1), the coil quality factors, Q_p and Q_s, need to be increased as well as the coupling factor, k. By studying (2) and (3), it can be noted that Q_p and Q_s can be increased by increasing the coil self inductances L_p and L_s, and reducing their ESRs, whereas k can be increased by increasing their corresponding mutual inductances. The mutual inductance between the primary and secondary coils varies with respect to the vertical distance between the coils, i.e., the air gap (z–distance), as well as their lateral and longitudinal misalignments (x- and y-displacements, respectively), assuming the EV's front-to-rear orientation is along the y-axis. The use of ferrite layer enhances the self and mutual inductances of the coils, and hence enhances their coupling performance and inductive link efficiency.

The coil quality factors are also directly related to the operating frequency of the inductive link, ω_0. The SAE J2954B standard [8] identifies 85 kHz as the nominal operating frequency for stationary RIPT EV chargers, ensuring that it complies with the maximum safety exposure limits identified in IEEE C95.1 standard [9]. Since the same secondary coil is expected to operate in stationary and dynamic charging modes, 85 kHz is recommended to be used for the design of dynamic wireless charging systems as well [10] and is selected as the link operating frequency in this work.

3 Design Specifications

3.1 Coil Construction

As explained in Sect. 2, the inductive link efficiency depends on the coils' inductances and their ESRs, which are determined according to the coil material and geometry. For EV charging coils, Litz copper wires are recommended in the literature as their stranded structure helps reduce the coils' ESRs [11, 12]. For this work, an 8 mm diameter wire is simulated to model a 4140 strands/AWG 38 Litz wire with a DC resistance of $0.18\Omega/1000\text{ft}$ and an AC resistance of $0.3623\Omega/1000\text{ft}$ [13]. As for the coil dimensions, the secondary coil area is restricted to the available space in the bottom chassis of the vehicle, estimated to be around 0.48 m^2 for a typical sedan vehicle [4]. While the size of the primary coil can be more freely selected, the use of comparable coil dimensions helps minimize variations in the resonant frequency and the power transfer efficiency [11]. For this work, the primary and secondary coils are designed to be identical rectangular coils with outer dimensions of $800 \times 600\text{ mm}^2$ and 11 turns of conductor wire, since typical values for the number of turns are reported in the literature to be between 6 and 20 turns [4, 14]. This guarantees a reasonable simulation time and a fair comparison between the ferrite geometries.

3.2 Ferrite Layer

At the high power transfer levels expected in an EV wireless charging system, the use of a ferrite layer guides the magnetic field lines to the area between the two coils and thereby reduces flux leakage to any surrounding magnetic objects [5]. This layer acts as a core of ferrite material that is placed below the primary coil and on top of the secondary coil with proper electrical insulation. The relative magnetic permeability of ferrite, μ_r, is significantly higher than that of copper, typically ranging from 1000 to 3000 [15]. However, a generic ferrite material with $\mu_r = 1000$ is used in this work as the main focus is to compare the geometry of the ferrite layer rather than studying the effect of the ferrite material.

The most generic structure of a ferrite layer is to use an extended ferrite sheet that spans the entire coil area. While this expectedly provides best field guidance, it introduces the risk of ferrite cracking, and in some cases, breaking, due to the brittleness of ferrite sheets [15]. In addition, the cost of ferrite depends on the volume of the material used. Moreover, since the weight of the EV affects its consumed power [16], the higher the ferrite volume used, the larger the weight of the inductive link and the higher the power consumed by the EV during its motion, hence translating to faster battery depletion. As a result, different ferrite configurations need to be studied in which smaller ferrite pieces are distributed over the coil area to reduce the overall ferrite volume required [17, 18], and the impact of each configuration on the coupling performance of the coils and their inductive link efficiencies needs to be investigated. In this work, three ferrite arrangements are studied using long and short ferrite bars of equal widths, thickness, and edge-to-edge bar spacing, while having different lengths depending on the configuration. The ferrite geometries under investigation in this work are presented in Fig. 2, and a summary of their construction details is presented in Table 1.

By studying the geometries presented in Fig. 2, it can be noted that the ferrite volume used for each of the first three structures is significantly smaller than that used for the ferrite sheet. This can be also observed by comparing the *percentage ferrite occupancy* of each structure, as shown in Table 1, which is the ratio occupied by the ferrite material of each geometry in comparison to the ferrite sheet arrangement. This can be used in conjunction with results of coupling performance to provide recommendations on the most effective ferrite geometry to be used with rectangular EV coils.

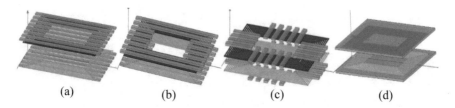

Fig. 2 Ferrite geometries simulated in this work: **a** long ferrite bars across entire structure, **b** long ferrite bars along the length and short bars along the width, **c** short ferrite bars perpendicular to the current conduction path, and **d** reference ferrite sheet

Table 1 Ferrite dimensions for simulated geometries

Parameter	Geometry 2(a)	Geometry 2(b)	Geometry 2(c)	Geometry 2(d)
No. of ferrite bars	8	12	18	1
Ferrite length	8 × 900 mm	4 × 900 mm + 8 × 250 mm	18 × 250 mm	900 mm
Ferrite width	37.5 mm	37.5 mm	37.5 mm	800 mm
Edge–edge spacing	75 mm	75 mm	75 mm	N/A
Ferrite thickness	16 mm	16 mm	16 mm	16 mm
Net ferrite area	270,000 mm²	210,000 mm²	168,750 mm²	720,000 mm²
Percent. occupancy	37.5%	29.2%	23.4%	100%

4 Simulation and Results

In order to compare the coupling performance of the four ferrite geometries in Fig. 2, FEM simulations are conducted on ANSYS Maxwell simulator over a realistic range of air gaps and lateral and longitudinal misalignments. The air gap between the coils is varied from 10 to 30 cm in steps of 5 cm and is then held constant at 20 cm to vary the lateral and longitudinal displacements, each at a time, in steps of 100 mm from a complete no overlap at $(x = +/- 600$ mm, $y = 0)$ and $(x = 0, y = +/- 900$ mm), to full overlap at $(x = 0, y = 0)$. The variations of the coupling factor versus air gap, lateral and longitudinal misalignments are presented in Figs. 3, 4, and 5, respectively.

By studying the graphs in Figs. 3, 4, and 5, few observations are made. First, Fig. 3 demonstrates that the coupling coefficient decreases with increasing air gap for all four ferrite geometries, which is expected due to the increase in the leakage of magnetic flux to the surrounding environment as the vertical distance between the coils increases. It is also observed that the performance of the ferrite sheet is superior to the other three geometries at different vertical distances, providing higher values of coupling coefficients over the complete range of air gaps. On the other hand, geometries 2(a) and 2(c) in Fig. 2 provide almost similar behavior for the coupling factor variations and are both superior in performance to geometry 2(b) in Fig. 2.

Figures 4 and 5 show bell-shaped curves for the coupling coefficients versus lateral and longitudinal misalignments, respectively. As expected, the maximum coupling coefficient occurs at zero misalignment in both the x- and y-directions, and the coupling coefficient decreases as the secondary coil is misaligned to either sides of the perfectly

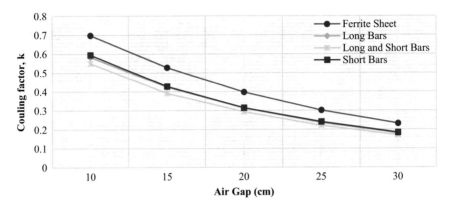

Fig. 3 Coupling factor versus air gap for ferrite geometries under study

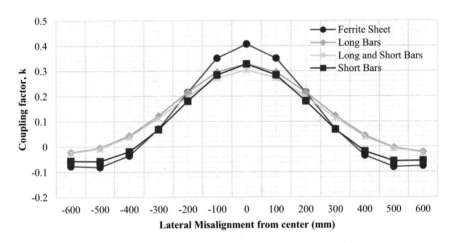

Fig. 4 Coupling factor versus lateral misalignment for geometries under study

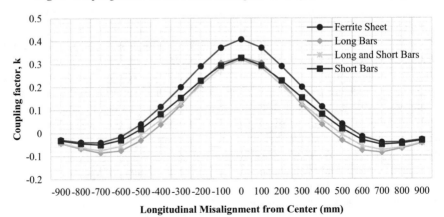

Fig. 5 Coupling factor versus longitudinal misalignment for geometries under study

aligned position. The same superior performance of using a ferrite sheet is also observed in the longitudinal misalignment variation results in Fig. 5 demonstrating better coupling factors over the entire range of longitudinal misalignments simulated. Figure 5 also reveals that the use of short ferrite bars, i.e., geometry 2(c), provides better longitudinal misalignment tolerance in comparison with geometries 2(a) and 2(b) accordingly. Figure 4, however, demonstrates that while the ferrite sheet performs best over a lateral misalignment range from −200 to 200 mm, the long ferrite bars make the inductive link more tolerant to lateral misalignments beyond 200 mm, providing higher values of coupling factors in comparison with all other geometries.

Since lateral misalignments are inevitable in dynamic wireless charging systems, the inductive link performance at larger lateral misalignments can be used as the key metric for the choice of the most recommended ferrite geometry for dynamic charging systems. Accordingly, long ferrite bars are recommended, due to their higher lateral misalignment tolerance compared with all other geometries. On the other hand, since lateral misalignments can be more effectively controlled in static charging, the decision can be based on other factors including ferrite volume and cost. While the ferrite sheet provides better coupling performance versus air gap and longitudinal misalignments, it adds significantly to the overall inductive link volume and hence increases the EV power consumption and its battery depletion rate. As a result, the use of short ferrite bars is a better option for stationary EV chargers as they provide similar air gap performance to that of long bars with better longitudinal misalignment tolerance, while utilizing a smaller ferrite volume than all other geometries. The least recommended topology is geometry 2(b) in which a combination of long and short ferrite bars is used.

5   Conclusions and Future Work

The presented work focuses comparing multiple designs of the ferrite layer of inductive wireless EV chargers in order to enhance the coupling behavior and inductive link efficiencies. Three ferrite bar arrangements have been designed and simulated with rectangular charging coils, and the results are compared to those obtained using a flat ferrite sheet that covers the entire coil surface area. Based on the conducted simulations, it is concluded that using long ferrite bars provides superior lateral misalignment performance above all other geometries including the ferrite sheet, and hence is highly recommended for dynamic EV charging systems in which lateral misalignments are inevitable. On the other hand, short ferrite bars shall be more effective for static charging systems in which lateral misalignment can be more easily controlled. Future work shall study the impact of the ferrite material itself on the lateral and misalignment tolerance of the inductive link in order to further optimize the design of the ferrite layer.

Acknowledgements. This work is supported by the American University of Sharjah through SCRI Grant No. SCRI 18-CEN-10.

References

1. Patil D, McDonough MK, Miller JM, Fahimi B, Balsara PT (2018) Wireless power transfer for vehicular applications: overview and challenges. IEEE Trans Transport Electr 4(1):3–37
2. Jang YJ (2018) Survey of the operation and system study on wireless charging electric vehicle systems. Transport Res Part C Emerg Technol 95:844–866
3. Ahmad A, Alam MS, Chabaan R (2018) A comprehensive review of wireless charging technologies for electric vehicles. IEEE Trans Transport Electr 4(1):38–63
4. Nguyen T, Li S, Li W, Mi CC (2014) Feasibility study on bipolar pads for efficient wireless power chargers. IEEE Appl Power Electron Conf Exposit APEC 2014:1676–1682
5. Ongayo D, Hanif M (2015) Comparison of circular and rectangular coil transformer parameters for wireless power transfer based on finite element analysis. In: IEEE 13th Brazilian power electronics conference and 1st southern power electronics conference, pp 1–6
6. Tavakoli R, Pantic Z (2018) Analysis, design, and demonstration of a 25-kW dynamic wireless charging system for roadway electric vehicles. IEEE J Emerg Selected Topics Power Electron 6(3):1378–1393
7. Zargham M, Gulak P (2012) Maximum achievable efficiency in nearfield coupled power-transfer systems. IEEE Trans Biomed Circuits Syst 6(3):228–245
8. Wireless power transfer for light-duty plug-in/electric vehicles and alignment methodology (2019) SAE J2954
9. IEEE approved draft standard for safety levels with respect to human exposure to electric, magnetic and electromagnetic fields, 0 Hz to 300 GHz (2019) PC95.1/D3.5, pp 1–312
10. Kabalan HS, Elghanam EA, Hassan MS, Osman A (2019) The impact of coupling and loading conditions on the performance of S-S EV dynamic wireless charging systems. In: IEEE international conference on electrical and computing technologies and applications (Accepted for publication)
11. Bosshard R (2015) Multi-objective optimization of inductive power transfer systems for EV charging. Ph.D. dissertation, ETH ZURICH
12. Ramezani A, Farhangi S, Iman-Eini H, Farhangi B, Rahimi R, Moradi GR (2019) Optimized LCC-series compensated resonant network for stationary wireless EV chargers. IEEE Trans Indus Electron 66(4):2756–2765
13. Litz Wire. Technical information. New England Wire Technologies
14. Vu V, Tran D, Choi W (2018) Implementation of the constant current and constant voltage charge of inductive power transfer systems with the double-Sided LCC compensation topology for electric vehicle battery charge applications. IEEE Trans Power Electron 33(9): 7398–7410
15. Soft ferrites and accessories (2013) Data handbook. Ferroxcube
16. Gesrou A (2019) Optimization of electric vehicles wireless charging. Masters thesis, American University of Sharjah, Sharjah
17. Elghanam EA, Kabalan HS, Hassan MS, Osman A (2019) Design and modeling of ferrite core geometry for inductive wireless chargers of electric vehicles. In: IEEE international conference on electrical and computing technologies and applications (Accepted for publication)
18. Strauch L, Pavlin M, Bregar V (2015) Optimization, design and modeling of ferrite core geometry for inductive wireless power transfer. Int J Appl Electromagn Mech 49:145–155

Comparative Study for Robust STATCOM Control Designs Based on Loop-Shaping and Simultaneous Tuning Using Particle Swarm Optimization

Syed F. Faisal[1(⊠)], Abdul R. Beig[1], and Sunil Thomas[2]

[1] ECCE Department, Khalifa University of Science and Technology,
Abu-Dhabi, UAE
syed.faisal@ku.ac.ae
[2] Electrical and Electronics Engineering Department, Birla Institute of
Technology and Science Pilani, Dubai Campus, Dubai, UAE

Abstract. Synchronous static compensator (STATCOM) can also be used to improve the dynamic performance of a power system apart from being used for reactive power compensation. This article performs a comparative study for robust STATCOM control designs based on graphical loop-shaping and simultaneous tuning using Particle Swarm Optimization (PSO) for a single machine infinite bus system (SMIB) equipped with STATCOM. The power system working at various operating conditions is considered as a finite set of plants. Fixed parameter robust controllers were designed considering the voltage magnitude of the voltage source converter (VSC), a part of the STATCOM system, as the input and speed deviation of the generator as the system output. Simulation studies are conducted on a simple power system which indicates that the designed robust controllers by the two methods provide very good damping properties over a wide range of operating conditions, but the simultaneous tuning method using particle swarm optimization is easy to implement compared to the cumbersome graphical loop-shaping technique.

Keywords: Power system damping · STATCOM · Voltage source converter (VSC) · Robust control · Loop-Shaping · Simultaneous tuning · Particle swarm optimization (PSO)

1 Introduction

Flexible AC transmission system (FACTS) devices are well known for their ability to improve the power system dynamic and transient performances. FACTS devices vary the admittance between the two points in a transmission network and hence dynamically control the power flow through the transmission network. With the advent of the Voltage Source Converters (VSC) static synchronous compensator (STATCOM) is made possible. STATCOM is a controllable synchronous voltage source and is increasingly being used in power systems to control the real-time power flow and for rotor oscillation damping. The STATCOM is similar to the static var compensator in a

© Springer Nature Singapore Pte Ltd. 2020
N. Goel et al. (eds.), *Modelling, Simulation and Intelligent Computing*,
Lecture Notes in Electrical Engineering 659,
https://doi.org/10.1007/978-981-15-4775-1_45

way that it provides shunt compensation, but STATCOM uses a voltage source converter instead of shunt capacitors and reactors as in the case of static var compensator [1–5]. STATCOM provides a large number of performance benefits. The control circuit in the STATCOM has auxiliary signals that can be used for damping enhancement of a power system in terms of its rotor angular stability [3–5]. Many control strategies for dynamic performance improvement of power system using STATCOM control have been reported in the literature. A rule-based control strategy for STATCOM is proposed in [4]. The multivariable sampled regulator was designed for STATCOM to control the rotor angle oscillations in [5]. Some control techniques for STATCOM control that show robust performance have also been studied [6–8]. In [8] robust control of SVC and STATCOM using the H∞ design approach is presented. Designing robust controllers using H∞ approach is a tedious task. A simple loop-shaping technique, that produces a robust controller with fixed parameters, has been reported in [6, 7]. The design of robust controllers using evolutionary optimization techniques for damping enhancement of power system oscillations by the simultaneous stabilization technique has been presented in [9, 10]. Simultaneous stabilization is known to be a very challenging problem and no general systematic solution is obtainable. Iterative optimization techniques are proven to be the best alternatives in the absence of analytic solutions. A single set of PSS parameters can be found using a simultaneous stabilization method to guarantee the stabilization of the power system over the wide range of operations as reported in [9, 10].

In this paper, robust controllers were designed by the graphical loop-shaping technique and simultaneous stabilization approach using PSO for a single machine infinite system installed with STATCOM. In the graphical loop- shaping technique changes in the system loading conditions were included as a structured uncertainty model. In the simultaneous stabilization technique, the power system operating at different operating points is considered as a finite set of system transfer functions. The problem of selecting a fixed parameter robust controller is transformed into a simple optimization problem which is solved by Particle Swarm Optimization (PSO) and an eigen value-based objective function. The effectiveness of both methods is compared through nonlinear simulations of the power system installed with STATCOM.

2   The Power System Model Installed with STATCOM

Figure 1 shows a SMIB with a STATCOM installed in the middle of the transmission line. The generator is represented in terms of two electromechanical swing equations and an internal voltage equation. The dynamic model of the power system installed with STATCOM at the mid-point of the transmission line can be represented by a set of nonlinear equations given by (1) [11]. In (1), δ and ω are the rotor angle and generator speed whereas E_{fd} and e_q' are the field voltage and internal voltage, respectively.

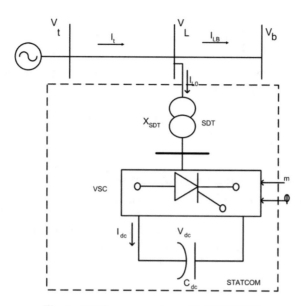

Fig. 1 SMIB power system with STATCOM

$$\dot{\delta} = \omega - \omega_o$$

$$\dot{\omega} = -\frac{1}{M}[P_m - P_e - D(\omega - \omega_o)]$$

$$\dot{e}'_q = \frac{1}{T'_{do}}[E_{fd} - e_q] \tag{1}$$

$$\dot{E}_{fd} = -\frac{1}{T_A}(E_{fd} - E_{fdo}) + \frac{K_A}{T_A}(V_{to} - V_t)$$

The STATCOM is connected to the transmission line through a step-down transformer (SDT) with a leakage reactance equal to X_{SDT}. A three-phase GTO-based voltage source converter, and a D.C. capacitor is an inherent part of the STATCOM system. The dynamic relationship that exists between STATCOM current (I_{dc}) and the capacitor voltage (V_{dc}) can be represented by (2),

$$\frac{dV_{dc}}{dt} = \frac{I_{dc}}{C_{dc}} = \frac{m}{C_{dc}}\left(I_{Lod}\cos\psi + I_{Loq}\sin\psi\right) \tag{2}$$

The direct and quadrature axes components of STATCOM current I_{Lo} are represented as I_{Lod} and I_{Loq}. The output voltage phasor is given by (3),

$$\bar{V}_o = mV_{dc}\angle\psi \tag{3}$$

In (2) and (3), ψ is the phase and m is the modulation index.
If the state and control vectors are taken as,

$$\left[\Delta\delta\ \Delta\omega\ \Delta e'_q\ \Delta E_{\mathrm{fd}}\ \Delta V_{\mathrm{dc}}\right]^{\mathrm{T}} \&\ [\Delta m\ \Delta\psi]^{\mathrm{T}}$$

Then the nonlinear state equations given by (1) and (2) for the SMIB with STATCOM can be represented as in (4),

$$\dot{x} = f(x, u). \tag{4}$$

3   The Design of Robust Controller for STATCOM Using Graphical Loop-Shaping Technique

The design of robust controller for the SMIB with STATCOM, begins by linearizing the nonlinear model of (4), around a nominal operating point

$$\begin{aligned} \dot{x} &= Ax + Bu \\ y &= Hx \end{aligned} \tag{5}$$

In the state space model of (5), control u is the modulation index m which is a measure of the STATCOM voltage magnitude. The nominal plant transfer function is

$$P = \mathrm{H}[sI - A]^{-1}B \tag{6}$$

Perturbations in the plant operating points is incorporated by a structured uncertainty model as

$$\tilde{P} = (1 + DW_2)P \tag{7}$$

Equation (7) is the multiplicative uncertainty model where, W_2 is a fixed stable transfer function, D is a variable transfer function satisfying $\|D\|_\infty < 1$. If $\|D\|_\infty < 1$ then

$$\left|\frac{\tilde{P}(j\omega)}{P(j\omega)} - 1\right| \leq |W_2(j\omega)|, \forall\omega \tag{8}$$

Therefore, $|W_2(j\omega)|$ is the uncertainty profile and in the frequency domain is the upper boundary of all the normalized plant transfer functions away from 1. For a controller function C in series with the plant P, the robustness criteria are

1. The nominal performance measure is $\|W_1 S\|_\infty < 1$
2. C provides robust stability if

$$\|W_2 T\|_\infty < 1 \tag{9}$$

3. Necessary and sufficient condition for robust nominal and robust performance is $\||W_1S| + |W_2T|\|_\infty < 1$.

It is to be noted here that W_1 is a real, rational, stable and minimum phase function. T is the input-output transfer function, S is the complement of the sensitivity function, and is given by

$$T = 1 - S = \frac{1}{1+L} = \frac{1}{1+PC} \tag{10}$$

Loop-shaping is a graphical method to design a robust controller C satisfying performance criteria and robust stability criteria as given in (9). The method relies on graphically constructing the open-loop transfer function, $L = PC$ to satisfy the robust performance criterion approximately, and then finally the robust controller is obtained from the relationship $C = L/P$. The constraints of this graphical method are internal stability of the plants and properness of C. Open-loop transfer function L should be constructed such that PC should not have any pole zero cancellation. An essential condition for robustness is that either or both $|W_1|$, $|W_2|$ must be less than 1 [7]. For a monotonically decreasing function W_1, it can be shown that at low frequency the open-loop transfer function L should satisfy

$$|L| > \frac{|W_1|}{1 - |W_2|} \tag{11}$$

while for high frequency

$$|L| < \frac{1 - |W_1|}{|W_2|} \approx \frac{1}{|W_2|}. \tag{12}$$

4   The Particle Swarm Optimization (PSO)

4.1   Introduction to PSO

The PSO is a heuristic optimization algorithm based on the social behavior of bird flocking and fish schooling and it was developed by Abdel-Magid et al. [9]. Similar to other evolutionary algorithms like Genetic Algorithms, PSO is also an optimization technique based on population. Unlike GA no operations in PSO are inspired by evolution to find new generations of candidate solutions, but instead, PSO depends on the exchange of information between individuals known as particles of the population [12–14]. The particles in the swarm update their own velocity and position according to the following equations given by (13) and (14), respectively.

$$V_i = QV_i + K_1\text{rand}_1(\,)(X_{\text{pbest}} - X_i) + K_2\text{rand}_2(\,)(X_{\text{gbest}} - X_i) \tag{13}$$

$$X = X_i + V_i \tag{14}$$

In (13) K_1 and K_2 are two positive constants, rand$_1$ ( ) and rand$_2$ ( ) are random numbers in the range [0, 1], and Q is the inertia weight. X_i represents the position of the ith particle in the swarm and V_i is particle velocity in Eq. (14).

4.2 PSO Flow Chart

The PSO flow chart used in this paper is shown in Fig. 2.

4.3 Simultaneous Tuning Using PSO

In this approach, the controller structure is pre-selected as two lead-lag blocks and a washout given by

$$C(s) = K \left(\frac{sT_w}{1 + sT_w} \right) \left(\frac{1 + sT_1}{1 + sT_2} \right) \left(\frac{1 + sT_3}{1 + sT_4} \right) \tag{15}$$

where K is the stabilizer gain, T_1, T_2, T_3, and T_4 are stabilizer time constants and T_w is the washout time constant. In this controller structure, T_w is usually a pre-selected value. The constraints on the values of K and the time constants are given by

$$
\begin{aligned}
K^{\min} &\le K \le K^{\max} \\
T_1^{\min} &\le T_1 \le T_1^{\max} \\
T_2^{\min} &\le T_2 \le T_2^{\max} \\
T_3^{\min} &\le T_3 \le T_3^{\max} \\
T_4^{\min} &\le T_4 \le T_4^{\max}
\end{aligned}
\tag{16}
$$

The objective function J to be minimized in the PSO algorithm is an eigen value-based objective function chosen to place the real part of the complex eigen values of the mechanical modes at a pre-determined location in the s-plane for all the operating points considered. The performance index is expressed as

$$J = \sum_{i=1}^{N} (\sigma_s - \sigma_i)^2 \tag{17}$$

where σ_s and σ_i are pre-selected value and real part of eigen value of the mechanical mode for the ith plant respectively and N is the number of operating points considered during the design.

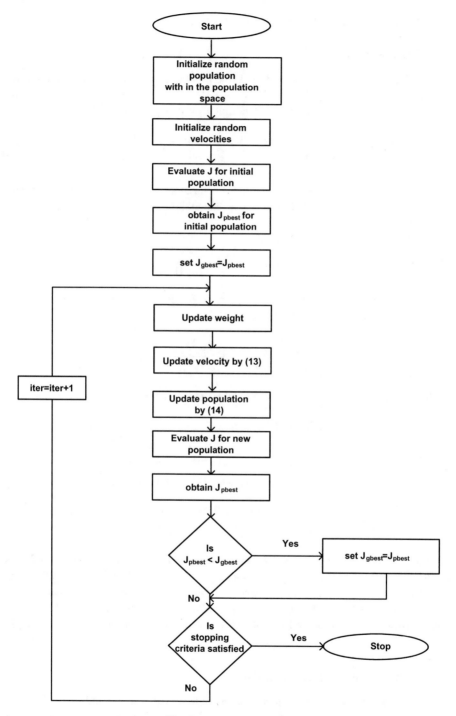

Fig. 2 Flow chart of PSO

5   Implementation of Robust Controllers

5.1   Graphical Procedure

For the SMIB with STATCOM of Fig. 1, the input is chosen as the modulation index m whereas changes in angular speed, $\Delta\omega$ of generator is chosen as the system output. For the chosen nominal operation, the nominal plant is given by (18)

$$P = \frac{0.2104s^2(s+100.827)(s-0.234)}{(s+99.17)(s+1.10)(s+0.05)(s^2+0.68s+21.63)} \tag{18}$$

Output power ranges between 0.4 and 1.4 pu and the range of power factor considered is 0.8 lagging to 0.8 leading. The uncertainty profile is fitted to the weight function, W_2 given by

$$W_2(s) = \frac{0.9s^2+15s+27}{s^2+5s+31} \tag{19}$$

$W_1(s)$ is chosen as a Butterworth filter that satisfies all the properties of the uncertainty profile and is given by (20)

$$W_1(s) = \frac{K_d f_c^2}{s^3+2s^2f_c+2sf_c^2+f_c^3} \tag{20}$$

In (20) $K_d = 0.01$ and $f_c = 0.1$ are chosen, and L is given by (21) and C by (22)

$$L = \frac{5.173s^2(s+100.827)(s-0.234)(s+1)(s+1.094)}{(s+99.17)(s+1.10)(s+0.05)(s^2+0.68s+21.63)} \\ \times \frac{(s+0.0476)(s+0.001)(s+99.174)}{(s+10)(s+0.1)(s+0.01)(s+0.03)(s+0.05)} \tag{21}$$

$$C = \frac{24.583(s+1)(s+1.094)(s+0.0476)(s+0.001)(s+99.174)}{(s+10)(s+0.1)(s+0.01)(s+0.03)(s+0.05)} \tag{22}$$

The bode plots for W_1, W_2, and L, which were used to find this robust controller, is given in Fig. 3a. The open-loop function L given by (21) is chosen to satisfy the boundaries set by (11–12). Figure 4 shows the plots for nominal and robust performance criteria. As evident from Fig. 3b, the nominal performance measure W_1S is well-satisfied whereas the combined robust stability and performance measure has a slight peak.

5.2   Simultaneous Tuning Using PSO

The proposed approach is implemented on the power system shown in Fig. 1. The PSO algorithm was used to minimize the objective function given by (17) where σ_s was chosen as −0.6 as it gives the best damping properties for the power system. The PSO

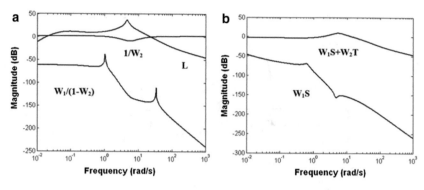

Fig. 3 **a** Graphical loop-shaping plots relating W_1, W_2, and L. **b** The robust performance criteria

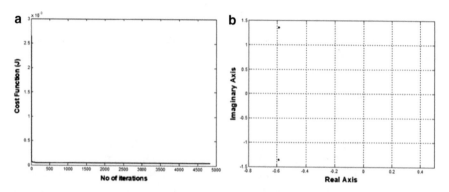

Fig. 4 **a** Convergence rate of the objective function. **b** Pole placement for four different operating points

algorithm converged to give the following settings for the lead-lag controller blocks: $K = 27.4$, $T_1 = T_3 = 0.0120$, $T_2 = T_4 = 0.0222$. The washout time constant T_w was selected as 2 s.

The convergence rate of the objective function J is shown in Fig. 4a. It is evident from Fig. 4a that the objective function reaches the minimum value very fast. Figure 4b shows the placement of poles for four different operating points. It can be observed clearly that the real parts of the poles (σ_i) are exactly placed as desired at -0.6 for the four operating points considered during the design. A comparison of the simulation results with the graphical loop-shaping and simultaneous tuning using PSO is given in Fig. 5.

The rotor angle response for the four loading conditions are presented in Fig. 5. It can be deduced that both the graphical loop-shaping technique and simultaneous tuning using PSO produce robust controllers that give almost identically good transient performance.

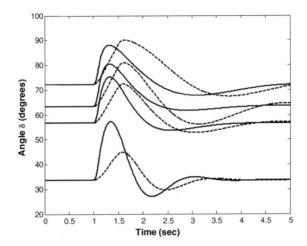

Fig. 5 Comparison of rotor angle changes upon a 50% disturbance on torque input (solid line is for loop-shaping technique and dotted line for simultaneous tuning using PSO)

6 Conclusions

In this paper, a comparative study was performed for designing robust STATCOM controllers by the graphical loop-shaping method and simultaneous tuning using Particle Swarm Optimization (PSO). Uncertainties in system loading conditions were considered for the purpose of design. However, the methods can also be applied with equal ease to include the uncertainties in the generator parameter variations. Both the graphical loop-shaping method and the simultaneous tuning method give almost similar transient responses. It was observed that the transient response for light conditions was found to be much better in the case of simultaneous tuning when compared to the graphical loop-shaping method. It is to be noted that compared to simultaneous tuning using PSO, the graphical loop-shaping method is more mathematically involved and requires graphical construction of weight function W_2 which is a bit time-consuming procedure for the amateur designer.

Acknowledgements. The authors would like to acknowledge the facilities provided at the Electrical Engineering and Computer Science Department at Khalifa University of Science and Technology and Birla Institute of Technology and Science Pilani, Dubai Campus, UAE towards this research.

References

1. Gyugi L (1994) Dynamic compensation of AC transmission lines by solid state synchronous voltage sources. IEEE Trans Power Deliv 9(2):904–911
2. Machowski J (1997) Power system dynamics and stability. Wiley
3. Wang H, Li F (2000) Multivariable sampled regulators for the coordinated control of STATCOM AC and DC voltage. IEEE Proc Gen Trans Distrib 147(2):93–98

4. Li C, Jiang Q, Wang Z, Rezmann D (1998) Design of a rule-based controller for STATCOM. In: Proceedings of the 24th annual conference of the IEEE industrial electronics society, vol 1, IECon'98, pp 467–472
5. Wang HF, Li F (2000) Design of STATCOM multivariable sampled regulator. In: International conference on electric utility deregulation and restructuring and power technologies, City University, London
6. Rahim AHMA, Al-Baiyat SA, Kandlawala FM (2001) A robust STATCOM controller for power system dynamic performance enhancement. In: IEEE PES summer meeting, Vancouver, pp 887–892
7. Rahim AHMA, Kandlawala FM (2004) Robust STATCOM voltage controller design using loop-Shaping technique. Electr Power Syst Res 68:61–74
8. Farasangi MM, Song YH, Sun YZ (2000) Supplementary control design of SVC and STATCOM using H∞ optimal robust control. In: Proceeding international conference on electric utility deregulation, City University, London, pp 355–360
9. Abdel-Magid YL, Abido MA, Al-Baiyat S, Mantawy AH (1999) Simultaneous stabilization of multi-machine power system via genetic algorithms. IEEE Trans Power Syst 14(4):1428–1439
10. Abdel-Magid YL, Abido MA (2003) Optimal multi-objective design of robust power system stabilizer using genetic algorithms. IEEE Trans Power Syst 18(3):1125–1132
11. Wang HF (1999) Philips-Hefron model of power systems installed with STATCOM and applications. IEEE Proc Gen Trans Distrib 146(5):521–527
12. Kennedy J, Eberhart R (1995) Particle swarm optimization. IEEE Int Conf Neural Netw 4:1942–1948
13. Shi Y, Eberhart RC (1999) Empirical study of PSO. In: IEEE Proceedings of the 1999 congress on evolutionary computation, vol 3, pp 6–9
14. Kennedy J (1997) The particle swarm optimization: social adaptation of knowledge. In: International conference of evolutionary computation, Indianapolis, pp 303–308

Real-Time Implementation of PID Controller for Cylindrical Tank System Using Short-Range Wireless Communication

Thulasya Naik Banoth$^{(\boxtimes)}$, Ravi Kumar Jatoth,
and Seshagiri Rao Ambati

National Institute of Technology Warangal, Warangal, Telangana, India
thulasyramsinghnaik@gmail.com, {ravikumar,seshagiri}
@nitw.ac.in

Abstract. Nowadays, wireless technology is developing at a great rate, as the process industries looking for the way to reduce the cost of the system, to improvise the performance of the system, and to comply with regulatory requirements. Especially in the short-range network, wireless technology addresses many operational challenges when compared with wired technology in oil, gas, and many other process industries. The wired network always needs real-time support for security, availability, and reliability in harsh industrial environmental conditions. These conditions will be overcome by introducing wireless technology in process industries, which in turn requires limited observation and maintenance. In this paper, a cylindrical tank is considered as single-input single-output (SISO) system, which is controlled wirelessly. Here, the plant is modeled as the first-order system by mathematical approach. After, IMC and direct synthesis tuning methods were used to tune the designed PID controller gain parameters.

Keywords: Arduino Mega · Bluetooth module HC-05 · Mobile application · MATLAB · Cylindrical tank modeling · PID controller

1 Introduction

In industrial control systems, the process control theory is the most commonly used modeling technique for controlling liquid levels in the tanks. The common problem in such systems is the variation of liquid flow rate at the input and output causes the irregular tank level. To make the liquid level constant, there is a need for a controller for the system. Various types of PID controllers were discussed in the literature [1–5]. Dighe Y. N. et al. applied the direct synthesis method approach for the design of PID controller [6]. To control the liquid level in a single tank system, Mostafa A. Fellaini et al. have developed PID, P, PI controllers. The simulations were done using MATLAB, and the responses of these controllers were compared for the parameters rise time, settling time, steady-state error, and overshoot [1]. Another method to control

© Springer Nature Singapore Pte Ltd. 2020
N. Goel et al. (eds.), *Modelling, Simulation and Intelligent Computing*,
Lecture Notes in Electrical Engineering 659,
https://doi.org/10.1007/978-981-15-4775-1_46

the liquid level using the fuzzy controller was proposed by Miral Changela, Ankit Kumar in which they discussed liquid level in tank and flow between the tanks. Later they compared the fuzzy controller with the PID controller and their performances were observed. And the performance of both is compared for their systems [2]. A fractional order PID controller was proposed by Janarthanan et al. [3] for a conical tank system. They presented different tuning methods to find the parameters of PID. Maxim et al. [4] used a model-based controller for MIMO process. Here, authors used IMC for the MIMO system. Finally, they tested the PID controller and IMC controllers for parameters consisting of set points tracking and disturbance rejection. Coupled tank controller was developed by Abbas [5] that consists of modeling and simulation of nonlinear SISO system. The sliding mode control method was used to control the liquid level in the SISO system, and this controller performance was compared with the PID controller.

The organization of the article is as follows: Sect. 2 describes the proposed work and Sect. 3 describes design of mobile application for Bluetooth. Section 4 includes mathematical modeling of the cylindrical tank system. Section 5 deals with the schematic diagram of the proposed work, followed by the results, conclusion, and future scope in Sect. 6.

2 Proposed Work

So far the plant is controlled through MATLAB and Arduino over a metered connection. The proposed work is to replace the metered connection with Bluetooth module and control the plant over a wireless connection as shown in Fig. 1, for that a mobile application is developed which is compatible with Bluetooth module and Arduino.

Fig. 1 Experimental setup

3   Application Design Using MIT App Inventor

Steps to Build a Mobile APP:

Step 1: Browse MIT App Inventor in Google and create a new account with a Gmail account [7].

Step 2: Then click on the new project and name the project, click ok as shown in Fig. 2.

Step 3: After creating a new project, we will get the *design screen*, which is shown in Fig. 3. The page consists of *component Palette*, which consists of components like *user interface, layout, sensors, storage, connectivity*, etc. The components can be dragged from the component palette and placed or dropped to the *viewer* to add them to the app and to see how the app looks like. For example, a button is dragged and placed on the phone in Fig. 3, which is shown in the red mark. Dragged components can appear in components which are shown in blue mark. The selected components can be renamed by clicking the *Rename* button which is shown in black mark. The properties of components can be changed in properties which are shown in the pink mark. The yellow mark consists of a *designer* and *blocks*, wherein the designer page is used for the screen design of the app. We can select the components from the palette based on our application and we can arrange them in a systematic view, which is shown on the left side of Fig. 3. The components which are highlighted in green mark in Fig. 3 are invisible components. With these invisible components, we will program the visible components based on our application. All the components properties can be edited as per the convenient of the appearance of the app in properties block as shown in Fig. 3.

Step 4: The event handler blocks describe how the phone should respond to certain events, which is programmed in Fig. 4. All the components selected in the screen designer page can appear in *Blocks*. If you click on the component, then the functionality or instructions can be shown in the *Viewer*.

Fig. 2   Home screen

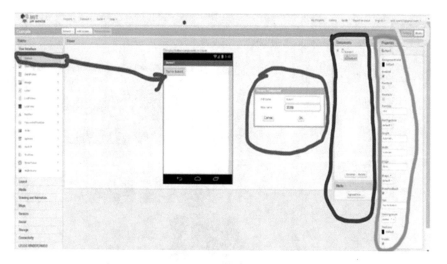

Fig. 3 Design screen

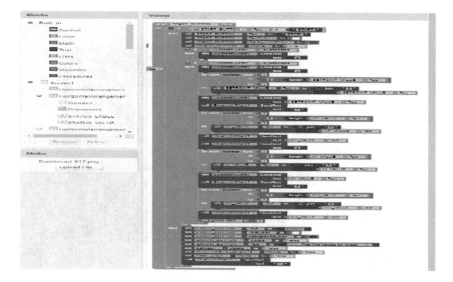

Fig. 4 Programming the data

When the Connect button has two options, that is when it is clicked then the blue-toothclint1 picks the available Bluetooth near its maximum range, which is shown in Fig. 5 as before picking the block, and then connects to the selected Bluetooth, which is shown in Fig. 5 as after picking block. If it is pressed again, then the Bluetooth is disconnected. When the app is opened, means app screen is initialized then the stored data in the excel sheet will be deleted. The file is the non-visible component for

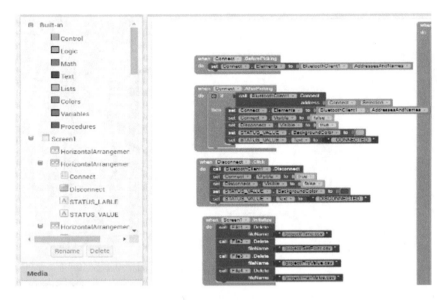

Fig. 5 Programming bluetooth

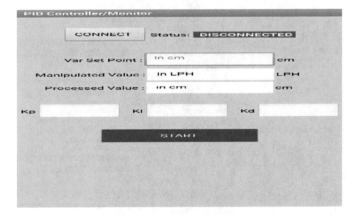

Fig. 6 Mobile view

storing data and retrieving data. This component writes or reads the data on our device. If the file path starts with a slash (/), then the file is created relative to /sdcard., i.e., it follows as /sdcard/project/name.csv (Figs. 6 and 7).

The clock is a non-visible component that provides an instant time using the mobile clock. At regular time intervals, the timer in the clock will perform time calculations, manipulations (process variable and manipulated variable in this project), and conversions. Operations on dates and times can be created using Make Instant methods,

Fig. 7 Non-visible components

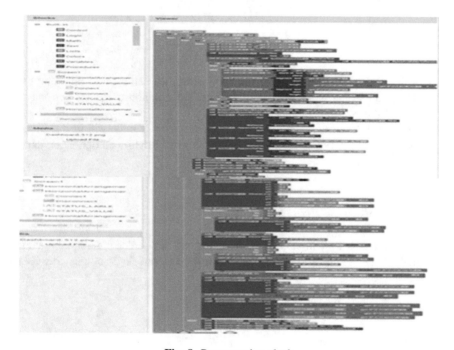

Fig. 8 Programming clock

and also appending data to created files using properties of the clock (Append to file) which is shown in Fig. 8.

Step 5: For the Android device through wireless Internet connection, we need to install App Inventor Companion App on the device.

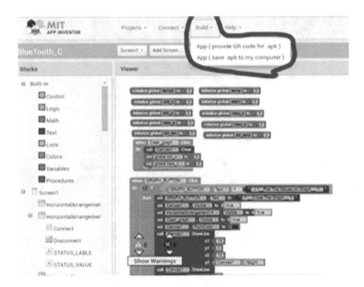

Fig. 9 Building APK file

Step 6: After the development of the application, we can create ".apk" file using Build option which is shown in Fig. 9. The Build option generates two ways to download the .apk file.

- It provides QR code, which can be scanned and then .apk file will be downloaded
- .apk file can be saved on a personal computer as it provides an option as "save to my computer."

4 Identification of Process Model

The cylindrical tank system can be modeled using a first-order differential equation. The liquid level in a tank is to be simulated as a single-input single-output (SISO) tank system (Fig. 10).

According to mass balance equation

$$q_i = q_o + \frac{dV}{dt}$$
$$q_i = q_o + \frac{d}{dt} A * h \tag{1}$$
$$q_i = q_o + A \frac{d}{dt} h$$

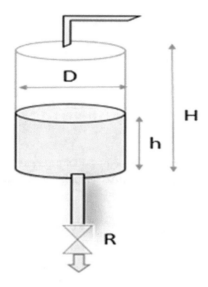

Fig. 10 Single-input single-output tank system

At steady state

$$q_{is} = q_{os} \tag{2}$$

The flowrate and liquid head relationship is

$$q_{os} = \frac{h_s}{R} \tag{3}$$

Therefore, the resistance of the outlet pipe is calculated as

$$R = \frac{h_s}{q_{is}}$$

$$q_i - q_{is} = Q_i$$

$$h - h_s = H$$

From Eqs. (1) to (2)

$$(q_i - q_{is}) = \frac{h - h_s}{R} + A\frac{dh}{dt}$$

$$Q_i = \frac{H}{R} + A\frac{dH}{dt} \tag{4}$$

By applying the Laplace transform on both sides

$$Q_i(s) = H(s)\left[1 + \frac{ARs}{R}\right]$$

$$\frac{H(s)}{Q_i(s)} = \frac{R}{(\tau s + 1)} \tag{5}$$

where $\tau = A * R$.

Where gain depends on the resistance of outlet pipe and time constant (τ) depends on the resistance of the outlet pipe and cross-sectional area. The transfer function of the system with delay time is

$$\frac{(H(s))}{(Q_i(s))} = \frac{\left(e^{(-7*s)} * 0.20999\right)}{((619.078s + 1))}$$

5   Schematic Diagram for Hardware Implementation

HC-05 Bluetooth module uses IEEE 802.15.1 standardized protocol to assemble wireless personal area networks and utilizes serial communication to communicate with Mobile and Serial Port Protocol (SPP) to communicate with Arduino Mega, so we need to begin the serial communication at 9600 baud rate. Using SPP, Bluetooth module, and Arduino Mega can send and receive data just as if there were RX and TX lines connected between them. After receiving the data from mobile for motor switching, two types of data string were used for switching from HIGH to LOW and LOW to HIGH. This signal is taken from one of the digital pins of the Arduino Mega, and controller performs the PID manipulations then the error signal is taken as process response from one of the digital pin of the Arduino Mega and fed to the V/I converter which is then sent to actuator which controls the inflow rate of liquid. As the height of liquid rises in the +cylindrical tank, the DPM sensor senses the differentiation in the pressure and converts it to the current is then converted to voltage and given as feedback signal to the analog pin on Arduino Mega is 10-bit resolution which converts analog signal to digital signal, i.e., it acts as A/D converter. The input given to the controller is an analog signal on the analog pin of Arduino Mega which is considered as the process variable (Fig. 11).

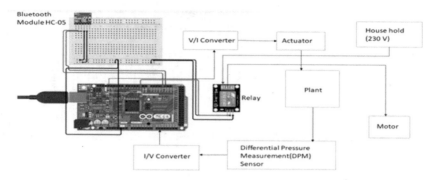

Fig. 11 Schematic diagram

6 Simulations and Results

Direct synthesis method for first-order time-delay system provides PI controller which is tuned and the gain values are tabulated in Table 1. The simulation result for controller gives better settling time without peak overshoot and zero steady-state error when compared to the IMC-based controller simulation result (Figs. 12 and 13; Tables 2 and 3).

Direct synthesis method gives the best results in the sense of system response and settling time and peak overshoot and considerable steady-state error when compared to the IMC-based controller. The system response characteristics are tabulated in the Table. The system performance is depicted in Fig. 14.

From the model Eq. (1) of the process, at the steady state of the liquid level in the tank, the liquid inflow rate varies from 350 LPH to 650 LPH. The above figure illustrates the performance of the controller.

Disturbance can be occurred by a sudden overflow of liquid and sudden closing of the bottom valve. Here, the disturbance is occurred by pouring the liquid manually. From Fig. 15, the disturbance is rejected but the settling time is slow because of derivative gain in the IMC-based controller, and from Fig. 15, the disturbance rejection and the settling time is fast for direct synthesis-based controller compared to IMC-based controller from Fig. 15.

Table 1 Controller gain parameters

Method	Filter coefficient (λ)	K_p	K_i	K_d
IMC	73.6	38.330	0.06176	133.40
Direct synthesis	73.6	36.129	0.05835	000

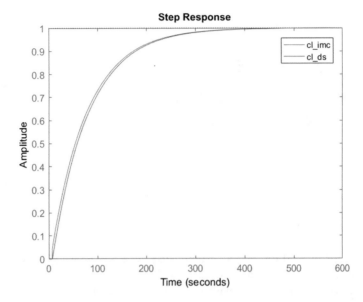

Fig. 12 Simulation response

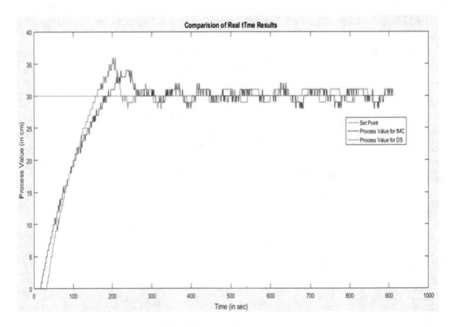

Fig. 13 Real-time response

Table 2 Observation from simulation response of the system:-

Method	Height (cm)	Rise time (sec)	Settling time (sec)	Steady-state error (%)
IMC	30	162	291	0
DS	30	161	294	0

Table 3 Observation from real-time response of the system:-

Method	Height (cm)	Rise time (sec)	Settling time (sec)	Peak overshoot (%)	Steady-state error (%)
IMC	30	159	695	13.333	5
DS	30	149	337	20	3.1

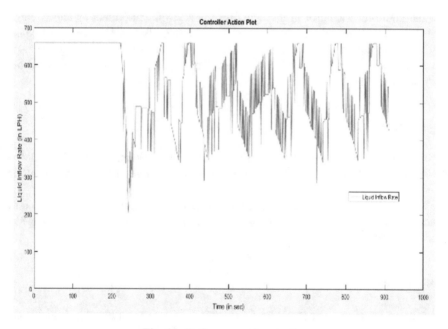

Fig. 14 Performance of controller

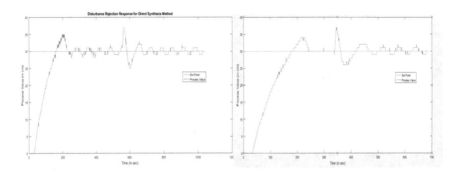

Fig. 15 Disturbance rejection response for DS method and IMC method

7 Conclusion

For the liquid-level control in the cylindrical tank, IMC, and direct synthesis control schemes, direct synthesis results fast response time and fast settling time and considerable steady-state error. As the liquid entering into the tank is splashing and turbulence, it results in noisy-level measurement. This is the reason, derivative controller does not reject the disturbance.

8 Future Work

In order to increase the access of wireless range of system, Bluetooth module is replaced with WI-FI module and the system can be controlled and monitored through Internet.

To improve the performance of the system, that is to decrease steady-state error, faster response, faster settling time, and PID tuning methods have to be changed to modern tuning methods.

References

1. Fellani MA, Gabaj AM (2015) PID controller design for two tanks liquid level control system using MATLAB. Int J Electr Comput Eng 5(3):436–442
2. Changela M, Kumar A, Designing of controller for two tank system. Int J Sci Res. ISSN 2319-7064
3. Janarthanan S, Thirukkuralkani KN, Vijayachitra S, Performance analysis of non-integer order PID controller for liquid level control of conical tank system. IEEE. ISBN No. 978-1-4799-3834-6/14
4. Maxim A, Ionescu CM, Copot C, De Keyser R, Multivariable model-based control strategies for level control in a quadruple tank process. IEEE. 978-1-4799-2228-4/13

5. Abbas H, Asghar S, Qamar S, Sliding mode control for coupled-tank liquid level control system. In: 2012 10th international conference on frontiers of information technology. IEEE. 978-0-7695-4927-9/12
6. Dighe YN, Kadu CB, Parvat BJ, Direct synthesis approach for design of PID controller. Int J Appl Innovation Eng Manag (IJAIEM) Volume 3, Issue 5, May 2014
7. http://appinventor.mit.edu/explore/library

Performance Enhancement in Stainless Steel Pressure Sensor

Sujan Yenuganti[(✉)] [iD]

Department of Electrical Engineering, Birla Institute of Technology and Science,
Pilani Campus, Pilani, Rajasthan, India
sujan.1053@gmail.com

Abstract. A piezoelectric resonant pressure sensor fabricated with stainless steel with a modified design to improve its performance is proposed in this work. The sensor consists of a stainless steel diaphragm, inclined trusses, hinged vertical mounts, and a resonating doubly clamped beam. The deflection of the diaphragm with applied pressure is transferred to the resonating beam via a stress transmission mechanism comprising of inclined trusses and vertical mounts. The sensor is fabricated with SS 304 grade stainless steel using electrical discharge machining (EDM) and wire-cut EDM process. The sensor was tested for its characteristics for an input pressure of 0–25 bar. The experimental results demonstrate that the proposed sensor was found to have better linearity, higher sensitivity, and low hysteresis as compared to a similar pressure sensor existing in the literature. Sensor design is simple; fabrication involves well-known machining process, self-packed, and hence cost effective.

Keywords: Pressure · Resonating beam · Piezoelectric · Diaphragm

1 Introduction

Pressure is an important physical parameter to be measured in almost all fields of engineering and industrial applications. Pressure measurements are not only important for monitoring and control, but also for measuring other parameters like flow and level. The development of pressure sensors is one of the well-established areas in sensor technology. Pressure sensors which depend on vibrating structures for sensing work on the principle of resonance frequency shift with applied pressure and are advantageous over other conventional sensing techniques [1–3]. The key advantage is that the frequency output is essentially in digital form and therefore can be easily integrated with digital electronics and instrumentation systems. In addition, the resonant measurement principle is usually based on the mechanical properties of the structure rather than the electrical properties and can be shown to offer stable performance, high resolution, reliability, and response time [4]. Among the excitation and detection methods used in resonant sensors, piezoelectric excitation and detection is gaining importance in recent years as the piezoelectric excitation offers advantages like strong forces, low voltage, high energy efficiency, linear behavior, high acoustic quality, and high speed [5–7].

Pressure sensors with stainless steel as sensing diaphragm are widely used to measure high pressures in corrosive environments. A complete self-packaged stainless

© Springer Nature Singapore Pte Ltd. 2020
N. Goel et al. (eds.), *Modelling, Simulation and Intelligent Computing*,
Lecture Notes in Electrical Engineering 659,
https://doi.org/10.1007/978-981-15-4775-1_47

steel high-pressure capacitive pressure sensor was developed and studied for its performance in harsh environments [8, 9]. A capacitive pressure sensor was designed and fabricated by Chang and Allen [10] which consists of stainless steel substrate, stainless steel diaphragm, and surface micro-machined back electrode for low pressure measurement. A piezoresistive pressure sensor using metallic thin film as a strain gauge bonded on to a stainless steel diaphragm was designed and fabricated for rail fuel injection systems [11]. A stainless steel pressure sensor using rigid center diaphragm and a fixed guided beam with FBG sensor was fabricated and tested for a pressure range of 0–20 bar [12]. In a similar fashion, a stainless steel-based pressure sensor using rigid center diaphragm and a fixed guided beam along with reduced graphene oxide-filter paper-based strain gauges was also fabricated and showed high sensitivity [13]. A stainless steel resonant pressure sensor with a new design was proposed with piezoelectric excitation and detection [14]. A complete analytical model of the sensor was developed using Ritz method and direct stiffness method. The sensor was also fabricated with three different grades of stainless steel, namely SS 304, SS 431, and 15–5 PH and tested for a pressure range of 0–25 bar.

In this work, a resonant pressure sensor with a modified design made up of stainless steel is designed, fabricated, and tested. The sensor was fabricated with SS 304 grade stainless steel and tested for its input–output characteristics. All parts of the sensor including the sensing diaphragm, stress transfer mechanism to the resonating beam, resonating beam, and the pressure port are completely fabricated using stainless steel. The advantage of the proposed sensor is that the stress transmission mechanism provides maximum stress to the resonating beam and isolates the resonating beam from the diaphragm; no hermetic package is required; sensor is fabricated from a single metal die and provides improved linearity and hysteresis with good repeatability as compared to the similar resonant pressure sensor reported by the author [14].

2 Sensor Design and Principle

The sensor comprises of three parts (I, II & III) as shown in Fig. 1. The part I is a cylindrical stainless steel (SS) rod with a certain diameter and thickness. Part II consists of a resonating beam which is supported by an elongated 'V'-shaped stress transmission mechanism secured on the top surface of the SS disk. The stress transmission mechanism comprises of two inclined trusses and two vertical mounts. The two trusses are inclined at an angle of 45° to the horizontal plane with their base resting on the square diaphragm and their other end to the vertical mount as shown in Fig. 1. The base of the two vertical mounts is firmly fixed on the top surface of the SS disk and hinged at other end where the inclined trusses unite. The inclined trusses and the vertical mounts together hold a resonating doubly clamped beam, on the top surface of which two piezoceramic patches are bonded for resonant actuation and sensing as shown in Fig. 1. Part III is also a stainless steel tube fitting for pressure inlet. The stainless steel disk is grooved from its bottom face to realize a square diaphragm of non-uniform thickness, with its edges fixed in all the directions acting as a primary sensing element for the input pressure as shown in Fig. 2.

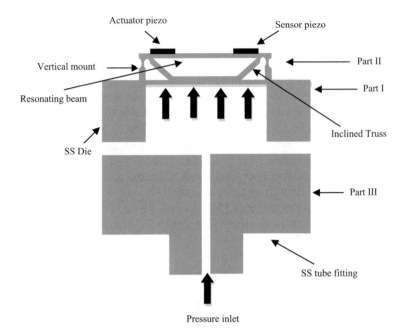

Fig. 1 Cross-sectional view of the pressure sensor

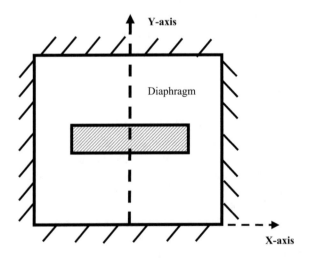

Fig. 2 Top view of the sensor diaphragm

The doubly clamped beam is made to vibrate at its first mode resonance frequency by the two piezoelectric patches under no pressure load conditions. When an input pressure P is applied through the SS tube fitting (Part III) to the square diaphragm in part I, the diaphragm deforms and exerts a force on the inclined trusses of part II. This force is transferred as an axial force to the resonating beam by the restriction offered by

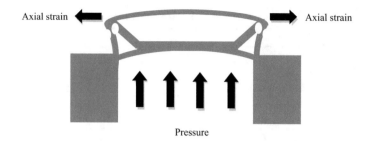

Fig. 3 Sensor under applied pressure

the hinged vertical mounts of Part II, thereby elongating the beam. This elongation produces an axial strain on the resonating beam as shown in Fig. 3 which shifts its resonance frequency from the initial value. The shift in the resonance frequency is proportional to the applied pressure P.

3 Fabrication and Testing

The pressure sensor design shown in Fig. 1 was fabricated using SS 304, and its dimensions are given in Table 1. Part I and Part II of the sensor were fabricated from a single SS cylindrical rod. The square diaphragm of Part I was realized by grooving from the bottom side of SS rod using electrical discharge machining (EDM) process. The resonating beam, inclined trusses, and vertical mounts with hinges described in Part II were realized by wire-cut EDM process. The pressure port in Part III was also made of SS 304 with M14 × 1.5 connector fabricated through conventional machining process. The fabricated resonating beam with inclined trusses and vertical mounts, square diaphragm machined on a single SS cylindrical rod and a pressure port of the sensor are shown in Fig. 4. The sensor was assembled by welding the pressure port (part III) to the SS rod (part I and part II) using LASER welding. Two piezoceramic patches (PZT 5H) of size 3 mm × 3 mm × 0.3 mm were bonded on the top surface of the resonating beam: One is used to excite the beam and the other one for sensing. The photograph of the fabricated sensor with bonded piezoelectric patches is also shown in Fig. 4.

The sensor was tested for its input–output characteristics using a dead weight tester (Make: FLUKE, Model: P3125-BAR) by applying pressure in the range of 0–25 bar in steps of 5 bar. The photograph of the test setup is shown in Fig. 5. During testing, the resonating beam was made to vibrate at its first mode frequency by applying a sinusoidal signal with an amplitude of 10 V_{P-P} to the piezoelectric actuator from an arbitrary waveform generator (Make: GW INSTEK, Model: AFG-3081) and the output of the piezoelectric sensor was measured using digital storage oscilloscope (Make: GW INSTEK, Model: GDS-2012A). In the absence of applied input pressure, the resonating beam remains to vibrate at resonance. When pressure is applied, the resonance frequency of the beam shifts and made to vibrate at its new resonance frequency by tuning the input signal frequency applied to the piezoelectric actuator. The input–output characteristics, namely nonlinearity, sensitivity, and hysteresis, of the pressure sensor obtained from testing were tabulated and plotted.

Table 1 Dimensions of the fabricated pressure sensor

Description	Value
Diaphragm length (mm)	13
Diaphragm thickness (mm)	0.3
Length of resonating beam (mm)	11
Thickness of resonating beam (mm)	0.3
Width of resonating beam (mm)	3
Length of the vertical mount (mm)	2.5
Thickness of the vertical mount (mm)	1
Width of the vertical mount (mm)	3
Length of the inclined truss (mm)	3.53
Cross-sectional area of the vertical mount (mm^2)	3
Cross-sectional area of the inclined truss (mm^2)	3

Fig. 4 Photographs showing the fabricated sensor

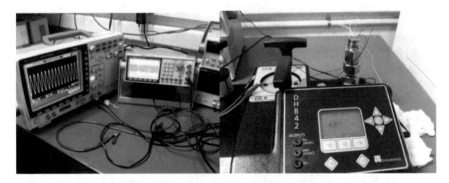

Fig. 5 Experimental setup for sensor testing

4 Results and Discussion

The experimental results show that the nonlinearity, sensitivity, and maximum hysteresis of the sensor were found to be 4.3% full-scale deflection (FSD), 23.12 Hz/bar, and 2.27% FSD, respectively. The sensor was also tested for repeatability by applying ten cycles of loading and unloading pressure and found to have good repeatability. The input–output characteristics of the sensor are shown in Fig. 6a, b. The resolution of the sensor was found to be 0.5 bar. The presence of the hinged vertical mounts in the design reduced the nonlinearity and hysteresis and improved the sensitivity as

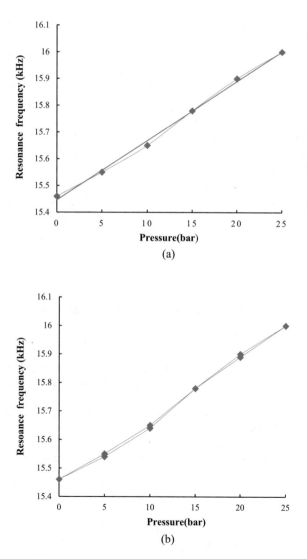

Fig. 6 Input–output characteristics of the sensor **a** nonlinearity and **b** hysteresis

compared to a similar pressure sensor of same dimensions which was fabricated and tested by the authors in their previous work. The existing sensor was found to have a nonlinearity of 7.7% FSD, sensitivity of 18.24 Hz/bar and a maximum hysteresis of 6.67% FSD, respectively [14].

Moreover, the sensor data can also be collected using a suitable DAC card and the sensor signal can be compensated for external noise, temperature etc., using a closed-loop control which will be taken up as a future work by the author.

5 Conclusions

A piezoelectric resonant pressure sensor has been designed and fabricated from stainless steel with a hinged vertical mount which was presented in this work. The sensor prototype is fabricated with 304 grade stainless steel and tested for its input–output characteristics to show the improvement in its performance for an applied pressure in the range of 0–25 bar. The maximum nonlinearity and maximum hysteresis of the proposed sensor were found to be 4.3% FSD and 2.27% FSD as compared to the sensor existing in the literature [14] (7.7% FSD & 6.67% FSD). The experimental results show a drop in nonlinearity by 44%, drop in maximum hysteresis was by 65% and improvement in sensitivity was by 21%. The ease of fabrication, low cost, SS being as a wetted part and frequency output, it can be used in noisy, corrosive and harsh environments like oil, petrochemical and aerospace industries. The diaphragm design and the stress transmission mechanism can be modified further for high pressure and corrosive environments. The similar design can also be micro-fabricated to improve its sensitivity further where the operating frequency will be in Mega hertz.

References

1. Melvas P, Kalvesten E, Stemme G (2001) A surface micro machined resonant beam pressure sensing structure. J Microelectromech Syst 10(4):498–502
2. Tang Z, Fan S, Cai JC (2009) A silicon micro machined resonant pressure sensor. JPCS 188 (012042):1–4
3. Tang Z, Fan S, Xing W, Guo Z, Zhang Z (2011) An electro thermally excited dual beam silicon resonant pressure sensor with temperature compensation. Microsyst Technol 21:1481–1490
4. Lanngdon RM (1987) Resonant sensors—a review. Current advances in sensors. Adam Hilger, Bristol, PA, pp 19–31
5. Suresh K, Uma G, Santhosh Kumar BVMP, Varun Kumar U, Umapathy M (2011) Piezoelectric based resonant mass sensor using phase measurement. Measurement 44 (2):320–325
6. Suresh KG, Uma G, Umapathy M (2011) A new resonance based method for the measurement of nonmagnetic conducting sheet thickness. IEEE Trans Instrum Measur 60 (12):3892–3897
7. Olfatnia M, Xu T, Miao JM, Ong LS, Jing XM, Norford L (2010) Piezoelectric micro diaphragm based pressure sensor. Sens Actuators, A 163:32–36

8. Ho SS, Rajgopal S, Mehregany M (2013) Media compatible stainless steel capacitive pressure sensors. Sens Actuators, A 189:134–142

9. Ho SS, Rajgopal S, Mehregany M Stainless steel capacitive pressure sensor for high pressure and corrosive media applications sensors. In: Proceedings of 23rd IEEE international conference on MEMS. Hong Kong, China, pp 647–650 (2010)

10. Chang SP, Allen MG (2004) Capacitive pressure sensors with stainless steel diaphragm and substrate. J Micromech Microeng 14:612–618

11. Stoetzler A, Dittmann D, Henn R, Jasenek A, Didra HP, Metz M (2007) A small size high pressure sensor based on metal thin film technology. In: Proceedings of 2007 IEEE sensors international conference. Atlanta, USA, pp 825–827

12. Manjunath MS, Nagarjuna N, Uma G, Umapathy M, Nayak MM, Rajanna K (2018) Design fabrication and testing of reduced graphene oxide stain gauge based pressure sensor with increased sensitivity. Microsyst Technol 24:2969–2981

13. Manjunath M, Uma G, Umapathy M, Nayak MM (2017) Design fabrication and testing of fiber bragg grating based fixed guided beam pressure sensor. Optic 158:1063–1072

14. Sujan Y, Uma G, Umapathy M (2016) Design and testing of piezoelectric resonant pressure sensor. Sens Actuators, A 205:177–186

Source/Drain (S/D) Spacer-Based Reconfigurable Devices-Advantages in High-Temperature Applications and Digital Logic

Abhishek Bhattacharjee[1] and Sudeb Dasgupta[2(✉)]

[1] Tripura Institute of Technology, Narsingarh, Tripura, India
[2] Indian Institute of Technology Roorkee, Roorkee, Uttarakhand, India
sudebfec@iitr.ac.in

Abstract. This paper explores source/drain (S/D) spacer technology-based reconfigurable field-effect transistors (RFETs) and a detailed physical insight toward the advantages of using spacer oxide in RFETs for applications involving rapid temperature fluctuations and reduction of circuit delay in contrast to conventional ambipolar FETs and other devices based on band-to-band tunneling (BTBT) such as TFETs. Temperature-based DC, analog and RF performance of gate-all-around (GAA), heterogeneous gate dielectric GAA, SiGe, and full silicon TFETs are compared. Moreover, it is also shown that the propagation delay in logic circuits is reduced for the proposed DG-RFET resulting in more robust and improved circuit performance.

Keywords: RFET · Temperature · BTBT · Source/drain · Spacer first section

1 Introduction

As the technology is advancing toward the sub-nm node [1–15], devices which can perform multiple functionalities are highly desirable. Reconfigurable nanowire field-effect transistor (RFET) which came into limelight in 2011 [1] attracted the eyes of many device researchers because of its ability to enhance the functional diversity by allowing the same transistor to behave as n- and p-FET. It can drastically reduce the number of functional units per logic block as well as provide a suitable alternative to the classical trend of downscaling [1, 2]. In order to further accelerate this performance trend of RFET, we demonstrate the *underlap* ambipolar FET architecture in this paper and also show the use of S/D spacers to improve the temperature-dependent DC, analog and RF performance in comparison to conventional *non-underlap* RFET and various other TFET configurations using well-calibrated TCAD tool [9]. Moreover, the reduction of propagation delay using spacer technology is also illustrated here for the first time.

© Springer Nature Singapore Pte Ltd. 2020
N. Goel et al. (eds.), *Modelling, Simulation and Intelligent Computing*,
Lecture Notes in Electrical Engineering 659,
https://doi.org/10.1007/978-981-15-4775-1_48

2 3D Device and Simulation Setup

Figure 1 shows the proposed S/D spacer-based *underlap* RFET architecture. Spacer length optimization is done through a series of TCAD simulations by taking I_{ON} and I_{ON}/I_{OFF} as the performance metrics for evaluation as shown in Fig. 2a, b. To validate the correctness and accuracy of our simulation setup, we have calibrated the BTBT and TCAD mobility models with experimental data for RFETs [1] as shown in Fig. 2c.

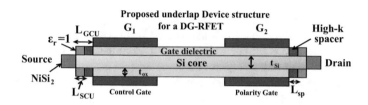

Fig. 1 Proposed underlap S/D spacer-based RFET

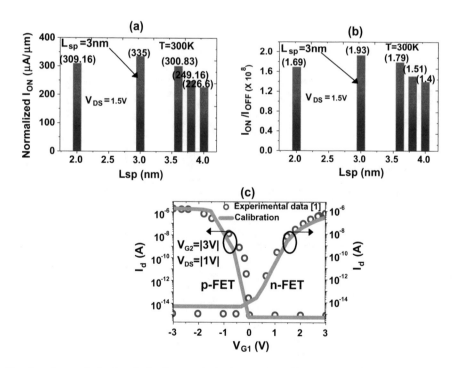

Fig. 2 **a** I_{ON} optimization, **b** I_{ON}/I_{OFF} optimization, and **c** calibration against experimental data [1]

3 Results and Discussion

The energy bandgap reduction with temperature (Fig. 3a) which has an inverse dependence on the tunneling current results in the rise in normalized I_{ON} with temperature (Fig. 3b, c).

Higher generation rate of EHP at elevated temperature results in an increase of I_{OFF} (Fig. 3b). But it is found to be lesser than TFET which shows that it is thermally more stable than TFET. S/S shows a linear dependence on the temperature at lower current values but this variation is slightly nonlinear when the device enters the off state (Fig. 3d).

A higher value of g_m and g_m/I_d (Fig. 4a, b) in case of the proposed device as compared to SiGe TFET [3] and Si abrupt TFET [3] is because of the BTBT dominant current conduction in the former case and TAT dominance in the later ones. The variation of output conductance (g_d) for the three devices is shown in Fig. 4c. Due to a larger drain voltage impact, the BTBT dominated device again shows a comparatively higher value of g_d. To verify this assumption, the maximum BTBT rate for the proposed device is also plotted in Fig. 4d. Moreover, for thermal-based analog applications, the proposed device is expected to be more suitable because of higher value of voltage gain, A_v (Fig. 4e).

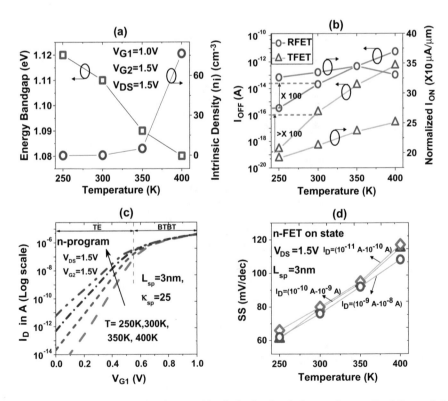

Fig. 3 Variation of **a** energy bandgap and intrinsic density, **b** I_{OFF} and normalized I_{ON}, **c** I_d in log scale, and **d** S/S with temperature

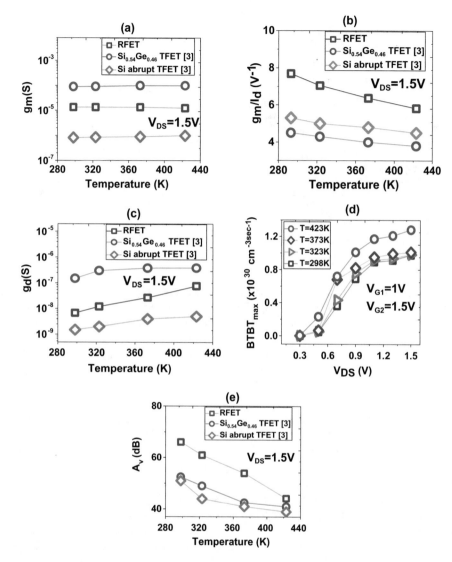

Fig. 4 Variataion of **a** g_m, **b** g_m/I_d, **c** g_d, **d** BTBT max, and **e** A_v with temperature for SiGe TFET [3], Si abrupt TFET, [3] and the proposed device

The total gate capacitance C_{GG} of the RFET is found to rise by 60.30% with a temperature increase from 250 to 400 K (Fig. 5a). It may be noted that the magnitude of C_{GG} is found to be much lower than GAA TFET [4]. It is because RFET has an almost intrinsic channel region as compared to the other cases. Secondly, TFET has a negligible potential drop near the drain–channel junction due to a larger reverse bias leading to a comparatively large gate–drain capacitance (C_{GD}). Now, C_{GG} is contributed equally by C_{GS} and C_{GD} in case of a GAA TFET. So, even a small change in C_{GS} causes C_{GG} to change by a great margin. In comparison to GAA TFET [4] and

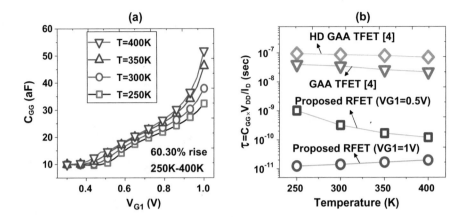

Fig. 5 Variataion of **a** C_{GG}, **b** τ with temperature for GAA TFET [4], HD GAA TFET [4], and the proposed RFET

HD GAA TFET [4], a considerable reduction in intrinsic gate delay (τ) in case of the proposed RFET can be seen from Fig. 5b, mainly because of higher I_{ON} in case of the proposed device which makes it more reliable for high-speed low/high-power applications.

f_T is the frequency at which the short-circuit current gain drops to unity. The variation in f_T with temperature is shown in Fig. 6a. The device shows its superior performance in the form of higher f_T. The difference in the values of C_{GS} and g_m values of the proposed and conventional RFET [1] and TFET devices is mainly reflected in the difference in f_T values. The larger f_T values of the proposed RFET also results in almost two orders of magnitude reduction in the transit time (τ_t) (Fig. 6b) which is very important in determining the speed of any device. An increase in GBW for the proposed device (Fig. 6c) is due to smaller capacitances in AF range whereas in case of both SiGe TFET [3] and Si abrupt TFET [3] it is in the order of fF. For bias point optimization in RFIC design, IIP3 is an important RF figure of merit. A better gate control over the channel for the proposed device is mainly responsible for an improved IIP3 (Fig. 6d). VIP2 decides the distortion characteristics for various dc parameters. The proposed architecture is found to be quite immune to variation in VIP2 with respect to temperature (Fig. 6e). A slightly higher value of VIP2 is displayed by the device under consideration as compared to the conventional RFET and slightly lower value as compared to SiGe TFET mainly because g_{m1} is higher for the proposed RFET as compared to the conventional ambipolar architecture but lower than that of SiGe TFET (Fig. 6e). 18–20 dB reduction in the third-order harmonics (IMD3) is exhibited by the proposed reconfigurable device as compared to TFETs thus ensuring lowered distortion in wireless applications (Fig. 6f).

Lastly, the improved electrical performance of the proposed ambipolar FET is reflected in a reduced NM_H and NM_L of 440 mV and 552 mV at $V_{DD} = 1$ V (Fig. 7)

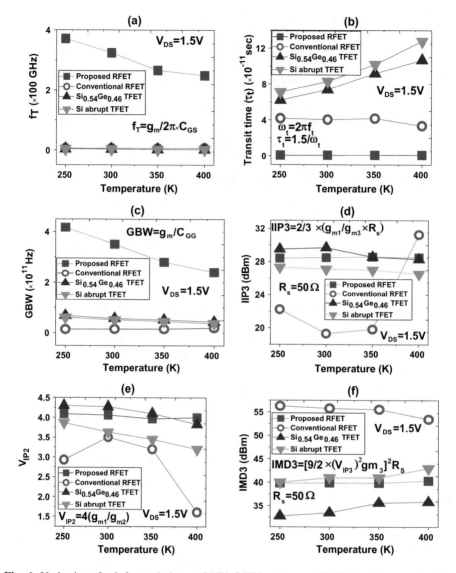

Fig. 6 Variataion of **a** f_T, **b** transit time, **c** GBW, **d** IIP3, **e** V_{IP2}, and **f** IMD3 with temperature for the proposed RFET, conventional RFET [1], SiGe TFET [3], and Si abrupt TFET [3]

and τ_{PHL}, τ_{PLH}, τ_P of 0.256 ns, 0.278 ns and 0.267 ns, respectively (Fig. 8), as compared to the conventional *non-underlap* RFET device [2]. On a whole, the overall improvement in the figure of merits of the proposed device may be attributed to better gate control over the channel due to spacers, improved on current and reduced capacitance.

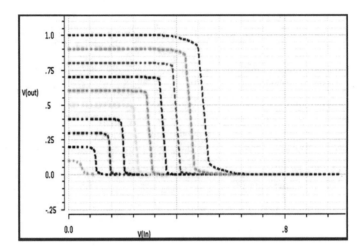

Fig. 7 Inverter VTC for various V_{DD} for the proposed RFET

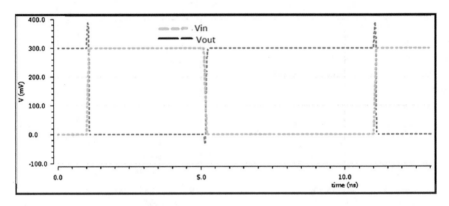

Fig. 8 Transient analysis for the proposed RFET

4 Conclusion

We evaluated the temperature-dependent DC, analog and RF performance of a spacer-based DG-RFET device and compared it with other devices which depend on BTBT for their on current generation such as TFETs. Unlike the TAT-dependent current conduction mechanism in SiGe and full silicon TFET, the BTBT-dependent on-state current is found to be beneficial for the proposed transistor owing to a superior analog performance. Further, we show that the device under consideration shows lower intrinsic delay and improved RF figure of merits in terms of cutoff frequency (f_T) and gain–bandwidth product (GBW) mainly because of enhanced I_{ON}, higher g_m, and lower intrinsic capacitance C_{GS}. Finally, the improved electrical performance is also reflected in terms of lower inverter delay as compared to the conventional RFET device.

Moreover, since the device FOMs are found to be quite immune to temperature variations, the S/D spacer-based RFET can be used in all those applications which demand rapid fluctuations in temperature.

References

1. Heinzig A, Slesazeck S, Kreupl F, Mikolajick T, Weber WM (2011) Reconfigurable silicon nanowire transistors. Nano Lett 12(1):119–124
2. Heinzig A, Mikolajick T, Trommer J, Grimm D, Weber WM (2013) Dually active silicon nanowire transistors and circuits with equal electron and hole transport. Nano Lett 13 (9):4176–4181
3. Martino MDV, Neves FS, Agopian PGD, Martino JA, Vandooren A, Rooyackers R, Simoen E, Thean A, Claeys C (2015) Analog performance of vertical nanowire TFETs as a function of temperature and transport mechanism. Solid-State Electron 112:51–55
4. Madan J, Chaujar R, Temperature associated reliability issues of heterogeneous gate dielectric—gate all around—tunnel FET. IEEE Trans Nanotechnol 99:1. https://doi.org/10. 1109/tnano.2017.2650209
5. Bhattacharjee A, Saikiran M, Dutta A, Anand B, Dasgupta S (2015) Spacer engineering-based high-performance reconfigurable FET with low off current characteristics. IEEE Electron Device Lett 36(5):520–522
6. Bhattacharjee A, Dasgupta S (2016) Optimization of design parameters in dual-κ spacer based nanoscale reconfigurable FET for improved performance. IEEE Trans Electron Devices 63(3):1375–1382
7. Bhattacharjee A, Dasgupta S (2017) Impact of gate/spacer-channel underlap, gate oxide EOT, and scaling on the device characteristics of a DG-RFET. IEEE Trans Electron Devices 64(8):3063–3070
8. Bhattacharjee A, Saikiran M, Dasgupta S (2017) A first insight to the thermal dependence of the DC, analog and RF performance of a S/D spacer engineered DG-ambipolar FET. IEEE Trans Electron Devices 64(10):4327–4334
9. Sentaurus TCAD (ver. 2012.06) Manuals. Synopsys Inc., Mountain View, CA
10. Trommer J, Heinzig A, Baldauf T, Slesazeck S, Mikolajick T, Weber WM (2015) Functionality-enhanced logic gate design enabled by symmetrical reconfigurable silicon nanowire transistors. IEEE Trans Nanotechnol 14(4):689–698
11. Migita S, Fukuda K, Morita Y, Ota H (2012) Experimental demonstration of temperature stability of Si-tunnel FET over Si-MOSFET. In: 2012 IEEE silicon nanoelectronics workshop (SNW), Honolulu, HI, pp 1–2
12. Boucart K, Ionesco AM (2007) Double-gate tunnel FET with high-κ gate dielectric. IEEE Trans Electron Devices 54(7):1725–1733
13. Marchiori C, Frank MM, Bruley J, Narayanan V, Fompeyrine J (2011) Epitaxial SrO interfacial layers for HfO$_2$-Si gate stack scaling. Appl Phys Lett 98(5):052908
14. Marchi MD, Zhang J, Frache S, Sacchetto D, Gaillardon P-E, Leblebici Y, De Micheli G (2014) Configurable logic gates using polarity-controlled silicon nanowire gate-all-around FETs. IEEE Electron Device Lett 35(8):880–882
15. Bhattacharjee A, Dasgupta S (2018) A compact physics-based surface potential and drain current model for an S/D spacer based DG-RFET. IEEE Trans Electron Devices 65(2):448–455

RoadNurse: A Cloud-Based Accident Detection and Emergency Relief Response Infrastructure

Aditya Rustagi, Vinay Chamola$^{(\boxtimes)}$, and Dheerendra Singh

Department of Electrical and Electronics Engineering, Birla Institute of
Technology and Science Pilani, Pilani Campus, Pilani 333031, India
{vinay.chamola,dhs}@pilani.bits-pilani.ac.in

Abstract. Casualties of roadside accidents often die due to the delayed arrival of rescue groups. This is because there is an interval between the accident occurring and the authorities being notified. In some cases, the authorities are failed to be notified due to the absence of any bystanders and the incapability of the victim to call for help themselves. We propose a compact system called the RoadNurse, to provide location-based emergency service which locates an accident quickly and notifies the emergency services and the loved ones of the victim. It also provides the live location of the accident to accelerate the transfer of the victim to the medical centers. The system contains vibration sensors which detect a value greater than certain threshold, determining the possible severity of injury and then utilizes the GPS module to determine the precise location of the accident. This location is sent to the cloud server, which contains the details (name, location, contact, and severity-level capability of treatment) of hospitals in the city. The server processes the optimal hospital with respect to proximity to the accident and the severity of injury. The hospital and the victims loved ones are messaged the details of the accident by the GSM module, and a phone call is initiated with the hospital. This procedure serves as a lifeline to the victim and might be the difference between life and death in the future.

Keywords: Cloud · GPS module · GSM module · Vibration sensor · NodeMCU

1 Introduction

Many cars manufactured in India fail the United Nation's minimum crash test regulations [1] and lack airbags that inflate during a mishap. Many reputed brands cars failed global New Car Assessment Program (NCAP) test [1], which is a reason to worry. Road accidents have a large stake in the annual mortality statistics with over 1.35 million people killed and 50 million injured [2]. In many accidental cases, drivers and passengers are rendered unconscious or are unable to reach out for any help. Many of these victims failed to receive any medical assistance due to the authorities being notified too late [3], or not at all. The victim is often at the mercy of luck, for if any on-goer will stop to help. It is often the bystanders themselves who bring the victim to the hospital [4].

© Springer Nature Singapore Pte Ltd. 2020
N. Goel et al. (eds.), *Modelling, Simulation and Intelligent Computing*,
Lecture Notes in Electrical Engineering 659,
https://doi.org/10.1007/978-981-15-4775-1_49

Such situations warrant policies be developed and strengthened to ensure compliance of international vehicle standards for manufacturers, thus limiting the sale of substandard vehicles in the country. Vehicles should further be equipped with automated alert systems. In that situation, a quick and automated call for help or emergency alarm is provided. Such implementations will help to reduce the growth in road accident deaths in developed countries [5] and developing countries where the injury-to-death ratio is maximum [6].

Many systems available in the market relies on emergency contacts to alert the hospitals, doctors, and ambulances, which adds a middle man leading to delay and the possibility of death of the victim. To overcome the above predicaments, we have proposed a cloud-based solution to contact the nearest hospitals with respect to the severity of the accident. Once the location and other relevant data is collected, it will be transmitted to the cloud server. Data will be processed, and the database will be searched to identify an appropriate medical facility in minimal time.

In this paper, we develop an accident detection system as well as a cloud-based infrastructure for emergency relief. This includes development of the prototype for vehicle accident detection and response unit in the vehicle, as well as developing the infrastructure (cloud services/database/message exchange system) for the emergency relief.

2 Related Works

Automatic accident detection implementations are present, but they offer some restrictions. The OnStar system [7] hosts provisions of emergency services via GPS along with roadside assistance. However, this is a rather bulky system that must inbuilt to the car itself, requiring the purchase of a new car to avail these services, which is neither available in countries like India, nor is cheap to the general public where available. Ki et al. [8] proposed a system of cameras placed at the intersections to detect accidents along with the velocity and direction. Clearly, such a system is suitable only at that particular intersection and fails to detect crashes at any other location.

Many implementations focus only on accident detection like Tushara et al. [9], where the authors propose an accident detection model where messages are sent to predefined numbers, but the system does not include any rescue system. Maleki et al. [10] proposed a location-based tracking system which collects the information of the accident through crash sensor readings. It communicates the information via SMS to the concerned emergency department. Few works such as [11–14] include contacting emergency services post detection. PoP et al. [15] built a smartphone application that utilizes a built-in gyroscope and accelerometer. The system depends on the fact that the smartphone will always be in the custody of the user and will detect accidental falls, reporting them to the nearest hospital. However, in the proposed system design, we propose a comprehensive system that account for the entire accident management including informing the relatives of the victim as well as hospital/ambulance selection/notification and sending periodic updates.

3 Requirements

This section discusses the various hardware components and software technologies used for building the proposed RoadNurse system.

3.1 Hardware Components

NodeMCU. The ESP8266 NodeMCU (CP2102) is the micro-controller used, due to its small size and connectivity to Wi-Fi without the need of external shields like the Arduino. NodeMCU is kept in a state of modem sleep, which shuts off the Wi-Fi connectivity, to consume less energy (using only 15 mA), since the GSM module allows Internet connectivity for lesser power. NodeMCU's Internet connectivity is kept in the hope of improving upon and adding more functionality to the system in the future. An energy-efficient system is a requirement for an emergency service provisioning product.

GPS Module. The NEO-6M-0-001 is included in the setup to receive information from GPS satellites and to acquire the device's geographical position. This data is used to locate the nearest optimal hospital from the dataset stored in the cloud, while simultaneously sending the data to the emergency services.

GSM Module. A GSM module is used to establish communication between the system and mobile devices, and the cloud and system (through the sim with Internet service provision). It serves a critical role since it is the mode of communication to other devices. The module used is the SIM900A which is a dual-band GSM/GPRS engine that works on frequencies EGSM 900 MHz and DCS 1800 MHz.

Vibration/Shock Sensor. The vibration sensor in use is the SW-420, which detects if there is any vibration beyond the threshold. The threshold is adjustable by the on-board potentiometer. The value of the sensor is also used to assign critical levels to the accident that has occurred, to warrant appropriate hospital selection based upon the possible severity of injury. Higher values imply greater severity of injury. For low vibrations, the module outputs logic LOW, while giving HIGH if a vibration is greater than the threshold.

Speaker and Microphone. This is used to communicate with the emergency services on their way to the accident site. The communication can include the current state of the victim along with their personal diagnosis of the injuries, so that the medical experts can provide on-the-phone treatment until the ambulance arrives.

3.2 Software Technologies

Structured Query Language. SQL is used for the database built in Microsoft Azure. SQL makes moving databases between the RoadNurse system and the cloud quick and easy. Other benefits include the ability to scale and its ability to quickly retrieve large amount of records from a database.

Azure Cloud Server. Azure Cloud is a cloud computing service created by Microsoft for building, testing, deploying, and managing applications and services, through an international array of Microsoft managed data centers. This cloud service contains a database of hospitals along with their location (latitude and longitude), contact details, and an associated level. This level is a number from a scale of 1–3 where these numbers indicate the ability to handle different severity of cases (determined by the specializations provided by the hospital). The severity is determined by the vibration sensor values. The database further contains a comprehensive list of all the registered users with the contact details of two friends or relatives whom they would like to be notified in the case of an accident.

4 Proposed Architecture

In the case of an accident, the vibration sensor senses a shock of value greater than a certain threshold, it will return a HIGH value to the NodeMCU. The value of the vibration determines the severity of the impact which is decided by preset values. The NodeMCU will then trigger the GPS module to instantaneously find the location of the victim. The location and severity will then be sent to the cloud via the GSM module. With regard to both location and severity, a suitable hospital is decided. The hospital database kept in the cloud contains its contact details, which will be contacted along with the loved ones of the victim, whose data is also fed into the database. The concerned people are contacted via the GSM module, which sends a SMS containing the time of accident and GPS location of the victim (so as to support the emergency relief).

The speaker and microphone installed in the accident detection and response unit will give updates to the victim on the emergency response and will receive input from the user (as well as those who might have come to help). An alarm deactivate button can also be pressed by the victim in case the accident is not serious and he wants to avoid the emergency relief action. This alarm would notify the concerned authorities with a pre-decided message stating "I am fine, please ignore the last message".

The whole unit is powered by rechargeable batteries which further draw their power from the vehicle's power unit. This way, in case the vehicle power unit fails during the accident, the batteries can power the micro-controller to carry on the post-accident functions (Fig. 1).

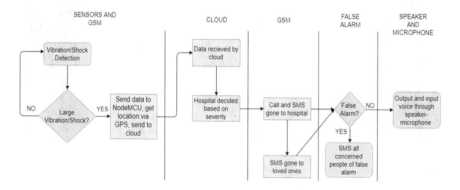

Fig. 1 Flow diagram of functions performed

5 Circuit Connections

All components described in Fig. 2 with the addition of bread board, jumper wires, BJT's, resistances, and power adapters will be used.

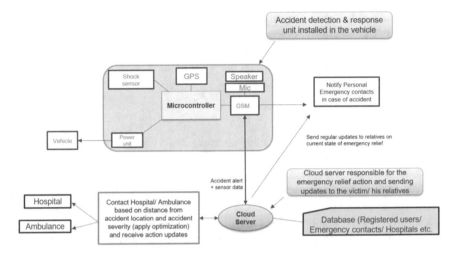

Fig. 2 Proposed system is presented in the figure. This would be a unit installed in the vehicle of the user and would have sensors for accident detection

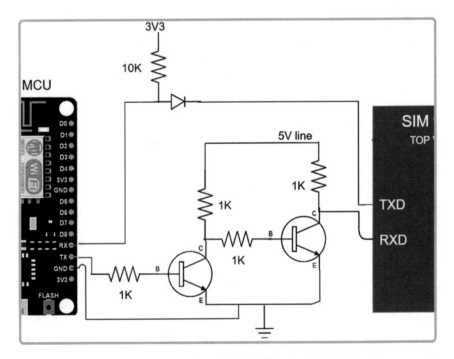

Fig. 3 Connection between GSM module and NodeMCU using TTL levels to form connections between 3.3 and 5 V input/outputs

Table 1 NodeMCU to GPS module and vibration/shock sensor connections

NodeMCU	GPS module	Vibration/shock sensor
3V3	Vcc	Vcc
GND	GND	GND
D1	RX	D0
D2	TX	Not applicable

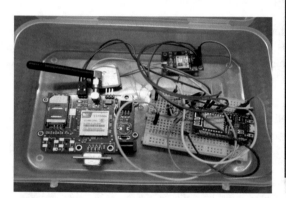

Fig. 4 LEFT: Compact implementation of the system to send GPS coordinates to the cloud and transmit processed data to concerned authorities and individuals. RIGHT: Automated message received by the hospitals and the loved ones of the victim

Figure 3 shows the implementation of a TTL logic, which allows the connection of TX of NodeMCU (functioning at 3.3 V) to form connections with the RXD of GSM which requires 5 V. Similarly, with the TXD of GSM and RX of NodeMCU, all resistors connecting the TPN2222A NPN BJT's are of 1 kΩ. The resistor between the 3V3 and RX of NodeMCU is of 10 kΩ (Table 1; Fig. 4).

6 Economic Viability

Expected cost of the RoadNurse response system (that is installed in the vehicle) would be around Rs. 2000 (based on the unit prices of the individual components given in Table 2). Note that if manufactured on mass scale, the cost could be brought down considerably lower and that this would be just a one-time investment for the user.

Table 2 Price tally in Indian Rupee

NodeMCU	GPS module	GSM module	Vibration/shock sensor
265	535	1000	200

Life/Vehicle insurance companies can mandate the low-cost accident detection and emergency relief system to be installed in the vehicle of the owner registered to their services. After an accident, third-party companies can provide services for the accident detection and relief system. Such a system serves as a complete revenue model as the customer is getting services and the companies are gaining a platform to provide these services quickly and efficiently.

7 Challenges in the Design of the System

With the aim to produce an industry-grade low-cost prototype, we also want to work out the infrastructure to implement this idea on large scale (i.e., working out the cloud service infrastructure). Following research/implementation challenges are relevant for designing such a system:

- The system has to be compact.
- The system should be rugged (should not get damaged during accident).
- The system should be waterproof.
- The battery powering unit should be properly designed to draw power from the vehicle.
- The system should be designed to typically run on low power, and if possible other key components (like GPS/GSM) should be powered up only after accident so as to reduce the power consumption.
- Maintaining a huge database of hospitals (nationwide) and their contact details.
- Analysis of the severity of the accident and choosing hospital accordingly (it might be possible that the hospital nearest to the victim might be a general hospital and not having specialized facilities to treat emergency cases).
- Low latency operations.
- Data security.

8 Future Scope

As this system uses GPS and GSM, it can be used to track vehicle thefts and to track the location of a vehicle which has gone missing (due to some accident, etc.). The registered user would have the facility to ping the cloud services so as to fetch the current location of the vehicle.

A low-cost display can be added to this device which has a dedicated GUI for displaying the shortest path to the hospital with respect to time. The victim can follow this path, given the severity of the accident is low and the victim is not in requirement of immediate emergency services. This would certainly be possible, and we look forward to such implementations.

9 Conclusion

We present an accident detection and response system for fast reporting and precise positioning of road accidents. The proposed system (RoadNurse) is a low-cost, low energy consumption system which can be the difference between life and death. The advantage of this system is the usage of a cloud-based architecture which not only reduces the reporting time, but also provides required medical facilities according to the severity of the accident. The vibration/shock sensor serves as the accident detection, relaying its information to the GPS and GSM module, which can utilize the database to provide information to concerned authorities. We hope that the RoadNurse will save many lives once it is commercialized and deployed on vehicles nationwide.

References

1. Popular Car Models in India Failed Global NCAP Crash Test. https://www.sayingtruth.com/indian-cars-failed-miserably-global-ncap-crash-test/. Last accessed 06 Nov 2019
2. World Health Organization (2018) Global status report on road safety 2018
3. Chotani HA, Razzak JA, Luby SP (2002) Patterns of violence in Karachi, Pakistan. Inj Prev 8(1):57–59
4. Waseem H, Naseer R, Razzak JA (2011) Establishing a successful pre-hospital emergency service in a developing country: experience from rescue 1122 service in Pakistan. Emerg Med J 28(6):513–515
5. Khan A, Bibi F, Dilshad M, Ahmed S, Ullah Z, Ali H (2018) Accident detection and smart rescue system using android smartphone with real-time location tracking. Int J Adv Comput Sci Appl 9(6):341–355
6. Jackson L, Cracknell R (2018) Road accident casualties in Britain and the world. London, UK, House of Commons Library
7. Carrigan R, Milton R, Morrow D (2005) Advanced automatic crash notification (AACN). Computerworld honors case study
8. Ki YK, Lee DY (2007) A traffic accident recording and reporting model at inter-sections. IEEE Trans Intell Transp Syst 8(2):188–194
9. Tushara DB, Vardhini PH (2016) Wireless vehicle alert and collision prevention system design using Atmel microcontroller. In: 2016 international conference on electrical, electronics, and optimization techniques (ICEEOT). IEEE, pp 2784–2787
10. Maleki J, Foroutan E, Rajabi M (2011) Intelligent alarm system for road collision. J Earth Sci Eng 1(3)
11. Rajkiran A, Anusha M (2014) Intelligent automatic vehicle accident detection system using wireless communication. Int J Res Stud Sci Eng Technol 1(8):98–101
12. Prabha C, Sunitha R, Anitha R (2014) Automatic vehicle accident detection and messaging system using GSM and GPS modem. Int J Adv Res Electr Electron Instrum Eng 3(7):10723–10727

13. Yee TH, Lau PY (2018) Mobile vehicle crash detection system. In: 2018 international workshop on advanced image technology (IWAIT). IEEE, pp 1–4
14. Khot I, Jadhav M, Desai A, Bangar V (2018) Go safe: Android application for accident detection and notification. Int Res J Eng Technol 5:4118–4122
15. POP AF, Puscasiu A, Folea S, Vălean H (2018) Trauma accident detecting and reporting system. In: 2018 IEEE international conference on automation, quality and testing, robotics (AQTR). IEEE, pp 1–5

A Compact Low-Loss Onchip Bandpass Filter for 5GnR N79 Radio Front End Using IPD Technology

V. Raghunadh Machavaram$^{(\boxtimes)}$ and Bheema Rao Nistala

Department of Electronics and Communication Engineering, National Institute of Technology Warangal, Warangal, India
{raghu,innbr}@nitw.ac

Abstract. It is reported in this paper, a compact very low-loss onchip bandpass filter meeting the requirements of 5GnR N79 radio frequency front end (RFFE), using 0.18 μm CMOS on Si substrate IPD technology. A series LC onchip BPF structure is designed and simulated by combining a passive multilayer (ML) inductor and a spiral capacitor in high frequency structural simulator (HFSS) at component level. The filter exhibited a quality factor (Q) value of 9.28, with a fractional bandwidth of 10.85% (<20%). It had exhibited very good insertion loss of −0.8 dB and also excellent return loss of −32.89 dB, at a center frequency of 4.5 GHz. The physical dimensions of the inductor, capacitor, and bandpass filter are 380×240 μm^2, 240×240 μm^2, and 480×240 μm^2, respectively. It had produced an excellent passband loss with a narrow passband characteristics, still occupying very small chip area. Hence, this proposed compact resonator filter definitely suits the 5G radio RFFE applications. We simulated the filter by focusing around 4.5 GHz, as this spectral band is being considered for the upcoming 5GnR N79 radio band trials and installations across several countries.

Keywords: RFFE · HFSS · IPD · BPF · Multilayer · Quality factor

1 Introduction

Worldwide explosion of smart systems had profound influence on the architectural evolutions for radio front end electronics. 5G radio receivers supporting beyond 1 Gbps data rates are being deployed in the 3–5 GHz (sub 6 GHz) band compatible to existing LTE network frequencies. Demand for onchip BPF with high-speed, small-size, and low-cost passive components is increased due to rapid growth in radio frequency integrated circuits (RFIC). Miniature 5G BPF design research focusses on silicon substrate CMOS ICs based on integrated passive device (IPD) technology. This is mainly because of minimal footprints, less consumed power, flexible bandwidths, and easy integration possible with IPD. Low-cost spiral inductor and capacitors are most successful silicon IPD passive components. Together, they occupy major portion of RFIC chip space (60–70%).

Minimum insertion loss and high selectivity with smallest onchip area pose big challenging task for passive bandpass filter (BPF) designs [1]. Power handling,

© Springer Nature Singapore Pte Ltd. 2020
N. Goel et al. (eds.), *Modelling, Simulation and Intelligent Computing*,
Lecture Notes in Electrical Engineering 659,
https://doi.org/10.1007/978-981-15-4775-1_50

selectivity, center frequency, and onchip area are usually compromised [2]. High-performance low-cost small-size passive components are realized using silicon IPDs [3]. Compact passive BPFs at high frequency bands are possible with LTCC technology, but they suffer from heat and size problems [4]. Recently, passive IPD BPF and LNA circuits are being embedded as CMOS chips as co-located system in package (SIP). These RFIC circuits yielded significant savings of chip area plus superior RF performance [5]. Hence, passive BPFs using Si IPD CMOS technology are being selected to meet with 5G radio filtering needs.

An open-loop resonator microstrip BPF showed very small insertion loss of only 0.8 dB at 3.6 GHz but occupied a large area of 99 mm^2 [6]. Tapered layout inductors enhanced Q_{max} from 15.9 to 22.8 and further to 26.7 with reverse excitation. However, Q_{max} frequency had increased from 5 to 8 GHz and SRF from 12 to 14.6 GHz [7]. A 2.31 GHz BPF using inter-wined spiral inductor and inter-digital capacitor showed a maximum return loss of 26.1 dB but occupied large area of 0.63 mm^2 [8]. A compact quad BPF was reported with 22–33 dB return loss and 0.2–1.2 dB insertion loss in the passband of 1.8–4.2 GHz, but occupied larger chip space of 545 mm^2 [9]. Another compact 5G NRN78 WLAN 3.3–3.8 GHz band BPF based on Si IPD had yielded low insertion loss of <1.8 dB but had large chip area of 1.28 mm^2 [10]. A 2.45 GHz LC silicon IPD BPF showed an insertion loss of 3–4 dB with an area of 0.75 mm^2 [11]. A compact 1.35 GHz GaAs substrate IPD BPF resulted in 0.26 dB insertion loss and return loss of 25.6 dB but consumed 1.25 mm^2 chip area [12].

In this paper, we report the design and simulation of a spiral ML inductor and a metal insulator metal (MIM) capacitor connected in series to form an LC resonator BPF. We found its S parameters in HFSS by concentrating the design for 4–5 GHz, as this radio band is being heavily explored for 5GnR N79 radio reception.

This paper comprises of Sects. 2, 3, and 4 as follows.

2 Design and Analysis of BPF

The inductance L, quality factor Q, and self-resonant frequency (SRF) depend on: conductor width, spacing, spiral diameter, number of turns, ring shapes, metal ring thicknesses, type and thickness of substrate, etc. These are traded off for cost and performance [13]. Selection of thick substrate and thin oxide layers not only reduces losses for passive components but also enhances the filter selectivity. Decreasing permittivity of metal, both the oxide capacitance and inductor Q value can be increased. This leads to an increase of maximum Q frequency and resonant frequency [14]. Analytical optimization techniques are attempted by varying the conductor width, shape, spacing, number of turns, outer diameters, and thickness to accomplish high performance [15].

A compact onchip BPF is designed and simulated with multilayer rectangular spiral inductor and a single-layer rectangular spiral capacitor connected in series configuration, on the Si substrate. Rectangle shape is chosen for the spiral structure of ML inductor and capacitor, because of its uniform current distribution and fabrication simplicity [2]. They are simulated in HFSS, and the filter analysis is performed with equivalent lumped model having with series RL and shunt RC components. S parameters are employed as they are good at high frequency analysis of RF circuits.

2.1 A Multilayer Onchip Spiral Inductor

Series stacked multi-turn spiral inductor structures showed higher inductance and Q values around 10. The proposed inductor has copper conductor on a thick Si substrate to reduce its parasitic resistance to enhance Q value. Its outer diameter is 380 μm and occupies 380 μm × 240 μm chip area. It is designed with three turns (each of 6 μm width) running onto three layers. The values of Q and impedance L are obtained via a lumped model in the HFSS simulation. The basic expression to extract inductance for rectangular spiral is taken from the greenhouse method [16]. The overall (net) inductance is [2]

$$L_{\text{Total}} = L_{\text{Self}} + \sum M_{+\text{ve}} + \sum M_{-\text{ve}} \tag{1}$$

The shunt and substrate capacitances in the lumped model are [2]

$$C_{\text{Substrate}} = \frac{wlC_0}{2} \tag{2}$$

$$C_{\text{Shunt}} = C_{\text{oxide}} \left\{ \frac{1 + w^2(C_{\text{Si}} + C_{\text{oxide}})C_{\text{Si}}R_{\text{Si}}^2}{1 + w^2(C_{\text{Si}} + C_{\text{oxide}})^2 R_{\text{Si}}^2} \right\} \tag{3}$$

In above expressions, w is width, l is length, and C_0 is unit area substrate capacitance. S_{ubstrate}, S_{hunt}, and O_{xide} are substrate, shunt, and oxide capacitances, respectively.

$$M_{21} = \frac{\mu_0 l_1}{4\pi l_2} \left\{ \frac{\varphi_{21}}{e^{-j\beta_1 Z}} \right\} \tag{4}$$

Self and mutual inductance values are fed into the greenhouse method to evaluate the desired inductance [2]. Quality factor for inductor is determined from

$$Q = \frac{\omega L_s}{R_s} \frac{1}{1 + \frac{R_s}{R_p} \left\{ \left(\frac{\omega L_s}{R_s}\right)^2 + 1 \right\}} \left[1 - \frac{R_s^2(C_s + C_p)}{L_s} - \omega^2 L_s(C_s + C_p) \right] \tag{5}$$

The proposed inductor structure has three layers with one turn per layer. The first turn of the conductor is placed in first layer. Next conductor turn is put into layer 2. Final turns are put in layer 3. The rectangular spiral planar view and 3D view structures of the designed inductor are depicted in Fig. 1.

This geometry is referred to as 3D structure. Same conductor width is chosen for all the three layers. Spacing between layers is selected so as to minimize the negative mutual inductance. The Q and L values for this 3D inductor are found from the

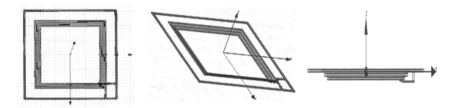

Fig. 1 Planar and 3D views of the onchip multilayer spiral 3D inductor

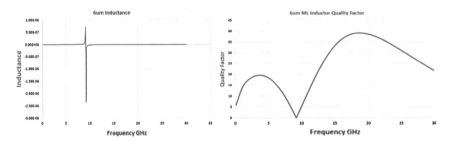

Fig. 2 Variation of inductance (nH) and quality factor with frequency for onchip ML spiral inductor

S parameters employing Eq. (6). Y parameters are found from simulated S parameters. Inductance change w.r.t. frequency is computed from the imaginary component of Y parameter.

$$Q = \frac{\mathrm{Im}[Y_{11}]}{\mathrm{Re}[Y_{11}]} \text{ and } L = \frac{-1}{2\pi f\{\mathrm{Im}[Y_{11}]\}} \tag{6}$$

Inductance and quality factor variation against frequency for the simulated inductor structure are given in Fig. 2.

Structural dimensions of this onchip ML inductor are: conductor width—6 μm, thickness—2 μm, and spacing—2 μm, with 2 μm layer spacing. It has three turns. Onchip occupied area is 380×240 μm^2. The inductor HFSS simulation showed higher Q value of 20 in filter passband and 8.64 GHz resonant frequency (SRF). This simulated inductor surely matches the upcoming 5G lower band (3–5 GHz), as it has smaller onchip area.

2.2   Onchip Rectangular Spiral Capacitor

Proposed single-layer rectangular spiral capacitor is designed and simulated in HFSS. Silicon material is chosen for substrate, and the copper metal is selected for conductor. All turns are placed in same layer. Every turn has same conductor width. The capacitance value majorly depends upon material used and conductor dimensions in the annular geometry of spiral. The geometry of spiral capacitor is given in Fig. 3.

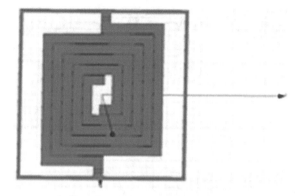

Fig. 3 Planar view of onchip spiral capacitor

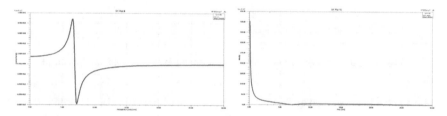

Fig. 4 Variation of the capacitance (pF) and the quality factor with frequency for onchip spiral capacitor

Capacitance and the quality factor Q for the capacitor are extracted from the S parameters' as given in Eq. (7) [2].

$$C = \frac{-\text{Imaginary}[Y_{21}]}{\omega} \text{ and } Q = \frac{\text{Imaginary}[Y_{11}]}{\text{Real}[Y_{11}]} \tag{7}$$

Change in the values of the capacitance and the quality factor w.r.t. frequency for this proposed capacitor is shown in Fig. 4.

The structural dimensions of this onchip capacitor are: Conductor width is 6 μm, thickness is 2 μm, and spacing is 1 μm. Number of turns is two and half. This capacitor occupied a chip space of only 240 × 240 μm². The capacitor HFSS simulation showed higher Q value of 10 in filter passband and 6.8 GHz resonant frequency (SRF). Simulated response of this capacitor on 0.18 μm CMOS technology showed a significant enhancement of the quality Q and capacitances. Therefore, this capacitor is sure to satisfy the 5G radio lower band applications.

2.3 Onchip Passive Band Pass Filter

BPF play crucial role in receiver performance and locate at the RF front end (antenna–BPF–LNA). The frequency response of a BPF significantly influences the selectivity of

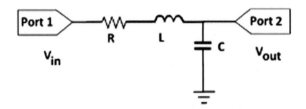

Fig. 5 Schematic of an LC resonant BPF used in typical radio receiver

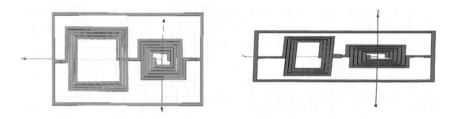

Fig. 6 Planar and 3D views of the LC resonant BPF component connection

a radio receiver. We have chosen a simple low-cost series LC resonator design approach to implement the BPF. This enables for easier analysis and implementation to prove the filter performance. The simulation of BPF replaces the circuit elements L and C with the designed multilayer inductor and the spiral capacitor. The LC resonant bandpass filter configuration is shown by the following equivalent circuit in Fig. 5.

The filter optimization and simulation were performed in the HFSS tool. The simulated values of the inductance and capacitance are very much stable during the entire passband. Figure 6 shown below depict the 3D and single-layer structural views for the simulated series LC filter configurations.

2.4   Simulation

We chose the first-order series resonant LC passive BPF for the filter design using lumped LC model and simulated in HFSS to find out the scatter parameters. After repeated simulations for different component dimensions and material type combinations, the filter losses got reduced to match the spectral characteristics of the 5G radio front end requirements. The obtained filter simulation results matched well with that of the theoretical values for the proposed BPF. The return loss characteristics of the proposed bandpass filter are shown in Fig. 7.

The above simulation results prove very good passband and outband performance for the proposed bandpass filter. The superior loss performance of this proposed filter yields good quality RF signals suitable for a 5G radio. Also, it possesses very good outband rejection capability.

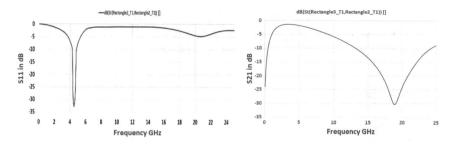

Fig. 7 Return loss characteristics of the proposed series resonator LC passive BPF

3 Results and Discussion

From the simulation results shown above, the quality factor for the inductor and capacitor has its maximum value as 39 and 405, respectively. HFSS simulation of this compact low-loss 4.5 GHz BPF had shown 485 MHz bandwidth from 4.3 to 4.785 GHz. The loss characteristics in Fig. 7 for this proposed BPF demonstrates very good return loss of −32.89 dB (maximum). It showed an excellent insertion loss of −0.8 dB around its center frequency of 4.5 GHz. It clearly shows that both of these important losses would definitely satisfy the performance requirements of a 5G bandpass filter.

BPF bandwidth = $f_{High} - f_{Low}$ = 4.785 − 4.3 = 485 MHz
BPF loaded $Q = f_0/\Delta f$ = 4.5/0.485 = 9.28
Fractional bandwidth = $\Delta f/f_0$ = 0.485/4.5 = 10.85%

The performance parameters like the insertion and return losses, Q, bandwidth, and fractional bandwidth obtained from HFSS simulation results for the proposed filter are summarized in Table 1.

If the fractional bandwidth is less than 20%, then a BPF is said to be narrow band filter. Proposed BPF had exhibited only 10.85% fractional bandwidth. Hence, this proven narrow inband response will surely match the stringent spectral demands of a

Table 1 Summarized BPF parameters

3.5 GHz BPF	Design specifications	Simulation results
Center frequency f_0—GHz	4–5	4.5
Bandwidth Δf—MHz	500	485
Fractional bandwidth—%	10–20	10.85
Quality factor Q	10–15	9.28
Return loss S_{11} dB	−20 to −30	−32.89
Insertion loss S_{12} dB	<−1.0	−0.8
Onchip area mm^2	0.1	0.1152

5G radio front end BPF. This filter also posses slowest occupied chip area. Due to several such prime parameter enhancements, proposed 4.5 GHz BPF will definitely suit for the realization of efficient RFICs for the 5G radio access filtering applications.

4 Conclusion

We have modeled and developed a simple 5G compatible miniature IPD BPF for the upcoming 3–5 GHz 5G radio band applications. The narrow band filter is successfully simulated by selecting a multilayer filter structure, and its analysis is done in HFSS. Along with the enhanced performance, the fabrication complexity, cost and time savings are easily achievable with this IPD filter structure laid onto multiple layers.

The return loss of −32.89 dB proposed filter shows very feeble returned signal in its pass band. The insertion loss of −0.8 dB is a considerable improvement, compared to many of the reported filters. Thus, the present filter design exhibited an excellent loss performance during entire filter passband, with stronger signal propagation. Also the filter size is below 400 × 400 μm, which is way smaller than the many of the reported BPFs. With a view by considering all the above proven performance merits, this proposed BPF is very much suitable to the fast emerging 5G radio access requirements. Proposed work has clearly demonstrated the feasibility of developing a highly selective low-loss compact 5G BPF integrated on the Si substrate in IPD technology.

References

1. Lin C-C, Huang CZ, Chen C-Y (2011) Compact and highly selective mm-wave meandered bandpass filter in 0.18-μm CMOS technology. In: Proceedings of Asia-Pacific microwave conference, pp 49–52. 978-0-85825-974-4©2011 Engineers Australia
2. Nagesh Deevi BVNSM, Bheema Rao N (2016) Miniature on-chip band pass filter for RF applications. Springer Microsyst Technol 23(3). https://doi.org/10.1007/s00542-016-3052-7
3. Xinhai B et al Fabrication and measurement of BPF using IPD technology. In: 2014 15th International conference on electronic packaging technology, pp 36–38. 978-1-4799-4707-2/14/©2014 IEEE
4. Lin Y-C, Huang H-H, Homg T-S Synthesis of LTCC multi-band bandpass filter using reflection and transmission zeros. 9781-4799-3869-8/14/®2014 IEEE
5. Wang S, Cho K-F (2016) CMOS/IPD switchable bandpass circuit for 28/39 GHz fifth-generation applications. IET Microw Antennas Propag 10(14):1461–1466. https://doi.org/10.1049/iet-map.2015.0806
6. Al-Yasir YIA et al (2019) Design, simulation and implementation of very compact dual-band microstrip bandpass filter for 4G and 5G applications. In: SMACD 2019, pp 41–44. 978-1-7281-1201-5/19/©2019 IEEE
7. Vanukuru VNR, Chakravorty A (2014) High-Q characteristics of variable width inductors with reverse excitation. IEEE Trans Electron Devices 61(9)
8. Chuluunbaatar Z, Adhikari KK, Wang C, Kim N-Y (2014) Micro-fabricated bandpass filter using intertwined spiral inductor and interdigital capacitor. Electron Lett 50(18):1296–1297
9. Chen Y-W, Tai T-C, Wu H-W, Su Y-K, Wang Y-H Design of compact multilayered quad-band bandpass filter, pp 1815–1818. 978-1-5090-6360-4/17©2017 IEEE

10. Shin KR, Eilert K Compact low cost 5GnR N78 band pass filter with silicon IPD technology. 978-1-5386-1267-5/18©2018 IEEE
11. Mao C, Zhu Y, Li Z Design of LC bandpass filters based on silicon based IPD technology. In: 19th international conference on electronic packaging technology, pp 238–240. 978-1-5386-6386-8/18/©2018 IEEE
12. Quan C-H, Wang Z-J, Lee J-C, Kim E-S, Kim N-Y (2019) A highly selective and compact bandpass filter with circular spiral inductor and an embedded capacitor structure using an integrated passive device technology on a GaAs substrate. Electronics 8(73):1–8 MDPI. https://doi.org/10.3390/electronics-8010073
13. Wang C, Kim N-Y (2012) Analytical optimization of high performance and high-yield spiral inductor in integrated passive device technology. Microelectron J 43:176–181
14. Buyuktas K, Koller K, Mller K-H, Geiselbrechtinger A (2010) A new process for on-chip inductors with high Q-factor performance. Int J Microw Sci Technol
15. Nagesh Deevi BVNSM, Bheema Rao N (2016) Analysis of miniature on-chip 3-D inductor for Rf circuits. ARPN J Eng Appl Sci 11(5). ISSN 1819-6608
16. Greenhouse HM (1974) Design of planar rectangular microelectronic inductors. IEEE Trans Parts Hybrids Packag 10(2):101–109. https://doi.org/10.1109/TPHP.1974.1134841

Supervised Feature Selection Methods for Fault Diagnostics at Different Speed Stages of a Wind Turbine Gearbox

Vamsi Inturi$^{(\boxtimes)}$ (ID), P. Ritik Sachin, and G. R. Sabareesh (ID)

Department of Mechanical Engineering, Birla Institute of Technology and
Science Pilani, Hyderabad Campus, Hyderabad, India
p20160025@hyderabad.bits-pilani.ac.in

Abstract. Individual condition monitoring (CM) strategies are capable to diagnose 30–40% of the defects, when they are performed individually. However, combining two or more individual CM strategies can provide more reliable information which will enhance the ability of fault detection. In this investigation, two intrusive CM strategies (vibration and lubrication oil analysis) and one non-intrusive CM strategy (acoustic signal analysis) are combined to form an integrated CM scheme. Experiments are performed on a miniature wind turbine gearbox bench top and the raw data is acquired and the defect sensitive features are extracted using discrete wavelet transform. Feature level fusion is accomplished to achieve integrated feature data set and the selection of optimal subset of significant features is done by various supervised featured selection methods. Finally, the obtained optimal feature subset is classified using SVM algorithm in order to diagnose the local defects of bearings as well as gears present in different stages of the wind turbine gearbox.

Keywords: Condition monitoring · Fault diagnosis · Integrated CM scheme · Supervised feature selection · Wind turbine gearbox

1 Introduction

As the wind turbines are operating under varying wind loads and subjected to expeditious atmospheric conditions, they are vulnerable to fail often. The gearbox of wind turbine is regarded as critical component since it is having highest failure rates [1]. Condition monitoring (CM) is executed to monitor the condition of the wind turbine gearbox, and it is capable of diagnosing and quantifying the defect levels. Various authors have exploited the vibration, thermography, lubrication oil, and acoustic signal analysis to diagnose the defects of gears as well as bearings of a single-stage gearbox [2–4]. Individual CM strategies have few limitations such as diagnosing a limited type of defects, inadequate at low operating speeds and positioning the sensors, etc. [5]. Because of these, the probability of diagnosing the defect is about 30–40%, when they are performed individually. However, combining two or more individual CM strategies can provide more reliable information which will enhance the ability of fault detection. Few authors have described about combining the vibration and temperature analyses

© Springer Nature Singapore Pte Ltd. 2020
N. Goel et al. (eds.), *Modelling, Simulation and Intelligent Computing*,
Lecture Notes in Electrical Engineering 659,
https://doi.org/10.1007/978-981-15-4775-1_51

and combining the vibration, oil debris, and acoustic emission analyses in order to monitor the condition of rotors and single-stage gearbox [5, 6].

Signal processing approaches are accomplished to extract the defect-sensitive features hidden in the acquired raw signatures. Eventually, the extracted features are encrypted as input to the feature classification algorithms. Discrete wavelet transform (DWT) is one of the extensively implemented multi-domain techniques for evoking the features as it is able to process non-stationary signals and having good resolution capability [2]. All the extracted features may not be equally significant, and hence, the feature selection is performed to determine the most discriminating features among the extracted. The feature subset achieved from the feature selection algorithms is further subjected to feature classifiers which work on machine learning algorithms. Owing to its ability to solve nonlinear data, high precision, and better generalization while classifying small number of observations, the application of support vector machine (SVM) is wider as a feature classifier [2]. Various authors have applied SVM algorithm to categorize the various defect severity levels of bearings as well as gears subjected to stationary operating speeds [3, 4].

Majority of the research investigations are performed by analyzing the data of a single CM strategy, and the studies related to the integration of CM strategies are limited. The feature selection algorithm that devises the optimal feature subset from a single CM strategy may not be the adequate while it is subjected to an integrated CM scheme. In order to bridge these gaps, the present investigation attempts to combine the features from three individual CM strategies (vibration, acoustic signal, and lubrication oil analysis) in order to devise a multi-variable integrated feature set. Further, selection of optimal subset of most significant features is done by various supervised feature selection algorithms and the number of features that are required to avoid the overfitting is investigated. Finally, the obtained optimal feature subset is classified using SVM algorithm in order to diagnose the local defects of bearings as well as gears present in different stages of the wind turbine gearbox subjected to non-stationary speeds.

2 Experimentation and Data Acquisition

2.1 Experimental Test Rig and Simulation of Defects

In order to mimic the operation of a wind turbine gearbox, a miniature three-stage spur gearbox was fabricated with an overall gear ratio of 48:1. The high speed stage (HSS) of the gearbox was driven by a 0.75 kW motor. The HSS drives the subsequent intermediate speed stage (ISS) and low speed stages (LSS) of the gearbox. Gear oil (80W-90) was used to lubricate the mating components. The bearing close to the pinion was test bearing, and all the experiments related to bearing were performed on the test bearing. Similarly, the bearing stays far from the pinion was attributed as support bearing, refer Fig. 1. The experiments related to gears were performed on the pinion. A slot (axial direction) on the inner race and outer race was seeded with a thickness of 0.25 mm using wire-cut electro-discharge machine. The depth of cut on the inner race was varied from 1.1 to 4.4 mm, which corresponds to the propagation of defect from 25 to 100%. On the other hand, the depth of cut on the outer race was varied from 1 to

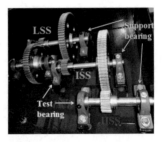

Fig. 1 Wind turbine gearbox bench top

Fig. 2 Seeded defects of bearing and gear

2.2 mm to represent the severity of defect from 25 to 100%. Similarly, a transverse cut was seeded at the root of the pinion tooth, and the depth of cut was varied from 1 to 4 mm so as to represent the propagation of defect from 25 to 100%. Figure 2 displays the defects seeded on the bearing race and pinion tooth.

2.2   Data Acquisition

One uniaxial accelerometer to the support bearing and one triaxis accelerometer to the test bearing were stud mounted to achieve the vibration response. Two microphones, one above the test bearing (Mic 1) and one above the support bearing (Mic 2), were situated to record the acoustic signal (sound pressure) data. One oil particle counting sensor was used to record the iron (Fe) particle deposition rate. All these sensors were connected to a data logger which further connected to a computer. The data was acquired with the sampling rate of 16 kHz, and the sample length was maintained as 4096. In the present study, a smooth fluctuating speed profile was generated by scaling the actual wind turbine data, and these speed profiles were given as input to the motor through a variable frequency drive in order to regulate the speed of the motor, ref. Fig. 3. The HSS, ISS, and LSS are subjected to speed profile 1, 2, and 3, respectively. While the different stages of the gearboxes are operating under the simulated speed profiles, the data in the form of vibration, sound pressure, and particle deposition rate was acquired and subjected to further processing.

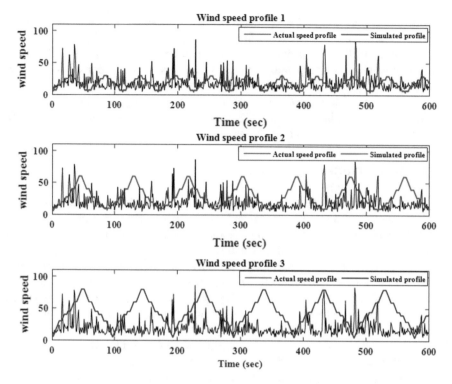

Fig. 3 Simulated wind speed profile

3 Feature Extraction and Integrated CM Scheme

3.1 Wavelet-Based Feature Extraction

Discrete wavelet transform (DWT) decomposes the signal into detail (cD_i) and approximation (cA_i) wavelet coefficients based on their frequency band. The approximation coefficients (cA_i) are subjected to next level of decomposition, and this process iterates further. The acquired vibration and acoustic signals are decomposed to fourth level of decomposition using Haar wavelet [4, 5]. Later, thirteen statistical features for vibration and thirteen statistical features for acoustic signal analysis are computed from the fourth-level approximation coefficients.

3.2 Integrated CM Scheme

The information from the individual CM strategies is integrated in order to enhance the probability of diagnosing the defect. The information from the two intrusive CM strategies (vibration and oil particle analysis) and one non-intrusive CM strategy (acoustic signal analysis) are combined to form an integrated CM scheme. Feature level data fusion is accomplished as the sensor data is non-commensurate. Therefore, 13 statistical features computed for vibration analysis, 13 statistical features computed for acoustic signal analysis, and one feature (particle deposition rate) for lubrication oil analysis are combined to achieve an integrated feature set. For each test condition of bearing (either of HSS, ISS, and LSS), 150 observations are considered and 27 features (thirteen of vibration, thirteen of acoustic signal, and one of lubrication oil analysis) are combined. Thus, the original integrated feature set has the order of 27 rows (features) and 750 columns (5 test cases * 150 observations). Similarly, the original integrated feature set of gear (either of HSS, ISS, and LSS) has 27 rows (features) and 600 observations (5 cases * 120 observations).

4 Feature Selection

All the features contained by the original integrated feature set are not equally significant. Analysis of high-dimensional and multi-variable data set is quite challenging in the field of machine learning, and feature selection algorithms are intended to remove the redundant and outlier features [7]. Feature selection is a process of achieving an optimal subset from the original data set, which selects the most significant features of the original data set. The supervised feature selection methods can be classified as filter, wrapper, and embedded models. As the accuracy of wrapper models is high when compared with filter models [7], we have exempted filter models in this study. Three wrapper model algorithms, namely Naïve Bayes, forward greedy, and backward greedy algorithm, and one embedded model algorithm (decision tree) are considered in the current investigation. The feature selection abilities of these algorithms are compared so as to select the feature selection algorithm that best suits to identify the informative features from the integrated feature data set.

4.1 Wrapper Models

Forward greedy algorithm (FGA) considers the whole integrated feature set as input and returns the subset of specified size as output. For example, the original feature set consists of 27 dimensions ($s = 27$) which is given as input to the algorithm, and a subset of ten features is returned as output along with the corresponding classification accuracy. The algorithm also verifies the possibility of improvement of the accuracy if some feature is excluded and discards the feature having the least accuracy. The step-wise procedure is as explained below [8]:

Input: original feature set, $Y = \{y_1, y_2, y_3, \ldots \ldots, y_s\}$
Output: subset of features, $X_j = \{x_i | i = 1, 2, \ldots \ldots, s; x_j \in Y\}$, where $j = (0, 1, 2, \ldots \ldots, s)$
Initialization: Initialize the algorithm with a null set *theta* so that size of subset
$j = 0$
$X_0 = theta, j = 0$

Step 1 – Inclusion: $x + = arg \ max \ J \ (x_j + x)$, where $x \in Y\text{-}X_j$
$X_{j+1} = X_j + x_+$
$j = j + 1$

Go to step 2

Step 2 – Conditional exclusion:

$x-= arg\ max\ J\ (x_j - x)$, where $x \in X_j$

if $J(xj - x) > J(xj)$

$Xj-1 = Xj - x-$

$j = j-1$

Go to step 1

Backward greedy algorithm (BGA) follows the similar structure; it performs exclusion followed by conditional inclusion. Naïve Bayes algorithm is a probability-based learning algorithm based on Bayes theorem. It assumes that the features of a data set are independent of each other and computes the posterior probability for each class. The subset suggested by the algorithm contains the crest factor (vibration analysis), maximum (vibration analysis), kurtosis (vibration analysis), skewness (acoustic signal), crest factor (acoustic signal), kurtosis (acoustic signal), median (vibration analysis), standard error (vibration analysis), mean (acoustic signal), median (acoustic signal), and standard error (acoustic signal).

4.2   Embedded Models

Decision tree (DT) is a tree-based knowledge demonstration which works on the principle of information gain. The existence of a feature in the tree gives the information about the significance of the associated feature. The features contributing towards the classification will appear in the decision tree, and the less contributing features do not appear. J48 algorithm (WEKA implementation of *c4.5*) is used to construct the decision tree. The integrated feature set is given as input to algorithm so as to measure the significance of features based on the information gain which depends on the statistical parameter called entropy. For a given data set of "*S*," the information gain of a feature "*a*" is given by Eq. (1).

$$Gain(S, a) = Entropy(S) - \Sigma(|Sv|/|S|)Entropy(Sv), \text{where } v \in \text{values of } a \quad (1)$$

Therefore, J48 algorithm examines the information gain of all features and expands the nodes of the DT based on their order of features which are following the descending order of information gain [9]. The features suggested by the decision tree are root mean square (RMS) value (vibration), skewness (vibration), mean (acoustic signal), RMS (acoustic signal), standard error (vibration signal), sample variance (acoustic signal), minimum (vibration signal), kurtosis (acoustic signal), crest factor (acoustic signal), and maximum (vibration signal).

5   Defect Classification Through Support Vector Machine

Support vector machine (SVM) is a supervised machine learning classifier which works on the principle of structural risk minimization. In this investigation, we have used *c-svc* based SVM model with tenfold cross-validation for defect classification, and the sample of obtained overall classification accuracies for HSS, ISS, and LSS is

Table 1 Classification accuracies of statistical features

Speed stage	No. of features	FGA	BGA	Naïve Bayes	DT	Defect type
HSS	4	91.4	92	92.8	<u>94.8</u>	IRF
		89.4	89.4	92.1	91.3	ORF
		86.5	89.3	93.4	94.2	RC
	5	92.4	<u>92.8</u>	92.8	92.5	IRF
		90.4	90.4	91.2	89.3	ORF
		86.1	86.1	92.6	91	RC
	7	**95.3**	**95.3**	89	88.3	IRF
		91.6	91.6	90.2	86.3	ORF
		86	81.7	90.6	88.6	RC
	10	<u>95</u>	<u>95</u>	84.4	82.7	IRF
		91.6	91.7	83.8	82.9	ORF
		85.9	82.3	84.2	86.3	RC
ISS	4	88.9	88.3	91.4	92.4	IRF
		89.3	87.1	91.5	<u>93</u>	ORF
		90.7	90	92.4	92.7	RC
	5	92	91.9	92.4	89.4	IRF
		89.9	88.5	92	93.3	ORF
		92	91.1	92.4	<u>93.8</u>	RC
	7	93.9	92.9	89.6	83.8	IRF
		91.3	91.3	88.9	85.1	ORF
		94.3	92.1	89.4	88.3	RC
	10	93.3	<u>93.9</u>	88.4	84.3	IRF
		91.3	91.9	87.5	84.7	ORF
		93.8	92.2	87.1	84.1	RC
LSS	4	91.2	91.6	88.2	<u>94.1</u>	IRF
		91.3	89.2	89.4	93.5	ORF
		86.8	86.8	88.7	91.6	RC
	5	91.7	91.7	92.1	88.6	IRF
		91.9	90.8	93.1	90	ORF
		89.3	89.3	<u>93.7</u>	92.1	RC
	7	92.1	**93.8**	89.2	85.3	IRF
		92.2	91.7	89	87.1	ORF
		91.7	91.7	89.3	88.6	RC
	10	92.1	<u>93.3</u>	87.7	84.5	IRF
		92.8	93.2	88.3	87.4	ORF
		<u>93.3</u>	<u>93.3</u>	88.1	82.7	RC

displayed in Table 1. It can be observed that DT algorithm in conjunction with SVM has shown better classification results while diagnosing the defect severity levels of IRD, ORD, and GTC when the number of features are less (four). In contrast to that, the classification efficiencies of greedy algorithm in conjunction with SVM are higher when the number of features are more (greater than six). It is worth noting that the overall classification efficiencies are decreasing, beyond certain number of features (after seven). Therefore, FGA/BGA consists of seven input features in conjunction with SVM which has yielded better classification accuracies while diagnosing the defect present at different stages of the wind turbine gearbox. The classification accuracies are 95.3%, 94.3%, and 93.8% at high, intermediate, and low speed stages, respectively.

6 Conclusions

In this investigation, two intrusive CM strategies (vibration and lubrication oil analysis) and one non-intrusive CM strategy (acoustic signal analysis) are combined to form an integrated CM scheme. Feature-level fusion is accomplished to achieve integrated feature data set, and selection of optimal subset of significant features is done by various supervised feature selection algorithms. Finally, the obtained optimal feature subset is classified using SVM algorithm in order to diagnose the local defects of bearings as well as gears present in different stages of the wind turbine gearbox. It can be concluded that forward/backward greedy algorithm consists of seven input features in conjunction with SVM which yielded better classification accuracies while diagnosing the defect present at different stages of the wind turbine gearbox.

References

1. Stetco A, Dinmohammadi F, Zhao X, Robu V, Flynn D, Barnes M, Nenadic G (2018) Machine learning methods for wind turbine condition monitoring: a review. Renew Energy 133:620–635
2. Vamsi I, Sabareesh GR, Penumakala PK (2019) Comparison of condition monitoring techniques in assessing fault severity for a wind turbine gearbox under non-stationary loading. Mech Syst Signal Process 124:1–20
3. Amarnath M, Sugumaran V, Kumar H (2013) Exploiting sound signals for fault diagnosis of bearings using decision tree. Measurement 46(3):1250–1256
4. Radhika S, Sabareesh GR, Jagadanand G, Sugumaran V (2010) Precise wavelet for current signature in 3ϕ IM. Expert Syst Appl 37(1):450–455
5. Inturi V, Sabareesh GR, Supradeepan K, Penumakala PK (2019) Integrated condition monitoring scheme for bearing fault diagnosis of a wind turbine gearbox. J Vib Control 25 (12):1852–1865
6. Nembhard AD, Sinha JK, Pinkerton AJ, Elbhbah K (2014) Combined vibration and thermal analysis for the condition monitoring of rotating machinery. Struct Health Monit 13(3):281–295
7. Cai J, Luo J, Wang S, Yang S (2018) Feature selection in machine learning: a new perspective. Neurocomputing 300:70–79

8. Raschka S (2018) MLxtend: providing machine learning and data science utilities and extensions to Python's scientific computing stack. J Open Source Softw 3(24):638
9. Konar A (2000) Artificial intelligence and soft computing: behavioral and cognitive modeling of the human brain. CRC Press, Florida

A Survey on Beamwidth Reconfigurable Antennas

Vikas V. Khairnar[1]([✉]), C. K. Ramesha[1], and Lucy J. Gudino[2]

[1] Department of Electrical and Electronics Engineering, Birla Institute of Technology and Science Pilani, K K Birla Goa Campus, Goa, India
p20130409@goa.bits-pilani.ac.in
[2] WILP Division, Department of Computer Science and Information Systems, Birla Institute of Technology and Science Pilani, Pilani, Rajasthan, India

Abstract. Next-generation wireless communication systems need antennas with multi-functionality, adaptability, and flexibility to provide efficient utilization of power and electromagnetic spectrum. Reconfigurable antenna can fulfill these demands by delivering multiple functionalities in a single antenna structure. These antennas can dynamically adapt to changing system requirements by altering their operating parameters. Reconfigurable antennas are classified as frequency, pattern, polarization, and compound reconfiguration. Compound reconfigurable antenna involves simultaneous reconfiguration of two or more parameters such as frequency and pattern, frequency and polarization, pattern and polarization and frequency, pattern, and polarization. This paper presents a comprehensive survey on reconfigurable antenna designs, realizing beamwidth reconfiguration with single or dual orthogonal polarization. Furthermore, this paper also investigates the performance comparison of multi-functional reconfigurable antennas achieving beam steering and beamwidth reconfiguration in a single antenna structure. The challenges and future research directions in beamwidth reconfigurable antennas are also discussed in detail.

Keywords: Reconfigurable antenna · Beamwidth · Beam steering · Polarization

1 Introduction

To address the challenges of modern wireless communication technologies, antennas capable of achieving multiple functionalities in a single structure are required. The concept of reconfigurable antenna (RA) is proposed in [1] to overcome the limitations of a conventional fixed performing antenna. RA technology is one of the hardware solutions developed to enhance wireless device connectivity. These new classes of radiating elements can adapt their physical characteristics to variations in environmental changes or user density and location. In contrast to the conventional fixed performing antenna, where energy is spent around the surrounding space, the use of RA enables smarter management of radiated energy as the beam can be focused in specific directions. As a result, it is possible to improve data throughput between two devices and significantly reduce interference between adjacent networks. RA incorporates an internal mechanism to redistribute radio frequency (RF) currents over its surface, which causes modifications in impedance and radiation characteristics [2]. RA can change its performance features such as resonant frequency, radiation pattern, beamwidth, and polarization by changing its architecture mechanically or electrically. Pattern reconfiguration is accomplished by changing the source current distribution on the antenna structure. Pattern reconfigurable antenna (PRA) avoids noise sources by directing null toward interference and radiating the main beam in the desired direction to improve coverage. PRAs can be classified according to the radiated beam shape and direction of the main beam. PRA produces various radiation patterns such as omnidirectional, conical, broadside, backfire, and endfire. The main beam of PRA can be discretely switched or continuously scanned in a certain angular direction. PRA also involves varying 3-dB beamwidth in narrow and wide beam modes. The research work on PRA has been carried out in areas by changing the main lobe direction, varying 3-dB beamwidth or combining these two characteristics. PRA designs reported in the literature are classified according to their radiation capabilities such as one-dimensional (1-D) beam steering, two-dimensional (2-D) beam steering, beamwidth RA, and beam steering and beamwidth RA.

In this paper, a comprehensive literature survey of beamwidth RAs is presented. The existing state-of-the-art techniques used to achieve beamwidth reconfiguration are described. In addition to this, the RA designs realizing beam steering and beamwidth reconfiguration in a single antenna structure are also discussed. The paper is organized as follows: Section 1 presents a detailed classification of RAs. Section 2 describes different techniques used to achieve beamwidth reconfiguration. The advantages and limitations of the existing beamwidth RAs are also discussed. Section 3 presents multifunctional RAs capable of achieving beam steering and beamwidth variability in a single antenna structure. Section 4 presents conclusion and future research directions to improve the performance of beamwidth and compound RAs.

2   Beamwidth Reconfigurable Antenna

Recently, a great deal of attention has been given to beamwidth RA. Dynamic control over beamwidth of the antenna improves coverage and traffic capacity of the wireless networks. The beamwidth RAs are highly suitable for cellular base station applications, which demand antennas of different beamwidth for diverse environments [3]. Beamwidth RAs proposed in the literature are based on magnetoelectric (ME) dipole [4–8], partially reflecting surface (PRS) [9–11], tunable parasitic elements [12–14], array structure [15–17], slotted patch [18], frequency selective surface (FSS) [19], and dipole [20].

The concept of ME dipole antenna is proposed in [21]. The ME dipole antenna is comprised of a planar electric dipole and a quarter-wave patch antenna. The ME dipole antennas are attractive for base stations in wireless communication systems due to their several advantages such as wide impedance bandwidth, stable unidirectional radiation pattern, and low back radiation. A linearly polarized three-element ME dipole antenna array is reported in [4] to accomplish beamwidth reconfiguration in the H-plane. The 3-dB beamwidth of this antenna can be discretely switched between 37° and 136° in the H-plane. Limitations of this antenna in terms of discrete beamwidth tuning and beamwidth reconfiguration with single polarization are overcome in [5] by achieving continuous beamwidth tuning in the H-plane from 80° to 160° with linear polarization (LP) and from 72° to 133° with dual orthogonal polarization. A ME dipole antenna in [6] uses tunable strip grating reflector to attain beamwidth reconfiguration in H-plane from 81° to 153°. The ME dipole antenna array developed in [7] realizes beamwidth reconfiguration in the E-plane and H-plane from 24° to 97° and 22° to 100°, respectively. It is observed that the earlier reported ME dipole antennas achieve beamwidth reconfiguration in either E-plane or H-plane individually. In [8], ME dipole antenna with the capability of individual and simultaneous beamwidth reconfiguration is proposed. Beamwidth of this antenna can be continuously tuned in the E-plane, H-plane, and both the E-plane and H-plane from 65° to 120°, 80° to 120°, and 65° to 130°, respectively. However, this antenna realized a narrow impedance bandwidth as compared with the other reported ME dipole antennas.

The antenna design proposed in [9] uses cylindrical electromagnetic band gap (EBG) structure to achieve reconfigurable elevation beamwidth. In [10], the reflection magnitude of the PRS antenna is controlled to achieve beamwidth reconfiguration. A PRS antenna design enabling one-bit dynamic beamwidth control is presented in [11]. This antenna achieves a 18° and 23° variation of 3-dB beamwidth in the E-plane and H-plane, respectively. The parasitic antennas generally make use of microstrip Yagi principle to achieve beamwidth reconfiguration. The tunable parasitic elements are placed symmetrically along the E-plane or H-plane of the driven element. Limitation of this technique is that it is very difficult to maintain impedance matching in the H-plane, due to the strong mutual coupling between driven and parasitic elements. In [12], the parasitic elements are used to control azimuth beamwidth with single or dual LP. In [13] and [14], tunable parasitic elements are used to achieve continuous beamwidth tuning in H-plane and both the principal planes, respectively. The antenna performance in [13] shows a dynamic control over radiation beamwidth that ranges from 50° to 112° with a capacitance tuning range of 0.5 to 2.5 pF. In [14], beamwidth of the RA is continuously tuned from 50° to 141° and 53.8° to 149° in the E-plane and H-plane, respectively.

The RA designs presented in [15–17] utilize array configuration to realize beamwidth variability. A variable beamwidth 2×2 antenna array that can tune its beamwidth approximately from 65° to 100° is proposed in [15]. A planar microstrip series-fed slot antenna array is proposed in [16], which provides 2-D beamwidth switching capability. The 3-dB beamwidth in E-plane and H-plane can be tuned from (35°, 65°) to (15°, 31°). The antenna design proposed in [17] consists of a coupler, three reconfigurable output ports, and 1×3 rectangular patch array. This antenna achieves broad beamwidth of 66.2° and a narrow beamwidth of 24.1°. A coplanar slotted-patch antenna is presented in [18] to realize broadside radiation with H-plane 3-dB beamwidth of 130° ± 10° and 55° ± 1°. A FSS-based antenna controls the E-plane beamwidth from 13.2° to 31.1° at an operating frequency of 5.5 GHz [19]. A reconfigurable dipole antenna presented in [20] achieves beamwidth reconfiguration in the E-plane. The 3-dB beamwidth is switched in three different modes narrow (77.6°), middle (90.7°), and wide (168.3°). Table 1 presents a detailed performance comparison of beamwidth RA designs reported in the literature. It is noted that,

Table 1 Performance comparison of the RA designs achieving beamwidth reconfigurability

Ref.	Antenna type	−10 dB bandwidth (%)	3-dB beamwidth (degree)	Plane	Peak gain (dBi)	Switches (number)	Polarization
Ge and Luk [4]	ME dipole	15	37–136	H	9.8	Switch (2)	Linear
Ge and Luk [5]	ME dipole	10	80–160 (LP) 72–133 (Dual LP)	H	7.1	Varactor (2)	Dual linear
Ge and Luk [6]	ME dipole	40	81–153	H	6.4	PIN (15)	Linear
Feng et al. [7]	ME dipole	78.4	22–100, 24–97	E, H	11.5	Switch (2)	Dual linear
Shi et al. [8]	ME dipole	4.87	65–120 80–120 65–130	E H E, H	5.8	Varactor (4)	Linear
Edalati and Denidni [9]	PRS	8	25–83	E	6.3	PIN (68)	n/a
Debogovic et al. [10]	PRS	5	21–29.5, 24–37	E, H	15.1	Varactor (100)	Linear
Debogovic et al. [11]	PRS	n/a	16–34, 16–39	E, H	n/a	MEMS	Dual linear
Khidre et al. [13]	Tunable parasitic	2	50–112	H	8.6	Varactor (2)	Linear
Saitoh et al. [14]	Tunable parasitic	n/a	50–141, 53.8–149	E, H	n/a	Varactor (4)	Linear
Lee et al. [15]	Array	n/a	65–100	H	7.36	Copper strips (2)	Linear
Chu and Ma [16]	Slot antenna array	2.5 5.8 10.8	(15, 31) (20, 45) (35, 65)	E, H	14.5 12.1 7	Varactor (1)	Linear
Kim and Oh [17]	Array	n/a	24.1, 66.2	H	11.3	PIN (4)	n/a
Tsai and Row [18]	Slotted patch	16.24	130, 55	H	5.3, 8.6	PIN (6)	Linear
Wang et al. [19]	FSS	n/a	13.2–31.1	E	19	Varactor (96)	n/a
Zhang et al. [20]	Dipole	4.8	77.6, 90.7, 168.3	E	n/a	PIN (4)	n/a

- The beamwidth RA designs suffer from either narrow bandwidth [8, 10, 13, 20] or narrow tunable beamwidth [11, 16, 19].
- Beamwidth RA designs based on ME dipole are non-planar and large in size [4–8].
- The methodologies reported in [4, 6, 7, 9, 11, 15, 17, 18, 20] do not provide continuous tuning of beamwidth.

- The antenna designs presented in [9, 10, 19] need many switches that make biasing circuit integration challenging.
- In some of the designs, beamwidth reconfigurability is achieved only in H-plane [4–6, 13, 15, 17, 18] or E-plane [9, 19, 20].
- There are limited number of antenna designs reported in the literature, which achieves beamwidth reconfiguration in both the principal planes with dual orthogonal LP [7, 11].

3   Beam Steering and Beamwidth Reconfigurable Antenna

As discussed in Sect. 2, most of the beamwidth RAs achieve unidirectional radiation characteristics. However, main beam of the antenna cannot be steered. Several antenna designs are reported in the literature to realize continuous beam scanning and tunable beamwidth in a single antenna structure [22–28]. Techniques used to achieve these objectives are PRS [22], metamaterial [23], antenna with metal walls [24], tunable parasitic elements [25–27], and parasitic pixel layer [28].

A PRS antenna is proposed in [22] to obtain independent beam scanning and dynamic beamwidth control. Beam scanning is realized from 15° to 20°, and beamwidth is tuned from 18.7° to 22.4°. A metamaterial-based leaky wave antenna (LWA) is proposed in [23] to attain tunable radiation angle and beamwidth functionalities. This antenna achieves continuous beam scanning from −49° to 50° at an operating frequency of 3.33 GHz. Continuous beam scanning is accomplished by uniformly biasing varactor diodes. Varactor diodes are non-uniformly biased to realize beamwidth tuning. An aperture-fed PRA with metal walls is presented in [24]. This antenna achieves boresight radiation with narrow and wide beamwidth. Main beam of the antenna is discretely switched in the E-plane and H-plane from −51° to 54° and −20° to 20°, respectively. The reflector and director properties of the parasitic elements are used in [25] to achieve beamwidth tuning from 60° to 130° and main lobe scanning from −20° to 20° in the H-plane. A low profile dual-polarized PRA is presented in [26]. Main beam of the antenna is directed to 14°, −17°, and −3° with narrow beamwidth. This antenna achieves narrow beamwidth of 49° and a wide beamwidth of 105°, when the main beam is steered in broadside direction. A cross parasitic antenna proposed in [27] consists of varactor-loaded tunable parasitic elements placed in the E-plane and H-plane of the driven element. This antenna provides continuous beam scanning in the elevation plane and covers complete azimuth plane. The beamwidth is continuously tuned in the E-plane and H-plane from 65° to 152° and 64° to 116°, respectively. A parasitic pixel layer-based RA design is developed in [28] to provide pattern and beamwidth variability on a single antenna structure. This RA generates nine beam steering and three beamwidth variable modes. The antenna covers complete azimuth plane, and the main beam is discretely switched to 40° in the elevation plane. This antenna achieves narrow and broad beamwidth of 40° and 100°, in $\phi = 45°$, 90° and −45° planes. Detailed performance comparison of beam steering and beamwidth RA designs is presented in Table 2. It can be concluded that,

Table 2 Performance comparison of the RA designs achieving beam steering and beamwidth variability in a single antenna structure

Ref.	Antenna type	−10 dB bandwidth (%)	360° azimuth coverage	Beam steering (degree)	3-dB beamwidth (degree)	Peak gain (dBi)	Switches (number)
Debogovi and Perruisseau-Carrier [22]	PRS	3	No	±10 (H-plane)	18.7–22.4 (H-plane)	14.7	Varactor (100)
Lim et al. [23]	Metamaterial	n/a	No	−49–50 (E-plane)	48, 37 (E-plane)	18	Varactor (90)
Yang et al. [24]	Antenna with metal walls	10.8	No	−51, 54 (E-plane) −20, 20 (H-plane)	Narrow Wide	6	PIN (2)
Wang et al. [25]	Tunable parasitic	6.4	No	±20 (H-plane)	60–130 (H-plane)	8.8	Varactor (2)
Deng et al. [26]	Parasitic	5.56	No	−17, 14	49, 105	7	PIN (8)
Khairnar et al. [27]	Tunable parasitic	1.63	Yes	10.8 (E-plane), 40 (H-plane)	65–152 (E-plane), 64–116 (H-plane)	3.78	Varactor (8)
Towfiq et al. [28]	Parasitic pixel layer	4	Yes	±40	40 (E-plane), 100 (H-plane), 100 (D-plane)	8	PIN (6)

- There are very few RA designs that achieve beam steering and beamwidth tuning in a single antenna structure.
- The RA design presented in [28] is capable of achieving complete 360° coverage in the azimuth plane and beamwidth variability. However, the main beam and 3-dB beamwidth of the antenna are discretely switched.

4   Conclusion

This paper presents a detailed survey on beamwidth RAs. The techniques used to realize beamwidth reconfiguration are reviewed in detail along with advantages and limitations of each technique. Moreover, this paper also discusses compound RAs realizing beam steering and beamwidth reconfiguration in a single antenna structure. This paper helps to choose suitable design technique to achieve beam steering and beamwidth reconfiguration. The challenges and future research directions in beamwidth and compound RAs can be summarized as follows:

- The RA should be designed with a minimum number of active switches as this will help to reduce cost, power consumption, and complexity of the DC biasing network.
- It is noted that the earlier reported RA designs attain beamwidth reconfiguration in either E-plane or H-plane. The major challenge is to accomplish beamwidth reconfiguration in both the principal planes with dual orthogonal LP.

- There are very few antenna designs reported, which achieve continuous beam scanning and tunable beamwidth in a single antenna structure. The compound RA capable of providing beam steering and beamwidth reconfiguration in a single antenna structure has the strong potential to improve the performance of wireless communication systems.

References

1. Schaubert D, Farrar F, Sindoris A, Hayes S (1981) Microstrip antennas with frequency agility and polarization diversity. IEEE Trans Antennas Propag 29(1):118–123
2. Bernhard JT (2007) Reconfigurable antennas. Morgan & Claypool
3. Li Y, Luk K (2014) A linearly polarized magnetoelectric dipole with wide H-plane beamwidth. IEEE Trans Antennas Propag 62(4):1830–1836
4. Ge L, Luk KM (2015) A three-element linear magneto-electric dipole array with beamwidth reconfiguration. IEEE Antennas Wirel Propag Lett 14:28–31
5. Ge L, Luk KM (2016) Linearly polarized and dual-polarized magneto-electric dipole antennas with reconfigurable beamwidth in the H-plane. IEEE Trans Antennas Propag 64 (2):423–431
6. Ge L, Luk KM (2016) Beamwidth reconfigurable magneto-electric dipole antenna based on tunable strip grating reflector. IEEE Access 4:7039–7045
7. Feng B, Tu Y, Chung KL, Zeng Q (2018) A beamwidth reconfigurable antenna array with triple dual-polarized magneto-electric dipole elements. IEEE Access 6:36083–36091
8. Shi Y, Cai Y, Yang J, Li L (2019) A magnetoelectric dipole antenna with beamwidth reconfiguration. IEEE Antennas Wirel Propag Lett 18(4):621–625
9. Edalati A, Denidni TA (2009) Reconfigurable beamwidth antenna based on active partially reflective surfaces. IEEE Antennas Wirel Propag Lett 8:1087–1090
10. Debogovic T, Perruisseau-Carrier J, Bartolic J (2010) Partially reflective surface antenna with dynamic beamwidth control. IEEE Antennas Wirel Propag Lett 9:1157–1160
11. Debogovi T, Bartoli J, Perruisseau-Carrier J (2014) Dual-polarized partially reflective surface antenna with MEMS-based beamwidth reconfiguration. IEEE Trans Antennas Propag 62(1):228–236
12. Korisch IA, Rulf B (2000) Antenna beamwidth control using parasitic subarrays. In: IEEE-APS conference on antennas and propagation for wireless communications, pp 117–120
13. Khidre A, Yang F, Elsherbeni AZ (2013) Reconfigurable microstrip antenna with tunable radiation beamwidth. In: IEEE antennas and propagation society international symposium (APSURSI), pp 1444–1445
14. Saitoh M, Honma N, Murakami T (2016) Impact of radiation pattern control of MIMO antenna on interfered multicell environment. IEEE Antennas Wirel Propag Lett 15:666–669
15. Lee S-N, Kim J, Yook J-G, Hu YC, Peroulis D (2007) A variable beamwidth antenna for wireless mesh networks. In: IEEE antennas and propagation society international symposium, pp 493–496
16. Chu HN, Ma T (2017) Beamwidth switchable planar microstrip series-fed slot array using reconfigurable synthesized transmission lines. IEEE Trans Antennas Propag 65(7):3766–3771
17. Kim D-W, Oh S-S (2017) Design of a coupler with three reconfigurable output ports and a beamwidth reconfigurable antenna. Int J Antennas Propag:1–8

18. Tsai C, Row J (2016) Beamwidth reconfigurable slotted-patch antennas. In: IEEE 5th Asia-Pacific conference on antennas and propagation (APCAP), pp 149–150
19. Wang M, Huang C, Chen P, Wang Y, Zhao Z, Luo X (2014) Controlling beamwidth of antenna using frequency selective surface superstrate. IEEE Antennas Wirel Propag Lett 13:213–216
20. Zhang J, Zhang S, Pedersen GF (2018) E-plane beam width reconfigurable dipole antenna with tunable parasitic strip. In: 12th European conference on antennas and propagation (EuCAP), pp 1–3
21. Luk K-M, Wong H (2006) A new wideband unidirectional antenna element. Int J Microw Opt Technol 1(1):35–44
22. Debogovi T, Perruisseau-Carrier J (2014) Array-fed partially reflective surface antenna with independent scanning and beamwidth dynamic control. IEEE Trans Antennas Propag 62 (1):446–449
23. Lim S, Caloz C, Itoh T (2005) Metamaterial-based electronically controlled transmission-line structure as a novel leaky-wave antenna with tunable radiation angle and beamwidth. IEEE Trans Microw Theory Tech 53(1):161–173
24. Yang G, Li J, Wei D, Zhou S, Xu R (2019) Pattern reconfigurable microstrip antenna with multidirectional beam for wireless communication. IEEE Trans Antennas Propag 67 (3):1910–1915
25. Wang J, Yin J, Wang H, Yu C, Hong W (2017) Wideband U-slot patch antenna with reconfigurable radiation pattern. In: 11th European conference on antennas and propagation (EuCAP), pp 611–615
26. Deng W, Yang X, Shen C, Zhao J, Wang B (2017) A dual-polarized pattern reconfigurable Yagi patch antenna for microbase stations. IEEE Trans Antennas Propag 65(10):5095–5102
27. Khairnar VV, Kadam BV, Ramesha CK, Gudino LJ (2019) A reconfigurable microstrip cross parasitic antenna with complete azimuthal beam scanning and tunable beamwidth. Int J RF Microw Comput-Aided Eng 29(1):e21472
28. Towfiq MA, Bahceci I, Blanch S, Romeu J, Jofre L, Cetiner BA (2018) A reconfigurable antenna with beam steering and beamwidth variability for wireless communications. IEEE Trans Antennas Propag 66(10):5052–5063

A Low Power CMOS Variable True Random Number Generator for LDPC Decoders

Jamel Nebhen[(⊠)]

College of Computer Engineering and Sciences, Prince Sattam Bin Abdulaziz
University, P.O. Box 151, Alkharj 11942, Saudi Arabia
j.nebhen@psau.edu.sa

Abstract. This paper presents a new structure of a variable integrated noise source (VINS) implemented in a commercial CMOS technology. This VINS circuit is based on the new technique of dual-drain MOS transistor. It consists of one dual-drain NMOS transistor and one dual-drain PMOS transistor with a particular innovation; the two drains have different lengths. The VINS circuit has been simulated in a CMOS FDSOI 28-nm process. It can produce good quality bit streams without any post-processing. It has a typical low power dissipation of 100-μW. This novel circuit is a promising unit for LDPC decoders. The new VINS circuit can be used in a CMOS system-on-chip (SoC) for a variety of applications ranging from the data encryption and mathematical simulation to the built-in-self test (BIST) of RF receivers.

Keywords: Random number generator · Low noise · Jitter · Voltage controlled oscillator

1 Introduction

The design of noise sources is widely used in several applications such as electronic tests and measurement and RF receivers [1, 2]. For the RF receiver, an idea is very interesting to carry out the main technique of its calibration. This idea consists in injecting at its inputs a quantity of known noise [3]. Therefore, with this methodology, it is very easy to measure the RF receiver power gain. As a result, we can accurately correct its variations [4]. For the RF receiver, if a noise power is injected with two different levels at its inputs, we can determine the level of power gain and the quality of the noise that represents its equivalent noise temperature. Consequently, the noise source makes it possible to carry out a complete calibration of the receiver parameters. In many examples of measurements, there is a need for a broadband signal with precisely known properties such as autocorrelation function, rms value, or amplitude probability density [5]. The random noise generator makes it possible to generate these useful signals. These test signals have statistical parameters that are easily manipulated and are known in advance. This shows well the use of the noise generated by these random noise generators in these techniques which are very accurate as insertion loss, measurement of impulse response, in the radar measuring trajectories modulated by noise and in the determination of intermodulation and linearity of communication systems. No work has been successfully design a variable integrated noise source in a

© Springer Nature Singapore Pte Ltd. 2020
N. Goel et al. (eds.), *Modelling, Simulation and Intelligent Computing*,
Lecture Notes in Electrical Engineering 659,
https://doi.org/10.1007/978-981-15-4775-1_53

CMOS technology until now. Random number generators (RNG) are an important part of modern communication system, modern cryptographic system, and statistical simulation system. TRNGs are mechanical or electrical devices for extracting random numbers from a source of physical hazard. If properly designed and implemented, even an absolute knowledge of their architecture and internal operation should not be able to predict their output bits. In the state of the art, three types of integrated RNGs have been reported: resistor-thermal-noise RNG [6, 7], oscillator-based RNG [8], and discrete-time chaotic RNG [9]. The most practical source of randomness in digital circuits is the jitter because it is ubiquitous and easily accessible. Combinations of these three techniques are often adopted to design RNG with better performances [10].

The performance obtained from an efficient implementation of error control codes is one of the key elements that make the difference within competition. The important features that make a product successful can be low complexity, low energy consumption, or low error probability performance. Several scientists have recently tackled the performance evaluation of iterative decoders in stochastic architectures. In the next generation of integrated circuits with transistor size below 40 nm, every single gate can temporarily output a wrong value due to transient defects [11]. One of the first proposed trends has been to evaluate, both theoretically and practically, the performance degradation induced by a stochastic architecture [12], then using wisely the redundancy to reduce the negative effects introduced by the transistor noise [13]. Through these research endeavors, an unexpected spin-off was identified: the noise inside the decoders is not necessarily an enemy to combat, but it can be used as an ally. Indeed, recent works have shown that the controlled injection of noise in an iterative error control decoder can significantly enhance the error correction performance, and thus, contribute to mitigate the effect of the transmission channel perturbations [14]. In other words, and even if it may appear as a paradox at first glance, noise in an iterative decoder can help to combat the channel noise. The objective of this work is to improve the performance of an iterative decoder by a "smart injection" of randomness inside the decoding process.

In this paper, we present a new VINS structure suitable for generating variable noise levels useful for use in many analogue or digital microelectronic applications. The idea is to use the noise properties of the MOS transistors in a CMOS FDSOI 28-nm technology. The rest of the paper is organized as following: the principle of the new VINS circuit is described in Sect. 2. The simulation results are presented in Sects. 3, and 4 concludes the paper.

2   Principle of the New VINS Circuit

The structure of the VINS circuit is composed by a dual-drain NMOS and a dual-drain PMOS, as shown in Fig. 1. This special MOS transistor is composed by two drains, one gate and one source. To fabricate this dual-drain MOS transistor in a standard CMOS process, we make a modification in the design of the layout of the MOS transistor. Indeed, we change the rectangular shape of the implantation mask of the source-drain to another concave [15]. Each dual-drain MOS transistor has a width size of W/N for the first drain and $W \cdot (N - 1)/N$ for the second drain with $N = 2, 3, 4, \ldots,$ n. When carrying out the layout of dual-drain MOS transistor, the mask concave form

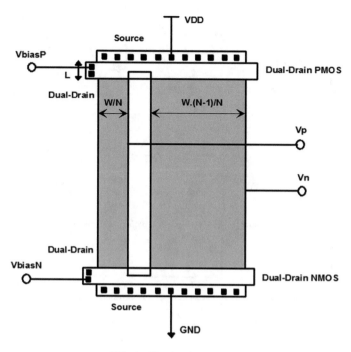

Fig. 1 Circuit of the VINS

two drains. By modifying the layout of the MOS transistor, we create a current circulation shed from the source to the two drains. If the dual-drain MOS transistor is polarized, then the majority carriers are derived from the source to the drain. In fact, the majority carriers meet the two sheds. Therefore, they enter stochastically into one of the two drains. Hence, they form the equivalent quantity of current that flows through each corresponding drain. In this case, the probability of a carrier entering one of the two drains is the same. But there is an uncertainty that a carrier makes the choice of one of the two drains. The uncertainty of crossing of a carrier in one of the two drains makes it possible to create a fluctuation through each drain. Therefore, if we compare a single-drain MOS transistor with a dual-drain MOS transistor, this fluctuation becomes an extra-noise source. The two drain shares the same active field so its two currents are inevitably correlated with each other. When two constant voltages are injected at the terminals of the source and of the gate of the dual-drain MOS transistor and when their two drains are polarized, respectively, with the two voltages at the same values, then the current in the source is constant. The value of this current is equal to the sum of the two currents. Therefore, when a drain is affected by a fluctuation of the current, this same phenomenon causes in the other drain an opposite phase fluctuation of the current. Accordingly, if one of the two drains of the dual-drain MOS is through by a current that increases stochastically, then the other drain undergoes the inverse operation and it is through by a current that decreases. Therefore, there is a correlation between the fluctuations of the two drains. The structure of dual-drain MOS transistor forms a differential architecture. The $1/f$ and thermal noise are the noise sources of the

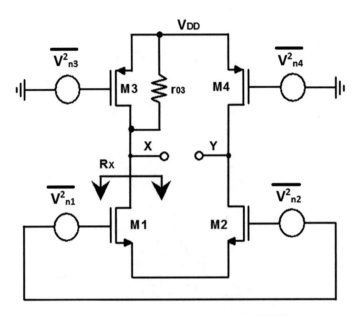

Fig. 2 Equivalent-input referred noise of the VINS circuit

dual-drain MOS transistor. We can design the dual-drain MOS transistor as a four-port circuit as shown in Fig. 2. It is, therefore, possible to model the overall noise. We only need to refer the noise of transistors $M3$ and $M4$ to the input because we can represent the model of the $1/f$ and thermal noise of transistors $M1$ and $M2$ as a voltage sources in series with the input. We first calculate the output noise contributed by transistor $M3$. It produces a noise voltage V_{nA} at node X with respect to ground as

$$V_{nA} = g_{m3} V_{n3} r_{03} \frac{r_{04} + 2r_{01}}{2r_{04} + 2r_{01}} \qquad (1)$$

where g_{m3} denotes the transconductance of transistor $M3$ and r_{01}, $r_{02,}$ and r_{04} denote the resistance of transistors $M1$, $M2$, and $M4$, respectively. Whereas the other part of circuit which includes transistors $M1$, $M2$, and r_{04} can generate a noise voltage V_{nB} at node Y as

$$V_{nB} = g_{m3} V_{n3} r_{04} \frac{r_{03}}{2r_{04} + 2r_{01}} \qquad (2)$$

Therefore, the transistor $M3$ generates a total differential output noise which can be written as

$$V_{nXY} = g_{m3} V_{n3} \frac{r_{03} r_{01}}{r_{03} + r_{01}} \qquad (3)$$

Applying (3) to transistor $M4$ and adding the resulting powers, the total output-referred noise can be expressed as

$$\overline{V^2_{n,out/M3,M4}} = g^2_{m3}(r_{01} \parallel r_{03})^2 \overline{V^2_{n3}} + g^2_{m4}(r_{02} \parallel r_{04})^2 \overline{V^2_{n4}} \tag{4}$$

The gain of this circuit is

$$G = \frac{g_{m1}(r_{01} \parallel r_{03}) + g_{m2}(r_{02} \parallel r_{04})}{2} \tag{5}$$

We divided (4) by G^2 to refer the noise to the input. Therefore, the total input referred noise can be expressed as:

$$\overline{V^2_{n,in}} = \overline{V^2_{n1}} + \overline{V^2_{n2}} + \frac{g^2_{m3}}{g^2_{m1}}\overline{V^2_{n3}} + \frac{g^2_{m4}}{g^2_{m2}}\overline{V^2_{n4}} \tag{6}$$

If we substitute V_{n1} and V_{n2} and we pose $X = r_{01} \parallel r_{03}$ and $Y = r_{02} \parallel r_{04}$ in (6), the total input referred noise can be expressed as:

$$\overline{V^2_{n,in}} = 8kT\left(\frac{1}{3g_{m1}} + \frac{1}{3g_{m2}} + \frac{4}{3}\left(\frac{g_{m3}X^2 + g_{m4}Y^2}{(g_{m1}X + g_{m2}Y)^2}\right)\right) + \frac{K_n}{W_1 L_1 C_{ox}f} + \frac{K_n}{W_2 L_2 C_{ox}f} \tag{7}$$

where k denotes the Boltzmann constant, T denotes the temperature in Kelvin, K_n denotes the flicker noise coefficient, C_{ox} denotes the gate oxide capacitance, f denotes the frequency, and W and L denote the channel width length of the MOS transistor. Now, if we replace $W_1 = W_3 = W/n$ and $W_2 = W_4 = W \cdot (n-1)/n$, we can write the transconductance of transistors $M1$, $M2$, $M3$, and $M4$ as:

$$g_{m1} = g_{m3} = \frac{g_m}{n}; \quad g_{m2} = g_{m4} = \frac{(n-1)g_m}{n} \tag{8}$$

We substitute g_m in (7) with his new value and we suppose $X \approx Y$, we can written the total input referred noise associated to g_{m1} and g_{m3} as:

$$\overline{V^2_{n,in}} = 8kT\frac{1}{3}\left(\frac{n^2}{(n-1)g_{m1}} + 4\frac{g_{m3}}{g^2_{m1}}\right) + \frac{n^2 \cdot K_n}{(n-1)W_2 L_2 C_{ox}f} \tag{9}$$

3   Simulation Results

The noise source circuit consists of a PMOS transistor with dual drain and a NMOS transistor with dual drain [16], as shown in Fig. 1. The dual-drain MOS transistor can be fabricated by only changing the source-drain implanting mask shape from a rectangle to a concave in a standard CMOS process. Each dual-drain MOS transistor has a width size of W/N for the first drain and $W \cdot (N-1)/N$ for the second drain. The special dual-drain MOS transistor structure can bring an extra noise to the currents through the double drains of the MOS transistor. This concave mask forms two drains of MOS transistors and results in a shed of the current that flows from the source to the dual drain of the MOS transistor. When the majority carriers drift from the source to the drain and meet the shed, the carriers enter one of the two drains stochastically and form the current through the corresponding drain. Although the probability that a carrier enters one of the two drains is the same as that a carrier enters the other drain. There is an uncertainty of the carriers' choosing of the drains of the transistor. This uncertainty creates the current fluctuation through each drain of the dual-drain MOS transistor. This noise characteristic of the dual-drain MOS transistor can be used as the noise source of a VINS circuit. In order to convert the noise current of dual-drain MOS transistors into the noise voltage signal, the noise source device is designed to be a dual-drain MOS transistor circuit with different width sizes for each drain. The fluctuation of the drain currents in the dual-drain MOS transistors is converted into a differential voltage signal between Vp and Vn. A differential amplifier follows the noise source device. It amplifies the differential noise voltage signal and transfers it to a single-ended noise voltage signal $Vout$, as shown in Fig. 3. The noise generator generates a proper noise signal and provides enough drive ability.

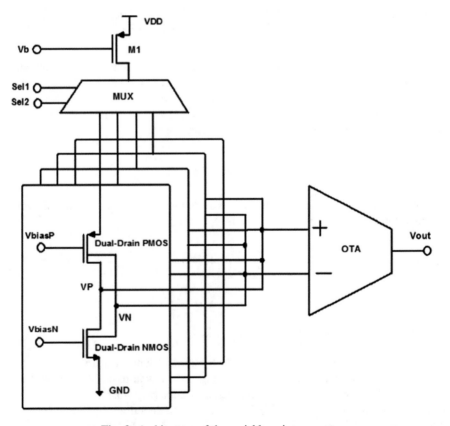

Fig. 3 Architecture of the variable noise generator

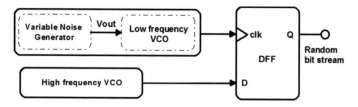

Fig. 4 Architecture of the VINS circuit

Table 1 Noise PSD simulation results of the VINS circuit

N	2	3	4	5
$(W/L)_{\text{dual-drain-NMOS}}$ (μm/μm)	5/2	3.33/2	2.5/2	2/2
$(W/L)_{\text{dual-drain-PMOS}}$ (μm/μm)	5/2	6.66/2	7.5/2	8/2
$\overline{V^2_{n,\text{in}}}$ $\left(\mu V/\sqrt{Hz}\right)$	10	11	11.7	12.7

The above results show that the noise generator can output a larger noise signal with a smaller DC offset. We use the noise generator to design a variable true random number generator based on the oscillator sampling architecture [17], as shown in Fig. 4. It consists of a noisy oscillator, a high-frequency oscillator, and a D flip–flop. The noisy oscillator consists of a dual-drain MOS transistor noise generator and a voltage-controlled oscillator. The output noise voltage signal of the noise generator controls the oscillator and makes it oscillate with a much large jitter.

To demonstrate the proper working of our new VINS circuit, we carried out various simulations for $N = 2$, 3, 4 and 5. Table 1 presents the simulation results. The simulation of the input referred noise of the VINS circuit under cadence is presented in Fig. 5. This figure clearly shows that if N increases, the input referred noise of the VINS circuit also increase which is in coincidence with the previous equations. There is a conversion of the fluctuation of the two drain currents into a differential voltage between Vp and Vn. The VINS circuit has several advantages. Firstly, the input referred noise of the dual-drain MOS transistor increases if N increase. If $N = 2$, it is equal to two times for a dual-drain MOS transistor than that for a normal MOS transistor. Secondly, the MOS transistor has an output impedance of about 1-GΩ, a small variation of 1-pA of drain current results in a 1-mV variation in the differential voltage (Vp, Vn). Finally, the VINS circuit has naturally a low power due to its operation in the sub-threshold state of the double drain MOS transistors.

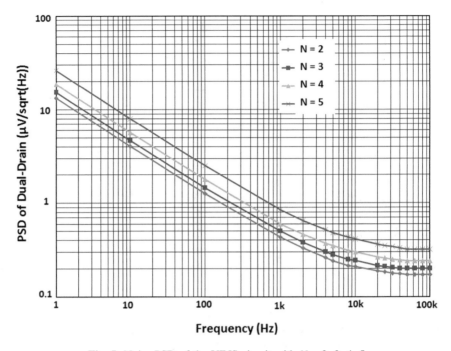

Fig. 5 Noise PSD of the VINS circuit with $N = 2, 3, 4, 5$

4 Conclusion

In this paper, we have shown a new structure of a VINS circuit implemented in a standard CMOS technology. The idea is inspired by the new technique of dual-drain MOS transistor with a particular innovation; the two drains have different lengths. The results of the simulation for the VINS circuit demonstrated the validity of our new architecture. The VINS circuit has been simulated in a CMOS FDSOI 28-nm process. It can produce good quality bit streams without any post-processing. It has a typical low power dissipation of 100-µW. The VINS circuit is a promising structure for the BIST of RF receivers, low power system, data security, encryption, and communication applications.

Aknowledgements. This project was supported by the Deanship of Scientific Research at Prince Sattam Bin Abdulaziz University under the research project 2019/01/11709.

References

1. International Technology Roadmap for Semiconductors. 2013 ITRS Editions Reports. http://public.itrs.net/reports.html
2. Ngassa CK, Savin V, Dupraz E, Declercq D (2015) Density evolution and functional threshold for the noisy min-sum decoder. IEEE Trans Commun 63(5):1497–1509

3. Gupta M (1975) Applications of electrical noise. Proc IEEE 63(7):996–1010
4. Tiuri ME (1964) Radio astronomy receivers. IEEE Trans Antennas Propag 12(7):930–938
5. Ulaby FT, Moore RK, Fung AK (1981) Microwave remote sensing. Artech House, Norwood, MA, USA
6. Jarosik N (1996) Measurement of the low-frequency-gain fluctuations of a 30-GHz high-electron-mobility-transistor cryogenic amplifier. IEEE Trans Microw Theory Techn 44 (2):193–197
7. Adam JA (1992) Data security-cryptography = privacy? IEEE Spectr 29(8):29–35
8. Tang Y, Boutillon E, Winstead C, Jego C, Jezequel M (2013) Muller C-element based decoder (MCD): a decoder against transient faults. In: IEEE International symposium on circuits and systems (ISCAS), pp 1680–1683
9. Sundararajan G, Winstead C, Boutillon E (2014) Noisy gradient descent bit-flip decoding for LDPC codes. IEEE Trans Commun 62(10):3385–3400
10. Fry PW, Hocy SJ (1969) A silicon MOS magnetic field transducer of high sensitivity. IEEE Trans Electr Dev ED 16:35–39
11. Sheng-hua Z, Wancheng Z, Nan-Jian W (2008) An ultra-low power CMOS random number generator. J Solid-State Electron 52:233–238
12. Mathew SK, Johnston D, Satpathy S, Suresh V, Newman P, Anders MA, Kaul H, Agarwal A, Hsu SK, Chen G, Krishnamurthy RK (2016) μRNG: a 300–950 mV, 323 Gbps/W all-digital full-entropy true random number generator in 14 nm FinFET CMOS. IEEE J Solid-State Circuits 51(7):1695–1704
13. Dongsheng L, Zilong L, Lun L, Xuecheng Z (2006) A low-cost low-power ring oscillator-based truly random number generator for encryption on SMART cards. IEEE Trans Circuits Syst II Exp Briefs 63(6):608–612
14. Karakaya B, Çelik V, Gülten A (2017) Chaotic cellular neural network-based true random number generator. Int J Circuits Theory Appl 45:1885–1897
15. Tomasi A, Meneghetti A, Massari N, Gasparini L, Rucatti D, Xu H (2018) Model, validation, and characterization of a robust quantum random number generator based on photon arrival time comparison. J Lightw Technol 36(18):3843–3854
16. Avesani M, Marangon DG, Vallone G, Villoresi P (2018) Source-device independent heterodyne-based quantum random number generator at 17 Gbps. Nature Commun 9 (1):5365-1–5365-6
17. Ray B, Milenković A (2018) True random number generation using read noise of flash memory cells. IEEE Trans Electron Dev 65(3):963–969

FPGA Implementation of Random Feature Mapping in ELM Algorithm for Binary Classification

Prabhleen Kaur Gill[1] ⓘ, Shaik Jani Babu[1](✉) ⓘ, Sonal Singhal[1] ⓘ, and Nilesh Goel[2] ⓘ

[1] Shiv Nadar University, Greater Noida, India
skjanibabu786@gmail.com
[2] Birla Institute of Technology and Science Pilani, Dubai Campus, Dubai, UAE
goel.nilesh@gmail.com

Abstract. Extreme learning machine (ELM) is a single layer feedforward neural network algorithm used for classification problems due to its accuracy and speed. It provides a robust learning algorithm, free of local minima, suitable for high-speed computation along with fast learning speed. In this paper, ELM algorithm implementation on hardware and software is discussed. A low-cost hardware implementation of 16-bit H-matrix generation on FPGA is discussed in the paper. Hardware implementation is carried out on Nexys-4 board using MATLAB and hardware description language (HDL). Generation of H-matrix is carried out using two activation functions, piecewise log-sigmoid and piecewise tan-sigmoid. This paper aims at optimizing the hardware implementation of ELM algorithm by minimizing the utilized resources of the FPGA. Finally, the ELM algorithm accuracy and hardware utilization for both activation functions are compared.

Keywords: Extreme learning machine (ELM) · H-matrix · Machine learning · Field-programmable gate array (FPGA)

1 Introduction

Supervised learning for classification involves learning from input data and subsequently using it for classifying new observation. There are several machine learning algorithms for the purpose of classification such as linear classifiers, neural networks (NN), support vector machines (SVM), and decision trees [1]. In feedforward NN, neurons are arranged in layers, each layer takes an input applies a nonlinear function to it and passes its output to the next layer. Weights are associated with each layer and these are tuned in the training phase to adapt itself to the particular problem in hand. This suffers from the bottleneck of slow-gradient-based learning algorithms and iterative tuning of the parameters [2]. Extreme learning machine (ELM) algorithm overcomes such issues. A single-hidden layer feedforward NN (SLFN) randomly chooses the hidden weights and analytically determines the output weights. It provides good results for most of the cases and extremely fast learning speed. ELM is proved to be

© Springer Nature Singapore Pte Ltd. 2020
N. Goel et al. (eds.), *Modelling, Simulation and Intelligent Computing*,
Lecture Notes in Electrical Engineering 659,
https://doi.org/10.1007/978-981-15-4775-1_54

better over other classification algorithms like conventional backpropagation feedforward neural networks and support vector machines (SVM) in various aspects [3].

Hardware or on-chip implementation of an algorithm is done to overcome the challenge to find an architecture that minimizes hardware costs, while maximizing performance, accuracy, and parameterization [4]. ELM provides a high speed, small size, low power consumption, autonomy, and true capability for real-time adaptation (i.e., the learning stage is performed on-chip) solution due to random mapping of features. Two real-world problems to which this algorithm can be extended are Landsat and driver identification system for smart car applications.

Field-programmable gate array (FPGA) is a robust hardware that offers features such as high configurability, low power, and high parallelism to meet the demand of high computational speed of ELM. FPGA performs better than general purpose processors in machine learning algorithms [5]. The computational time and accuracy are the main focus of research interest for implementing any machine learning algorithm on hardware, which motivates present research work to implement low-cost and high-speed ELM algorithm. Sigmoid activation function provides better accuracy compared to other activation functions like sinc, hard limit, radial function, etc., in ELM algorithm [6].

In this work, low-cost hardware implementation of ELM algorithm is carried out on Nexys-4 using MATLAB and hardware description language (HDL). This paper aims at optimizing the hardware implementation of ELM algorithm by minimizing the utilized resources of the FPGA. The activation function used for hardware implementation of a 16-bit H-matrix is a user-defined piecewise sigmoidal function. Final accuracy calculation is carried out in MATLAB environment for comparing the two methods of implementation, i.e., hardware and software, drawing appropriate conclusions. The paper discusses and explains the background of extreme learning machine (ELM) algorithm in Sect. 2. Section 3 describes the hardware implementation of ELM on FPGA. Results and conclusions are presented in Sects. 4 and 5, respectively.

2 ELM Algorithm

This section describes the ELM architecture and algorithm in detail. Figure 1 illustrates the ELM algorithm architecture. ELM is a two-layer neural feedforward network with 'L' hidden neurons and 'n' input nodes with activation function $g(\ )$.

The output, y is considered to be scalar here, for classification problem. The output of the network, y is given by the Eq. (1):

$$y = \sum_{i}^{L} \beta_i H_i = \sum_{i}^{L} \beta_i g\left(a_i^T x + b_i\right) a_i, x \in R^d, \beta_i, b_i \in R \quad (1)$$

where β denotes the output weights, H_i is the output of the ith hidden layer neuron. a_i denotes the input weight and b_i is the bias for the ith neuron.

An appropriate activation function is assumed to be $g(\ )$ [7, 8]. Unlike, conventional backpropagation learning rule that modified all the weights, the ELM algorithm uses

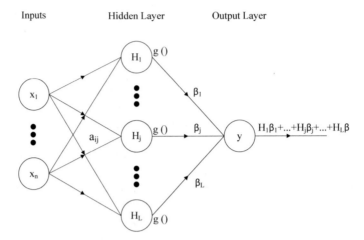

Fig. 1 ELM architecture

random feature mapping. It allows a_i and b_i to be random numbers drawn from any continuous distribution while only the output weights, β needs to be tuned based on the training data T. For N samples (x_k, t_k), the hidden layer output matrix H is defined as in Eq. (2):

$$
H = \begin{bmatrix} g\left(a_1^T x_1 + b_1\right) & \cdots & g\left(a_L^T x_1 + b_L\right) \\ \vdots & \cdots & \vdots \\ g\left(a_1^T x_N + b_1\right) & \cdots & g\left(a_L^T x_N + b_L\right) \end{bmatrix} \tag{2}
$$

The desired output weights β are then the solution of the following optimization problem:

$$
\text{Minimize}_\beta \|H\beta - T\|^2 \tag{3}
$$

where output weight matrix, β, and training data target, T, are given by the following Eq. (4),

$$
\beta = \begin{bmatrix} \beta_1 \\ \vdots \\ \beta_L \end{bmatrix}, \quad T = \begin{bmatrix} t_1 \\ \vdots \\ t_N \end{bmatrix} \tag{4}
$$

The ELM algorithm proves that the optimal solution β is given by

$$
\beta = H^+ T \tag{5}
$$

where H^+ denotes the pseudo-inverse or the Moore–Penrose generalized inverse of a matrix [9]. This method is used because it removes the need for iterative tuning and gives a simple formula to calculate the weights. The orthogonal projection method can be efficiently used to find H^+ as $(H^TH)^{-1}H^T$ if H^TH is non-singular or as $H^T(HH^T)^{-1}$ if HH^T is non-singular.

3 Implementation of ELM Algorithm on FPGA

This section describes methodology for the implementation of the algorithm. Figure 2 shows the implementation flow of the ELM algorithm. Sample data points taken as input are used to find the corresponding H-matrix. Following this, output weight, β, is evaluated. Finally, the accuracy was calculated for the obtained output (y). Generation of H-matrix is performed using both MATLAB script and hardware description language (HDL). Later, β-vector is evaluated using MATLAB script. In this work, 20 hidden neurons ($L = 20$) along with 10,000 sample data points ($N = 10000$) were used. The data set used is a skin data set, collected by randomly sampling RGB values from the face of images obtained from FERET database [10].

To generate H-matrix, random weights (a_i) are multiplied to the input data (x). The weighted inputs are then added with random bias values (b_i), which are passed through a soft limiting activation function. In this work, piecewise tan-sigmoid and log-sigmoid functions are created and used as activation functions $g(\ )$. H-matrices corresponding to the two activation functions are obtained using Eq. (4). The training and testing H-matrices are generated from respective input data points. The HDL code is generated for the obtained 16-bit H-matrix and was used to generate a bit stream which was dumped onto the FPGA. Since floating point is not compatible with FPGA, fixed point data type was used in the HDL code.

For calculating output weight vector (β), generated training H-matrix and training targets (T) are used. Pseudo-inverse of training H-matrix is multiplied with the training targets to evaluate output weights.

The predicted targets for classification are computed by multiplication testing H-matrix and output weight vector (β). Accuracy of the ELM algorithm is calculated using misclassification error [11] as represented in Eq. (6):

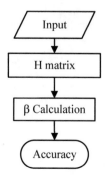

Fig. 2 Implementation and methodology of the algorithm

$$\text{Accuracy} = \frac{\text{Number of correct predictions}}{\text{Total number of predictions}} = \frac{\text{TP} + \text{TN}}{\text{TP} + \text{TN} + \text{FP} + \text{FN}} \quad (6)$$

where TP is true positive, TN is true negative, FP is false positive, and FN is false negative. TP and TN together constitute for correct predictions and all TP, TN, FP, FN constitute for all the predictions.

4 Results and Discussion

This section presents the hardware utilization of the algorithm on FPGA for the two activation functions. The accuracy of the algorithm having two modes, software and hardware, of implementation are compared.

Table 1 shows the hardware resource utilization report of the FPGA for H-matrix implementation using log-sigmoid and tan-sigmoid activation functions. For both activation functions about 25% of the available DSP processors are utilized to generate weighted inputs. It is observed that LUTs for tan-sigmoid are more compared to log-sigmoid. This is due to additional combinational logics created in tan-sigmoid for sign magnitude formation.

Equation (6), misclassification error, is used to compute the accuracy of ELM algorithm. It is computed for two modes of implementation, software and hardware, with activation functions, tan-sigmoid and log-sigmoid. Table 2 shows the accuracy obtained from both modes of implementation. In software mode implementation, the accuracy is found to be approximately 98% for both activation functions. Whereas, in hardware mode implementation, the accuracy for log-sigmoid exceeds that of tan-sigmoid by half a percent. The accuracy obtained by hardware implementation will further reduce upon increasing the sample size. This happens because of the linear piecewise approximation of the activation functions in the hardware implementation.

Table 1 Post synthesis utilization report for H-matrix generation

Resources	Tan-sigmoid		Log-sigmoid	
	Utilization	(In %)	Utilization	(In %)
LUT	4054	5.35	3392	6.39
FF	24	0.02	23	0.02
DSP	60	25	60	25
IO	32	15.24	32	15.24

Table 2 Accuracy obtained by various methods of implementation

Mode	Accuracy	
	Tan-sigmoid (%)	Log-sigmoid (%)
Software implementation using MATLAB	98.1	98.4
Hardware implementation using FPGA	97.1	97.7

Figure 3 shows the deviation of each testing sample output from the actual targets for hardware and software implementation of ELM algorithm with tan-sigmoid and log-sigmoid activation functions. The sample points with deviation 1 or −1 depict the error points, i.e., wrongly predicted sample points, whereas those with zero deviation are correctly predicted sample points. It can be observed from the plots that hardware implementation has more error points compared to software implementation, which is consistent with the accuracies obtained.

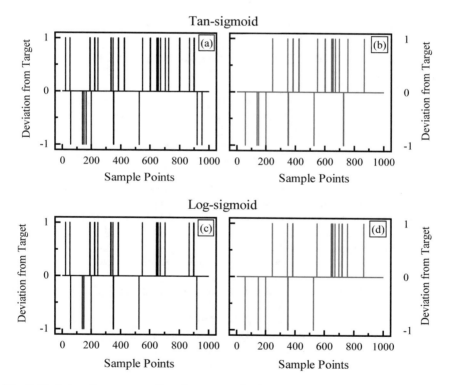

Fig. 3 Deviation from target of each test sample for **a, c** hardware and **b, d** software implementation with tan-sigmoid and log-sigmoid activation function, respectively. Here, in the figure black and red lines correspond to hardware and software mode of implementation, respectively

5 Conclusion

This paper presents hardware and software implementation of 16-bit H-matrix generation and used for ELM algorithm in binary classification problem. Hardware implementation of H-matrix is carried out using hardware description language (HDL). Both modes of implementations are carried out using two activation functions, log-sigmoid and tan-sigmoid. Piecewise model of the activation functions is used for hardware implementation. After generation of H-matrix from the two modes, misclassification error was used for accuracy calculation in MATLAB script. It was observed that the accuracy of log-sigmoid was slightly higher than that of tan-sigmoid in both modes of implementation. It can be concluded from hardware utilization synthesis report for the activation functions that tan-sigmoid has a higher resource usage compared to log-sigmoid. Therefore, log-sigmoid is more efficient compared to tan-sigmoid in terms of both accuracy and hardware utilization. This work can be further extended by implementing β and accuracy calculation on hardware.

References

1. Cortes C, Vapnik V (1995) Support-vector networks. Mach Learn 20(3):273–297
2. Huang G-B, Zhu Q-Y, Siew C-K (2006) Extreme learning machine: theory and applications. Neurocomputing 70(1–3):3–8
3. Finker R, del Campo I, Echanobe J, Martínez V (2014) An intelligent embedded system for real-time adaptive extreme learning machine. In: 2014 IEEE symposium on intelligent embedded systems (IES), Orlando, FL, pp 61–69
4. Gomperts A, Ukil A, Zurfluh F (2011) Development and implementation of parameterized FPGA-based general purpose neural networks for online applications. IEEE Trans Ind Inf 7 (1):78–89
5. Woods L, Teubner J, Alonso G (2011) Real-time pattern matching with FPGAs. In: Proceedings—international conference on data engineering, pp 1292–1295
6. Yeam TC, Ismail N, Mashiko K, Matsuzaki T (2017) FPGA implementation of extreme learning machine system for classification. In: TENCON 2017–2017 IEEE region 10 conference, Penang, pp 1868–1873
7. Huang G-H, Chen L (2007) Convex incremental extreme learning machine. Neurocomputing 70(16–18):3056–3062
8. Huang G-B, Chen L (2008) Enhanced random search based incremental extreme learning machine. Neurocomputing 71(16–18):3460–3468
9. Huang G-B, Zhou H, Ding X, Zhang R (2012) Extreme learning machine for regression and multiclass classification. IEEE Trans Syst Man Cybern B Cybern 42(2):513–529; Author F (2016) Article title. Journal 2(5):99–110
10. UCI machine learning repository: 'Skin Segmentation Dataset' 2010. http://archive.ics.uci.edu/ml
11. Gopal M (2018) Applied machine learning. McGraw Hill, New York

Design of Efficient Approximate Multiplier for Image Processing Applications

C. Sai Revanth Reddy, U. Anil Kumar, and Syed Ershad Ahmed[✉]

Birla Institute of Technology and Science Pilani, Hyderabad Campus,
Hyderabad, India
anilkumaruppugundur@gmail.com, syed@hyderabad.bits-
pilani.ac.in

Abstract. Approximate computing is an emerging paradigm to create energy-efficient computing systems. Most of the image processing applications are inherently error-resilient and can tolerate the error up to a certain limit. In such applications, energy can be saved by pruning the data path modules such as a multiplier. In this paper, we propose a new truncation scheme and an error correction term which are applied to recursive multiplier architecture. Further, truncation method and correction term that compensates the error in the proposed approximate multiplier significantly reduce the area, delay and power. Finally, the proposed multiplier is validated on an image sharpening algorithm. Simulations carried out clearly prove that the proposed multiplier performs better compared to the existing multipliers.

Keywords: Approximate computing · Multiplier · Image sharpening

1 Introduction

Approximate computing is a potential technique that yields savings in computation resources particularly in applications that can tolerate error. This approach is becoming increasingly popular in image and multimedia applications that have energy and speed constraints. Multiplication is an ubiquitous and power-hungry operation in most of these applications. Therefore, in recent years, approximate computing has become an important and emerging research area.

A typical multiplier includes steps such as partial product generation, partial product reduction and carry propagate addition. Most of the previous work focuses on simplifying the partial product reduction stage; however, approximations can be carried out in any of the steps. For instance, a well-known technique [1] in which the lower significant partial products are not formed and the error is compensated using a correction function. Other techniques try to simplify the multi-operand addition at partial product reduction stage by using approximate compressors [2].

In this work, we propose a technique with an objective to reduce the area, delay and power at partial product generation and reduction stage. In the partial product generation stage, the lower significant partial products are not formed, and the correction is done using a constant correction function. Finally, the proposed method aims to reduce area, delay and power while sacrificing accuracy to a tolerable limit.

© Springer Nature Singapore Pte Ltd. 2020
N. Goel et al. (eds.), *Modelling, Simulation and Intelligent Computing*,
Lecture Notes in Electrical Engineering 659,
https://doi.org/10.1007/978-981-15-4775-1_55

The rest of this paper is structured as follows. Section 2 reviews the existing literature, while the proposed approximate multiplier is explained in Sect. 3. A detailed comparison of error, synthesis results and mapping of proposed multiplier on image sharpening application is carried out in Sect. 4. Finally, the conclusions are drawn in Sect. 5.

2 Related Work

Work by Kulkarni et al. [3] presented a 2 * 2 or 4 * 4 sub-multiplier to simplify the partial product matrix. Most of the papers try to reduce the computation complexity at the partial product reduction stage. One such multiplier, broken-array multiplier (BAM), was proposed by Mahdiani et al. [4]. In this design, the optimization is achieved by truncating the carry save adders used at accumulation mode. An approximate Wallace tree multiplier (AWTM) was presented by Bhardwaj et al. [5] with carry-in speculation for building recursive blocks. The AWTM uses accurate multiplier blocks for MSB calculation, while the LSB portion is computed using in-exact modules. Consequently, AWTM design is faster, occupies less area and consumes less power. Yang et al. [6] try to achieve a low error rate by exploiting the probabilistic characteristic of AND gate output that is used at partial product generation stage. Ha and Lee [7] achieve better error rate by adding a simple error correction circuit to Yang design.

2.1 Preliminaries

Most of the multiplier architectures with truncation schemes are either tree-based, array-based or recursive-based approaches. However, due to regular structure, recursive and array multiplier schemes based on binary number system are considered in this work. Recursive multipliers [8] have the regularity of array multipliers and delay almost equal to tree multipliers. These characteristics make it the most suitable candidate for VLSI implementation of multiplication. Given the advantages of recursive multiplier architecture, in this work, we adopt recursive multiplication for our approximate multiplier. The recursive multiplication algorithm is described in the below subsection.

2.2 Existing Recursive Multiplier

A typical recursive multiplier architecture has a hierarchical structure composed of several sub-multipliers. Let N and M be two $2p$-bit unsigned binary numbers, where N is the multiplicand and M is the multiplier. They can be expressed as follows [9]:

$$N = \sum_{i=0}^{2p-1} n_i \cdot 2^i \tag{1}$$

$$M = \sum_{i=0}^{2p-1} m_i \cdot 2^i \qquad (2)$$

The recursive multiplication is performed by splitting each of the binary number into two equal individual parts of p-bit each. Accordingly, the operand N is split into N_H and N_L, while the operand M is split into M_H and M_L, respectively, as mentioned in Eqs. (3–4). Where the subscript H denotes the most significant portion and L denotes lower significant portion of the respective binary numbers.

$$N = N_H * 2^p + N_L \qquad (3)$$

$$M = M_H * 2^p + M_L \qquad (4)$$

The product (R) of N and M can be written in the following form:

$$R = N \cdot M = (N_H * 2^p + N_L)(M_H * 2^p + M_L)$$
$$= N_H M_H * 2^{2p} + (N_H M_L + N_L M_H) * 2^p + N_L M_L$$

Thus, the $2p * 2p$ bit multiplication of operands M and N is accomplished by four $p * p$ multiplications in parallel. The multiplication of these individual portions is carried out by four $p * p$ binary sub-multipliers namely $N_H * M_H$, $N_H * M_L$, $N_L * M_H$ and $N_L * M_L$. The resultant partial products from the individual sub-multipliers are reduced to final product (R) of $4p$-bit using a reduction tree [9].

3   Proposed Multiplier

In this work, we propose a new truncation scheme and a constant correction term to compensate the error in a recursive multiplier architecture with an objective to reduce the area, delay and power at partial product generation and reduction stage.

The proposed approach is explained using $8 * 8$ recursive multiplier architecture, and the same can be extended to higher width multiplier. Consider a typical $8 * 8$-bit recursive multiplier with sub-multiplier of size $4 * 4$. The sub-multiplier $N_H M_H$ forms the most significant portion (MSP), while $N_H M_L$ and $N_L M_H$ form intermediate significant portion (ISP). The sub-multiplier $N_L M_L$, the least significant portion (LSP), contributes the least compared to other sub-multipliers to the final product.

Since approximating $N_H M_H$ will result in large error, it is implemented using exact multiplier architecture [10]. The proposed work attempts to prune the sub-multipliers ($N_H M_L$, $N_L M_H$ and $N_L M_L$) based on the positional weight. The truncation of a portion of multiplier leads to error in the result and has to be compensated. Therefore, we propose a constant correction term for $N_L M_L$ after truncating the major portion of the sub-multiplier as discussed in Sect. 3.1. Similarly, the sub-multipliers ($N_H M_L$ and $N_L M_H$) are pruned, and a constant correction term is used as discussed in the below Sect. 3.2.

3.1 Constant Correction Applied to $N_L M_L$ Sub-multiplier

Figure 1a shows the 4 * 4 $N_L M_L$ sub-multiplier generating partial products (PPs) denoted with solid dots. These PPs can be reduced using an adder tree to obtain the result, $x_7 x_6 x_5 x_4 x_3 x_2 x_1 x_0$. In order to save hardware, the sub-multiplier $N_L M_L$ has to be truncated; however, it will result in large error. For this reason, in this work, we follow the below-mentioned approach. Since $x_7 x_6 x_5$ forms the most significant result and also requires less hardware, the respective PPs are formed. The partial product matrix enclosed in the dotted rectangular box consumes major area compared to the other partial products. Hence, truncating these will aid in reducing the hardware complexity. However, this will result in reducing the accuracy of the multiplier.

To compensate this, an appropriate correction term has to be added. The correction function is achieved in a systematic manner as discussed below.

3.2 Mathematical Analysis of Constant Correction Term

In order to estimate the constant term, let us consider the following procedure. Consider an event 'E' with all possible outcomes $[n_1, n_2, n_3, \ldots, n_i]$. Suppose we have to estimate a constant 'K' for the outcome of event 'E' so that the average error between estimated outcome and actual outcome will be zero.

$$\text{Average Error} = (K - n_1 + K - n_2 + \cdots + K - n_i)/i$$
$$= (i * K - (n_1 + n_2 + n_3 + \cdots + n_i))/i$$

Since the requirement is to achieve zero average.

$$\text{Accordingly, } i * K = n_1 + n_2 + n_3 + \cdots + n_i,$$
$$K = (n_1 + n_2 + n_3 + \cdots + n_i)/i$$

Therefore, K is the average of all possible outcomes.

The above-mentioned concept has been applied in obtaining the error correction term for highlighted dotted portion in $N_L M_L$, where the truncated part is the event E is,

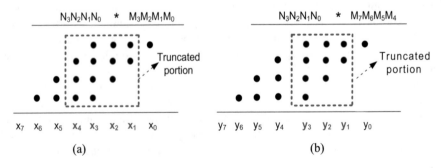

Fig. 1 **a** Partial product matrix of sub-multiplier $N_L M_L$ **b** Partial product matrix of sub-multiplier $N_L M_H$

and the sum generated from the truncation part for all possible input combinations from the sample space [0:255] are the outcomes. Therefore, the correction term to compensate the truncated part is the average value of the resultant sum of the truncated block considering all the possible input combinations. The calculated correction term in $N_L M_L$ is $(12)_{10}$ which is 1100 in binary. Accordingly, we replace $x_4 x_3 x_2 x_1$ with constant correction binary term 1100, while $x_7 x_6 x_5$ are calculated using exact adders. Further, including x_0 as part of truncation would require a non-integer $(24.25)_{10}$ as a correction term which is difficult to implement and hence that PP has not been truncated. The similar concept can be extended to obtain constant correction term for $N_L M_L$ sub-multipliers of larger width.

3.3   Constant Correction Applied to $N_L M_H$ and $N_H M_L$ Sub-multiplier

The truncation technique applied to $N_L M_L$ sub-multiplier is extended to $N_L M_H$ and $N_H M_L$. The truncation method in intermediate significant portion ($N_L M_H$ and $N_H M_L$) is explained using the sub-multiplier $N_L M_H$ which is shown in Fig. 1b, and the same can be applied to $N_H M_L$. It can be noted that the sub-multiplier $N_L M_H$ contributes more to the final product than the sub-multiplier $N_L M_L$. Therefore, PPs in $N_L M_H$ that contribute most significant result $y_7 y_6 y_5 y_4$ are retained, while partial product matrix enclosed in the highlighted in the rectangular box is eliminated, and the product terms $y_3 y_2 y_1$ are compensated using the constant correction function obtained by computing the partial products that are not formed. By following the approach mentioned in Sect. 3.2, the correction term for sub-multiplier $N_H M_L$ is computed which will be equal to 110 in binary.

Finally, the resultant products of $N_H M_H$, $N_L M_H$, $N_L M_H$, $N_L M_H$ sub-multipliers are further reduced using carry save adders into two rows and then finally to a product ($r_{15} - r_0$) using ripple carry adders.

4   Experimental Results

4.1   Error Analysis of Approximate Multipliers

Exhaustive error analysis has been carried out on various multiplier architectures including proposed multiplier using 65,536 and 1 million random cases for 8-bit and 16-bit multipliers, respectively. Error metrics such as error rate (ER), normalized mean error distance (NMED), mean relative error distance (MRED) and average error are computed and tabulated in Tables 1 and 2.

It is observed from Tables 1 and 2 that the average error of proposed multiplier is the lowest compared to all the existing multipliers. The NMED of proposed multiplier is low compared to existing multipliers except Yang et al. and Ha et al.

Though Yang et al. and Ha et al. multipliers achieve low NMED compared to proposed multiplier, they tend to occupy larger area, dissipate more power and are prone to higher delay as discussed in Sect. 4.2.

Table 1 Error analysis of various 8-bit approximate multipliers

Multiplier	Error rate (%)	MRED (%)	NMED (10^{-3})	Average error (10^4)
BAM-8 [4]	98.82	9.78	6.9	0.0451
BAM-9 [4]	99.51	16.34	13.9	0.0903
BAM-10 [4]	98.73	25.47	25.8	0.1677
BAM-12 [4]	99.15	53.01	77.4	0.5031
TruM-4 [11]	99.63	25.493	28.845	0.1870
AWTM-4 [5]	99.88	23.36	6.02	0.0385
Yang et al. [6]	82.91	0.76	0.48	0.00128
Ha and Lee [7]	82.91	0.78	0.43	0.00102
Proposed multiplier	99.69	16.35	2.8	0

Table 2 Error analysis of various 16-bit approximate multipliers

Multiplier	Error rate (%)	MRED (%)	NMED (10^{-3})	Average error (10^4)
BAM-16 [4]	99.99	0.21	0.06	24.6
BAM-17 [4]	99.99	0.36	0.11	49.2
BAM-18 [4]	99.99	0.63	0.22	95.0
BAM-20 [4]	100	1.79	0.79	337.5
TruM-8 [11]	100	2.85	1.94	834.1
AWTM-4 [5]	99.94	0.33	0.02	8.6
Yang et al. [6]	98.83	0.014	0.0049	3.44
Ha and Lee [7]	98.75	0.016	0.005	1.85
Proposed multiplier	99.99	0.58	0.017	0.0045

4.2 Hardware Synthesis

The area, delay and power of various existing multiplier schemes including the proposed one have been investigated at 180 nm technology using TSMC library and are tabulated in Table 3. An exhaustive comparison of these multiplier architectures in terms of area, delay and power is provided in Table 3. It can be observed that the proposed multiplier performs better in terms of area (31–40%), delay (21–66%) and power (40–43%) in comparison with existing designs [5–7]. Though the proposed design occupies larger area compared to BAM [4] and TruM-4 [11], they suffer from large average error and MRED. Finally, it can be concluded that proposed multiplier performs better in terms of area, delay and power compared to the best performing schemes [5–7] in terms of error metrics.

Table 3 Synthesis results of various 8-bit approximate multipliers

Multiplier	Area (μm²)	Delay (ps)	Power (nW)
BAM-8 [4]	521.4	2447	15,055
BAM-9 [4]	363.3	1722	9622
BAM-10 [4]	247.6	1691	5915
BAM-12 [4]	76.2	692	1484
TruM-4 [11]	234.2	1420	4989
AWTM-4 [5]	1052.7	3354	29,931
Yang et al. [6]	951.14	2637	31,093
Ha and Lee [7]	925	2557	30,171
Proposed multiplier	635	2018	17,418

4.3 Bench Marking Application

To validate the proposed multiplier in image processing applications, we chose image sharpening algorithm [12]. Accordingly, the exact multiplications are replaced with proposed approximate one, while all other operations are carried out using 'accurate' techniques.

Figures 2a–c and 3a–c depict the images sharpened using exact and proposed approximate multiplier. It can be observed that images processed using exact and proposed multiplier are almost similar.

(a) Input Image

(b) Exact Multiplier Output

(c)   Proposed Multiplier Output

Fig. 2 Lena output images of exact and proposed multiplier after image sharpening

(a) Input Image

(b) Exact Multiplier Output

(c)   Proposed Multiplier Output

Fig. 3 Cameraman output images of exact and proposed multiplier after image sharpening

5   Conclusions

This paper presents a new truncation scheme and an error correction term which is applied to recursive multiplier architecture. The correction term aids in obtaining better average error, while the truncation method reduces the area, delay and power. The design is implemented in 0.18 μm CMOS process, and the savings in area, delay and power up to 40%, 66% and 43%, respectively, are achieved compared to existing designs. Finally, proposed design is validated on image sharpening algorithm, and results prove that the quality of image obtained is almost similar to exact multiplier.

References

1. Biswas K, Wu H, Ahmadi M (2006) Fixed-width multi-level recursive multipliers. In: 2006 Fortieth Asilomar conference on signals, systems and computers. IEEE
2. Townsend WJ, Swartzlander E Jr, Abraham JA (2003) A comparison of Dadda and Wallace multiplier delays. In: Advanced signal processing algorithms, architectures, and implementations XIII, vol 5205. International Society for Optics and Photonics
3. Kulkarni P, Gupta P, Ercegovac M (2011) Trading accuracy for power with an underdesigned multiplier architecture. In: 2011 24th International conference on VLSI design. IEEE, pp 346–351
4. Mahdiani HR, Ahmadi A, Fakhraie SM, Lucas C (2009) Bioinspired imprecise computational blocks for efficient VLSI implementation of soft-computing applications. IEEE Trans Circ Syst I Regul Pap 57(4):850–862
5. Bhardwaj K, Mane PS, Henkel J (2014) Power-and area-efficient approximate Wallace tree multiplier for error-resilient systems. In: Fifteenth international symposium on quality electronic design. IEEE, pp 263–269
6. Yang Z, Han J, Lombardi F (2015) Approximate compressors for error-resilient multiplier design. In: 2015 IEEE international symposium on defect and fault tolerance in VLSI and nanotechnology systems (DFTS). IEEE, pp 183–186
7. Ha M, Lee S (2017) Multipliers with approximate 4–2 compressors and error recovery modules. IEEE Embed Syst Lett 10(1):6–9
8. Danysh AN, Swartzlander EE (1998) A recursive fast multiplier. In: Conference record of thirty-second Asilomar conference on signals, systems and computers (Cat. No. 98CH36284), vol 1. IEEE
9. Karatsuba A, Ofman Y (1963) Multiplication of multidigit numbers on automata. Sov Phys Dokl 7
10. Wallace CS (1964) A suggestion for a fast multiplier. IEEE Trans Electron Comput 1:14–17
11. Parhami B (2010) Computer arithmetic, vol 20. Oxford University Press, Oxford
12. Lau MSK, Ling K-V, Chu Y-C (2009) Energy-aware probabilistic multiplier: design and analysis. In: Proceedings of the 2009 international conference on compilers, architecture, and synthesis for embedded systems. ACM

Study of Performance of Ant Bee Colony Optimized Fuzzy PID Controller to Control Two-Link Robotic Manipulator with Payload

Alka Agrawal[1]($\boxtimes$), Vishal Goyal[1] (ORCID), and Puneet Mishra[2] (ORCID)

[1] Department of ECE, GLA University, 17km Stone, NH-2, Mathura 281406, India
alka.agrawal@gla.ac.in
[2] Department of EEE, Birla Institute of Technology and Science Pilani, Pilani Campus, Pilani, Rajasthan 333031, India

Abstract. Two-link robotic manipulator system with payload at tip is a highly complex and nonlinear system and faces a challenging task to control. Thus, a nonlinear proportional–integral–derivative (PID) controller is implemented in this paper using fuzzy logic where the parameters of the controller are optimized with a new metaheuristic algorithm based on the foraging behavior of the swarm of bees. The performance indices function to minimize the error between the reference signal, and the system's output is taken as the integral of absolute error (IAE). The implemented controller is compared with the conventional PID controller. From the simulation studies, it is found that the implemented fuzzy PID controller works more efficiently than the PID controller in terms of the trajectory tracking, in the presence of parametric uncertainties as well as disturbance rejection and the noise suppression.

Keywords: PID · Fuzzy · Optimization · IAE · ABC · Robotic manipulator system

1 Introduction

Robotic manipulators are mechanized devices to mimic human behavior in order to perform a continuous operation to fulfill various tasks like picking and placing of an object in industries, welding, assembling in automobile industries, handling of radioactive and bio-hazardous materials in nuclear plants, assistance in surgery in medical fields, etc. [1]. Since the dynamics of robotic manipulators is highly complex and nonlinear as well as it is associated with unavoidable structured and unstructured uncertainties like parameter variations, external disturbances, friction, noise, etc., it is a challenging task for researchers to effectively control the end effector of a manipulator to follow a desired trajectory [2]. Since the invention of proportional–integral–derivative (PID) controller in 1910 [3], the PID controller has been extensively used in industries due to its simple structure and effectively controls many real-world problems. In literature, different implementations of PID controller to control robotic manipulators have been cited by the researchers [4, 5]. In recent years, researchers showed that the nonlinearities must be included and modeled to study the system satisfactorily as they are inevitable and intrinsic for the system [6]. Now, researchers have been attracted toward the intelligent process control due to the progress in the area of fuzzy logic control (FLC), neural network (NN) and the genetic algorithm (GA) [7]. In 1965, Zadeh gave the concept of fuzzy logic based on fuzzy sets [8]. Not the requirement of exact mathematical modeling, incorporation of human expertise into the design and dealing with the uncertainties are the features, which make the FLC as an efficient controller to control the highly complex, nonlinear systems under parameter variations as well as external disturbances. Mamdani [9] reported a successful application of FLC in a laboratory scale process. Since then, different structures of the FLC have been developed by the researchers for the linear as well as nonlinear are presented in the literature. An improved fuzzy PI controller was introduced in [10], and it was proved that the tracking performance of the controller is much good as the conventional PI controller. A new hybrid fuzzy PI + conventional D was presented in [11] which was proved to be much robust as compared to a PID controller.

In addition to the implementation of the controller, it is very essential to properly determine the parameters of the FLC to make it more robust, and hence, a vast research has been done in the optimization of the parameters of a controller. As computational intelligence keeps on improving, the intelligent optimization technique has also been used for estimate the parameters of FLC controller. In this paper, a comparative study of the PID controller with fuzzy PID controller is done which is optimized with a highly efficient metaheuristic ant bee colony optimization technique to justify the superiority of fuzzy PID controller over PID controller. The implemented controllers are applied to highly complex and nonlinear two-link manipulator system with payload. The superiority of the fuzzy PID controller is also discussed in the case of parameter variations, disturbance rejection and noise suppression. It was found that the fuzzy PID controller overruled the PID controller in all the cases. This paper has six sections as follows: In the first section, introduction of the paper is given, second section shows the mathematical modeling of the system, third section gives the details of the PID controller, and fourth section explains the fuzzy logic controller design and optimization technique. Fifth section represents the simulation results, and sixth part gives conclusion of the paper.

2   Dynamic Model of the System

In this paper, the two-link robotic manipulator system with payload has been used as shown in Fig. 1. The mathematical model of the system is described in Eq. (1) given below [12].

$$\begin{bmatrix} S_{11} & S_{12} \\ S_{21} & S_{22} \end{bmatrix} \begin{bmatrix} \ddot{\theta}_{11} \\ \ddot{\theta}_{22} \end{bmatrix} + \begin{bmatrix} P_{11} \\ P_{21} \end{bmatrix} + \begin{bmatrix} f_{r1} \\ f_{r2} \end{bmatrix} + \begin{bmatrix} f_{n1p} \\ f_{n2p} \end{bmatrix} = \begin{bmatrix} \tau_1 \\ \tau_2 \end{bmatrix} \tag{1}$$

where θ_1 and θ_2 are the position of the links, τ_1 and τ_2 are the generated torques, and f_{r1} and f_{r2} represent dynamic friction. The parameters values are shown below in Table 1 [12].

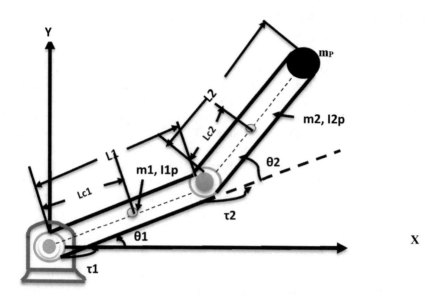

Fig. 1 Two-link planar robotic manipulator system with payload at tip

Table 1 Parameter values for link 1 and link 2 for the system

Link 1	Values	Link 2	Values
m_1	0.392924 kg	m_2	0.094403 kg
l_1	0.2032 m	l_2	0.1524 m
l_{c1}	0.104648 m	l_{c2}	0.081788 m
b_{1p}	0.141231 m/radian/s	b_{2p}	0.3530776 m/radian/s
I_{1p}	0.0011411 kg-m^2	I_{2p}	0.0020247 kg-m^2
m_p = 0.566699 kg			
g = 9.81 m/s^2			

3   PID Controller

Reference [13] gives the conventional PID controller in time domain as in Eq. (2) given below:

$$u(t) = k_\mathrm{p}e(t) + k_\mathrm{I} \int e(t)\mathrm{d}t + k_\mathrm{D}\frac{\mathrm{d}e(t)}{\mathrm{d}t} \tag{2}$$

where $u(t)$ is the output of the controller, and $e(t)$ is the instantaneous error signal obtained as the difference between the reference signal and the plant actual output. The three terms k_p, k_I and k_D are the proportional gain, integral gain and differential gain parameters of the controller.

4 Design of Fuzzy PID Controller for the Two-Link Manipulator System

In this section, implementation of fuzzy PID (FPID) controller has been described which is used to generate desired torque to effectively deflect the links of the system to move the end effector in a prescribed trajectory. Figure 2 shows the block diagram of the control system for the two-link manipulator system with payload using FPID. It depicts that the fuzzy PID is the combination of fuzzy PI and fuzzy PD controller. The system requires two inputs, i.e., error and rate of change of error. At the output of fuzzy logic controller (FLC), the output is integrated and added again with the output of the FLC to form fuzzy PID controller. K_{P1}, K_{D1}, K_{PI1} and K_{PD1} are the parameters of the controller for link 1, and K_{P2}, K_{D2}, K_{PI2} and K_{PD2} are the parameters of the controller for link 2 as shown in Fig. 2.

4.1 Design of Fuzzy Logic Controller

Fuzzy logic controller (FLC) performs in three steps. First step is the fuzzification in which the crisp input data is converted into the fuzzy data with the help of membership functions defined in the range $(-1, 1)$. In this work, seven membership functions with triangular shape as shown in Fig. 3 have been used. In the second step, a two-dimensional rule base is formed based on human knowledge using if–then, which is the principal part of a FLC. Table 2 shows the rules formed for the membership functions for the error and the derivative of the error. In addition to this, Mamdani inference mechanism is used which applies min–max composition for implication and aggregation using logical AND-OR operator. At last, in the third step, the fuzzified data is converted into again the crisp data by using the center of gravity method [14].

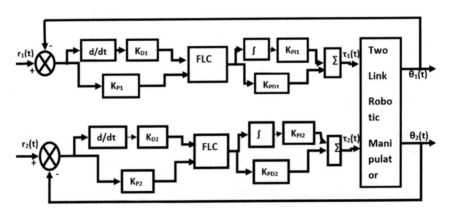

Fig. 2 Block diagram of PID controller using fuzzy logic for two-link robotic manipulator with payload

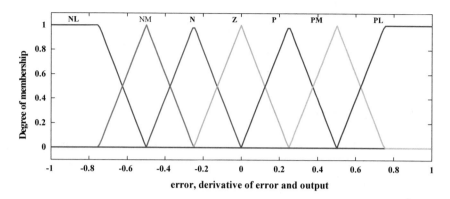

Fig. 3 Triangular membership functions for fuzzification of error, the derivative of error and output

Table 2 Input–output rule base for FLC [1]

E \ DE	NL	NM	N	Z	P	PM	PL
NL	NL	NL	NL	NM	N	N	Z
NM	NL	NM	NM	NM	N	Z	P
N	NL	NM	N	N	Z	P	PM
Z	NL	NM	N	Z	P	PM	PL
P	NM	N	Z	P	P	PM	PL
PM	N	Z	P	PM	PM	PM	PL
PL	Z	P	P	PM	PL	PL	PL

4.2 Optimization

In 2005, D. Karaboga having inspired by the foraging behavior of the honeybees introduced ant bee colony (ABC) algorithm, which is a swarm intelligence-based optimization algorithm [15]. In this work, the fitness function (fit) has been chosen as the sum of the integral of absolute error (IAE) of the links chosen given by the formula as shown in Eq. (3).

$$\text{fit} = \int_0^t |e_1(t)| \mathrm{d}t + \int_0^t |e_2(t)| \mathrm{d}t \tag{3}$$

where $e_1(t)$ is the instantaneous error signal between the reference signal $r_1(t)$ and the plant's output $\theta_1(t)$, and $e_2(t)$ is the instantaneous error signal between the reference signal $r_2(t)$ and the plant's output $\theta_2(t)$ as given in Eqs. (4) and (5).

$$e_1(t) = r_1(t) - \theta_1(t) \tag{4}$$

$$e_2(t) = r_2(t) - \theta_2(t) \tag{5}$$

5 Simulation Results

The parameters of the controller obtained by the ABC algorithm minimize the fitness function (fit). The desired trajectory to be followed is taken as given in Eqs. (6) and (7):

$$r_1(t) = -2\sin((\pi/3)t) \tag{6}$$

$$r_2(t) = 2\sin((\pi/2)t) \tag{7}$$

where $r_1(t)$ is the desired position for the link 1, and $r_2(t)$ is the desired position for the link 2. The settings of the parameters of the ABC for optimization are as follows: no. of population—40, food number—20, limit—100, max. cycle—100, lower bound—1 and upper bound—500. Figure 4 shows the convergence curve for the optimization technique which minimizes the value of the fitness function to follow the desired trajectory by both controllers. It is clear from the figure that the FPID controller acquires less value of the fitness function as compared to the PID controller and hence follows the desired path more effectively and efficiently than the PID. After optimization for trajectory tracking, the obtained parameters for the PID as well as fuzzy PID for both links are given in Table 3, and the overall fitness value of PID is found to be 3.38e−2 while of FPID is 13e−4. It depicts that the FPID controller is more efficient than the PID controller for trajectory tracking. Figure 5 shows the trajectory-tracking curve of the output of the link 1 and link 2.

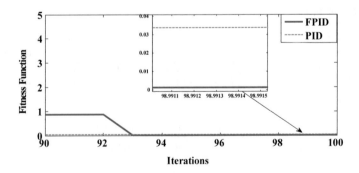

Fig. 4 Convergence curve for the fitness function (IAE) of the ABC optimized fuzzy PID and PID

Table 3 Parameter values for the controllers and the IAE values achieved after optimization

FPID		PID	
Link 1	Link 2	Link 1	Link 2
$K_{P1} = 48.66$	$K_{P2} = 106.61$	$K_{P1} = 500$	$K_{P2} = 500$
$K_{D1} = 0.1$	$K_{D2} = 0.1$	$K_{D1} = 2.15$	$K_{D2} = 0.1$
$K_{PI1} = 499.15$	$K_{PI2} = 231.22$	$K_{I1} = 500$	$K_{I2} = 500$
$K_{PD1} = 71.22$	$K_{PD2} = 61.8$		
IAE1 = 8.07e−4	IAE2 = 5.19e−4	IAE1 = 1.89e−2	IAE2 = 1.48e−2

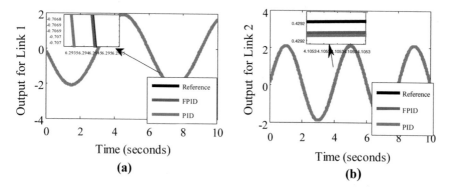

Fig. 5 Output curve during trajectory tracking for **a** link 1 **b** link 2

5.1 Testing Against the Parameters Uncertainties

The PID controller and the fuzzy PID controllers are simulated by varying the parameter values of the system like +5% change in the masses of the links individually, +5% variation in the lengths of the links as well as +5% variation in the payload mass. The obtained IAE values for the links are listed in Table 4. The IAE values show that the FPID controller acquires less value of error as compared to PID even the parameters of the system are varying while the parameters of the controller are kept constant.

Table 4 IAE and fitness function values for link 1 and link 2 during parameters variations

Parameters variations	IAE for link 1		IAE for link 2		Fitness value (fit)	
	PID	FPID	PID	FPID	PID	FPID
m_1	1.91e−2	8.1955e−4	1.48e−2	4.9953e−4	3.39e−2	13e−4
m_2	1.90e−2	8.4432e−4	1.48e−2	5.7382e−4	3.38e−2	14e−4
m_p	1.96e−2	8.4377e−4	1.50e−2	5.1144e−4	3.46e−2	14e−4
l_1	1.94e−2	8.8069e−4	1.48e−2	5.6941e−4	3.42e−2	15e−4
l_2	1.92e−4	8.2917e−4	1.50e−2	5.1405e−4	3.42e−2	13e−4

5.2   Testing for the Disturbance Rejection and Noise Suppression

For robustness testing of the controller, a disturbance signal as given in Eq. (8) is applied to the controller outputs of both links simultaneously for 2 s, a noise signal as shown in Eq. (9) is accumulated at the output, and the results are shown in Table 5.

Table 5 IAE and fitness function values for link 1 and link 2 during disturbance and noise accumulation

	IAE for link 1		IAE for link 2		Fitness value (fit)	
	PID	FPID	PID	FPID	PID	FPID
Disturbance	2.71e−2	11e−4	2.96e−2	8.2277e−4	5.68e−2	19e−4
Noise	1.93e−2	37e−4	1.50e−2	30e−4	3.44e−2	67e−4

$$d(t) = 5(u(t-5) - u(t-7)) \tag{8}$$

$$n(t) = r(t) * \text{random} * 0.001 \tag{9}$$

6   Conclusion

Aim of the work done in this paper is to study and verify the efficacy and robustness of a nonlinear PID controller using fuzzy logic as well as an appropriate optimization technique to control a highly complex and nonlinear two-link robotic manipulator system. In this paper, it is shown that the FPID is more accurate as the IAE values of the links of the manipulator controlled with FPID are having very much less value as compared to the PID controller. In addition to this, the ant bee colony optimization technique used in this paper proved to give optimum parameters to get better results. Further, in the study of the controller in the presence of parametric uncertainties, disturbance and noise, it is found that the FPID overruled the PID controller by giving lower IAE values. The feasibility of the implemented controller in this work could be checked for more complex nonlinear systems like three-link manipulator systems, etc., in the future.

References

1. Kumar J, Kumar V, Rana KPS (2019) Fractional-order self-tuned fuzzy PID controller for three-link robotic manipulator system. Neural Comput Appl
2. Refoufi S, Benmahammed K (2018) Control of a manipulator robot by neuro-fuzzy subsets form approach control optimized by the genetic algorithms. ISA Trans
3. Ang K, Chong G, Li Y (2005) PID control system analysis design and technology. IEEE Trans Control Syst Technol 13(4):559–576

4. Ayala HVH, Coelho LDS (2012) Tuning of PID controller based on a multi objective genetic algorithm applied to a robotic manipulator. Expert Syst Appl 39:8968–8974
5. Li X, Yu W (2011) A systematic tuning method of PID controller for robot manipulators. In: 9th IEEE international conference on control and automation, pp 274–279
6. Goyal V, Mishra P, Deolia VK (2019) A robust fractional order parallel control structure for flow control using a pneumatic control valve with nonlinear and uncertain dynamics. Arab J Sci Eng 44(3):2597–2611
7. Goyal V, Mishra P, Shukla A, Deolia VK (2019) A fractional order parallel control structure tuned with meta-heuristic optimization algorithms for enhanced robustness. J Electr Eng 70 (1):16–24
8. Zadeh LA (1973) Outline of a new approach to the analysis of complex systems and decision processes. IEEE Trans Syst Man Cybern 3(1):28–44
9. Mamdani EH (1977) Applications of fuzzy logic to approximate reasoning using linguistic synthesis. IEEE Trans Comput 26(12):1182–1191
10. Tang W, Chen G (1994) A robust fuzzy PI controller for a flexible joint robot arm with uncertainties. In: Proceedings FUZZ-IEEE, Orlando, FL, pp 1554–1559
11. Kumar J, Kumar V, Rana KPS (2018) A fractional order fuzzy PD + I controller for three-link electrically driven rigid robotic manipulator system. J Intell Fuzzy Syst 35(5):5301–5315
12. Sharma R, Bhasin S, Gaur P, Joshi D (2019) A switching based collaborative fractional order fuzzy logic controllers for robotic manipulators. Appl Math Modell
13. Agrawal A, Goyal V, Mishra P (2019) Adaptive control of a nonlinear surge tank-level system using neural network-based PID controller. In: Malik H, Srivastava S, Sood Y, Ahmad A (eds) Applications of artificial intelligence techniques in engineering. Advances in intelligent systems and computing, vol 698. Springer, Singapore
14. Behera L, Kar I (2010) Intelligent systems and control principles and applications. Oxford University Press Inc., Oxford
15. Gao W, Liu S (2011) A modified artificial bee colony algorithm. Comput Opera Res

Investigating the Impact of BTI and HCI on Log-Domain Based Mihalas–Niebur Neuron Circuit

Shaik Jani Babu[1]([⊠]) , Anish Vipperla[1] ,
Haarica Vinayaga Murthy[1] , Chintakindi Sandhya[1] ,
Siona Menezes Picardo[2] , Sonal Singhal[1] , and Nilesh Goel[2]

[1] Shiv Nadar University, Greater Noida, India
skjanibabu786@gmail.com
[2] Birla Institute of Technology and Science Pilani, Dubai Campus, Dubai, UAE
goel.nilesh@gmail.com

Abstract. Neuromorphic circuits are becoming quite popular due to their ability to mimic the structure and behavior of human brain. Current research focuses on approximating spiking biological neuron behavior. Various neuron models have been proposed in the past that aid in investigating the behavior of neuronal systems mathematically. Mihalas–Niebur (MN) neuron model is one among them. In this paper log-domain based MN neuron model is implemented at 45 nm technology node. The paper studies the effects of process-temperature variations and also investigates the impact of Hot Carrier Injection (HCI), Bias Temperature Instability (BTI) on the performance of MN circuit. Average power consumption and spiking frequency are chosen as key performance measures to analyze the circuit performance before and after degradation.

Keywords: Neuromorphic circuits · Mihalas-Niebur neuron · Process corner · Bias temperature instability · Hot carrier injection

1 Introduction

Neuromorphic computing has been referred to a variety of brain-inspired computers, devices, and models that differentiate the predominant Von Neumann architecture. Neuromorphic engineering aims to emulate human cognition enabling architectures to deal with problems such as uncertainty and ambiguity. Mimicking the brain would give them the ability to adapt and learn from unstructured stimuli with the energy efficiency of the human brain [1]. Neuromorphic chips have potential to accomplish tasks such as image and pattern recognition quite efficiently which is still challenging for modern Von Neumann based hardware architectures. Bulk of the current research in neuromorphic circuitry is aimed at approximating spiking biological neuron behavior. Mathematical descriptions of neural dynamics have been introduced by Hodgkin and Huxley model. Although it is capable of describing several types of neuron behaviors, it suffers from the requirement of a large number of parameters resulting in increased circuit complexity. In order to overcome this obstacle, the generalized Leaky Integrate-and-Fire (LIF) model is

© Springer Nature Singapore Pte Ltd. 2020
N. Goel et al. (eds.), *Modelling, Simulation and Intelligent Computing*,
Lecture Notes in Electrical Engineering 659,
https://doi.org/10.1007/978-981-15-4775-1_57

often utilized as it captures fundamental properties of neurons such as integration, threshold adaptation, etc. using computationally efficient topologies [2–4].

Mihalas-Niebur neuron is a popular neuron model that replicates the bursting and spiking behaviors of a biological neuron [4–7]. Many architectures based on Hodgkin–Huxley model, in an attempt to provide an accurate description of the neuron, turn out to be quite complex in architecture. As a result, they occupy a large silicon area on a neuromorphic chip [5]. On the other hand, Leaky-Integrate and Fire models have simplistic architecture but fail to reproduce a few characteristics of biological neurons. MN neuron aims to find a balance between both of them. It tries to mimic biological neurons as closely as possible while having a simplistic architecture of a LIF neuron.

These advantages of MN-neuron model are the key motivations of paper to implement the circuit at 45 nm technology node and study the performance degradation due to aging. Once the design functionality is satisfied, device-level simulations are needed to be performed for device degradation due to temperature and aging caused by stress from Bias-Temperature Instability (BTI), Hot Carrier Injection (HCI), etc. in n-channel and p-channel MOSFET devices. Shifting towards lower technology nodes, factors such as degradation due to HCI/BTI stress become more prominent for analog circuits [8, 9].

Reliability of analog circuits fabricated using CMOS in the deep-submicron range is also significantly affected by Process, voltage, and temperature (PVT) variations. Process variations occur due to deviations in semiconductor fabrication process. This would result in variations of key parameters of the circuit like threshold voltage, oxide thickness, etc. [10]. PVT variations along with aging mechanisms lead to lifetime degradation of device performance and deviate from intended values. The increasing variability of technology parameters causes mismatches and yield problems. Accurate prediction of aging-induced performance degradation is a significant aspect that needs to be taken into consideration right from the design stage in order to avoid catastrophic chip failures and expensive design re-spins.

The paper is organized as follows: Section 2 briefly describes the MN Neuron model and its CMOS implementation. Section 3 outlines an approach for the reliability analysis of the circuit. Section 4 presents the simulation results obtained. Section 5 provides conclusions.

2 Mihalas–Niebur (MN) Neuron Model

This section discusses Mihalas–Niebur (MN) neuron model and its CMOS based implementation using first-order differential equations.

Mihalas and Niebur introduced a neuron model that produces a wide range of firing patterns that closely mimic the spiking in real neurons [4]. This model uses simple first-order differential equations to describe each of the state variables such as membrane potential (V_M) and spiking related currents (I_k) [6, 7]. All the complexity of the MN model derives from reset rules that are applied when a spike is generated. Simple MN neuron model is described by Eqs. (1) and (2).

$$\frac{dI_k}{dt} = -A_k I_k; \quad k = 1, 2, 3, \ldots, N \tag{1}$$

$$\frac{dV_M}{dt} = \frac{1}{C_M}\left(I_{ex} + \left(\sum_k I_k - GV_M\right)\right) \tag{2}$$

In Eq. (1), I_k represents spiking related currents to incorporate different behaviors as well as model synaptic dynamics. In Eq. (2), V_M represents the membrane potential which exhibits a typical leaky integration behavior of the excitatory input current I_{ex} and the spiking related currents I_k [6].

A log-domain based CMOS implementation of MN neuron by using tau-cell as first-order low-pass filter is shown in Fig. 1. Transistors M_1–M_4, C-$_{MEM}$ and bias currents I_O and $2I_O$ constitute the tau-cell which model membrane current [6]. Transistor M_4 in the tau-cell is kept in sub-threshold region in order to obtain an exponential relationship between current and voltage [7]. Membrane potential (V_{MEM}) is developed

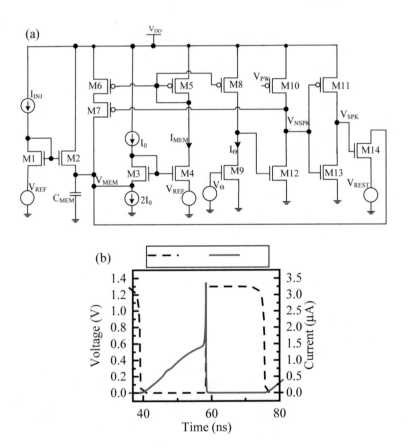

Fig. 1 a Log-domain based MN neuron circuit **b** simulation waveform of Voltage spike (V_{SPK}) and Membrane current (I_{MEM})

due to current (I_{M2}) flowing through transistor M_2. By trans-linear principle in tau-cell model, membrane current I_{MEM} through transistor M_4 has an exponential relationship with V_{MEM}. Membrane current I_{MEM} is copied by transistors M_5 and M_8 due to current mirror effect. A threshold current I_Θ is compared with I_{MEM} resulting in a negative voltage spike (V_{NSPK}). Therefore, an inverted spike (V_{SPK}) is generated using CMOS based inverter consisting of transistors M_{11} and M_{13}. A steep charging of V_{MEM} occurs through positive feedback transistors M_5–M_7 using voltage spike (V_{NSPK}). The discharge path of membrane capacitance (C_{MEM}) is created through transistor M_{14}. The refractory period can be controlled by an external voltage V_{REST}.

3 Simulation Framework

Figure 2 shows the simulation flow for investigating the impact of MN neuron circuit with BTI, HCI and process-temperature variations. Industry standard 45 nm technology library from foundry is used to create a circuit netlist. Process and temperature variations are included in the circuit netlist along with the technology library at time-zero. Aged circuit netlist is generated by industry standard reliability simulator ©Cadence RelXpert. These circuit netlists are provided to the SPICE simulator for both time-zero and aged simulation. For SPICE simulations, time-zero and aged circuit netlists are referred to as time_zero and aged simulations respectively. Time_zero simulations are carried out for different process corners (TT, SS, FF, SF, FS) at temperatures 300 K and 358 K, respectively. Aged simulations are performed for three degraded cases (i) only HCI, (ii) only BTI, and (iii) combination of both HCI and BTI; by considering a lifetime span of ten years. Frequency of spiking events and average power consumed by the neuron circuit is considered to be the most critical parameters. SPICE simulations are carried out in order to observe the spiking frequency and average power by varying inject current (I_{INJ}) from 2 to 50 μA. Performance of spiking

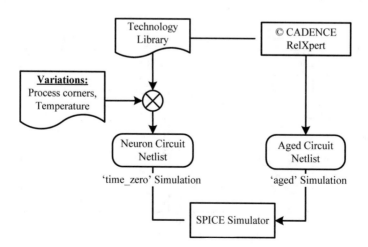

Fig. 2 Simulation framework to evaluate performance of MN-neuron circuit

neuron is characterized by calculating the percentage deviation in spiking frequency (F_{SPIKE}) and average power of aged and time-zero circuit. Percentage change is calculated by using Eq. (3).

$$\text{Percentage change} = \frac{(\text{aged} - \text{time\_zero})}{\text{time\_zero}} * 100 \tag{3}$$

4 Results and Discussion

This section discusses the effect of process-temperature variations and the impact of HCI, BTI degradation on spiking frequency, and average power consumption.

4.1 Process-Temperature Variations

Figure 3 shows spike frequency with injecting current (I_{INJ}) for different process corners at two temperatures 300 and 358 K. For TT corner, as I_{INJ} increases, frequency of spikes increases at both temperatures. At higher I_{INJ}, the change in frequency reduces, and frequency saturation is observed for both temperatures. Similar trend is observed at other process corners. At SS process corner, lower magnitude of currents flow through transistors in the circuit due to rise in threshold voltage. This results in an increased charging time of membrane capacitance (C_{MEM}). Therefore, a lower frequency range is obtained at SS process corner for both temperatures. Faster PMOS transistors are responsible for generating high-frequency spikes. Hence at SF process corner, frequency of spiking tends to dominate for lower inject currents while FF process corner dominates for higher I_{INJ}.

Figure 4 shows average power consumption of the circuit by varying inject current (I_{INJ}) for different process corners at two temperatures 300 and 358 K. For TT corner, average power consumption by the circuit increases with raise in I_{INJ} for both

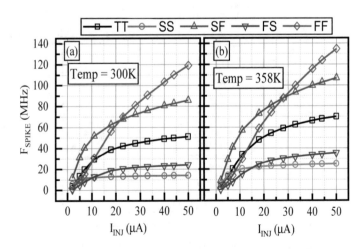

Fig. 3 Spiking frequency by varying I_{INJ} for different process corners at **a** 300 K and **b** 358 K

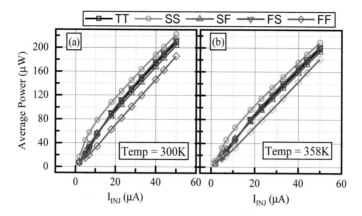

Fig. 4 Average power by varying I_{INJ} for different process corners at **a** 300 K and **b** 358 K

temperatures. Similar trend is observed at other process corners. It is observed from the figure that average power consumption is minimum for FF corner while maximum power is consumed at SS process corner for both temperatures. This is because of higher spike turn-ON period causing current to flow through the transistors for an increased amount of time and thus higher power consumption is observed. Whereas for FF corner vice versa trend is observed. Unlike frequencies of spike outputs which were affected significantly by process variations; average power consumption variation amongst the five corners was minimal at both temperatures.

4.2 Reliability Analysis

Figure 5 shows the percentage change in spiking frequency (F_{SPIKE}) by varying inject current (I_{INJ}) for considered degraded cases. It is observed that for lower current range (<15 uA), MN circuit does not generate spike. With an increase in I_{INJ} the percentage change in spiking frequency decreases for HCI and combination of HCI and BTI. In case of BTI the percentage change in frequency degrades for all I_{INJ}. As discussed in Sect. 2, current through transistor M_2 charges membrane capacitance $C_{MEM,}$ and membrane potential is developed at V_{MEM}. As membrane current I_{MEM} reaches the threshold current I_Θ, a resultant spike is generated. Therefore, current I_{M2} flowing through transistor M_2 is largely responsible for spike generation in MN circuit. For HCI degraded case, M_2 transistor degrades with aging. At lower inject current range (<15 µA), lesser magnitude of current flows through degraded M_2 transistor which is insufficient for the membrane potential V_{MEM} to develop. This results in lower I_{MEM} compared to I_Θ and therefore circuit does not generate spikes. Whereas with an increase in I_{INJ}, I_{M2} increases due to current mirror formed by M_1 and M_2 transistors. Hence, increase of spiking frequency is observed with increase in I_{INJ}. As a result, reduction in percentage deviation is observed with increase in I_{INJ}. For BTI degraded case, since NMOS transistors M_1 and M_2 do not degrade, the spiking frequency largely depends on PMOS transistor M_{11}. Degradation in M_{11} results in slower V_{SPK} generation which in turn leads to M_{14} transistor conduction slowing down. This causes the Reset duration of membrane

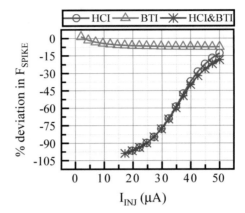

Fig. 5 Percentage deviation in spiking frequency (F_{SPIKE}) by varying I_{INJ} for considered degraded cases

potential (V_{MEM}) to increase. As a result, spike frequency decreases with increase in inject current. For combined effect of HCI and BTI, frequency variation tends to follow HCI curve as the contribution of BTI stress is minimal.

Figure 6 shows the percentage change in average power consumption of MN neuron circuit with variation in I_{INJ} for considered degraded cases. It is observed that for lower current range (<15 uA), percentage change in average power increases up to ~70%. With increase in I_{INJ} the percentage change in average power decreases for HCI and combination of HCI and BTI. While for BTI the percentage change in average power consumption degrades for all I_{INJ}. For HCI degraded case, degradation in M_2 transistor increases with aging. At lower range of I_{INJ}, reduced current I_{M2} flows through the degraded transistor M_2 which is unable to charge the membrane capacitance (C_{MEM}). Therefore, small constant static current flows through transistors M_{10} and M_{12} resulting in increased power consumption. For higher I_{INJ}, spikes are

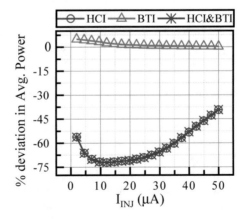

Fig. 6 Percentage deviation in average power by varying I_{INJ} for considered degraded cases

generated due to which V_{NSPK} switches between high and low potential. The magnitude of the current flowing through transistors M_{10} and M_{12} now depends on the switching frequency of V_{NSPK}. Hence, reduced average power consumption is observed by accounting the switching frequency of V_{NSPK} which is shown in the figure. For BTI degraded case, it is observed that as I_{INJ} increases degradation in transistors M_6 and M_{11} increases. The average power consumption by the circuit is largely responsible due to the current flowing through transistor M_6. As I_{INJ} increases, current through M_6 reduces resulting in lower power consumption. For combined effect of HCI and BTI, average power deviation tends to follow HCI curve as the contribution of BTI stress is minimal.

5   Conclusions

In this paper, the effects of process-temperature (PT) variations and impact of HCI and BTI on simple MN-neuron circuit have been presented. For all process corners, as I_{INJ} increases, frequency of spikes and average power consumption increases at both temperatures. Higher magnitudes of spike frequencies are observed at FF process corner with the maximum being 120 MHz and 134 MHz at 300 K and 358 K respectively. Whereas the lowest spike frequency range is observed at SS process corner up to 14 MHz and 25 MHz at 300 K and 358 K respectively. Power consumed is least at FF corner while maximum power was consumed at SS process corner. The average power consumption variation amongst the five corners was minimal. For HCI degraded case, with increase in I_{INJ}, deviation in spiking frequency decreases. Maximum percentage deviation in average power consumption is found to be 75% at $I_{INJ} = 12$ µA. For BTI degraded case, with an increase in I_{INJ} percentage change in frequency increases while percentage change in average power decreases. However, the contribution of BTI stress on the circuit was minimal. Hence, for combined effect of HCI and BTI, the frequency and average power deviation curves tend to follow HCI curve.

Acknowledgements. The authors are grateful to the Electrical Engineering department of Shiv Nadar University for providing simulation resources. The authors would also like to acknowledge the support from Electrical and Electronics department of BITS Pilani-Dubai Campus for providing reliability tools used in this work.

References

1. Neuromorphic computing—beyond today's AI—Intel, https://www.intel.in/content/www/in/en/research/neuromorphic-computing.html
2. Indiveri G, Linares-Barranco B, Hamilton T, van Schaik A, Etienne-Cummings R, Delbruck T, Liu S-C, Dudek P, Hafliger P, Renaud S, Schemmel J, Cauwenberghs G, Arthur J, Hynna K, Folowosele F, Saighi S, Serrano-Gotarredona T, Wijekoon J, Wang Y, Boahen K (2011) Neuromorphic silicon neuron circuits. Frontiers Neurosci 5
3. Thakur CS, Molin JL, Cauwenberghs G, Indiveri G, Kumar K, Qiao N, Schemmel J, Wang R, Chicca E, Olson HJ, Seo J, Yu S, Cao Y, van Schaik A, Etienne-Cummings R (2018) Large-scale neuromorphic spiking array processors: a quest to mimic the brain. Frontiers Neurosci 12

4. Niebur ŞME (2009) A generalized linear integrate-and-fire neural model produces diverse spiking behaviors. J Neural Comput 21(3):704–718
5. Varghese V, Molin J, Brandli C, Chen S, Etienne-Cummings R (2015) Dynamically reconfigurable silicon array of generalized integrate-and-fire neurons. In: 2015 IEEE biomedical circuits and systems conference (BioCAS), 1–4
6. van Schaik A, Jin C, McEwan A, Hamilton TJ, Mihalas S, Niebur E (2010) A log-domain implementation of the Mihalas-Niebur neuron model. In: Proceedings of 2010 IEEE international symposium on circuits and systems. Paris, pp 4249–4252
7. Thanigaivelan B, Postula A, Jin CT, Schaik AV, Hamilton TJ (2010) Symbolic analysis of the Tau Cell log-domain filter using affine MOSFET models. In: 2010 IEEE Asia pacific conference on circuits and systems, 1095–1098
8. Alam MA, Mahapatra S (2005) A comprehensive model of PMOS NBTI degradation. Microelectron Reliab 45(1):71–81
9. Sengupta D, Sapatnekar SS (2017) Estimating circuit aging due to BTI and HCI using ring-oscillator-based sensors. IEEE Trans Comput-Aided Design Integr Circ Syst 36:1688–1701
10. Weste NHE, Harris D (2005) CMOS VLSI design: a circuits and systems perspective. 3rd edn. Addison-Wesley

Time Series Prediction of Weld Seam Coordinates for 5 DOF Robotic Manipulator Using NARX Neural Network

Abhilasha Singh[1(✉)] ⓘ, V. Kalaichelvi[1] ⓘ, and R. Karthikeyan[2] ⓘ

[1] Department of Electrical & Electronics Engineering, Birla Institute of
Technology and Science Pilani, Dubai Campus, Dubai, UAE
{p20180906, kalaichelvi}@dubai.bits-pilani.ac.in
[2] Department of Mechanical Engineering, Birla Institute of Technology and
Science Pilani, Dubai Campus, Dubai, UAE
rkarthikeyan@dubai.bits-pilani.ac.in

Abstract. In general, welding is a process in which two workpieces are joined together. The edge interface of the two halves are called weld seam. The main scope of this paper is to perform prediction analysis of 3D weld seam coordinates based on Non-Linear Auto Regressive with Exogeneous Input (NARX) Neural Network using various training functions and training ratios. Because developing a model for such complex processes using analytical techniques is time-consuming and prerequisite knowledge of the process is needed. Training NARXNN with the appropriate combination of learning rate, training-testing ratios, momentum coefficient and training function for the prediction of robot coordinates is a challenging task in Neural Networks. This work investigates Gradient Descent based Back Propagation, Scaled Conjugate Gradient method, Resilient Back Propagation, Levenberg–Marquardt algorithms in determining the 3D coordinates of weld seam. The proposed work compares the training algorithms based on Mean Absolute Error (MAE), Mean Square Error (MSE), and Root Mean Square Error (RMSE) for the real-time experimental data of weld shape. Experimental analysis is performed using the data obtained from real-time weld seam detection using 5 DOF robotic manipulators.

Keywords: Robotic manipulator · Weld seam · Nonlinear Auto Regressive with Exogenous Input (NARX) Neural Network · Mean Absolute Error (MAE) · Mean Square Error (MSE) · Root Mean Square Error (RMSE)

1 Introduction

With the introduction of intelligent techniques in robotic manipulators, robotic welding is one of the active research areas in which any welding process needs manual human intervention to identify the weld seams and also develop accurate models from analytical approach and complex mathematical calculations to program the robot to weld them precisely irrespective of shapes of weld pieces. This inherently takes time and cost to program the robot to weld new parts which is a great disadvantage for mass production environment. Hence using intelligent techniques like Artificial Neural

© Springer Nature Singapore Pte Ltd. 2020
N. Goel et al. (eds.), *Modelling, Simulation and Intelligent Computing*,
Lecture Notes in Electrical Engineering 659,
https://doi.org/10.1007/978-981-15-4775-1_58

Networks makes modeling easier and it does not need knowledge about the process thereby it is easy to automate the welding process. The active research in the field of Artificial Neural Network (ANN) started since the beginning of the 1980s and a lot of the survey for performance evaluation based on different ANN architectures, different training-testing ratios, hardware implementations received large attention [1]. Furthermore, learning phase is very crucial in Neural Networks because proper choice of activation functions also plays a major role in efficient training of networks [2]. For the past few decades, ANN has been a better tool for time series prediction, pattern reorganization, and classification [3]. To predict any data, several ANN types are used namely Nonlinear Autoregressive Exogenous Neural Network (NARXNN) [4, 5], Nonlinear Autoregressive Neural Network (NARNN), and Recurrent Neural Network (RNN). In this paper, NARXNN has been used to evaluate the performance of coordinates of 3D weld seam and also to perform step ahead prediction of robot coordinates to track the weld seam. One of the major benefits of this structure is that they can accept dynamic inputs represented by time series datasets.

The objectives of this paper are as follows:

1. To analyze various training functions with different training and testing ratios.
2. To validate network performance for different numbers of hidden nodes and learning rates.
3. To evaluate the prediction data obtained from real-time by calculating MAE, MSE, and RMSE for various training algorithms as well as training-testing ratios.

The paper is organized into five sections. Section 2 describes the literature survey of NARX Neural Network (NARXNN). The problem statement and experimental approach to automatically predict weld seam coordinates are discussed in Sect. 3. Section 4 describes the results obtained in this work followed by a conclusion and future scope in Sect. 5.

2   Related Works

This paper builds on prior research in NARX Neural Network for learning as well as prediction in robotic manipulators. The approach presented here is motivated by the performance evaluation of robotic welding tasks in advanced manufacturing. Xue et al. [6] proposed the prediction of weld bead geometry using BackPropagation (BP) neural network and improved BP neural network for wire and arc welding with optimized weights and threshold values based on Genetic Algorithm thereby avoiding premature convergence. Saldanha et al. [7] developed a multi-layered feed forward neural network for prediction analysis of weld bead geometry of TIG welding. The algorithms used for training are traimlm, trainrp, and trainscg. The performance was better with 60/40 training-testing ratios and LM algorithm showed good convergence.

Zhou et al. [8] proposed nonlinear autoregressive neural network with exogenous inputs (NARX) for surface Electromyography (EMG) to predict the natural guidewire rotation to estimate the motion states of the guidewire in complex vascular environment for robotic surgery. Rafiei et al. [9] has performed system identification and modeled dynamics based on NARX for Ferdowsi University of Mashhad (FUM) SCARA robot using Conjugate Gradient, Levenberg–Marquardt, and Bayesian Regularization training algorithms to train the Neural Networks with varying hidden nodes and results reveal that with a large number of hidden layer neurons, overfitting occurs and NARX with Bayesian showed higher accuracy. Raglend et al. [10] analyzed inverse kinematics of 5 DOF robotic manipulators using NARX based neural network to remove the difficulty of traditional mathematical calculation of inverse kinematics. Vrushali et al. [11] proposed extraction and classification of weld seam from 3D point clouds and since inverse kinematics cannot be implemented in TAL BRABO manipulator, an Artificial Neural Network (ANN) was implemented to convert the image coordinates to robot coordinates for real-time tracking.

3 Prediction Analysis Using NARXNN

To obtain the prediction model for real-time weld image coordinates, NARXNN which is a recurrent dynamic neural network with feedback connections to map the time series modeling is used. The overall NARXNN architecture is shown in Fig. 1.

In the neural network, there are several learning algorithms but "traingda", "trainrp", "trainscg", and "trainlm" are used [12–15] for the present work to predict robot coordinates for 3D weld seam. The weld datasets of various shapes are obtained from Kinect depth camera and the image coordinates are sent to NARX network for prediction of data accurately based on current and past values of robot coordinates.

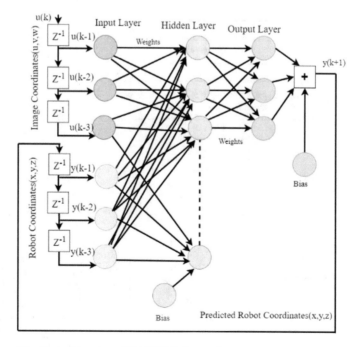

Fig. 1 Architecture of NARXNN for prediction of robot coordinates

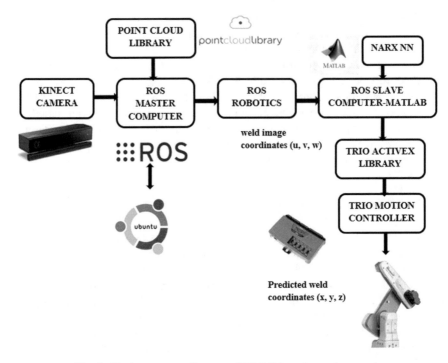

Fig. 2 Hardware setup diagram of NARX based neural network

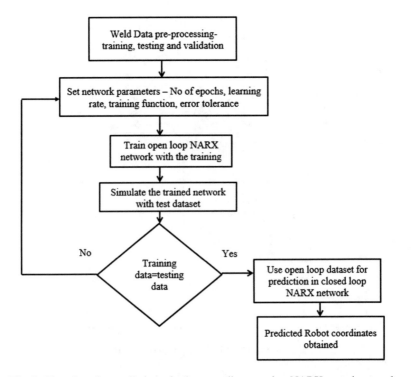

Fig. 3 Flowchart for prediction of robot coordinates using NARX neural network

Finally, robot coordinates are sent to TAL BRABO robotic manipulator via Trio Controller to trace the weld seam. The hardware setup diagram for the preparation of weld datasets for prediction analysis is shown in Fig. 2.

The overall workflow followed in this paper is explained using a simple flow chart in Fig. 3.

4   Results and Discussion

4.1   NARXNN Model Configuration

The coordinates of weld seam (u, v, w) were extracted for three different shapes namely radially straight, inclined, and curved as mentioned in [11]. These coordinates are fed as inputs $x(t)$ of size 91×3 to NARXNN model which is implemented using MATLAB and predicted robot coordinates $y(t)$ is obtained. The model is in two phases —open-loop and closed-loop. The open-loop model consists of three layers namely the input layer with three inputs, hidden layer with varying number of neurons from 10 to 60, and output layer with three neurons as shown in Fig. 4. In closed-loop structure, the parameters obtained from open-loop are used to train the model, and output is fed back to reduce the error between actual and predicted outputs thereby increasing the performance of the network and reduces error. The closed-loop configuration of NARXNN is shown in Fig. 5. The activation functions for hidden and output layers used are "tansig" and "purelin", respectively.

4.2   NARXNN with Different Training, Testing and Validation Ratios and Training Algorithms

Once the model is developed, the performance analysis of NARX Neural Network is evaluated by varying the training, testing, and validation ratios and training algorithms as shown in Fig. 6 and Mean Absolute Error, Mean Square Error, and Root Mean Square Error is calculated. It is seen that from the graph that network performed better when training data is 60%, testing data is 20% and validation data is 20%. With this ratio, low MAE value of 5.85, MSE value of 67.46, and RMSE value of 8.21 was obtained compared to other training-testing ratios which might fail to produce better results. It is also observed that MSE value for 50-25-25 data split ratio saw the highest

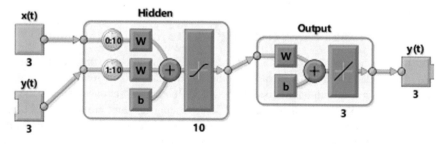

Fig. 4 Open-loop configuration of NARXNN in MATLAB

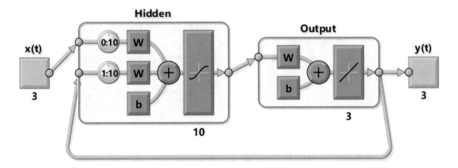

Fig. 5 Closed-loop configuration of NARXNN in MATLAB

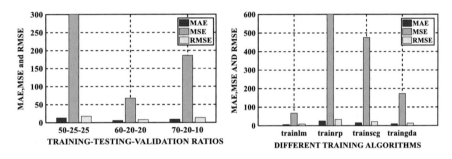

Fig. 6 Error plot of NARXNN for various training-testing ratios and training algorithms

MAE, MSE, and RMSE values of 13.07, 313.41 and 17.70, respectively which will make the network performance poor. The number of neurons in hidden layer was fixed to 50 and learning rate to 0.01. Next, the network performance is analyzed by fixing dataset ratios as 60-20-20 for implementing different training algorithms. It was observed from the graph that trainlm has outperformed with minimum MAE, MSE and RMSE values of 5.85,67.46, and 8.21, respectively, and trainrp performed worst with high error values and the order of performance can be derived in such a way that traimlm > traingda > trainscg > trainrp. Therefore, it can be inferred that Levenberg–Marquardt method can be used to test the real-world weld seam data to achieve exact tracing of weld seam using robot.

4.3   NARXNN with Different Nodes in Hidden Layer and Learning Rates

To further enhance the analysis, the network was validated by changing the number of nodes in the hidden layer without varying other parameters, and the results illustrated in Fig. 7 shows that optimum number of nodes in hidden layer was found to be 10 neurons leading to better performance with less training time and prevents overfitting. The right way to choose the size of hidden layer is no of neurons should be two-thirds of input layer size plus size of output layer [15]. Finally, the network is further trained

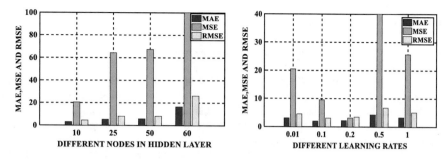

Fig. 7 Error plot of NARXNN for different no of nodes in hidden layer and learning rates

with different learning rates varying from lower value of 0.01 to maximum value of 1 with optimum nodes in hidden layer of 10. For learning rate of 0.2 and "trainlm", MAE, MSE, and RMSE were 2.24, 3.16, and 3.49, respectively which is comparatively low than other learning rates because with higher learning rate training quickly converges to local minima or sometimes diverge and with very low learning rate network converges slowly to local minima [16, 17]. Hence from the above analysis, the network was trained with optimum values and it was found that for the training-testing-validation ratio of 60-20-20 with learning rate of 0.2 and no of neurons of 10, MAE, MSE, and MSE was 2.24,3.16 and 3.49 respectively.

5   Conclusion and Future Scope

In this paper performance evaluation of weld seam coordinates obtained from real-time experimental data was analyzed based on MAE, MSE, and RMSE values. The network was trained using different training algorithms, training-testing-validation ratios, and different learning rates with a varying number of neurons. It can be concluded that NARX NN performed better with 60-20-20 dataset ratio using trainlm with faster convergence and learning was better with value of 0.2. The optimal number of hidden neurons was found to be 10 which plays a vital role in overfitting and under fitting of data. The only disadvantage is for larger datasets it leads to overfitting or under fitting leading to slower convergence and also it increases computational complexity because of large number of hidden layers. The future works include implementation of proposed techniques for real-time validation in robotic manipulator. Also, analysis can be further enhanced with the use of machine learning and deep learning techniques for larger datasets.

Acknowledgements. The authors are extremely grateful to the authorities for all the hardware and software facilities provided to us for carrying out this research work in Birla Institute of Technology and Science Pilani, Dubai Campus, Dubai, UAE.

References

1. Fausett LF (2006) Fundamentals of neural networks-architectures, algorithms and its applications. 1st Impression. Pearson, New Jersey
2. Sharma B, Venugopalan K (2014) Comparison of neural network training functions for hematoma classification in brain CT images. IOSR J Comput Eng (IOSR-JCE) 16(2):31–35
3. Sarkar R, Julai S, Hossain S, Chong WT, Rahman M (2019) A comparative study of activation functions of NAR and NARX neural network for long-term wind speed forecasting in Malaysia. Mathe Probl Eng 1–14. https://doi.org/10.1155/2019/6403081
4. Ashok Kumar D, Murugan S (2018) Performance analysis of NARX neural network backpropagation algorithm by various training functions for time series data. Int J Data Sci 3 (4):308–325
5. Jaikumar R, Nagendra SS, Sivanandan R (2018) Development of NARX based neural network model for predicting air quality near busy urban corridors. In: Recent developments and the new direction in soft-computing foundations and applications. Studies in Fuzziness and Soft Computing 361581-593. https://doi.org/10.1007/978-3-319-75408-6_45
6. Xue Q, Ma S, Liang Y, Wang J, Wang Y, He F, Liu M (2018) Weld bead geometry prediction of additive manufacturing based on neural network. In: 11th International symposium on computational intelligence and design (ISCID), vol 2. Hangzhou, China, pp 47–51
7. Saldanha SL, Kalaichelvi V, Karthikeyan R (2017) Prediction analysis of weld-bead and heat affected zone in TIG welding using Artificial Neural Networks. In: IOP conference series: materials science and engineering, vol 346. IMMT, Dubai, UAE, pp 1–8
8. Zhou XH, Bian GB, Xie XL, Hou ZG, Hao JL (2017) Prediction of natural guidewire rotation using an EMG-based NARX neural network. In: 2017 international joint conference on neural networks (IJCNN). Anchorage, AK, USA, pp 419–424
9. Rafiei H, Hosseini AA, Tootoonchi AA (2018) Modeling the dynamic of SCARA robot using nonlinear autoregressive exogenous input neural network model. In: Iranian conference on electrical engineering (ICEE). Mashhad, Iran, pp 994–999
10. Raglend GJ, Anand MD, Prabha DMMSR (2016) Inverse kinematics solution of a five joint robot using NARX algorithm. JChem Pharm Sci 9(4):2677–2687
11. Vrushali P, Indraneel P, Kalaichelvi V, Karthikeyan R (2019) Extraction of weld seam in 3D point clouds for real time welding using 5 DOF robotic arm. In: PROCEEDINGS IN IEEE, 5th international conference on control, automation and robotics (ICCAR). Beijing, China, pp 727–733
12. Anastasiadis AD, Magoulas GD, Vrahatis MN (2005) New globally convergent training scheme based on the resilient propagation algorithm. Neurocomputing 64:253–270
13. Beale MH, Hagan MT, Demuth HB (2015) Neural network toolbox TM user's guide
14. Mustafidah H, Hartati S, Wardoyo R, Harjoko A (2014) Selection of most appropriate backpropagation training algorithm in data pattern recognition. Int J Comput Trends Technol (IJCTT) 14(2):92–95
15. Panchal FS, Panchal M (2014) Review on methods of selecting number of hidden nodes in artificial neural network. Int J Comput Sci Mobile Comput 3(11):455–46

16. Li Y, Lee TH, Wang C (2018) An artificial neural network model for predicting joint performance in ultrasonic welding of composites. In: 7th CIRP conference on assembly technologies and systems, vol 76. Tianjin, China, pp 85–88

17. Atakulreka A, Sutivong D (2007) Avoiding local minima in feedforward neural networks by simultaneous learning. In: Orgun MA, Thornton J (eds) AI 2007, Advances in artificial intelligence. AI 2007. Lecture notes in computer science, vol 4830. Springer, Berlin, Heidelberg, pp 100–109

Chaotic Aspects of EMG Signals in Normal and Aggressive Human Upper Arm Actions

K. M. Subhash$^{(\boxtimes)}$ (iD) and K. Paul Joseph

Department of Electrical Engineering, National Institute of Technology Calicut,
Calicut, Kerala 673601, India
subhash@nitc.ac.in

Abstract. The aim of this research work is to demonstrate a standardized procedure for extracting subtle chaotic features to study the dynamics of aggressive and non-aggressive human muscle actions in the chaotic domain. The relevant features present in the electromyogram (EMG) signals are analyzed by exploiting the chaotic characteristics of the signal. Degree of Self-Similarity (DoSS), Largest Lyapunov Exponent (LLE), Correlation Dimension (CD), Approximate Entropy (ApEn) and Katz Fraction Dimension (KFD) are the features, extracted to study the chaotic aspects of normal and aggressive human upper arm muscles. This chaotic feature vector is utilized for signal characterization, which is fruitfully extended for classification of the EMG signals into aggressive and normal. The proposed extraction and classification technique was experimentally verified for validating the findings, using EMG signals available from the UCI machine learning repository database. The features are statistically categorized into three significant levels, applying ANOVA technique. The inferences lead us to conclude that the extracted chaotic features qualify as a distinguishing multi-feature set for EMG signals of different classes. Five different classifications approached were used for classification by using tenfold cross-validation. The maximum classification accuracy achieved was 97.5% with two of the most significant chaotic features.

Keywords: Electromyogram · Feature extraction · Chaos · Classification · SVM · KNN · Naive Bayes

1 Introduction

Surface electromyographic (sEMG) signals are one of the best known bioelectric signals which are extensively applied in the estimation of neurophysiological characteristics of skeletal muscles. The EMG signals are records of electrical signals generated due to polarization of skeletal muscle fibers [1]. EMG provides information regarding overall muscle function and conduction of skeletal muscles. The most significant functional unit for the neural control of the muscular contraction process is called a motor unit. During the innervation process, every muscle fiber of a given motor unit acts as a single system. Hence, an EMG signal is the result of the summation of action potentials of numerous motor units that travel through tissues. Unlike needle EMG (nEMG), surface EMG is not a localized signal, as it is recorded by placing

© Springer Nature Singapore Pte Ltd. 2020
N. Goel et al. (eds.), *Modelling, Simulation and Intelligent Computing*,
Lecture Notes in Electrical Engineering 659,
https://doi.org/10.1007/978-981-15-4775-1_59

surface electrodes on the skin. Though nEMG gives a more localized recording, the wide acceptance of sEMG over nEMG is because of its noninvasive and painless recording procedure. Nearly after 35 years, since the first description of sEMG by Hermes et al. [2] research in the computerized analysis of EMG is still in progress to completely explain the actual applicability of SEMG in clinical settings. The limited special resolution of sEMG makes it inferior to nEMG in clinical applications [3].

Numerous computer-aided feature extraction algorithms have been extensively investigated in Time Domain (TD), Frequency Domain (FD), Time-Frequency Domain (TFD) and in Fractal Domain (FD). The gold standard methodology for estimating the neurophysiological characteristics of skeletal muscles is recommended as conducting NEMG analysis combined with nerve conduction studies [4]. Acknowledging the recent developments in computer-aided feature extraction and classification, it is pertinent to state that, even in this era of big data analytics and deep learning, the most critical factor in analyzing the EMG signals is the experience and expertise of the electrophysiologists, who evaluates them visually and audibly to interpret the neuromuscular condition of the nervous system. Neurologists extract sample epochs of the signal by visually inspecting the EMG signals in the time domain, and identify the discriminative inherent patterns.

EMG is one of the complex biological signals, highly corrupted by noise and is highly influenced by the introduction of nonlinearity by interference and muscle crosstalk. The underlying physiology of biological systems such as nervous systems and skeletal muscles cannot be completely described by traditional mathematical concepts of calculus due to the introduction of nonlinearity by interference and muscle crosstalk. To attain a better understanding of the dynamics and movement science of EMG signals and to quantify the complexity can be probed by using nonlinear and chaotic approaches [5, 6]. The key to the success of a classification task is an appropriate representation for the given data by a feature descriptor, which should be compact within a category and, far apart among different categories [7]. This work concentrates on developing a feature vector by extracting and analyzing the subtle chaotic features of the sEMG signals recorded from the upper arm muscles. Further to this, the signals are classified to aggressive and normal classes. The statistical significance of the extracted features was analyzed in order to achieve an efficient classification.

The rest of the paper is organized as follows: Sect. 2 describes materials and methods in detail, followed by the description of the data sets used. Section 3 discusses the experimental results, classifier evaluation and performance comparison. Finally, Sect. 4 draws the main conclusions.

2 Materials and Methods

The chaotic phenomenon was discovered by Lorenz [8]. With the recent developments in the theory of chaos and nonlinear dynamics have paved way in analyzing biomedical signals in a more effective and meaningful way. Researchers have begun to understand and restudy these random-looking bio-signals so as to unearth the inherent deterministic nature of these signals.

2.1   EMG Feature Extraction

The underlying fundamental principle of chaotic analysis is that a very complex process is a result of infinite iterations of a simple process. The exact periodicity of biological signals is not an indication of normal condition. The dynamics in the complexity and strength in muscle activation during various neural conditions are assessed from the chaotic features—Correlation Dimension (CD), largest Lyapunov Exponent (LLE), Degree of Self-Similarity (DoSS), Katz Fractal Dimension (KFD) and Approximate Entropy (ApEn). The following sections discuss the features in detail.

Correlation Dimension (CD). Chaotic behavior in some dynamical systems is found only in a subset of phase space. The chaotic motion leads to form a strange attractor with great detail and complexity. Chaotic attractors often have a complex fractal structure resulting in non-integer dimensions in the phase space called fractal dimensions. These dimensions can be estimated on the base of Grassberger-Procaccia algorithm [9].

Correlation dimension method is to estimate appropriate dimension D_2 in terms of correlation theorem [10] using Eq. (1) [11].

$$D_q = \lim_{r \to 0} \frac{\log[C(q, r)]}{(q - 1)\log[r]} \tag{1}$$

where $C(q, r)$ is the correlation function of the attractor, i.e., a measure of probability that two points on the attractor are separated by a distance of r. When q is 2, the resulting dimension is the correlation dimension D_2.

Degree of Self-Similarity (DoSS). The physical interpretation of self-similarity in complex bio-signals arises since a single motor unit statistically resembles the larger structure. When an electrode is placed on the skin surface to record sEMG, it picks up the signal which is a result of the aggregate of action potentials of various motor units that travel through tissues. The degree of self-similarity property of sEMG was computed using the procedure proposed by Kalden and Ibrahim [12]. If $y(k)$ be the discrete-time signal, the aggregated processes $y^m(k)$ in Eq. (2) are created with non-overlapping blocks of size m such that

$$y^m(k) = \frac{1}{m} \sum_{i=0}^{m-1} y(km - i) \tag{2}$$

The degree of self-similarity is measured from the variance of aggregated processes plotted against block size as given in Eq. (3). For this analysis, 25 aggregated processes are computed ($m = 1, 2 \ldots 25$) for computing degree of self-similarity [13].

$$\text{Var}\left(y^{(m)}\right) \approx m^{-\beta} \tag{3}$$

with $0 < \beta < 1$ and DoSS $= 1 - \beta/2$

where DoSS expresses the degree of self-similarity.

Katz Fractal Dimension (KFD). Katz fractal dimension can be directly computed from the discrete-time signal. According to Katz's algorithm [14], KFD is defined as in Eq. (4):

$$KFD = \frac{\log L}{\log d} \tag{4}$$

where L is the total length of the discrete-time signal and d is the distance between the first point in the time series and the furthest point.

Largest Lyapunov Exponent (LLE). The largest Lyapunov exponent, which quantifies the rate of the exponential divergence of nearby trajectories in the phase space is calculated from the slope of the least-square plot $y(n)$ [15–17], as given by Eq. (5), of the logarithms of the distance between the trajectory of the initial point and the trajectories of the neighboring points in the reconstructed phase space.

$$y(n) = \frac{1}{\Delta t} \langle ln(d_i(n)) \rangle \tag{5}$$

$\langle \cdot \rangle$ denotes the average of overall points in the phase space and $d_i(n)$ represents the distance between ith pair of nearest neighbors.

Approximate Entropy (ApEn). Approximate entropy quantifies the entropy of the system and uncovers the underlying episodic nature of the time series. The probability that the embedded vector $X(i)$ is similar to other vectors within a tolerance r can be indicated by the correlation integral $C_i^m(r)$ which is expressed as Eq. (6)

$$C_i^m(r) = \frac{N_i^r}{N - m + 1}, i = 1, 2, \dots N - m + 1 \tag{6}$$

where N is the number of data points of the signal and N_i^r is the number of vectors with distance $X(i)$ smaller than r, which is taken as 0.2 times the standard deviation. Following this ApEn is finally defined [18, 19] as Eq. (7)

$$ApEn(m, r, N) = \phi^m(r) - \phi^{m+1}(r) \tag{7}$$

where $\phi^m(r) = \frac{1}{N-m+1} \sum_{i=1}^{N-m+1} ln C_i^m(r)$

2.2 Data Sets Used

The input signals for probing the chaotic aspects of EMG signals were taken from the EMG physical action data set from the UCI machine learning repository [20]. The physical action data set includes 10 normal and 10 aggressive physical actions that measure human muscle activity. This data set contains eight-channel SEMG records of 10 aggressive and 10 normal actions for each of three male and one female subjects. The normal class has SEMG recordings correspond to bowing, clapping, handshaking, hugging, jumping, running, seating, standing, walking and waving actions. The aggressive class has SEMG recordings correspond to elbowing, front kicking,

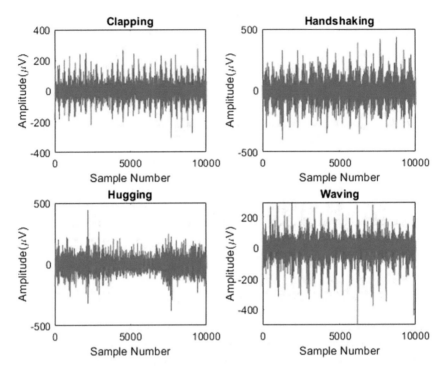

Fig. 1 Sample signals from UCI machine learning repository

hammering, hindering, kneeing, pulling, punching, pushing, side kicking and slapping actions. The signals are recorded using a Delsys EMG wireless apparatus in 8 channels. Some of the sample sEMG signals are shown in Fig. 1.

3   Results and Discussion

In this section, the results obtained after the experimental verification of the proposed algorithm are described and the proposed work is also compared with similar reported works which use EMG physical action data set from the UCI machine learning repository.

Chaotic features extracted from normal and aggressive upper arm EMG signals of Subject-1 are given in Tables 1 and 2, respectively.

3.1   Statistical Analysis

The statistical significance analysis was carried out by one-way ANOVA test [21] and is performed using IBM-SPSS v16. The performance analysis is shown in Table 3. The ANOVA procedure helped us to conclude that the differences observed in classification performance using the chaotic features are attributed to the advantages or disadvantages of each feature, not by chance.

Table 1 Chaotic features for normal action of Subject-1

Action class	KFD	DoSS	LLE	CD	ApEn
Clapping	1.4685	0.9975	3.5623	1.9622	0.6055
Hand shaking	1.4117	0.9965	3.6066	1.9938	0.6234
Hugging	1.5455	1.0003	3.6602	1.9834	0.6785
Waving	1.3274	0.9962	3.5566	1.9917	0.6915

Table 2 Chaotic features for aggressive action of Subject-1

Action class	KFD	DoSS	LLE	CD	ApEn
Elbowing	2.1242	0.9995	3.7015	1.8905	0.4652
Hammering	2.3693	0.9995	3.7356	1.9218	0.4963
Pulling	2.1907	0.9978	3.6649	1.9584	0.5994
Punching	2.2241	0.9965	3.6231	1.9537	0.4779
Pushing	2.3954	1.0024	3.6255	1.9266	0.4793
Slapping	2.1403	0.9994	3.6509	1.9547	0.4472

Table 3 One-way ANOVA for the chaotic features of human upper arm EMG signals

Features	Normal action Min/Max/Mean/SD	Aggressive action Min/Max/Mean/SD	p-value	F-score
KFD	1.283/1.873/1.459/0.157	1.815/2.563/2.144/0.183	$p < 0.005$[a]	148.885
DoSS	0.993/1.002/0.998/0.002	0.994/1.002/0.998/0.001	0.786[c]	0.075
LLE	3.524/3.779/3.631/0.066	3.493/3.735/3.630/0.064	0.952[c]	0.004
CD	1.614/2.098/1.938/0.113	0.928/1.996/1.703/0.354	$p < 0.05$[b]	6.566
ApEn	0.128/0.691/0.563/0.171	0.255/0.601/0.453/0.095	$p < 0.05$[b]	6.781

[a]High statistical significance ($p < 0.005$)
[b]Statistical significance ($p < 0.05$)
[c]No statistical significance

It is observed from Table 3 that two features, namely CD and ApEn have statistical significance (p-value < 0.05) and KFD is highly statistically significant (p-value < 0.005) and the features DoSS and LLE has no statistical significance. Hence, only KFD, CD and ApEn are used for classification of sEMG signals. The performance evaluation of different classifiers is presented in the next section.

3.2 Performance Evaluation of Classification

The features are separated into two clusters in the three-dimensional feature space as shown in Fig. 2. Classification is then performed using various classifiers, and the

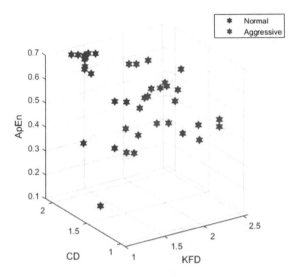

Fig. 2 3D feature space for normal and aggressive EMG signals

Table 4 Performance evaluation of classifiers

Classifier	Error Rate	Accuracy	Sensitivity	Specificity	Precision	MCC	*F*-Score
LDA	0.025	0.975	0.937	1	1	0.948	5.625
Naïve Bayes	0.025	0.975	0.937	1	1	0.948	5.625
Linear SVM	0.025	0.975	0.937	1	1	0.948	5.625
Q-SVM	0.025	0.975	0.937	1	1	0.948	5.625
Cubic SVM	0.025	0.975	0.937	1	1	0.948	5.625
Fine KNN	0.050	0.950	0.875	1	1	0.898	5.250

MCC Mathew's correlation coefficient

performance indices of best results are reported in Table 4. Ten cross-validation is adopted throughout the experiments. The experiment environment is MATLAB 2019b.

The comparison of performance of the proposed method with other reported research works in terms of accuracy is shown in Table 5.

Table 5 Performance comparison

References	Features	Limb separation	Classification accuracy (%)
[22]	Bi-spectrum analysis	NO	99.75
[23]	WPD	NO	92.1
[24]	Variational mode decomposition	NO	98.17
Proposed algorithm	**Chaotic features**	**YES (upper arm muscles)**	97.5

4 Conclusion and Future Work

The present research work demonstrated the efficiency of well-defined chaotic features in characterizing sEMG signals to various action classes. Normal and aggressive actions could successfully be modeled and classified by carefully choosing the classifier algorithm. The experimental results and findings revealed that the proposed multi-feature set is capable of distinguishing heterogeneous neuro-muscular conditions. Our future efforts are progressing toward extracting distinguishing nonlinear features for lower limb muscles, which out performs the current reported works.

References

1. Kimura M (2013) Anatomy and physiology of the peripheral nerve and types of nerve pathology. Electrodiagnosis Dis Nerve Muscle 49–73
2. Hermens G, Schlechl HJ, Zilvold MC (1989) The clinical use of surface EMG for quantitative evaluation in rehabilitation. Int J Rehabil Res 12(2):214
3. Hogrel JY (2005) Clinical applications of surface electromyography in neuromuscular disorders. Neurophysiol Clin 35(2–3):59–71
4. Pullman M, Goodin SL, Marquinez DS, Tabbal AL, Rubin S (2000) Clinical utility of surface EMG: report of the therapeutics and technology assessment subcommittee of the American Academy of Neurology. Neurology 55(2):171–177
5. Padmanabhan P, Puthusserypady S (2004) Nonlinear analysis of EMG signals—a chaotic approach. Annu Int Conf IEEE Eng Med Biol Proc 26(I, Mvc):608–611
6. Lei G, Meng M (2012) Nonlinear analysis of surface EMG signals. Comput Intell Electromyogr Anal 450
7. Subhash KM, Pournami PN, Joseph PK (2018) Census transform based feature extraction of emg signals for neuromuscular disease classification. In IEEE student conference on research and development: inspiring technology for humanity, SCOReD 2017—proceedings
8. Lorenz EN (1963) Deterministic nonperiodic flow. J Atmos Sci 20:130–141
9. Grassberger I, Procaccia P (1983) Measuring the strangeness of strange attractors. Phys D Nonlinear Phenomena 9(1–2):189–208
10. Ding M, Grebogi C, Ott E, Sauer T, Yorke JA (1993) Plateau onset for correlation dimension: when does it occur? Phys Rev Lett 70(25):3872–3875
11. Haykin S, Puthusserypady S (1997) Chaotic dynamics of sea clutter. Chaos 7(4):777–802

12. Kalden R, Ibrahim S (2004) Searching for self-similarity in GPRS. Lect Notes Comput Sci (including Subser Lect Notes Artif Intell Lect Notes Bioinformatics) 3015:83–92
13. Subhash KM, Pournami PM, Joseph PK (2018) Characterizing EMG signals using aggregated CENSUS transform. In: The 2018 biomedical engineering international conference (BMEiCON-2018)
14. Katz MJ (1988) Fractals and the analysis of waveforms. Comput Biol Med 18(3):145–156
15. Briggs K (1990) An improved method for estimating Liapunov exponents of chaotic time series. Phys Lett A 151(1–2):27–32
16. Wolf A, Swift JB, Swinney HL, Vastano JA (1985) Determining Lyapunov exponents from a time series. Phys D Nonlinear Phenom 16(3):285–317
17. Rosenstein MT, Collins JJ, De Luca CJ (1993) A practical method for calculating largest Lyapunov exponents from small data sets. Phys D 65(1–2):117–134. http://dx.doi.org/10.1016/0167-2789(93)90009-P
18. Pincus S (1995) Approximate entropy (ApEn) as a complexity measure. Chaos 5(1):110–117
19. Mesin L, Estimation of complexity of sampled biomedical continuous time signals using approximate entropy. Front Physiol 9, 1–15
20. Dua D, Graff C (2019) UCI machine learning repository. University of California, School of Information and Computer Science, Irvine, CA. https://archive.ics.uci.edu/ml/index.php
21. Castillo-Valdivieso PA, Merelo JJ, Prieto A, Rojas I, Romero G (2002) Statistical analysis of the parameters of a neuro-genetic algorithm. IEEE Trans Neural Netw 13(6):1374–1394
22. Sezgin N (2012) Analysis of EMG signals in aggressive and normal activities by using higher-order spectra Sci World J
23. Abdullah AA, Subasi A, Qaisar SM (2017) Surface EMG signal classification by using WPD and ensemble tree classifiers. IFMBE Proc 62:475–481
24. Sukumar N, Taran S, Bajaj V (2018) Physical actions classification of surface EMG signals using VMD. In: Proceedings 2018 IEEE international conference on communication signal process, ICCSP 2018, pp 705–709

Integration of Distributed PV System with Grid Using Nine-Level PEC Inverter

Shahbaz Ahmad Khan, Deepak Upadhyay, Mohammad Ali[ID],
Khaliqur Rahman, Mohd Tariq[✉], Adil Sarwar, and Anas Anees

Department of Electrical Engineering, ZHCET, Aligarh Muslim University,
Aligarh, India
mohad_ali92@yahoo.com, tariq.iitkgp@gmail.com

Abstract. In this paper, packed E-cell converter (MPEC) topology is investigated when integrated with three different PV sources of two different voltages and power ratings. Nine-level hybrid PWM with half parabola vertically shifted carriers is used which gives lesser THD when compared with triangular carrier waves. The advantage of the investigated topology is that the MPEC can continue a five-level operation even if a fault occurs on four-quadrant switch, without a change in topology. The modelling of the investigated system is done in MATLAB®/Simulink, and results obtained are presented and discussed in the paper.

Keywords: Distributed PV systems · Packed E-Cell · Modified packed E-Cell · Power quality · Renewable energyconversion

1   Introduction

Renewable energy demand is increasing day by day, and solar energy is the major contributor in this with capability of distributed solar photovoltaic systems. As such, distributed PV systems must be designed with resilience in mind and combined with other sources of generation or with grid [1, 2]. But, as the nonlinear loads on the grid and the penetration of intermittent renewable sources increases, the need for converters having lower THD, having lower number of components, the capability to incorporate different voltage rating sources and more number of levels arises. İntroduction of a new type of multilevel topology called packed U-cell converter with one DC source and less number of switches paved a way for using multilevel inverters in various applications [3, 4]. Supremacy of this type of inverter topology over well-established topologies like flying capacitor and cascaded H-bridge inverters in terms of compactness and reliability continued the research in this arena. Over many advantages, packed U-cell topology was having a problem of requirement of a complex controller for voltage balancing of capacitors [5, 6]. External controllers were proposed for controlling the voltage of capacitors, but requirement of separate sensors and tuning of weighting factors was still a hectic task and time consuming. İn literature, the Packed U-Cell has been replaced by Packed E-Cell topology with modular design. There is no requirement of DC link, and number of switches have been reduced to seven [7, 8]. Another distinguishing feature is high redundancy which leads to reduced switching frequency and automatic voltage

© Springer Nature Singapore Pte Ltd. 2020
N. Goel et al. (eds.), *Modelling, Simulation and Intelligent Computing,*
Lecture Notes in Electrical Engineering 659,
https://doi.org/10.1007/978-981-15-4775-1_60

regulation of capacitors. Among other benefits, PEC has one fascinating attribute, i.e. ability of operating as five-level otherwise seven-level during faulty state of four-quadrant switch [9]. These all advantages have made PEC topology a promising solution and replacement of PUC [10, 11].

In this paper, authors have used the modified PEC converter for connecting distributed PV system with the grid [4]. This modified topology is comparable with existing well-established topologies like NPC, CHB, FCC, PUC and PEC. The control strategy is much easier than the parent PEC topology [12, 13]. In this topology, three distributed PV sources are used in which two PV sources are having same power and voltage rating and quarter times the third voltage source (e.g. sources of voltage level E and E/4 for generating nine-level output) with low harmonic content and proper distribution of voltages between the levels. The peak output AC voltage level is equal to the maximum amplitude of the PV source to the inverter.

Simulation and comprehensive analysis of modified PEC topology and its control strategy is done in this paper. The integrated distributed PV system is described in Sect. 2. Afterwards, the PEC converter and its modification are explained in Sect. 3. Further, control strategy is explained in Sect. 4. Simulation results and full discussion are done in Sect. 5 of this paper.

2   PEC and Modified PEC Inverter Topology

It is to be noted that the upper switches (S_1 and S_4) are linked to the DC source switches at comparatively lower switching frequency, whereas the remaining switches (S_2, S_3, S_5, S_6) switch at the switching frequency. Also, the two switches at the top (S_1 and S_4) must be capable of withstanding the DC voltage at input side. The input voltage at the auxiliary DC link, which is actually half of the voltage of DC source at the top, employs identical switches which have lower voltage rating as their upper counterparts and operates at higher frequency.

The MPEC has also the ability to operate at multilevel voltage at output AC terminals, where different voltage levels can be obtained without any modification in the original topology and these features make this topology a very reliable and secure configuration. For instance, MPEC will not cease to function in case of a fault on four-quadrant switch. In such a situation, five levels output can be obtained if DC link input voltage is maintained at half of DC supply voltage at input, and seven levels can be attained in case DC link voltage is maintained at one third of input DC voltage.

In this paper, a modified PEC converter topology is proposed in which the replacement of the capacitors in the DC link is taken place by two PV modules having the voltage magnitude one fourth of the third PV module. In MPEC topology, there is no need of active voltage balancing across DC link to obtain nine levels. So, the major advantage associated with the modified PUC is that nine-level (Fig. 1).

Single-phase AC voltage can be obtained by incorporating three PV modules, having two different voltage and power rating on the grid.

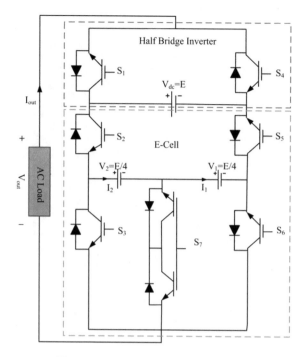

Fig. 1. Modified packed E-Cell inverter

3   System Description

Renewable energy, especially solar energy, demand has been increasing day by day, and for solar photovoltaic applications, inverter plays very important role of controlling voltage levels and grid integratation. Different control topologies are used to control the output of solar panels for optimizing the use of solar energy (Table 1; Fig 2).

Table 1. Modified PEC switching states

STATE	S1	S2	S3	S4	S5	S6	S7	Vout
1.	1	0	0	0	1	1	0	+Vdc
2.	1	0	0	0	1	0	1	+3Vdc/4
3.	1	0	0	0	1	0	0	+Vdc/2
4.	1	1	0	0	0	0	1	+Vdc/4
5.	0	0	1	1	1	1	0	0
6.	0	0	1	1	1	0	1	−Vdc/4
7.	0	1	1	1	0	1	0	−Vdc/2
8.	0	1	1	1	0	0	1	−3Vdc/4
9.	0	1	1	1	0	0	0	−Vdc

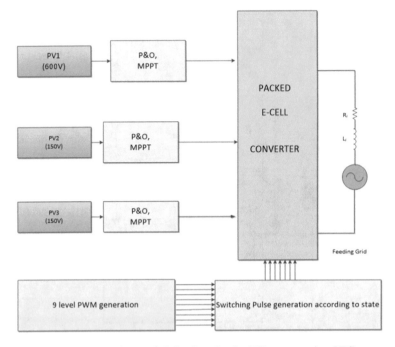

Fig. 2. Block diagram of distributed solar PV system using PEC

Packed E-cell inverter topology with slight modification is used in the photovoltaic application. In the system proposed, the different PV sources of different ratings are used as the replacement of two DC link capacitors and the DC source. Maximum Power Point Tracking (MPPT) techniques in conjunction with boost DC converters are used to control the DC voltages of PV arrays.

In the proposed system, three PV arrays, two of same power and voltage rating, i.e. one fourth of the third one is used. For simulation and experimental purpose, two 150 V and one 600 V PV arrays have been used. The P&O technique is used for drawing high power. This P&O technique is widely used in industries and well explained in literature. MPEC inverter delivers solar energy into the grid or loads.

Multiple controllers have been used to control the DC voltage and phase difference of load current and grid voltage. Waveforms and grid voltage also regulate voltage across capacitors to obtain nine-level voltage waveform having very less harmonic distortion, while Perturb & Observe MPPT algorithm extracts maximum power from PV array.

4   Control Structure

Control strategy for modified PEC is described as follows:

  I. Half parabola carrier and nine-level PWM generation wave generation.
 II. Switching function and pulse generation (Fig. 3).

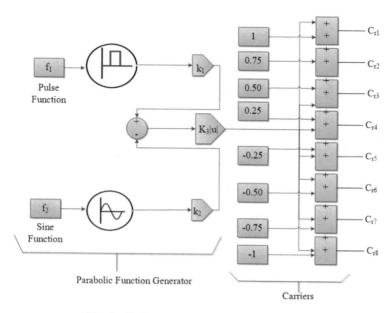

Fig. 3. Half parabola carrier wave generation

4.1 Half Parabola Carrier Wave Generation

Carrier wave which is half parabolic has been designed for or implementation of PWM technique for better performance and controlling the THD of voltage waveform at lower switching frequency as shown in figure. The half wave is generated by using sine and pulse function, considering pulse function f_1 and sine function f_2, where f_1 is two times greater than sinusoidal function f_2 ($f_1 = 2f_2$). The half parabola wave is obtained by those two waves f_1 and f_2. The factors k_1, k_2 and k_3 are also selected as 0.5, 1 and 0.25, respectively. Eight vertically shifted carrier waves are generated having amplitude of 0.25. In this stage, the carrier waves are compared with sinusoidal signal of fundamental frequency PWM pulses are generated by limiting the magnitude of signal generated in the previous stage for setting the magnitude of different voltage level.

4.2 Switching Function and Pulse Generation

In this stage, PWM signals for nine-level generation of specified magnitude are now summed up to form the final switching function generation. MATLAB program for switching pulse is generated by considering the states for corresponding nine levels according to the switching function generated in the previous stage. Seven switching signals are generated by this MATLAB functional block incorporating the effect of switching function.

Waveforms of load voltage and approximately in phase current are shown in above figure depicts that the output voltage consists of nine voltage levels, and the current waveform is approximately sinusoidal with very low harmonic distortion (Fig. 4).

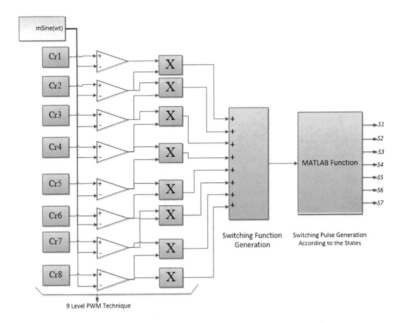

Fig. 4. Nine-level PWM and switching pulse generation

5   Simulation Results

Also, the peak value of the voltage appears at the output is equal to the voltage of that PV module having the maximum voltage rating. The above voltage across load and current waveforms clearly reveals that power factor is approximately unity (Fig. 5).

Low harmonic grid current and output load voltage of nine multilevel are recognizable in the figure. Cuurent harmonics in output are reduced because of the multilevel voltage waveform. A 40 mH inductor and 80 Ω resistor have been used in experiments. Harmonic spectrum of load current or grid feeding current clearly showing that the feeding current is under the harmonic limits laid down by IEEE 519 standard stating

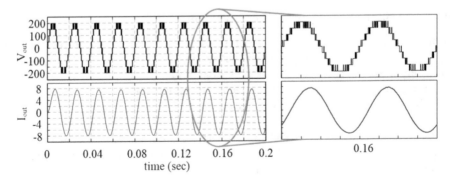

Fig. 5. Nine-level load voltage and current waveforms of PEC-9 converter topology

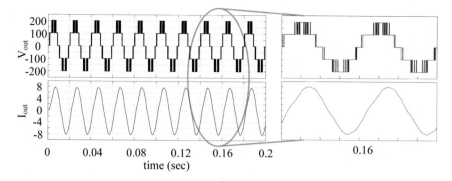

Fig. 6. Five-level load voltage and current waveforms of PEC9 running on five-level converter

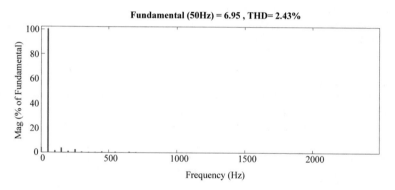

Fig. 7. Harmonic analysis of grid current injected by invereter from the PVs

current harmonics should be less than 5%. Also, it should be marked that the highest peak between harmonic order is on 2 kHz except the fundamental. Compared to the other well-established multilevel inverter topologies like modified PUC, the current THD in the MPEC inverter is very low, which is just 2.4% (Figs. 6 and 7).

6 Conclusions

In this paper, a novel modified packed E-cell converter topology has been presented in which three different PV sources of two different voltages and power ratings can be integrated together to give a single-phase AC which is close to sinusoidal waveform. The power factor obtained is approximately unity. The proposed inverter can generate nine-level load voltage waveform at the output having low harmonic contents. Also, the nine-level hybrid PWM with half parabola vertically shifted carriers is used which gives lesser THD when compared to triangular carrier waves. The proposed topology has multilevel terminal capability. MPEC can continue its operation even if the four-quadrant switch stops operating, without any change in power factor. In such cases, five- or seven-level output voltage can be achieved if the PV array of half or one third of rated voltage of the main PV array voltage rating is implemented.

References

1. Daher S, Schmid J, Antunes FL (2008) Multilevel inverter topologies for stand-alone PV systems. IEEE Trans Ind Electron 55(7):2703–2712
2. Nademi H, Das A, Burgos R, Norum LE (2016) A new circuit performance of modular multilevel inverter suitable for photovoltaic conversion plants. IEEE J Emerg Select Topics in PowerElectron 4(2):393–404
3. Sebaali F, Vahedi H, Kanaan HY, Moubayed N Al-Haddad K (2016) Sliding mode fixed frequency current controller designed for grid-connected NPC inverter. IEEE J Emerg Select Topics Power Electron 4(4):1397–1405
4. Sharifzadeh M, Vahedi H, Portillo R, Khenar M, Sheikholeslami A, Franquelo LG et al (2016) Hybrid SHM-SHE pulse amplitude modulation for high power four-leg inverter. IEEE Trans Ind Electron 63(11):7234–7242
5. Singh B, Chandra A, Al-Haddad K (2014) Power quality: problems and mitigation techniques. Wiley
6. Vahedi H, Al-Haddad K, Ounejjar Y, Addoweesh K (2013) CrossoverSwitches Cell (CSC): a new multilevel inverter topology with maximum voltage levels and minimum dc sources. In: IECON2013-39th annual conference on IEEE industrial electronics society, Austria, pp 54–59
7. Wang K, Zheng Z, Wei D, Fan B, Li Y (2017) Topology and capacitor voltage balancing control of a symmetrical hybrid nine-level inverter for high-speed motor drives. IEEE Trans Ind Appl 53(6):5563–5572
8. Sandeep N, Yaragatti UR (2017) Operation and control of an improved hybrid nine-level inverter. IEEE Trans Ind Appl 53(6):5676–5686
9. Sandeep N, Yaragatti UR (2018) Design and implementation of active neutral-point-clamped nine-level reduced device count inverter: An application to grid integrated renewable energy sources. IET Power Electron 11(1):82–91
10. Sharifzadeh M, Vahedi H, Sheikholeslami A, Labbé P-A, Al-Haddad K (2015) Hybrid SHM-SHE modulation technique for four-leg NPC inverter with DC capacitors self-voltage-balancing. IEEE Trans Ind Electron 62(8):4890–4899
11. Sebaali F, Vahedi H, Kanaan HY, Moubayed N, Al-Haddad K (2016) Sliding mode fixed frequency current controller designed for grid- connected NPC inverter. IEEE J. Emerg Select Topics Power Electron 4(4):1397–1405
12. Vahedi H, Sharifzadeh M, Al-Haddad K (2018) Modified seven-level pack U-cell inverter for photovoltaic applications. IEEE J Emerg Sel Topics Power Electron 6(3):1508–1516
13. Sharifzadeh M, Al-Haddad K (2019) Packed E-Cell (PEC) converter topology operation and experimental validation. In: IEEE Access, vol 7, pp 93049–93061

Using Sentiment Analysis to Obtain Plant-Based Ingredient Combinations that Mimic Dairy Cheese

Urvashi Satwani, Jaskanwar Singh, and Nishant Pandya[(✉)]

Birla Institute of Technology and Science Pilani, Dubai Campus, Dubai International Academic City, Dubai, UAE
{f20160222, f20160075, nishant}@dubai.bits-pilani.ac.in

Abstract. In this paper, crowdsourcing has been used to obtain meaningful conclusions and insights about consumer behavior with respect to plant-based cheese. A fivefold rise in the number of people that follow strict, plant-based diets (and those who are lactose intolerant), increase in conscious consumerism due to environmental awareness and high costs of natural cheese production are factors driving research to make plant-based food products accessible to the masses, with respect to taste similarity and cost-effectiveness. Previous research focused on the sustainability and cheaper costs associated with the production process. However, taste and textural similarity to dairy counterparts were found to be lacking. This paper aims to tackle the barriers attached with organoleptic properties (taste and texture) of food products by making use of widely available data from the online vegan community who are immersed in preparing versions of famously non-vegan foods. These recipes are then tried by thousands of others who leave reviews on their experiences. The underlying objective of this research was to analyze sentiments behind the reviews and comments left on each recipe. This was useful to analyze which base ingredient was responsible for the most positive sentiment. To recognize sentiments, Natural Language Toolkit (NLTK) Valence Aware Dictionary and Sentiment Reasoner (VADER) was put to use, which was able to score each review from -1 to 1, a compounded score based on the negativity, neutrality and positivity of the statement. These scores aided in the decision of raw material selection.

Keywords: Food technology · Opinion mining · Crowdsourcing

1 Introduction

1.1 Changing Food Trends

The research and development of plant-based food products is encouraged by compelling statistics of rising interest throughout the world. According to GlobalData, it has been published that the number of US consumers identifying as vegan grew from 1 to 6% between 2014 and 2017, a 600% increase [1]. Even though 6% is a small fraction of the overall population, there is significant evidence reporting growing interest by consumers who do not consider themselves vegetarian or vegan. For

N. Goel et al. (eds.), *Modelling, Simulation and Intelligent Computing*,
Lecture Notes in Electrical Engineering 659,
https://doi.org/10.1007/978-981-15-4775-1_61

example, sales of plant-based alternatives to animal-based foods, including meat, cheese, milk and eggs grew 17% over the year of 2017, while overall US food sales rose only 2%, according to data from Nielson and Good Food Institute [1].

This growing interest is due to many factors. With respect to dairy cheese, the need to find alternatives is because of rise in conscious consumerism. Injection of hormones in cows to meet demands, high greenhouse gas emissions, high energy requirement during processing and nutritional concerns over saturated fat content [2] are aspects discouraging consumers to consume dairy cheese.

1.2 Problems Encountered in Previous Research

Previous research was carried out to address concerns over cost-effectiveness and overall sustainability of the plant-based cheese production process. A home-scale production unit was set up, wherein three batches of cheese were produced, which varied in base ingredient (see Fig. 1).

However, challenges were posed with respect to taste and textural integrity of the cheese complexes, which are vital in food products. Achieving desirable sensory experiences that were comparable to dairy cheese proved difficult as it is not easy to quantify intangible qualities, and opinions vary from one subject to another. Bigger food research and developmental facilities typically study sensory experiences by testing on a 50–100 human member sensory panel [3]. Statistical tools are then used to analyze results and choose or pivot product formulations. Access to such a facility which is cost-intensive was not available. Secondly, there is no hard and fast rule with respect to ingredient selection so hand-picking ingredients to try out was a time-consuming process. Even after considerable time investment, no guarantee is given about taste and texture.

1.3 Proposed Methodology

Tracking sentiment in popular media has been of long interest to media and other economic channels. Rather than relying on polarity judgments from single expert, crowdsourcing techniques have been used in many fields to generate trend statistics from a number of non-expert users [4]. With the amelioration of technology and

Fig. 1 Cheese produced as part of previous research; left: cashew-based, middle: sunflower seeds-based and right: sweet potato and oats-based

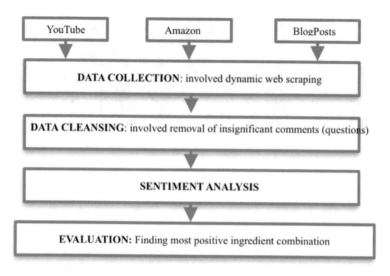

Fig. 2 Flow of work (proposed methodology)

mainstreaming of social media, the same principle can be used to tackle issues faced in food technology research. A promising alternative to the sensory panel employed by huge research facilities and would be crowdsourced opinions of the existing online vegan community on social media and mining those opinions to obtain meaningful insights about consumer behavior. There are a considerable number of people who post recipes online of plant-based alternatives to famous non-vegan foods, in the form of Blogposts or YouTube videos. They meticulously write about the serving size, ingredient composition and nutritional value. There also exists an equitable share of people who try and test these recipes, sharing their experiences as reviews. Since this information is easily available, this collective social intelligence can be harnessed by crowdsourcing this data and performing sentiment analysis over the reviews to aid in development of a product that would suit the target population, of whom the reviewers are an ideal sample set. Along with YouTube videos and Blogposts, consumer reviews of Amazon product listings of vegan cheese would also be considered. Figure 2 shows the flow of work.

2   NLTK: The Natural Language Toolkit, VADER

NLTK, the Natural Language Toolkit, is a suite of open-source program modules, tutorials and problem sets, providing ready-to-use computational linguistics courseware [5]. It is a library of python, which provides a base for building programs and classification of data. NLTK uses a simple bag of words model and returns for each text three probabilities: a probability of it being positive, one of it being neutral and one of being negative [6]. A compounded score is also shown that exhibits overall polarity (see Fig. 3).

```
In [85]: sid.polarity_scores('Aditya is good boy')
Out[85]: {'neg': 0.0, 'neu': 0.508, 'pos': 0.492, 'compound': 0.4404}

In [86]: sid.polarity_scores('Aditya is not good boy')
Out[86]: {'neg': 0.376, 'neu': 0.624, 'pos': 0.0, 'compound': -0.3412}

In [87]: sid.polarity_scores('The food here isn't really all that great')
Out[87]: {'neg': 0.0, 'neu': 0.616, 'pos': 0.384, 'compound': 0.6557}
```

Fig. 3 Negative, positive and neutral probability

Valence Aware Dictionary and Sentiment Reasoner (VADER), a tool of NLTK was put to use here. Social media and micro-blog content pose critical roadblocks to applications of sentiment analysis. This is due to the contextual sparseness resulting from shortness of the text as well as the tendency to use abbreviated language conventions and emoticons to express sentiments [7]. VADER is especially trained and tested on social media content, and thus it becomes natural choice for use. VADER uses a lexicon-based approach, where the lexicon contains the intensity of all the sentiment showing words. The intensities are fetched, the sentiment score is calculated and based on this sentiment score, the review is classified as either positive or negative [8].

2.1 Comparative Advantage

2.1.1 Other Sentiment Analysis Tools
When comparing VADER to seven other well-established sentiment analysis lexicons, it has been found that VADER performs exceptionally well in the social media domain and generalizes favorably [7] (see Fig. 4). In tweets, it shows greater accuracy than humans.

2.1.2 Machine Learning Approaches
VADER performs as well or better across domains than the machine learning approaches (like support vector machine and neural networks) do in the same domain for which they were trained [7, 8].

2.2 Inherent Advantages

It is also widely known that the simplicity of VADER makes it computationally efficient. Most traditional "opinion mining" tools work in a two-step process [9]. Furthermore, in this use case, it does not require any training data. It does not suffer from a speed-performance trade-off. It works well with social media text as it is able to understand intricacies of human involvement in social media where punctuation, such as addition of exclamation mark increases the intensity without changing the semantic orientation. Another example is where the capitalization of a whole word increases intensity too. It is also able to quantify sentiments behind emojis, emoticons and slangs.

	Correlation to ground truth (mean of 20 human raters)	3-class (positive, negative, neutral) Classification Accuracy Metrics			Ordinal Rank (by F1)		Correlation to ground truth (mean of 20 human raters)	3-class (positive, negative, neutral) Classification Accuracy Metrics		
		Overall Precision	Overall Recall	Overall F1 score				Overall Precision	Overall Recall	Overall F1 score
Social Media Text (4,200 Tweets)							**Movie Reviews (10,605 review snippets)**			
Ind. Humans	0.888	0.95	0.76	0.84	2	1	0.899	0.95	0.90	0.92
VADER	0.881	0.99	0.94	0.96	1*	2	0.451	0.70	0.55	0.61
Hu-Liu04	0.756	0.94	0.66	0.77	3	3	0.416	0.66	0.56	0.59
SCN	0.568	0.81	0.75	0.75	4	7	0.210	0.60	0.53	0.44
GI	0.580	0.84	0.58	0.69	5	5	0.343	0.66	0.50	0.55
SWN	0.488	0.75	0.62	0.67	6	4	0.251	0.60	0.55	0.57
LIWC	0.622	0.94	0.48	0.63	7	9	0.152	0.61	0.22	0.31
ANEW	0.492	0.83	0.48	0.60	8	8	0.156	0.57	0.36	0.40
WSD	0.438	0.70	0.49	0.56	9	6	0.349	0.58	0.50	0.52
Amazon.com Product Reviews (3,708 review snippets)							**NY Times Editorials (5,190 article snippets)**			
Ind. Humans	0.911	0.94	0.80	0.85	1	1	0.745	0.87	0.55	0.65
VADER	0.565	0.78	0.55	0.63	2	2	0.492	0.69	0.49	0.55
Hu-Liu04	0.571	0.74	0.56	0.62	3	3	0.487	0.70	0.45	0.52
SCN	0.316	0.64	0.60	0.51	7	7	0.252	0.62	0.47	0.38
GI	0.385	0.67	0.49	0.55	5	5	0.362	0.65	0.44	0.49
SWN	0.325	0.61	0.54	0.57	4	4	0.262	0.57	0.49	0.52
LIWC	0.313	0.73	0.29	0.36	9	9	0.220	0.66	0.17	0.21
ANEW	0.257	0.69	0.33	0.39	8	8	0.202	0.59	0.32	0.35
WSD	0.324	0.60	0.51	0.55	6	6	0.218	0.55	0.45	0.47

Fig. 4 VADER three-class classification performance as compared to individual human raters and seven established lexicon baselines across four distinct domain contexts [7]

3 Results and Discussions

Results were obtained, wherein the compounded score was taken into account to assess the overall polarity of each review. Overall picture of the sentiments can be seen from Fig. 5 and understanding the difference between positive and negative reviews can also be judged from the same. The next step was to label each review according to the base ingredient used, which is the most fundamental component of the cheese. (A recipe contains three main sets of ingredients: base, flavoring and textural) Then, the bar graph showing each of the recipes labeled by ingredient with respect to the percentage polarity portrays which ingredient combination is most favorable. If there was more than one recipe with the same base ingredient, percentage positivity of the common ingredient was calculated as the weighted mean of the corresponding individual. This same method of analysis can be extrapolated to understand water content, textural ingredient compositions as well as the range of flavoring ingredients that can be possible to prepare the most similar plant-based cheese. This is shown for each of the three datasets classified on data source: YouTube videos, Amazon product offerings as well as Blogposts.

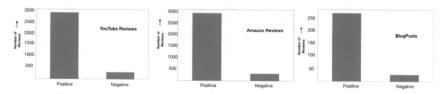

Fig. 5 Number of positive reviews against negative reviews for (left to right) YouTube comments, Amazon reviews and Blogposts

A common observation is that the number of negative reviews is significantly lower than positive (as shown in Fig. 5). This might be because of the demographics associated with the reviewers as most of the commenters were vegan, who continuously want to reinforce positivity throughout the small, but growing community. This can also be due to the inherent precision error with respect to negative sentiments by VADER. Also, product reviews and opinions on recipes should have been treated differently since they are of different intent [10].

3.1   YouTube

Around 3000 data points were taken into account. Cashew-soya milk was the most positive reviewed combination (refer Fig. 6).

3.2   Amazon Product Listings

Around 3000 data points were examined, here too. The cashew-based plant-based cheese performed well compared to the others. The other three products: soy, zucchini and walnut-based perform equally well. Walnuts and zucchini are that have not been studied earlier and show the potential for further research (see Fig. 7).

3.3   Recipe Blogposts

Only around 250 data points were collected because personal Blogposts cannot gain as much traction as YouTube videos or commercial Amazon product listings. When analyzed equal highest sentiment is attached to both almond-based and cashew-coconut oil (see Fig. 8). On the other hand, the combination of cashew-tapioca is seen to have the least percent positivity across all the reviews from the three data sources (i.e., less than 80%).

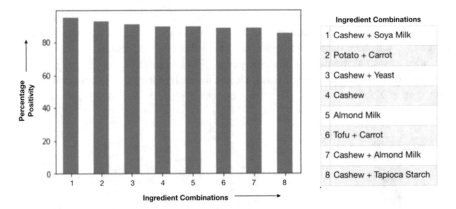

Fig. 6 Cashew and soya milk combination is most appreciated from YouTube videos

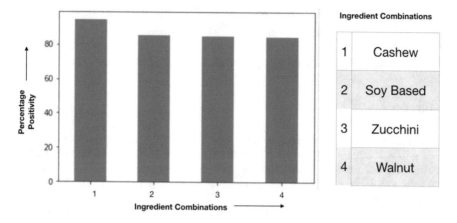

Fig. 7 Cashew is most appreciated from Amazon reviews

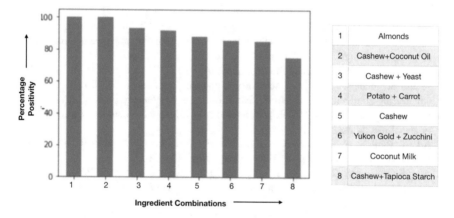

Fig. 8 Almond and cashew and coconut combinations perform well from all the recipes on Blogposts

4 Conclusion

It can be concluded that cashew-soy, cashew, cashew-coconut oil and almond-based cheese substrates can be produced by the home-scale method, as per the positive results obtained from the NLTK VADER tool that analyzed sentiments of reviews on plant-based cheese for three types of data sources. It can also be said that roughly, only 10% of the reviews were negative out of the total reviews on every dataset. However, some factors affect the clarity of the final base ingredient selection such as the number of reviews, the number of times the Web site was visited, security of pages (since insecure Web sites are prone to scrapers and automated scripts), quality of comments (in significant instances, they were questions about procedure, instead of review based on trial), click rates on Web sites and understanding of content on sites for different kinds of people.

The research done provides a novel way of transitioning into sustainable plant-based foods without compromising taste. This research concluded that by employing NLP tools and data crowdsourcing we can achieve accurate results rivaling those achieved in expensive laboratories which employ extensive taste improving techniques.

References

1. Forbes: the Growing Acceptance of Veganism (2018). https://www.forbes.com/sites/janetforgrieve/2018/11/02/picturing-a-kindler-gentler-world-vegan-month/#511f6fb92f2b
2. Animal production and health division, food and agriculture organisation of the Unites States: a report on greenhouse gase emissions from the dairy sector (2010). http://www.fao.org/3/k7930e/k7930e00.pdf
3. Peyvieux C, Dijksterhuis G (2001) Training a sensory panel forTI: a case study. Food Quality and Preferance 12(1):19–28
4. Brew A, Greene D, Cunningham P (2010) Using crowdsourcing and active learning to track sentiment in online media. In: ECAI 2010: 19th European conference on artificial intelligence. pp 145–150
5. Loper E, Bird S (2002) NLTK: the natural language toolkit. In: Proceeding of ACL-02 workshop on effective tools and methodologies for teaching natural language processing and computational linguistics, vol 1, pp 63–70
6. Jongeling R, Sarkar P, Datta S, Serebrenik A (2017) On negative results when using sentiment analysis tools for software engineering research. In: Empirical software engineering conference 2017, vol 22, No 5, Springer, Heidelberg, pp 2543–2584
7. Hutto C, Gilbert E (2014) VADER, a parsimonious rule-based model for sentiment analysis of social media text, AAAI
8. Adarsh R, Patil A, Rayar S, Veena KM (2019) Comparison of VADER and LSTM for sentiment analysis. Int J Recent Technol Eng. 7(6):540–543
9. Attardi G, Simi M (2016) Blog mining through opinionated words. In: Proceeding of 15th text retrieval conference (TREC). http://citeseerx.ist.psu.edu/viewdoc/download?doi=10.1.1.76.1128&rep=rep1&type=pdf
10. Nyugen H, Veluchamy A, Diop M, Iqbal R (2018) Comparative study of sentiment analysis with product reviews using machine learning and lexicon-based approaches. MU Data Sci Rev 1(4)

Real-Time Fog Removal Using Google Maps Aided Computer Vision Techniques

Ashlyn Selena DSouza, Rifah Mohamed, Mishika,
and V. Kalaichelvi[✉] [ID]

Birla Institute of Technology and Science Pilani, Dubai Campus, Dubai, UAE
{f20160003, f20160211, f20160261, kalaichelvi}@dubai.
bits-pilani.ac.in

Abstract. This paper aims to tackle the problem of impaired visibility for drivers on the road due to fog, which is a safety concern. This novel approach is a unique comparative algorithm through integration with Google Maps and has several embedded functionalities to reduce noise caused by fog. Real-time input is collected in the form of continuous video frames, on which image processing is carried out. This is a two-step process, first using dark channel prior and second using histogram matching with ideal weather Google Street View images. In order to measure the fogginess of the image at each step, horizontal variance is used. The results obtained show a drastic increase in variance during the two-step process, which is in line with the theory that the higher the variance, the lesser the fogginess. The fog-free images are retrieved and put together to form continuous frames of a video, which is displayed on the driver's screen in real time.

Keywords: Image processing · Driver's assistance system · Fog removal · Google street view

1 Introduction

Road safety is a perpetual concern, and car manufacturers constantly strive toward developing new technology and safety equipment. Yet dense fog reduces visibility drastically to much below 1 km [1] and is a perilous driving scenario that causes major pile-ups and accidents. Fog results in less light reaching the driver's eye as it has a whitish hue due to air light and causes a reduction in saturation and contrast of the view ahead as described in [1]. All these contributing factors make fog a serious concern on roads, yet no satisfactory and fail-safe system has been implemented.

Fog detection and removal are particularly complex as ambiguity with regard to image depth prevents differentiation of foreground and background essential to image processing. Image processing is the process of improving quality or extracting information from images. It can be used for fog reduction through restoration techniques and overall enhancement. Major drawbacks still persist though, such as imperfect data collection, inefficient real-time processing and low adaptivity of trained code to real scenarios and failure in case of unforeseen circumstances. This paper strives to come up with an innovative approach that existent issues. Advanced driver's assistance systems

(ADASs) are a growing technology that aims to increase driver comfort and safety on the road through techniques such as reverse parking aids, lane detection and sensory collision avoidance. Fog disrupts many such functions, an example is traffic sign detection as studied in paper [2], especially if they are camera dependent, as image degradation occurs with low contrast and color intensity.

2   Literature Review

There are various ways in which researchers have worked on image processing of fog in real time. K He and J Sun introduced an algorithm called dark channel prior (DCP) in paper [3] which cleared the haziness in an image in three major steps—dark channel construction, atmospheric light estimation and transmission map estimation. Several variations in the method were proposed. For dark channel construction, different local patch sizes were used—3 × 3 in [4], 11 × 11 in [5] and 15 × 15 in [3, 6, 7]. Atmospheric light is estimated in order to obtain the brightest pixel in the sky. Long [8] and Chen [5] used the pixel with the highest dark channel value directly, but this gave false outputs when there was a bright object in the image. Transmission map is then obtained which provides depth information. Further, the map estimated is refined with filters, to smoothen the pixels on the video frames. The methods like guided filter, soft matting and cross-bilateral filter [9, 10] use foggy image as a cross channel and give sharper output images as compared to other methods (Bilateral and Gaussian filters).

Earlier methods required the use of multiple images taken under different weather conditions. One such method was the polarization filter method proposed by Schechner et al. [11] that uses images taken through polarizers at different orientations and reverses the polarization of scattered atmospheric light caused by fog. DCP used a single input but the clearer output has a higher contrast as compared to natural light image as well as a halo effect. DCP also cannot enhance image patches that match air–light conditions and have no shadows cast on them. This paper focuses on overcoming limitations of DCP through a novel approach that makes use of the extensive road data readily available through the Google Street View Application Programming Interface (API). This algorithm uses just two images, the real-time input and a 360° panoramic image obtained from the API. In an advanced drivers' assistance system, street view images can be obtained at the driver's real-time location. Having an ideal Google Street View and the real-time foggy view, histogram matching is used to redistribute the pixels in the foggy image to match the histogram of the ideal case. This is a derived method of histogram equalization, which redistributes the pixels equally among each intensity level. Contrast Limited Adaptive Histogram Equalization (CLAHE) is a modification which applies histogram equalization to pixels clipped below a certain maximum [12]. In the next Sect. 3, there is an elaborate discussion on how histogram matching and DCP algorithm have been incorporated together and improved upon in this paper.

3   Analysis of Proposed Algorithm

3.1   Mathematical Description

3.1.1   Dark Channel Prior (DCP)

Dark channel prior-based defogging algorithm used in this paper has the following steps:

For clear images, there are pixels with low intensity that is close to zero in at least one color channel, in a local patch, everywhere except the sky region, this is called the dark channel. Thus, for any image J, it is dark channel, and J^{dark} can be obtained as,

$$J^{dark}(x) = min_{c \in \{r,g,b\}} \left(min_{y \in \Omega(x)} (J_c(y)) \right) \tag{1}$$

where J^c is the color channel of J, and $\Omega(x)$ represents the local patch centered at x.

Next, atmospheric map A needs to be estimated in order to obtain transmission map. For this, the most opaque hazed pixels are chosen where these constitute the $p\%$ dark channel values. In this scenario, p is the probability of a pixel value in a local patch x.

$$A = I \left(argmax_x \left(I^{dark}(x) \right) \right) \tag{2}$$

where I is the foggy input image, and I^{dark} is its dark channel image.

The transmission map is defined by the following equation,

$$\widehat{t}(x) = 1 - \omega_{y \in \Omega(x)} \left(min_c \frac{I^c(y)}{A^c} \right) \tag{3}$$

The image will not look natural if fog is removed completely; hence, a small amount of fog is retained through a constant ω $(0 < \omega < 1)$.

After considering a minimum transmission value t_{min} as 0.1 whenever t tends to 0, the final defogging model J is described as,

$$J(x,y) = (I(x,y) - A)/\max(t(x,y), t_{min}) \tag{4}$$

3.1.2   Histogram Matching

The probability distribution of an image histogram is given by,

$$p_r(r_j) = n_j/n \tag{5}$$

where p_r is the probability of the intensity level r_j, n_j is the number of pixels with the intensity level, and r_j is the total number of pixels in the image. If $p_z(z_j)$ is our desired probability distribution function, then $p_r(r_s)$ must be transformed to match it. The pdfs can be represented by the distribution function,

$$S(r_k) = \sum_{j=0}^{k} p_r(r_j) \tag{6}$$

where $k = 0, 1, 2, 3,...$

$$G(z_k) = \sum_{j=0}^{k} p_z(z_j) \tag{7}$$

where $k = 0, 1, 2, 3,..., L$

where L is the number of intensity levels. Then each r value is mapped to a z value, that has the same probability in the expected output pdf.

$$S(r_j) = G(z_i) \text{ or } z = G^{-1}(S(r)) \tag{8}$$

3.2   Two-Step Approach for Fog Removal

Figure 1 gives an overview of the working algorithm using a two-step approach. Foggy input in the form of video frames is obtained using a 180-degree-wide angle camera and may be mounted behind the rear-view mirror as this provides a road view closely matched to that of the driver. As per the discussion of feasibility of transferring data to a cloud processor in paper [13], it is proposed that since this is not a very heavy application and can be run at good speeds even without a GPU, all processing on the proposed system shall be on board. Any computational advantages offered by cloud processing would be negligible and as per the paper would require superior 5G network speeds.

For Step 1, these input frames are fed into a DCP algorithm which involves four processes. Initially, a dark channel is constructed using a patch size of 15×15 in Eq. (1) as a larger patch size enables faster processing of video frames. For the first process of atmospheric light estimation, in Eq. (2), the top 0.1% of the brightest pixels in the dark channel is used and pixel with the highest intensity value in the input image, among the selected pixels, is used as the atmospheric estimation. This is instead of directly using highest intensity pixel, which causes false values of atmospheric light.

Then, the second process, transmission map estimation is carried out as per Eq. (3), and fog percentage of $\omega = 0.95$ was retained as it gave the best results experimentally. The output is used for the third process—transmission map refinement, for which a guided filter is used for edge-preserving pixels smoothing and noise reduction. It is a highly efficient, linear-time transformation for color images and takes neighboring pixels into consideration. The same foggy image is used as guidance; thus, edges are preserved perfectly, and intensity similarities are maintained. The fourth and final step is image restoration given by Eq. (4).

Step 2 of the algorithm is enhancement of the obtained DCP output using a comparative algorithm, i.e., the histogram matching, and this utilizes Google Street View API. Images are extracted for the driver's location using GPS and the {latitude, longitude} is then set as described in the article [14]. Heading and pitch parameters are

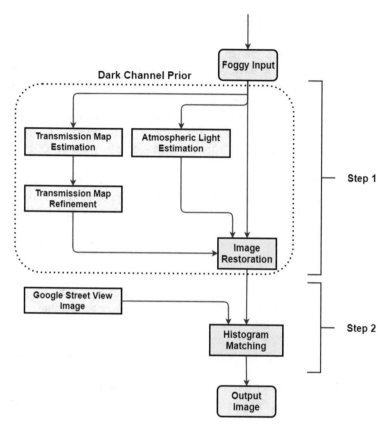

Fig. 1 Flowchart of the proposed methodology

set, so that obtained 360 Street View image can be constrained to the direction and angle that closely aligns with the drivers' view of the road. {left heading, right heading} is obtained to restrict the camera rotation. Images are extracted for a bracket of ±90 degree offset and matched with the 180-degree foggy input. Further pitch is left at the default value 0, as it provides sufficient upward and downward view for the driver. For continuous input of data with decent processing speed, the sequence of the road {sequence} is obtained every 10 meters ensuring sufficient overlap in the images. The images are downloaded over an internet connection. In order to hasten this process, we propose that they are downloaded in the background, based on the GPS route set by the driver. The images once used, can be discarded to save storage space.

The obtained Google Street View image is used as the reference in the histogram matching algorithm for the DCP output image. Since Google ensures that its street view images are taken in the best weather conditions, it fares well for this application. The histogram matching algorithm redistributes the pixel intensities in the DCP output frame, so that its new histogram matches the histogram of the ideal street view image. Thus, DCP frame is made to resemble the optimal weather image, with the benefit of brightened and natural-looking output frame.

4   Results and Discussion

All simulation studies were carried out using Python and MATLAB software to enhance the effectiveness of our proposed algorithm. The two-step algorithm has processing speed of 50 ms per frame and is effective in real time. Further, it processes 25 frames per second, and with the 180-degree foggy input, this ensures that the proposed system gives prior warnings of fast-moving objects in the lateral view of the driver that may enter the frames. To better meet the real-time threshold of 30–40 ms, at 25 fps as suggested in paper [13], a better processor and possibly GPUs may be considered in the future.

It can be clearly seen that applying DCP on foggy input image Fig. 2a gives a depth-restored, clearer output Fig. 2b. However, the image obtained is dark and has a halo effect on it. So, histogram matching was performed with the corresponding Google Street View image as reference, to improve image quality.

Figure 3a is the output image obtained after DCP that is used as input in histogram matching algorithm. It is dark, and hence, its image histogram Fig. 3b that depicts the frequency distribution of 'RGB' pixels over the image is highly concentrated toward the low-intensity region. Figure 3c is a brighter Google Street View image, and its histogram Fig. 3d is used as reference histogram. After histogram matching, we get the resulting image shown in Fig. 3e with a new histogram and Fig. 3f that closely resembles the reference histogram. Thus, the result is a brighter and sharper output.

According to [15], which discusses the use of statistical measures in digital image processing, the variance can be utilized to determine edge positions in images. Highly foggy images do not have clearly demarcated edges and therefore have very low horizontal variance. In [16], this property of images is used to estimate visibility range in foggy images. Figure 4 gives the horizontal variance ($\sigma 2$) for each stage of the algorithm. The drastic rise in horizontal variance is apparent, between the original foggy image ($\sigma 2 = 261$), the output of the DCP algorithm ($\sigma 2 = 1091$) and the final result after histogram matching with Google Street View ($\sigma 2 = 2739$). Since fogginess has an inverse proportionality to pixel variance, this incremental change correlates to a significant decrease in fogginess.

(a)   (b)

Fig. 2   **a** Foggy input image, **b** DCP output image

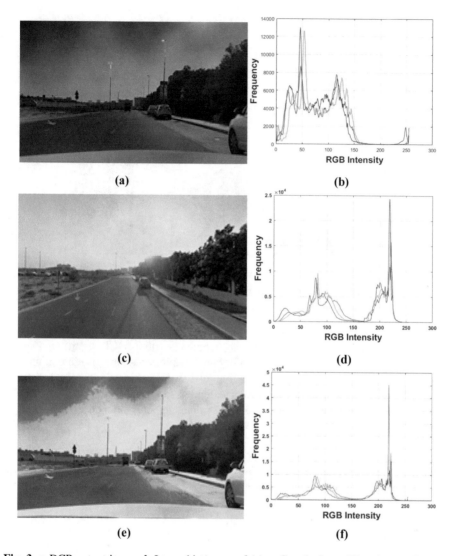

Fig. 3 **a** DCP output image, **b** Image histogram of (a), **c** Google Street View image, **d** Image histogram of (c), **e** Histogram matching output image, **f** Image histogram of (e)

5   Conclusion and Future Work

The results of the paper match the visual findings and imply an increase in horizontal pixel variance. RGB pixel variance from foggy image to DCP output shows a 4.19 times increase and from DCP output to histogram matched output shows a 2.51 times increase. This shows that the two-step proposed method is an improvement over traditional stand-alone DCP processing. The 180-degree input taken at a rate of 25 frames per second ensures safety on road through early warnings, and a processing speed of 50

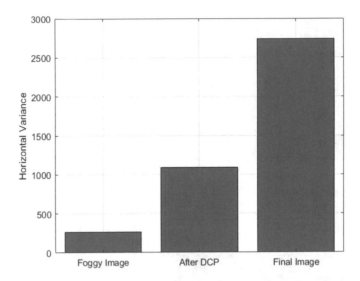

Fig. 4 Fogginess comparison using horizontal variance

ms ensures real-time information. The comparative algorithm results in realistic natural light images, where the halo effect is also avoided. Further, DCP algorithm may be unable to clarify patches that are whitish in hue and similar to air light. In conjunction with street view obtained images, through comparison, such details can be recovered by the algorithm. Research can be further extended to enhance other existent ADAS features such as lane maintenance and object detection in conjunction with the fog removal.

Acknowledgements. The authors are greatly appreciative for the support provided by Birla Institute of Technology and Science Pilani, Dubai Campus, Dubai, UAE. Further, the lab facilities provided a conducive environment for innovation.

References

1. Patel SP, Nakrani M (2016) A review on methods of image dehazing. Int. J. Comput. App. 133(12):44–49
2. Belaroussi R, Gruyer D (2014) Impact of reduced visibility from fog on traffic sign detection. In: IEEE Intelligent vehicles symposium proceedings, Dearborn, MI, pp 1302–1306
3. He K, Sun J, Tang X (2009) Single image haze removal using dark channel prior. In: Proceedings of the IEEE Computer vision and pattern recognition (CVPR '09), Miami, Fla, USA, pp 1956–1963
4. Cheng YJ, Chen BH, Huang SC, Kuo SY, Kopylov A, Seredin O, Mestetskiy L, Vishnyakov B, Vizilter Y, Vygolov O, Lian CR, Wu CT (2013) Visibility enhancement of single hazy images using hybrid dark channel prior. In: Proceedings of IEEE International conference on systems, Man, and Cybernetics, SMC, Manchester, pp 3267–3632

5. Lv X, Chen W, Shen IF (2010) Real-time dehazing for image and video. In: Proceedings of the 18th IEEE Pacific conference on computer graphics and applications, HangZhou, pp 62–69

6. Jeong S, Lee S (2013) The single image dehazing based on efficient transmission estimation. In: Proceedings of IEEE International conference on consumer electronics, Las Vegas, pp 376–377

7. Kil TH, Lee SH, Cho NI (2013) Single image dehazing based on reliability map of dark channel prior. In: Proceedings of IEEE 20th International conference on image processing, Melbourne, pp 882–885

8. Long J, Shi Z, Tang W (2012) Fast haze removal for a single remote sensing image using dark channel prior. In: Proceedings of international conference on computer vision in remote sensing, Xiamen, pp 132–135

9. Linan Y, Yan P, Xiaoyuan Y (2012) Video defogging based on adaptive tolerance. TELKOMNIKA Indonesian J. Elec. 10(7):1644–1654

10. Lin Z, Wang X (2012) Dehazing for image and video using guided filter. Appl. Sci. 2 (4B):123–127

11. Schechner YY, Narasimhan SG, Nayar SK (2001) Instant dehazing of images using polarization. In: Proceedings of the 2001 IEEE computer society conference on computer vision and pattern recognition, CVPR 2001, Kauai, HI, USA, pp 1–1

12. Xu Z, Liu X, Ji N (2009) Fog removal from color images using contrast limited adaptive histogram equalization. In: 2009 2nd International congress on image and signal processing, Tianjin, pp 1–5

13. Olariu C, Ortega J, Yebes J (2018) The role of cloud-computing in the development and application of ADAS. In: 26th European signal processing conference (EUSIPCO), Rome, pp 1037–1041

14. A Google street-view image processing pipeline. https://datasciencecampus.github.io/street-view-image-processing/. Accessed 10 Nov 2019

15. Kumar V, Gupta P (2012) Importance of statistical measures in digital image processing. Int. J. Emerging Technol. Adv. Eng. 2(8):6–12

16. Chincholkar S, Rajapandy M (2020) Fog image classification and visibility detection using CNN. In: Pandian A, Ntalianis K, Palanisamy R. (eds) Intelligent computing, information and control systems. ICICCS 2019. Adv. Intell. Syst. and Computi. vol 1039, Springer, Cham, pp 249–257

Experimental Verification of Shunt Active Power Filter for Harmonic Elimination

Neethu Elizabeth Michael[1](✉) ⓘ, Suhara E. M[2] ⓘ,
and Jayanand B[2] ⓘ

[1] Department of Electrical and Electronics Engineering, Birla Institute of
Technology and Science Pilani, Dubai Campus, Dubai, UAE
neethueliza@gmail.com
[2] Department of Electrical Engineering, Government Engineering College,
Thrissur, India
{suharanas, jayanandb}@gmail.com

Abstract. Active power filter (APF) is one of the effective means for harmonic current compensation in power grid. In this study, design and implementation of a shunt active power filter based on synchronous detection method is done. Harmonic and reactive current drawn by a nonlinear load are compensated by this filter. The proposed control technique has been simulated using MATLAB/Simulink and validated experimentally. Control algorithm is implemented in Real-Time Windows Target, along with a PCI 1711 card for data acquisition.

Keywords: Shunt active filter · Synchronous detection · Real-Time windows target

1 Introduction

Because of the widespread use of nonlinear loads, such as diode or thyristor rectifiers and a vast variety of power electronic appliances, harmonic current pollution of three-phase electrical power systems has become a serious issue. Conventionally, passive inductance–capacitance (LC) filters have been used to eradicate current harmonics. However, they have the following disadvantages: (a) Possibility of resonances with source impedance; (b) fixed compensation; (c) they are bulky, load dependent and inflexible [1]. They can also cause resonance to the system [2]. To meet all these problems, active power filters (APFs) have been considered. The evaluation of reference current extraction methods in active power filters has become more popular and has attracted great attention [3].

Figure 1 shows the basic compensation principle of shunt active filter (SAPF). It is designed to be connected in parallel with the nonlinear load to detect its harmonic and reactive current and then to inject a compensating current into the system. Thus, it can cancel the harmonic components drawn by the nonlinear load and keeps the utility line current sinusoidal. SAPF consists of an inverter with switching control circuit. The inverter generates the desired compensating harmonics. The SAPF injects harmonic

N. Goel et al. (eds.), *Modelling, Simulation and Intelligent Computing*,
Lecture Notes in Electrical Engineering 659,
https://doi.org/10.1007/978-981-15-4775-1_63

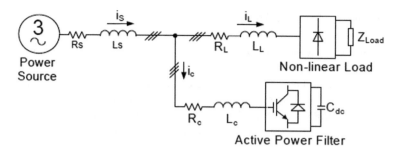

Fig. 1 Shunt active power filter

current required by the nonlinear load and makes the current at the source side purely sinusoidal.

This shunt active power filter is based on synchronous detection theory [4]. This work is motivated by the fact that the ability of a converter to perform effectively as an active filter is limited by the power and the frequency distribution of the distortion for which it must compensate. This system is comprised of a three-phase shunt active filter and the main objective is to simulate and implement a shunt active filter utilizing synchronous detection technique. It also substantiates the reliability and effectiveness of this model for the integration into power system network. The scope of the project based on the objective above is; firstly, to identify and determine the nonlinear load and its current waveform patterns. Secondly is to develop and design a harmonics extraction algorithm and current controller for a closed-loop power system. Thirdly is to simulate the designed harmonics extraction algorithm and controller by using MATLAB/Simulink environment. Finally, is to implement the shunt active filter for harmonic current compensation. Thus, in this paper, the simulation and hardware implementation of synchronous detection algorithm is presented. The simulation results are experimentally tested on a standard inverter to confirm the validity of the proposed theory and it verifies the sustainability and effectiveness of the shunt active filter.

Abaali, Lamchich, and Raoufi, "Shunt Power Active Filter Control under Non Ideal Voltages Conditions" [5] propose the modified synchronous detection (MSD) method for determining the reference compensating currents of the shunt active power filter under unbalanced voltages source conditions. Bhuvaneswari, Sathish Kumar Redd and Randhir Sigh "Simulation and Hardware Implementation of DSP Based Shunt Active Power Filter" [4] present the simulation and real-time hardware implementation of shunt active power filter in Analog Devices DSP (ADMC401) environment. The user manual for DIGIVAC AC Drive, Startup Manuel and Lab Experiment Manuel [6] explains how the DSP-based motor controller trainer kit can be used for harmonic elimination. The Math works user's guide-Embedded Target for the TI TMS320C6000™ DSP Platform [7] introduces the usage of Embedded Target DSPs with real-time workshop.

2 Harmonic Current Detection

The synchronous detection method is applied to calculate compensating currents while the three-phase source is feeding a highly nonlinear load [4]. The following assumptions are made in calculating the three-phase compensating currents using equal current distribution method of synchronous detection algorithm: (i) Voltage is not distorted; (ii) loss in the neutral line is negligible. In this algorithm, three-phase main currents are assumed to be balanced after compensation [2]. Therefore,

$$I_{am} = I_{bm} = I_{cm} \tag{1}$$

where I_{am}, I_{bm}, I_{cm} are the amplitudes of the three-phase main currents after compensation. The real power consumed by the load can be represented as

$$P = V_a V_b V_c \begin{bmatrix} I_a \\ I_b \\ I_c \end{bmatrix} \tag{2}$$

The real power P is sent to a low pass filter to obtain its average value P_{av} and is split into the three phases of the main supply as

$$P_a = (P_{av} V_a)/V_{tot} \tag{3}$$

$$P_b = (P_{av} V_b)/V_{tot} \tag{4}$$

$$P_c = (P_{av} V_c)/V_{tot} \tag{5}$$

where V_a, V_b, V_c are the amplitude of the mains voltages and V_{tot} is the sum of V_a, V_b, V_c. The desires mains currents are calculated as

$$I_{acc} = (2V_a P_a)/V_a^2 \tag{6}$$

$$I_{bcc} = (2V_b P_b)/V_b^2 \tag{7}$$

$$I_{ccc} = (2V_c P_c)/V_c^2 \tag{8}$$

The reference compensation currents can then be calculated as

$$i_{can}^* = i_{an} - i_{acc} \tag{9}$$

$$i_{cbn}^* = i_{bn} - i_{bcc} \tag{10}$$

$$i_{ccn}^* = i_{cn} - i_{ccc} \tag{11}$$

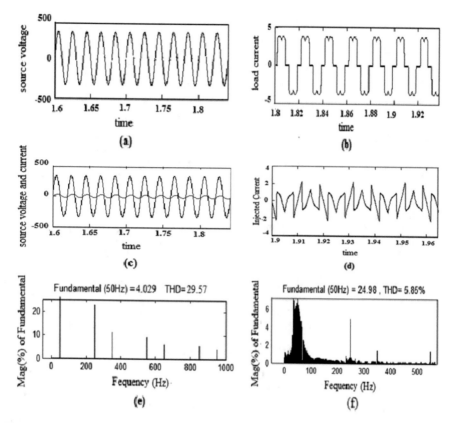

Fig. 2 a Voltage source, **b** line current before compensation, **c** source voltage and current after compensation, **d** injected current, **e** Spectrum of line current before compensation, **f** spectrum of line current after compensation

3 Simulation Results

MATLAB/Simulink was used in as simulation tool in this development, as it offered an integrated environment between designing control algorithms and the electrical network models. Here, the three-phase voltages source is balanced. The system parameters are phase voltage 415 V and frequency 50 Hz. The interface reactor used is 6.5 mH and the DC-link capacitor is 2000μF with the DC-link reference voltage being 680 V. Hysteresis current control is used for controlling the inverter so that an output current is generated which follows a reference current waveform [8, 9]. It decides the switching pattern of active power filters. Figure 2a, b shows the voltage source and the line current before compensation. Source voltage and current after compensation injected current spectrum of line current are shown in Fig. 2c–f. The line current total harmonic distortion (THD) before compensation is 29.57% which is reduced to 5.85% after compensation.

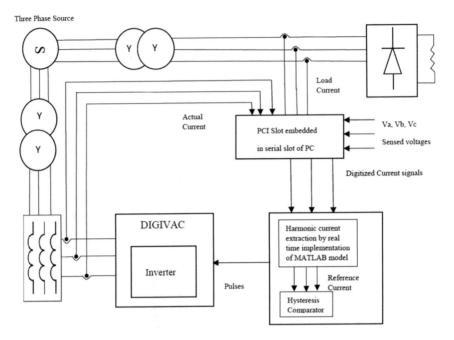

Fig. 3 Connection diagram

The switching logic is formulated as:

If $I_{ca} < (i^*_{can} - H_b)$, upper switch is off and lower switch is on for leg "a" (SA = 1). If $I_{ca} > (i^*_{can} - H_b)$, upper switch is on and lower switch is off for leg "a" (SA = 0). Where I_{ca} is the reference line current of the active power filter and i^*_{can} is the actual line current of the active power filter. The switching functions for three phases are determined similarly, using corresponding reference and measured currents and hysteresis bandwidth which is the error difference between them (Fig. 3).

4   Hardware Setup and Testing

The feasibility of hardware experimentation for the shunt active power filter was evaluated using a three-phase inverter available with a DSP-based motor controller trainer kit, DIGIVAC. The inverter drive consists of a selector switch by which the DSP mode or standard mode can be selected. Figure 4 shows the schematic of DIGIVAC setup. Here for experimental purpose, AC motor is disconnected and inverter output is connected back to the system under test for injecting back the harmonics required by the nonlinear load as shown in Fig. 3. When the selector switch is set to the standard mode in DIGIVAC, the MATLAB programs corresponding to the reference current generation are converted to C Code and loaded through the PCI card. Thus, the required pulses for the inverter are obtained from the PCI card. The current sensor here used is HE055T01 Hall effect sensor.

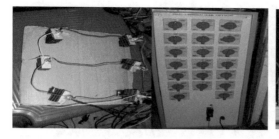

Fig. 4 Diode rectifier and PCI 1711/L Multifunction card

The voltage is sensed using step down transformers and opamp voltage follower circuit. The diode rectifier (12 A, 1200 V) with loading rheostat (10 KW, 220 V) is used as load. PCI-1711L is a multifunction card for the PCI bus to allow high speed data transfer from peripheral devices to processor. Shunt active filter considered for the experimental set is of 1 KVA rating.

5   Experimental Results

In order to verify the purpose of the shunt active filter with synchronous detection algorithm, a three-phase diode bridge rectifier module with resistive load is used as the nonlinear load in the system. The source current and voltage waveforms are witnessed using MATLAB/Simulink environment. The current waveform is found to be highly nonlinear without the compensator and the harmonic contents in the input currents are observed. The load currents and supply voltages are measured, and reference currents are calculated using synchronous detection method. Actual current is taken from DIGIVAC output. Gating signals are generated by hysteresis comparator, and finally, the inverter drive produces the actual compensating currents which are injected to the source side through an inductor. Details of experimental setup are shown in Figs. 4, 5. When we execute build action, code corresponding to the MATLAB Simulink model is automatically loaded to DIGIVAC, and then it runs the files. The selector switch is put in the standard mode; the pulses are obtained from PCI Card. An internal dead time

Fig. 5 Digivac inverter drive and experimental prototype

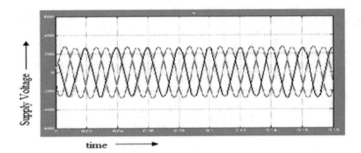

Fig. 6 Supply voltage measured

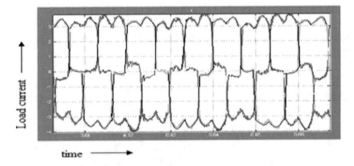

Fig. 7 Load current measured

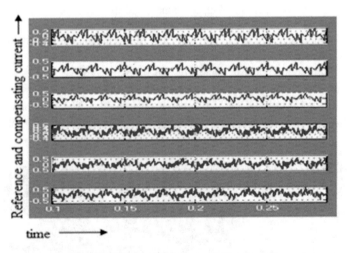

Fig. 8 Reference and compensating current

generator produces the required delays before feeding pulses. The provided drive will generate the required dead times of about 1.8–2 microseconds and convert three-phase pulses to six individual pulses to drive the converter circuit. An uncontrolled rectifier

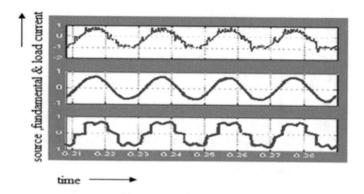

Fig. 9 Source current fundamental and load current

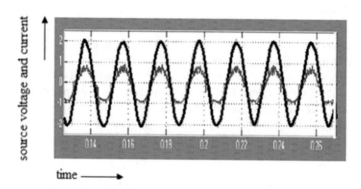

Fig. 10 source voltage and current

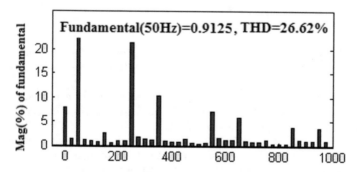

Fig. 11 Harmonic spectrum before compensation

and filter inside the DIGIVAC provide the DC-link voltage from the input single-phase supply. The experimental results are detailed in Figs. 6, 7, 8, 9, 10, 11, 12.

The line current total harmonic distortion (THD) before compensation is 26.6% which is reduced to 6.92% after compensation.

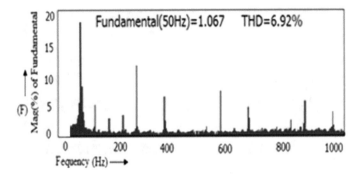

Fig. 12 Harmonic spectrum after compensation

6 Conclusion

The simulation has been carried out in MATLAB/Simulink environment and the hardware has been implemented using DIGIVAC trainer kit. The simulation results show that the shunt active filter model is suitable for use in current compensation. Hysteresis control for harmonic current compensation has been proposed. It is shown by simulation that the proposed scheme is successfully able to track the harmonic current required to be injected. The pulses are generated with the codes produced in real-time workshop along with MATLAB code. Thus, the designed system is also being realized using hardware.

Acknowledgements. I also take this opportunity to record my sincere thanks to Dr. Shazia Hasan, my current supervisor for her constant encouragement.

References

1. Ucak O, Kocabas I, Terciyanli A (2008) Design and implementation of a shunt active power filter with reduced dc link voltage. TUBITAK-Space Technol. Res. Inst., Power Electron, TURKEY
2. Rajagopal R, Palanisamy K, Paramasivam S (2018) A technical review on control strategies for active power filters. In: International conference on emerging trends and innovations in engineering and technological research (ICETIETR)
3. Vardar K, Akpınar E, Sürgevil T (2009) Evaluation of reference current extraction methods for DSP implementation in active power filters. Electr. Power Syst. Res. 79, 10:1342–1352
4. Bhuvaneswari G, Reddy SK, Singh R (2006) Simulation and hardware implementation of dsp based shunt active power filter. In: 2006 Annual IEEE India conference
5. Abaali H, Lamchich MT, Raoufi M (2006) Shunt power active filter control under non ideal voltages conditions. Int. J. info. Technol. 2:164–169
6. The user manual for DIGIVAC AC Drive, Startup manuel and lab experiment manuel, Integrated Electric Co. Pvt. Ltd, 2003–2004
7. The Math works user's guide-Embedded Target for the TI TMS320C6000™ DSP Platform (2002)

8. Bouzid AEM, Sicard P, Cheriti A, Chaoui H, Koumba PM (2017) Adaptive hysteresis current control of active power filters for power quality improvement. In: 2017 IEEE Electrical power and energy conference (EPEC), Saskatoon, SK, Canada
9. Biricik S, Komurcugil H (2016) Three-level hysteresis current control strategy for three-phase four-switch shunt active filters. IET Power Electron. 9(8):1732–1740

An Efficient Thermoelectric Energy Harvesting System

Tirth Lakhani$^{(\boxtimes)}$ and Vilas H. Gaidhane 🆔

Birla Institute of Technology and Science Pilani, Dubai Campus, Dubai, UAE
lakhanitirth@yahoo.com, vhgaidhane@dubai.bits-pilani.
ac.in

Abstract. This paper proposes an option to harvest energy by using the See-beck effect, which harvests energy through the temperature differences present. This energy harvesting tool/Peltier Module is fabricated using ceramic outer shell and the inner part made of bismuth telluride. The thermoelectric generator can supply low-power electronics and a combination of these TEGs can power much more than low-power electronics. The aim of this thermoelectric generator is to supply electricity of 5 V and 1 A to places where placing a Solar panel is not commercially viable. The devices that are aimed to be powered are the devices that are able to charge themselves using a USB port.

Keywords: Seebeck effect · Energy harvesting · Solar energy · Thermoelectric generator

1 Introduction

In the recent years, there has been a rise of interest in scavenging "waste energy" from high power consuming, non-efficient devices [1]. The use of batteries cannot guarantee an uninterrupted working round the clock, unless they are powered continuously through an electrical line which then reduces mobility of the device [2]. Such energy scavenging devices are rising in quantity in every field. Curiosity rover (on mars) uses a similar Peltier Generator [3]. These are used in conjunction with the heat produced by the nuclear elements that radiate heat to keep the rover's components warm during the Martian nights [3]. These devices are usually very small. TEGs are able to convert the ambient temperature into millivolts of electricity, with a few small cost friendly additions to increase this temperature difference more current can be produced. While most of the energy scavenging devices are still under research, thermoelectric energy scavengers were the first to hit the markets, mainly due to the ease in fabrication of solid-state devices and also presence of a well-known physical-theory to back it, The Seebeck Effect. In 1822, Thomas Seebeck observed that a magnetic needle was deflected when kept next to two dissimilar metals that were electrically connected in series and thermally in parallel, exposed to a temperature difference, this deflection proved the presence on flow of current (Ampere Law) [4]. Whenever warmed, electron/hole pairs are made at the hot end and ingest the warmth all the while [5]. The sets recombine and reject heat at the cool edges. The voltage potential, the Seebeck

© Springer Nature Singapore Pte Ltd. 2020
N. Goel et al. (eds.), *Modelling, Simulation and Intelligent Computing*,
Lecture Notes in Electrical Engineering 659,
https://doi.org/10.1007/978-981-15-4775-1_64

voltage, which drives the opening/electron stream, is made by the temperature distinction between the hot and cold edges of the thermoelectric components [6, 7].

To maximize the efficiency of thermoelectric generators, the Seebeck effect is to be maximized. This happens in the presence of highly doped semiconductors (Bismuth Telluride). The efficiency can also be increased if the materials used for fabrication are bad thermal conductors (to retain the heat) and good electrical conductors (reduced Joule heating effect).

The figure of merit (ZT) decides the performance of the Seebeck generator:

$$ZT = \frac{\alpha 2}{\rho \lambda} T \tag{1}$$

where α is the Seebeck coefficient, ρ is the electrical resistivity, λ is the thermal conductivity, and T is the temperature. The efficiency of a thermoelectric generator can be defined as:

$$\eta = \frac{\Delta T}{T} \tag{2}$$

where T is the temperature at the hot side of the generator.

2 Real-Time Applications

2.1 Curiosity Rover (Mars)

Curiosity Rover, which is currently on mars, uses a device called multi-mission radio-isotope thermoelectric generator (MMRTG) [3].

It was chosen for this mission due to its maintenance-free design. This device uses heat from the radioisotope (Plutonium-238) and Seebeck effect/generators to generate electricity to power the rover [3]. Although it is less efficient than solar arrays. It was selected on the rover due to the fact that the dust storms on Mars cannot affect this system. While the solar arrays could get covered in dust and gradually reduce the power output. The TEGs are heated on one side with heat from the radioactive material onboard. The temperature gradient is created on the cold side by using the radiators mounted on the circumference of the whole device as shown in Fig. 1. The heat from the cold side is transferred to the radiator panels using cooling tubes. The cooling tubes can either be made of solid metals or contain a coolant [3].

2.2 Offshore Deep-Sea Beds Projects

On offshore deep-sea beds, where sunlight does not reach the site, TEGs are used [8]. These places where a solar panel is not efficient enough, TEGs are used. Hydrothermal sites vent out hot liquids or sites where geothermal wells are drilled. The cold seawater and the hot liquids being vented or drilled out create a temperature difference. Hydrothermal vents are situated near volcanically active places. This guarantee's the presence of heat around the TEG, which increases the efficiency of the TEG [1].

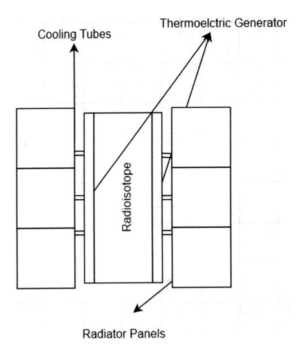

Fig. 1 Basic design of a MMRTG

3   Possible Uses in the Future

There is a vast scope for thermoelectric generator in the future. Places where a photovoltaic cell is not a viable option.

3.1   Vehicle Internal Combustion Engines

All of the internal combustion engines in vehicles these days waste the heat energy to as it is considered a resistance in the performance of the vehicle. The heat is transferred to the atmosphere using multiple radiators in the vehicle. With a small change, this waste energy can be utilized. When the heat from the engines is taken away by the coolant, the heat should be then transferred to hot side of the TEG. And the cold side on the heat sink/radiator. The system is shown in Fig. 2.

3.2   District Heating

Most of the European countries have this system for their residents. District heating provides heating to houses in a local area. The heat is transferred from the heat generation plant to the houses or buildings through heavily insulated pipes. Although no heat here is wasted, it can be used for multiple uses for which one is thermoelectric

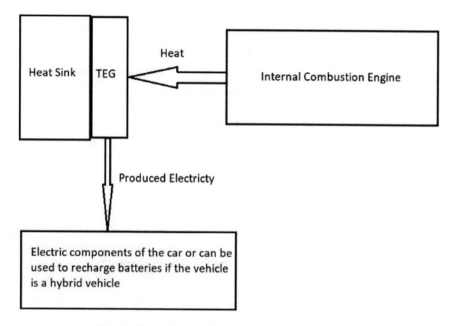

Fig. 2 Block diagram of using a TEG in an IC engine

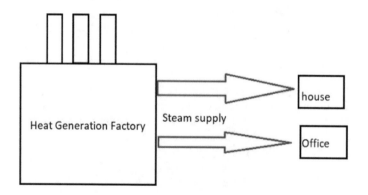

Fig. 3 Basic diagram of a district heating system

generation. The heat generation plants cogenerate heat by burning fossil fuels or bio-mass. As the ambient temperature in these countries during winters goes below 0 °C, a large temperature gradient can be maintained. The heat is transferred using steam which is usually at 100 °C. The system is shown in Fig. 3.

During summers, the same feeding lines convert into cooling stations as they feed cold water through these pipes. This assures us that power will be generated all around the year (Fig. 4).

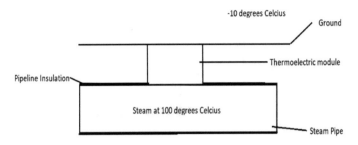

Fig. 4 Proposed use of thermoelectric module

3.3 Chiller Units

In countries like the United Arab Emirates, air conditioning is provided through liquid-cooled chiller units. In BITS Pilani Dubai Campus, a cold water supply runs from the chiller unit to different buildings in the campus. The temperature of the cold water ranges from 2 to 7 °C [9]. Thermoelectric generators can be placed in these conditions. A minor line from the main pipeline can be taken. The water enters a small tank in which the heat sinks are placed. The heat sinks are cooled which are in contact with the cold side. The average temperature in Dubai during summers is 45 °C. The hot side can be directly exposed to atmosphere or hot environment. The second option being place a metal box of sand on the hot side and place concave mirrors to heat the sand up to a higher temperature than the ambient temperature (45 °C). The second option can prove to be an efficient method due to the higher temperature gradient (Fig. 5).

3.4 Human Body Heat

The normal body temperature of a human is 37 °C [6]. The ambient temperature in cold countries during peak winter can range from −1 to −4 °C. During this time, the body heat can be utilized to charge smart watches. TEGs can be integrated into the

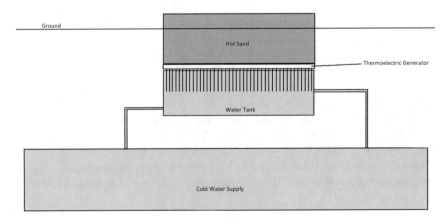

Fig. 5 Model design

gloves. Our palms which tend to be usually warmer than our body temperature will act as the heat source. The cold source would be the ambient temperature. As the temperature difference is almost 40°, enough power could be obtained from the TEG to charge smart watches. Even hearing aids can be charged using this as hearing aids use batteries with less voltage.

3.5 Passive Cooling

To maximize cost efficiency, we reduce the initial investment. As most of the other options require some changes to be made in the systems that exist already. That increases the initial cost. The heat sink can be directly placed in wet soil or even normal soil. The sand labelled as "hot sand" will be heated using concave mirrors or lenses. The concentrated heat of the sun on the sand increases the temperature of the sand higher than that of the surrounding environment. Only digging has to be done to install this device. Hence, a lower initial cost for this system is predicted.

3.6 Prototype

The design methodology used for this prototype was of passive cooling. The prototype was designed, manufactured, and tested in the facilities of Birla Institute of Technology and Science, Dubai Campus. The prototype consists of a $25 \times 15 \times 10$ cm aluminium box structure. The proposed prototype for the energy harvesting is shown in Fig. 6.

This aluminium structure is to hold the heated sand. The TEGs are to be held to the bottom of the aluminium structure using thermoconductive paste/tape (Silicone based or epoxy). An aluminium plate of 25×15 cm is to be placed on the cold side of the TEG are required to be made in the systems.

The prototype consists of a $25 \times 15 \times 10$ cm aluminium box structure. This aluminium structure is to hold the heated sand. The TEGs are to be held to the bottom of the aluminium structure using thermoconductive paste/tape (Silicone based or epoxy). An aluminium plate of 25×15 cm is to be placed on the cold side of the TEG again by using thermoadhesive paste/tape.

The prototype consists of 10 TEG SP1848-27145 SA connected in a series array. The combination of series and parallel can be decided based on the priority if more voltage is required or more current. The output wires from this setup are connected to a

Fig. 6 The prototype manufactured using the proposed method

Table 1 Results for presented system

Temperature gradient (°C)	Voltage (V)	Temperature range (°C)
12	1.65	25–13
45	5.08	55–10
65	7.39	70–5
68	7.68	74–5
69	7.87	75–5

multitester to record the readings. A voltage regulator can be attached after which a battery, LED light, a phone charger, etc. can be connected for practical/real-life purposes. The aluminium box is to contain the sand that would be heated due to sun.

3.7   Testing

The testing was conducted in the facilities of Birla Institute of Technology and Science, Dubai Campus. The tests were conducted during the winter time in Dubai. The test was simulated to resemble the summer conditions in the region. The cold sink taken during this test was cold moist garden soil. The following results were obtained. As we obtain a higher voltage, the combination of the TEG can be changed to obtain a 5 V with a current enough to charge a phone battery or power fairy lights. The various results are summarized in Table 1.

3.8   Scope for Improvement

More TEGs could be used in the design for a higher output. A glass pane put on top the aluminium box would increase the temperature of the sand. Instead of cold moist sand, the cold sink could be replaced of a water line (cold) which runs through gardens, roads, etc. Solar concentrators can be added to concentrate the heat onto the sand. This increases the temperature of the sand and also the temperature gradient giving a much higher output.

4   Conclusion

In this paper, a thermoelectric generator was designed for energy harvesting. It has been observed that the presented design can be used efficiently as a source of energy in deferent conditions, where the other sources like solar PV array, cannot use during winter time. Therefore, this model may have the scope in future for better and low-cost power generation. It can be extensively used in parks, street lighting, autonomous homes, etc.

References

1. Carrasco JM, García Franquelo L, Bialasiewicz JT, Galván E, Portillo Guisado RC, Martín Prats MDLÁ, León JI, Moreno-Alfonso N (2006) Power-electronic systems for the grid integration of renewable energy sources: a survey. IEEE Trans Ind Electron 53(4):1002–1016
2. Carmo JP, Gonçalves LM, Correia JH (2010) Thermoelectric micro converter for energy harvesting systems. IEEE Trans Industr Electron 57(3):861–868
3. Levihn F (2017) CHP and heat pumps to balance renewable power production: Lessons from the district heating network in Stockholm. Energy 137:670–678
4. Seebeck TJ (1825) Magnetische Polarisation der Metalle und Erzedurch Temperatur-Differenz (Magnetic polarization of metals and minerals by temperature differences). Abhandlungen der KöniglichenAkademie der Wissenschaftenzu Berlin (Treatises of the Royal Academy of Sciences in Berlin), pp 265–373
5. Stanford III HW (2016) HVAC water chillers and cooling towers: fundamentals, application, and operation, 2nd edn. CRC Press
6. Gaidhane VH, Mir A, Goyal V (2019) Energy harvesting from far field RF signals. Int J RF Microwave Comput Aided Eng 29(5):e21612
7. Mir A, Gaidhane VH (2017) Deriving energy from far field RF signal. In: IEEE international conference on electrical and computing technologies and applications (ICECTA), pp 1–4
8. Liu L (2014) Feasibility of large-scale power plants based on thermoelectric effects. New J Phys 16(12):123019
9. Hutchison JS, Ward RE, Lacroix J, Hébert PC, Barnes MA, Bohn DJ, Dirks PB, Doucette S, Fergusson D, Gottesman R, Joffe AR (2008) Hypothermia therapy after traumatic brain injury in children. N Engl J Med 358(23):2447–2456

FinFET Optimization in the Design of 6T SRAM Cell

Sreeja Rajendran[✉] and R. Mary Lourde

Department of Electrical and Electronics Engineering, Birla Institute of
Technology and Science Pilani, Dubai Campus, Dubai, UAE
sreejamanojnair@gmail.com, marylr@dubai.bits-pilani.ac.in

Abstract. To overcome the challenges in MOSFET scaling, FinFETs have
emerged as a probable candidate compatible with CMOS technology. Memory
forms an integral part of almost all IC chips and contributes to the major share of
power dissipated. Replacing MOSFET-based memory arrays with the quasi-
planar FinFET helps to lower the leakage currents and thereby the power dissipation. The important criteria in the design of an SRAM cell are cell stability
and cell area. The stability of the cell is determined by the static noise margin
(SNM). This paper describes the modelling and simulation of a double-gate n-
FinFET. It also discusses the effect of varying the gate material on the performance characteristics of the FinFET. The optimization of a 6T FinFET-based
SRAM cell has also been presented. The cell optimization is in terms of the fin
dimensions, namely fin width and fin pitch.

Keywords: FinFET · Static noise margin · 6T SRAM

1 Introduction

As scaling in technology proceeds, leakage and power are two areas which pose a
challenge for MOSFET technology. Replacing the traditional MOSFET with a FinFET
which can be fabricated using the same technology and at the same time reduce the
second-order effects like leakage, power dissipation is one of the solutions [1]. The
primary reason for incorporating memory in circuits is for on-chip data storage capabilities. Reduction in memory density has become very essential since memory plays
an important role in many designs. Memory arrays are classified into random access
memory (RAM), serial access memory (SAM) and content access memory (CAM).
Static RAM is one of the categories of random access memory. SRAM cell can hold
data only as long as power is available, and they also require a greater number of
transistors than dynamic RAM. But due to their high speed of operation, they are
preferred as memory on-chip [2].

Replacing existing SRAM cell design with FinFETs is gaining interest in recent
times. This is mainly because FinFETs are more scalable and at the same time reduce
second-order effects drastically. This paper focuses on the design of a double-gate

© Springer Nature Singapore Pte Ltd. 2020
N. Goel et al. (eds.), *Modelling, Simulation and Intelligent Computing*,
Lecture Notes in Electrical Engineering 659,
https://doi.org/10.1007/978-981-15-4775-1_65

n-FinFET and its application in a 6T SRAM cell. The paper also discusses the effect of varying gate materials on the FinFET performance characteristics. A 6T SRAM cell using FinFETs shows better static noise margin (SNM) and also reduced variability.

2 Basic Structure of FinFETs

FinFET is a three-dimensional structure where source, drain and channel are formed over a buried oxide (BOX). Fin is the connecting link between the source and the drain and is perpendicular to the substrate. Figure 1 shows a Multifin FinFET. The fin serves as the channel and the fin parameters, namely the fin width W_{fin}, fin height H_{fin} and fin pitch P_{fin} play key roles in the device performance. The short channel effects can be better controlled with scaling of fin width. This helps in the gate gaining better control over the channel potential [3]. The drive current increases with increased channel width which can be achieved by increasing the fin height. An increased drive current at the expense of gate area benefits memory arrays [4, 5]. However, integrating a FinFET into an SRAM cell requires a lot of precision since etching process to remove excess material deposited during gate stack and spacer formation can result in fin erosion.

SENTAURUS TCAD software tool has been used to model a double-gate n-FinFET with the following specifications. The dopant used is arsenic. The polysilicon gate has also been doped with arsenic to the same concentration of source region. The device model has been created using structure editor where a 3D model can be created specifying the coordinates and material of each layer. We can also select the type of doping profile we want to include as well as set the contacts on the desired faces. Furthermore, we also need to define the mesh for each layer differently since the structural geometry is complex. The simulated structure and doping profiles are shown in Figs. 2 and 3, respectively. Table 1 gives the design specifications.

$Si_{0.4}Ge_{0.6}$ has silicon along with 60% Ge mole fraction as the gate material. Using $Si_{0.4}Ge_{0.6}$ provides the advantages such as variability in work function by changing the Ge concentration and also compatibility with CMOS technology. It can be noted that using $Si_{0.4}Ge_{0.6}$ has provided lower leakage currents, and at the same time, slightly higher drain currents than the case when PolySi is used as the gate material. Table 2 shows the device parameters obtained using different gate materials.

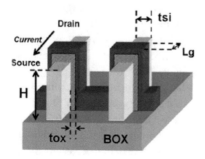

Fig. 1 Multifin FinFET [6]

n3_MYFINFET_bnd

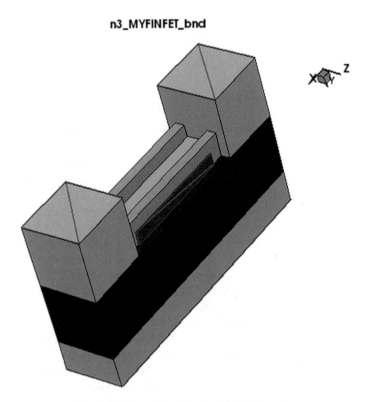

Fig. 2 Simulated double-gate FinFET structure

3 SRAM Cell Design

A memory array consists of rows and columns. The selection of a cell into which data is to be written or read from is done by using a row and column decoder. The transfer of data to and from a cell occurs using bit line. Select line is used to select a particular cell. Figure 4 shows the basic structure of a RAM cell. The delay involved in driving load on the word line and bit lines determines the performance of the SRAM array [5].

Stability, bit cell read and leakage currents are the various parameters that evaluate the performance of SRAMs. Data retention margin, access disturb margin and write margin constitute the stability metric. Apart from these, static noise margin (SVM) can also be used as a metric for SRAM stability [7]. With technology scaling and lower supply voltages, data retention of SRAM cell becomes a functional constraint and thereby limits the stability of the cell [8].

The 6T Static RAM cell as shown in Fig. 5 comprises of cross-coupled inverters and access/pass transistors. The access transistors provide access to the memory cell. When select = 1, data can be written into or read from the cell. When select = 0, the cell gets decoupled from the data lines. This is referred to as the hold state. For reasons of symmetry, it is important that the n FinFETs and the p-FinFETs have the aspect ratios β_n and β_p, respectively.

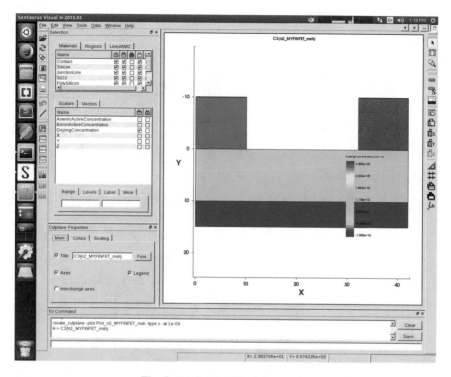

Fig. 3 Doping profile along z-axis

Table 1 Design specifications of double-gate n-FinFET

Process parameters	Gate length	Height of fin	Thickness of fin	Oxide thickness	Spacer length	Channel doping	Source doping
Values	18 nm	2 nm	4 nm	1 nm	1.5 nm	1e16/cm^3	1e20/cm^3

Table 2 Device parameters obtained for different gate materials

Gate Material	Vtgm (V)	Idsat (A)	Ioff (A)	SS
PolySi	0.259	1.121e-3	4.753e-8	98.522
Si$_{0.4}$Ge$_{0.6}$	0.239	1.148e-3	8.205e-8	103.184

3.1 Read Operation in a 6T SRAM Cell

Standby, read and write are the three operating modes in SRAM. In standby mode, word line (WL) is connected to the ground. For read operation, BL and BLB lines in Fig. 5 are initially high. Assume $Q = 0$ and $QB = 1$. When the word line is high, the bit should be pulled to zero through the transistors Mn1 and Ax1. Due to the current

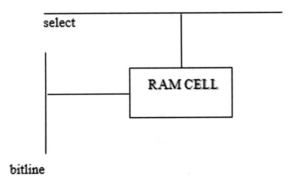

Fig. 4 RAM cell with bit line and select line

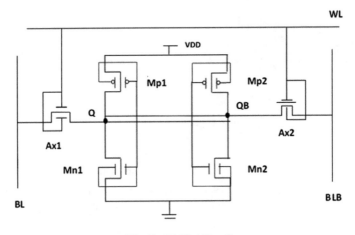

Fig. 5 6T SRAM cell

flowing through Mp1, the voltage at node Q may rise from zero. To avoid this, Ax1 should be stronger than Mp1. Hence, the transistors should be ratioed such that the node a remains below switching threshold of Mp1/Mn1 inverter. Therefore, the criterion for read stability is Mn1 ≫ Ax1 ≫ Mp1 [9]. The SNM during a read operation helps to quantify the read stability. Read access is the period when the cell becomes highly susceptible to noise [4]. This is due to the intrinsic disturbance produced by direct data access mechanism of the 6T SRAM cell [10].

3.2   Write Operation in a 6T SRAM Cell

Assume Q is 0, and we want to write 1 into the cell. Bit line is charged high. BLB is pulled low by a write driver. Due to read stability constraint, bit line will be unable to pull a high through Ax1. Hence, the cell is written forcing QB low through Ax2. Mp2 opposes this. So Mp2 ≪ Ax2. This constraint is called writeability. When QB becomes low, Mn1 turns off and Mp1 turns on pulling Q to 1 [9].

3.3 Static Noise Margin

SNM is the ability of a cell to hold a stable state. The intersection points on the V_{DD} axes gives the two stable states of the circuit. The ability of the cell to reject noise depends on the separation between the characteristics of the two inverters. Hence, wider the separation, greater will be the ability of the cell to reject noise. The butterfly curve is shown in Fig. 6. SNM is the length of the largest square that can be implanted within the lobes of the curve [11].

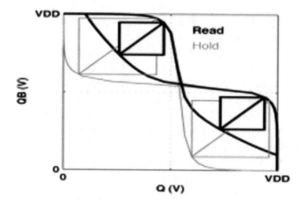

Fig. 6 Butterfly curve showing SNM [11]

3.4 Design Constraints of a Memory Cell

For a memory cell, the crucial properties are density and functionality. Functionality is ensured by optimal selection of supply voltage, threshold voltage and device sizing. Read stability, write stability, access time and power consumption also play key roles in the performance of the memory cell. The word line remains high only during the cell access time. If a read or write operation does not occur within this allocated time, then an access failure occurs [12]. The SRAM cell access time is directly proportional to the drain capacitance of the access transistor and inversely proportional to the device on current. The access time of a cell will always be lower than that of memory as a whole because it also needs to take the word line decoder delay into consideration. SRAM arrays contribute to a major share of power consumption due to its short active period and long idle periods. The static power dissipation can be taken care of by using sleep transistors and also through body biasing. The trade-off between delay and power consumption becomes more crucial in sub-100 nm technologies.

Transistor sizing will not aid in improving the robustness of the cell due to width quantization, process variations and device-to-device mismatch. In the design of SRAM systems, process variations and leakage currents have a great impact. The authors in [12] suggested optimizing the gate sidewall spacer thickness of FinFETs to improve the robustness of SRAM cell. Authors in [13] have discussed the application of a row-based forward bias above V_{DD} to the back gates of access transistors as well as pull-down transistors during read/write operations to improve performance. A reverse bias below GND is applied to these transistors during standby/sleep mode to reduce leakage. These biasing voltages are generated by a bias generator which provides the necessary bias (greater than V_{DD} or less than GND) using a simple inverter. The bias generator is driven by the row decoder. Selected cells will be connected to positive bias to enhance performance while the unselected ones will be connected to a negative bias to curtail leakage.

In an SRAM cell, due to cell sizing the cell area ($L_g * W_{ch}$) decreases and σV_t varies. So, there is an issue of V_t variability [3, 16]. A fin with ion implantation from two wafer orientations is referred to as a double-sided fin while an implantation in only one direction results in a single-sided fin. It has been shown that single-sided fin provides better control of threshold voltage variability as well as a lower DIBL [5].

In order to improve the readability as well as the writeability of the cell, in a double-gate FinFET, one gate of the access as well as the pull-up transistors is permanently grounded. Improved performance is availed by dynamically adjusting the second gate of the transistors during circuit operation. It has been demonstrated in [14] that the SNM is increased by 92% and leakage power consumption is reduced by 36% by using independent gate FinFETs compared to shorted gate ones in an SRAM cell. Authors in [15] demonstrated the use of double etch sidewall image transfer technique to produce SRAM cell with fin pitch of 40 nm. Merging pull-down transistors with tight pitch while keeping pull-up transistors with relaxed pitch separate is accomplished through epitaxy. This leads to differential fin pitch formation in SRAM cells. Uniform extensions were formed on the fins by the in situ-doped epitaxial films. This helps to reduce the external resistance of the FinFET by about 30%. Aligning the transistors along a particular orientation can be accomplished through fin rotation which is possible due to the quasi planar structure. The electron and hole mobility, effective mass, scattering rates are different along with different orientations [12]. Maximum read stability is obtained when pull-down transistors are oriented along <100> and the access transistors along <110>. Doping the source and drain unequally gives rise to Asymmetrically doped FinFETs (AD FinFETs). These FinFETs show lower currents when region with lighter doping is chosen as the source. The read/write conflict can be alleviated by connecting the lightly doped region to the storage nodes [13, 17].

4 Simulation Results

The 6T SRAM cell was simulated using H-Spice BSIM-CMG model. The static noise margin for read and write cycles was determined for different fin thickness and fin pitch as shown in Tables 3 and 4, Figs. 7 and 8, respectively.

Table 3 Read operation SNM values for various fin thickness and fin pitch

N_P: N_A: N_N	SNM (mV) Tfin = 10 nm	SNM (mV) Tfin = 13 nm	SNM (mV) Tfin = 15 nm
01:01:02	283.912	277.134	270.735
01:01:03	313.385	306.771	300.417
01:01:04	332.611	325.779	319.192
01:02:04	275.839	267.961	260.676
01:03:04	239.243	230.299	222.072
01:02:03	253.619	245.426	237.892
02:03:04	248.419	240.588	233.311
02:02:03	262.568	255.591	249.005
02:02:04	283.912	277.134	270.735

Table 4 Write operation SNM values for various fin thickness and fin pitch

N_P: N_A: N_N	SNM (mV) Tfin = 10 nm	SNM (mV) Tfin = 13 nm	SNM (mV) Tfin = 15 nm
01:01:02	451.681	441.739	432.716
01:01:03	451.548	441.253	431.95
01:01:04	450.5	439.887	430.292
01:02:04	450.5	439.887	430.292
01:03:04	450.5	439.887	430.292
01:02:03	451.548	441.253	431.95
02:03:04	451.681	441.739	432.716
02:02:03	450.794	441.011	432.083
02:02:04	451.68	441.739	432.716

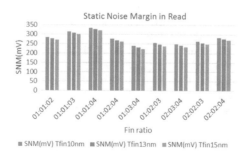

Fig. 7 SNM during read for different values of fin thickness

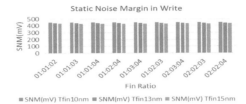

Fig. 8 SNM during write operation for different fin thickness

5 Conclusions

FinFETs are a promising replacement for traditional MOSFETs in various circuit designs. The simulation work carried out on double-gate n-FinFET has shown that using $Si_{0.4}Ge_{0.6}$ as the gate material provides better performance characteristics, namely higher drain and lower leakage currents. Further, the use of FinFETs in 6T SRAM cell is studied. The optimization of the FinFET-based SRAM with respect to its fin dimensions, namely fin width and fin pitch, has provided the following inferences:

- The highest value of static noise margin in the read cycle is obtained when the fin ratio is 1:1:4, and the fin thickness is 10 nm.
- In the case of write cycle, the static noise margin is almost insensitive to the fin ratio. Also, it can be observed that for a particular value of fin thickness, the SNM is almost constant.
- Hence, optimization of the SRAM can be done keeping the read SNM as top priority.

References

1. Rajendran S, Mary Lourde R (2015) Fin FETs and their application as load switches in micromechatronics. In: 2015 IEEE international symposium on nanoelectronic and information systems, pp 152–157. IEEE
2. Uyemura JP (1992) Circuit design for CMOS VLSI, Kluwer Academic Publishers
3. Hisamoto D, Lee WC, Bokor J et al (2000) FinFET—a self aligned double gate MOSFET scalable up to 20 nm. IEEE Trans Electron Dev
4. Carlson A, Guo Z, Balasubramanian S, Pang LT, King Liu TJ, Nikolic B (2006) FinFET SRAMS with enhanced read/write margins. In: IEEE international SOI conference proceedings, pp 105–106
5. Kawasaki H, Basker VS, Yamashita T, Lin C-H, Zhu Y, Faltermeier J, Schmitz S, Cummings J, Kanakasabapathy S, Adhikari H, Jagannathan H, Kumar A, Maitra K, Wang J, Yeh C-C, Wang C, Khater M, Guillorn M, Fuller N et al (2009) Challenges and solutions of FinFET integration in an SRAM cell and a logic circuit for 22 nm node and beyond, IEEE
6. Khandelwal S, Gupta V, Raj B, Gupta RD (2015) Process variability aware low leakage reliable nano scale double-gate-FinFET SRAM cell design technique. J Nanoelectron Optoelectron 10(6):810–817

7. Bhoj AN, Niraj KJ (2013) Parasitics-aware design of symmetric and asymmetric gate-workfunction FinFET SRAMs. IEEE Trans Very Large Scale Integr (VLSI) Syst 22(3):548–561

8. Grossar E, Stucchi M, Maex K, Dehaene W (2006) Read stability and write-ability analysis of SRAM cells for nanometer technologies. IEEE J Solid State Circuits 41(11):2577–2588

9. Weste N, Harris D, Banerjee A CMOS VLSI design-a circuits and system perspective, 3rd edn. Pearson Education

10. Guo Z, Balasubramanian S, Zlatanovici R, King T-J, Nikolić B (2005) FinFET-based SRAM design. In: Proceedings international symposium on low power electronics and design, pp 2–7

11. Birla S, Singh RK, Pattnaik M (2011) Static noise margin analysis of various SRAM topologies. Int J Eng Technol 3(3):304

12. Gangwal S, Mukhopadhyay S, Roy K (2006) Optimization of surface orientation for high-performance, low-power and robust FinFETSRAM. In: Proceedings IEEE custom integration circuits conference, pp. 433–436

13. Joshi RV, Kim K, Williams RQ, Nowak EJ, Chuang CT (2007) A high performance, low leakage and stable SRAM row-based back gate biasing scheme in FinFET technology. In: International conference on VLSI design, pp. 665–672

14. Tawfik SA, Liu Z, Kursun V (2007) Independent-gate and tied-gate FinFET SRAM circuits: design guidelines for reduced area and enhanced stability. In: International conference on microelectronics, pp. 171–174

15. Basker VS, Standaert T, Kawasaki H, Yeh C-C, Maitra K, Yamashita T, Faltermeier J, Adhikari H, Jagannathan H, Wang J, Sunamura H, Kanakasabapathy S, Schmitz S, Cummings J, Inada A, Lin C-H, Kulkarni P, Zhua Y, Kuss J, Yamamoto T, Kumara A, Wahl J et al (2010) A 0.063 μm² FinFET SRAM cell demonstration with conventional lithography using a novel integration scheme with aggressively scaled fin and gate pitch. In: IEEE symposium on VLSI technology, pp 19–20

16. Kawasaki H, Khater M, Guillorn M, Fuller N, Chang J, Kanakasabapathy S, Chang L, Muralidhar R, Babich K, Yang Q, Ott J, Klaus D, Kratschmer E, Sikorski E, Miller R, Viswanathan R, Zhang Y, Silverman J, Ouyang Q, Yagishita A, Takayanagi M, Haensch W, Ishimaru K (2008) Demonstration of highly scaled FinFET SRAM cells with high-κ/metal gate and investigation of characteristic variability for the 32 nm node and beyond. In: IEEE international electron devices meeting, pp 1–4

17. Ananthan H, Bansal A, Roy K (2004) FinFET SRAM—device and circuit design considerations. In: Proceedings international symposium on quality electronic design, pp 511–516

Analyzing the Impact of NBTI and Process Variability on Dynamic SRAM Metrics Under Temperature Variations

Siona Menezes Picardo[1(✉)] ⓘ, Jani Babu Shaik[2] ⓘ, Sakshi Sahni[2] ⓘ,
Nilesh Goel[1] ⓘ, and Sonal Singhal[2] ⓘ

[1] Birla Institute of Technology and Science Pilani, Dubai Campus, Dubai, UAE
{sionacmenezes, goel.nilesh}@gmail.com
[2] Shiv Nadar University, Greater Noida, India

Abstract. Continuous scaling of CMOS technology has led to reliability issues and process variability that affect the circuit performance of the SRAM cell. The dynamic behavior of SRAM cells are characterized by critical read-stability (Tread) and critical write-ability (Twrite) while the Static Noise Margins (SNMs) are deduced by the static metrics that are the key performance metrics. The work in this paper demonstrates the cumulative impact of process variability and Negative Bias Temperature Instability (NBTI) degradation on the dynamic metrics of the SRAM cell under varied temperature conditions. Degradation due to NBTI is incorporated by considering different activity factors (α) for the dynamic metrics. Time-zero or process variability is performed for fresh-case, symmetric and asymmetric degradation by Monte Carlo run simulations using foundry models in addition to examining the effect of correlation with their corresponding static metrics.

Keywords: SRAM · NBTI degradation · Dynamic metrics · Process variability

1 Introduction

Bias Temperature Instability (BTI) is a primary reliability issue that impacts CMOS-based devices and circuits with Negative Bias Temperature Instability (NBTI) and Positive Bias Temperature Instability (PBTI) affecting p-channel MOSFETs and n-channel MOSFETS, respectively. Predominantly, NBTI has been found to be a critical concern of the two issues that lead to the shift in device parameters like threshold voltage (ΔV_{TH}), drain current (ΔI_D), mobility ($\Delta\mu$), transconductance (Δg_m) and sub-threshold slope (ΔS) [1–4]. Threshold voltage (ΔV_{TH}) is most adversely affected among the stated parameters which in turn affect the performance of the circuits implementing these devices. Modern processors comprise of SRAM whose read-stability and write-ability are important dynamic metrics and thus been considered in device degradation study [5, 6]. Several static metrics are used to study the impact of NBTI on the SRAM cell [7–9] that are also employed for characterizing read-stability and write-ability. Static Noise Margin (SNM) is used to characterize read-stability by means of various static metrics such as Read Static Noise Margin (RSNM), Static Current Noise Margin

© Springer Nature Singapore Pte Ltd. 2020
N. Goel et al. (eds.), *Modelling, Simulation and Intelligent Computing*,
Lecture Notes in Electrical Engineering 659,
https://doi.org/10.1007/978-981-15-4775-1_66

(SINM) and Static Voltage Noise Margin (SVNM) whose standards are retrieved from the read N-curve [9]. Similarly, write-ability adopts Write Static Noise Margin (WSNM), Bit-line Write Trip Voltage (BWTV), and Word-line Write Trip Voltage (WWTV) [5, 8] as its static metrics.

In this paper, dynamic SRAM metrics namely critical read-stability (T_{read}) for read operations and critical write-ability (T_{write}) for write operations are studied for different conditions of device usage or activity factor (α) for varied temperatures and their correlation is examined with their corresponding static metrics. The correlation of SRAM dynamic metrics with static metrics has been reported recently [5] but the temperature variations were not taken into consideration. This study focuses on the impact of temperature on T_{read} and T_{write} against the RSNM and BWTV static metrics due to the proven best correlation [8]. Various studies on SRAM have been done at different temperatures and therefore a range of temperatures is considered for this work [5, 10, 11].

For circuit-level simulations, a compact model for NBTI [12] is implemented to derive ΔV_{TH} in P-MOSFETs of the SRAM cell. The equation of the compact model (1) takes into account the activity factor (α) to describe the extent of usage of individual P-MOSFETs of the two cross-coupled inverters and consists of an occupancy factor of "f_{of}". SRAM cell stores data at its nodes which are complementary, implying that at any given instance, one among the P-MOS devices will be in an absolute relaxed state while the other will be completely degraded due to NBTI.

$$\Delta V_{TH} = \Delta V_{TH\_MAX} * (\alpha)^{0.35} * (f_{of}) \tag{1}$$

The parameter "α" takes the value of "0" or "1" for extreme usage (asymmetric) and "0.5" for symmetric cases. ΔV_{TH_MAX} is the maximum possible shift in threshold voltage for extreme NBTI degradation. f_{of} is assigned 1 for extreme degradation and 0.5 for all other conditions.

The simulation set-up to characterize the critical read-stability (T_{read}) is represented in Fig. 1a. T_{read} is defined as the maximum width of the word-line that does not lead to a failure in the read operation. The nodes L_{OUT} and R_{OUT} have initial stored values of "0" and "1", respectively. Bit lines BL and BLB are fixed at V_{DD}. A finite value of T_{read} is achieved by connecting a noise source "V_n" between the node L_{OUT} of the left inverter and the node of the right inverter, likewise, another noise source "V_n" between R_{OUT} of the right inverter and the node of the left inverter. A similar simulation set-up is used to characterize the critical write-ability (T_{write}) shown in Fig. 1b without the noise sources. T_{write} is defined as the minimum width of the word-line needed to successfully complete the write operation. The bit lines BL are set to V_{DD} and BLB to GND for a write operation.

Critical read-stability (T_{read}) is measured as the duration from 50% of word-line to 50% transition of either of the internal nodes (L_{OUT} or R_{OUT}) [5, 8] and is displayed in Fig. 2a. Critical write-ability (T_{write}) is measured as the duration from 50% of word-line to 50% transition of the internal node (L_{OUT}) [5, 8] and is shown in Fig. 2b.

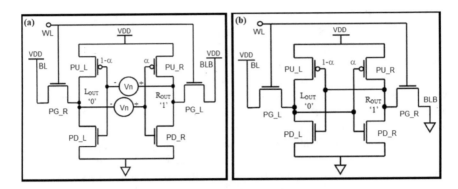

Fig. 1 a SRAM cell for T_{read} consisting of a CMOS inverter pair with pass transistors connecting the bit lines (BL and BLB) and the word-line (WL) and **b** SRAM cell for T_{write} consisting of a CMOS inverter pair with pass transistors connecting the bit lines (BL and BLB) and the word-line (WL)with activity factors for PU_L fixed at $(1 - \alpha)$ and PU_R for both the setups

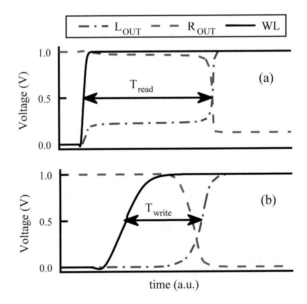

Fig. 2 a Waveforms for T_{read} and **b** waveforms for T_{write}

2   Simulation Method

The dynamic metrics of SRAM are evaluated at time-zero (no NBTI degradation) and degradation post-NBTI. SPICE simulations are carried out using the SRAM netlist with the augmented NBTI technology library, while time-zero variability is done with foundry models. The degradation with respect to alpha (activity factor) is varied for the two P-MOSFETs to perform simulations to incorporate NBTI degradation.

The mean values of the dynamic metrics are derived for various sets of alpha values with intermediate α = 0.05, 0.25, 0.5, 0.75, 0.95, and extreme α of "0" and "1". The left pull-up P-MOSFET (PU_L) is considered to have a fixed activity factor of $(1 - \alpha)$ and the right pull-up P-MOS device (PU_R), α. T_{read} has a stronger dependency on PU_R and thus when α takes the value of 1, it suffers from extreme degradation. Similarly, T_{write} depends on the voltage of L_{OUT} for a successful write operation which consequently, results in extreme degradation of the left pull-up P-MOS device when α = 0. Extreme NBTI degradation is considered over a period of 10 years with $\Delta V_{TH\ MAX}$ = 100 mV.

Monte-Carlo simulations are done for 1000 samples of T_{read} and T_{write} and their distributions are plotted along with correlation against the two static metrics, i.e., RSNM and BWTV for fresh, symmetric, and worst-cases of NBTI degradation.

3   Results and Discussions

This section focuses on the degradation effects due to NBTI on T_{read} and T_{write} by varying the activity factor for the P-MOS devices of the right and left inverters of SRAM cell. These results are discussed along with their corresponding correlation with the static metrics (RSNM and BWTV) for varied activity factor and temperatures.

3.1   Dynamic Metrics T_{read} and T_{write} for Different Activity Factor

Figure 3 indicates the effect of NBTI on T_{read} for different α and temperatures. The transistor (PU_R) controlling T_{read} for "fresh" and α = 0, does not incur NBTI degradation. For the above two stated cases, T_{read} depends on the initial voltage of L_{OUT}. The right inverter of SRAM cell is critical in deciding the trip point of T_{read} (known as R_{trip}) which leads to the flipping of the internal nodes when the voltage on L_{OUT} attains its critical value. This eventually consequences in a failure for T_{read}. The value of R_{trip} is found to reduce with an increase in activity factor (α) which implies that

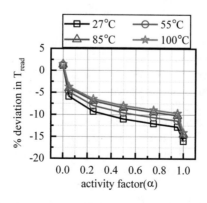

Fig. 3 Percentage deviation in T_{read} (w.r.t. fresh-case) due to NBTI by a varying activity factor (α) for different temperatures

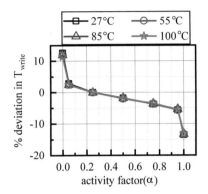

Fig. 4 Percentage deviation in T_{write} (w.r.t. fresh-case) due to NBTI by varying activity factor (α) for different temperatures

the critical value of L_{OUT} approaches at a faster rate. It is found that T_{read} degrades for all values of α with respect to fresh, except in a case of $\alpha = 0$ (improves by around 5%). The worst-case NBTI degradation is found to occur at $\alpha = 1$ due to the absolute degradation of the PU_R transistor with its node storing a value of "1". Temperature variations effects T_{read} constructively as it can be seen that with an increase in temperature, the worst-case degradation reduces from 16% to about 14%. As the temperature rises, the impact of NBTI on Tread reduces.

Figure 4 shows the degradation due to NBTI with different activity factor (α) and temperatures. T_{write} has significant dependence on L_{OUT} transition from "0" to "1" which highlights the importance of degradation of the left pull-up transistor for T_{write}. This dependency increases the duration needed to flip the state of internal nodes in comparison to the fresh-case. Therefore, for $\alpha = 0$ and $\alpha = 0.05$ degradation of T_{write} occurs. Conversely, an improvement for higher α is observed by virtue of L_{OUT} being able to flip the right pull-up transistor more rapidly owing to the decrease in R_{trip}, with almost 15% for $\alpha = 1$(best-case) and 2.5% for $\alpha = 0.5$. Increase in temperature to 55, 85 and 125 °C does not impact T_{write} for higher alpha values and there is a minimum deflection in T_{write} mean values for the fresh-case for $\alpha = 0.05$ and the worst-case degradation.

The distribution of T_{read} in the fresh-case for different temperatures is indicated in Fig. 5. The dynamic metrics attain least values for the extreme case showing the most amount of degradation of T_{read} as compared with the fresh-case, followed by $\alpha = 0.5$ for all temperatures. These results concur with the mean values obtained in Fig. 3. Additionally, at 27 °C temperature, the fresh values tend to deviate at higher values of T_{read} which remains consistent across all temperatures indicating sensitivity towards the applied noise voltages. With increasing temperatures, it is observed that the mean T_{read} achieves reduced values indicating an improvement in degradation.

The distribution of T_{write} for different temperatures is represented in Fig. 6 for fresh, $\alpha = 0.5$, and $\alpha = 0$. At 27 °C, T_{write} for symmetric case of degradation, shows improvement with respect to the fresh-case while in the extreme case it degrades which

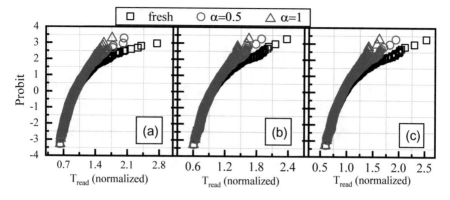

Fig. 5 Distribution of T_{read} for different α values at **a** 27 °C, **b** 55 °C and **c** 100 °C

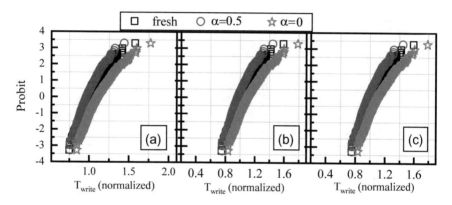

Fig. 6 Distribution of T_{write} for different α values at **a** 27 °C, **b** 55 °C and **c** 100 °C

validates the graph in Fig. 4. For higher temperatures, this trend remains consistent depicting minimal impact of temperature on the overall distribution of T_{write}.

3.2   Correlation of Dynamic Metrics with Static Metrics

The correlation of the dynamic metrics, T_{read} is done w.r.t. the static metrics RSNM and T_{write} against the static metrics BWTV with NBTI degradation considering fresh-case as a reference at different temperatures.

A good correlation is observed for T_{read} at 27 °C in all three cases considered ranging between 0.88 and 0.9. At higher temperatures, the correlation drops to 0.82 for the fresh-case and is found to improve for $\alpha = 0.5$ and 1 as indicated in Fig. 7b, c. Amongst the three different considered temperatures, Fig. 7a shows the best correlation of T_{read} with RSNM indicating improvement of R^2 for symmetric and worst-case degradation.

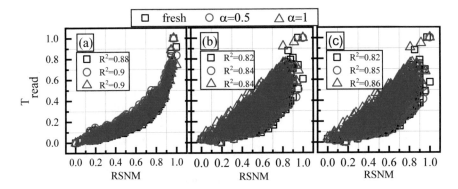

Fig. 7 T_{read} and RSNM correlation distribution at **a** 27 °C, **b** 55 °C and **c** 100 °C for fresh-case, symmetric, and worst-case

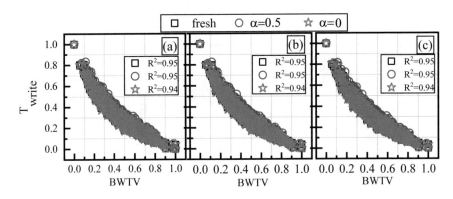

Fig. 8 T_{write} and BWTV correlation distribution at **a** 27 °C, **b** 55 °C and **c** 100 °C for fresh-case, symmetric and worst-case

T_{write} is correlated with BWTV that is shown in Fig. 8. The correlation is strong for the three cases with R^2 ranging from 0.94 to 0.95. For all temperatures, correlation in the fresh-case $\alpha = 0.5$, and 0 are found to be similar with R^2 dropping slightly for $\alpha = 0$.

4   Conclusions

SRAM comprises of two critical parameters namely T_{read} and T_{write} that have been considered in this study to investigate the effect of NBTI degradation and variability with various activity factors (α) and temperatures. It has been illustrated that at time-zero, T_{read} suffers from almost 16% when PU_R is completely degraded. However, with increasing temperatures, the degradation is found to improve from 16% to around 14% for worst-case degradation ($\alpha = 1$) due to NBTI. For T_{write}, the worst degradation ($\alpha = 0$) is a maximum of around 13% with no significant fluctuations in its values due to temperature variations. T_{read} projects sensitivity to increasing α, noise voltage, and temperatures while T_{write} degrades to a significantly lesser extent with increasing α and marginally impacted by temperature. The plots for correlation of the dynamic metrics with the two corresponding static metrics indicate that there is a fairly good correlation of T_{read} with RSNM obtaining an R^2 of 0.88–0.9 for fresh-case at the lower temperature and subsequent decrease in R^2 at elevated temperatures. An excellent correlation of 0.94–0.95 is obtained for T_{write} with BWTV across all considered temperatures. It can, therefore, be concluded that T_{read} is not drastically impacted under temperatures variations owing to the insignificant drop in correlation and minimum deviation in NBTI degradation over a period of 10 years. In the case of T_{write}, it projects an improvement during symmetric usage of P-MOS transistors of the SRAM cell and with increasing α values, PU_L does not undergo severe degradation irrespective of temperature variations.

Acknowledgements. The authors would like to acknowledge and thank the Department of Electrical and Electronics Engineering of BITS Pilani—Dubai Campus and Cadence Academic Network for providing resources to perform simulations required for the work conducted in this paper. The authors are thankful for the support extended towards simulation resources from the Electrical Engineering Department of Shiv Nadar University.

References

1. Parihar N, Goel N, Mukhopadhyay S, Mahapatra S (2018) BTI analysis tool—modeling of NBTI DC, AC stress and recovery time kinetics, nitrogen impact, and EOL estimation. IEEE Trans Electron Dev 65(2):392–403
2. Alam MA, Kufluoglu H, Varghese D, Mahapatra S (2007) A comprehensive model for PMOS NBTI degradation: recent progress. Microelectron Reliab 47(6):853–862. https://doi.org/10.1016/j.microrel.2006.10.012
3. Mahapatra S et al (2013) A comparative study of different physics-based NBTI models. IEEE Trans Electron Dev 60(3):901–916
4. Mahapatra S (2012) A physics-based model for NBTI in p-MOSFETs. In: IEEE 11th international conference on solid-state and integrated circuit technology, Xián
5. Jani Babu S, Chaudhari SP, Singhal S, Goel N (2018) Analyzing impact of NBTI and time-zero variability on dynamic SRAM metrics. In: 15th IEEE India council international conference (INDICON), Coimbatore
6. Seevinck E, List FJ, Lohstroh J (1987) Static-noise margin analysis of MOS SRAM cells. IEEE J Solid State Circuits 22:748–754. https://doi.org/10.1109/JSSC.1987.1052809
7. Guo Z, Carlson A, Pang LT, Duong KT, Liu TJK, Nikolić B (2009) Large-scale SRAM variability characterization in 45 nm CMOS. IEEE J Solid State Circuits 44:3174–3192
8. Chaudhari SP, Babu SJ, Singhal S, Goel N (2018) Correlation of dynamic and static metrics of SRAM cell under time-zero variability and after NBTI degradation. In: IEEE international symposium on smart electronic systems (iSES), Hyderabad
9. Toh SO, Guo Z, Liu TK, Nikolic B (2011) Characterization of dynamic SRAM stability in 45 nm CMOS. IEEE J Solid State Circuits 46(11):2702–2712
10. Mishra S, Mahapatra S (2016) On the impact of time-zero variability, variable NBTI, and stochastic TDDB on SRAM cells. IEEE Trans Electron Dev 63(7):2764–2770

11. Kang K, Kufluoglu H, Roy K, Ashraful Alam M (2007) Impact of negative-bias temperature instability in nanoscale SRAM array: modeling and analysis. IEEE Trans Comput Design Integr Circuits Syst 26(10):1770–1781
12. Goel N, Joshi K, Mukhopadhyay S, Nanaware N, Mahapatra S (2014) A comprehensive modeling framework for gate stack process dependence of DC and AC NBTI in SiON and HKMG p-MOSFETs. Microelectron Reliab 54(3):491–519

An Efficient Design of Multi-logic Gates Using Quantum Cellular Automata Architecture

Avinashkumar$^{(\boxtimes)}$, Anuj Borkute, and Nilesh Goel

Birla Institute of Technology and Science Pilani, Dubai Campus, Dubai, UAE
{avimkumar16,goel.nilesh}@gmail.com

Abstract. Quantum Cellular automata (QCA) is one of the promising next-generation technology which enables high performance and low energy Nano-electronic circuits. QCA presents a new dimension of ideas of designing the fundamental gates in digital electronics with minimum hardware. Here the logic level switching depends on the change in the polarization between the cells neglecting the current transfer which marks this viable technology as a promising candidate for upcoming generations. Moreover, the processing and transfer of information make use of quantum mechanics and cellular automata to deal with the disputes of CMOS transistor technology. In this manuscript, we have proposed the design of multi-logic gates using QCA architecture. The design layout has been simulated using QCA designer 2.0 and is in accordance with the desired logic. The results are compared with the existing design and found with less number of cells and less area.

Keywords: QCA gates · Nano electronics · Multi-logic gates · QCA designer 2.0

1 Introduction

Studying QCA is theoretically motivated by the view that information is the fundamental element of nature and computing is the fundamental process. The universe is then perceived as giant quantum cellular automaton [1]. The quest of futurists for a reliable technology at the end of CMOS era paved them towards QCA. From the experimental point of view, QCA has a significant advantage over CMOS circuits with respect to the current-switching paradigm where the cell-cell responses are based on the underlying physics of Columbic repulsion. Thence this makes the plausible approach to design more energy-efficient, smaller area, and high-density circuits. In this paper, we propose the design of multi-logic gates in QCA with the same inputs. The proposed design makes use of the eight neighborhoods and partial polarization. The proposed design is extendable and it can be used to implement any large size in QCA. Validation of the proposed design is done using QCA Designer 2.0.3 software.

© Springer Nature Singapore Pte Ltd. 2020
N. Goel et al. (eds.), *Modelling, Simulation and Intelligent Computing*,
Lecture Notes in Electrical Engineering 659,
https://doi.org/10.1007/978-981-15-4775-1_67

2 QCA Fundamentals

QCA is based on arrays of coupled quantum dots called *cell* which is the basic building block [2]. The Columbic exclusion phenomenon, which is exhibited by quantum dots when these cells are put together and the way they have Columbic effect on each neighbor cell leads to quantum confinement [2]. This effect leads to two stable state behavior of each cell which makes them usable in cellular arrays. The physical interaction between neighboring cells can be utilized to implement logic functions.

A basic QCA cell has four quantum dots and two dots which are diametrically opposite comprises of electron. Electron cannot occupy adjacent quantum dots according to the principle of repulsion of Coulomb. The junction between two quantum dots is called tunneling junction [3]. Electron can move to different quantum dots in the QCA cell by means electron tunneling which is completely controlled by a potential barrier that can be raised high or low between adjacent QCA cells by means of capacitive plates [4]. Figure 1a shows a basic QCA cell.

2.1 QCA Cells and Wire

As the Columbic repulsion aligns two mobile electrons in the diametrically opposite dots preventing escape from the confinement of the cell, it results in two polarizations which are $P = +1$ and $P = -1$. Cell polarization $P = +1$ denotes binary "1" and cell polarization $P = -1$ denotes binary "0" [5]. Figure 1b demonstrates the polarization of QCA cell and the quantum tunneling of charge between dots enables device switching.

$$P = \frac{P2 + P4 - (P1 + P3)}{P1 + P2 + P3 + P4} \tag{1}$$

Each adjacent QCA cell has its effect on others, so interaction between cells transmits data from input to output. Since information transfer takes place by polarized charge with the help of force of repulsion instead of flow of current QCA circuit has low energy and high processing speed [6]. Figure 2 represents the binary wire which propagates information from input to output.

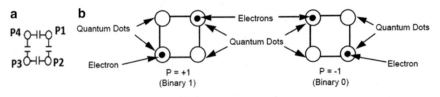

Fig. 1 **a** Basic cell with barriers. **b** QCA cell logic "1" and "0"

Fig. 2 Representation of binary wire with QCA cells

Table 1 QCA cell polarization at different clock phases

Clock phase	Potential barrier	Cell polarization state
Switch	Low to high	Polarized
Hold	Held high	Polarized
Release	Low	Unpolarized
Relax	Remain lowered	Unpolarized

2.2 QCA Clocking

To transfer information through QCA cells, clock is important and with the help of clock, information can be locked within specified cells. Clocking in QCA has four clock zones, clock zone 0 to clock zone 3 and each clock zone has four phases which are Switch, Release, Relax and Hold. Table 1 shows clock zones and position of potential barrier with corresponding polarization.

Based on potential barrier, QCA cells acquire different polarization. During "Switch" phase of clock potential barrier goes from low to high and QCA cell gets polarized. During "Hold" phase potential barrier is held high and polarization is preserved. In "Release" phase potential barrier is lowered and the cell becomes non-polarized. In the Relax phase barrier remain lowered and cells remain non-polarized. From this phase, cells are ready to switch again [5] (Fig. 3).

2.3 Majority Gate and Invertor

Logic circuits can be designed using array of QCA cells which are arranged in a specific manner with different, same, or fixed polarization according to required logic. In binary wire, array of cells are placed one after another in the same polarization (Fig. 2). Similarly, inverter logic can be designed by making cells to interact with each other at 45° or diagonally so that quantum dots get different polarization and electrostatic interaction between cells gets inverted. Additionally, QCA cells can form a majority gate in which output cell gets polarization according to the majority

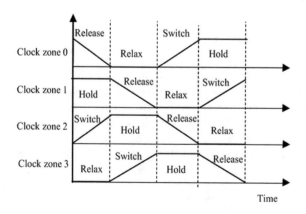

Fig. 3 QCA clocking zones and phases

Table 2 Basic QCA gates with number of cells existing in literature

Basic gates	No. of cells [reference]
NOT	2 [7]
AND	5 [5]
OR	5 [5]
NAND	14 [3]
NOR	14 [3]
XOR	10 [5]
XNOR	10 [5]

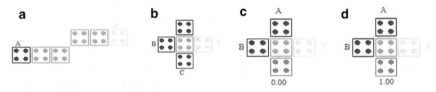

Fig. 4 a Inverter. **b** Majority gate. **c** AND gate from majority gate. **d** OR gate from majority gate

polarization at input side. The functioning of a 3-input majority gate with three inputs A, B, C can be defined by the Boolean logic as given in (2). Figure 4a shows the inverter and Fig. 4b shows the majority gate.

$$M(A, B, C) = AB + BC + AC \tag{2}$$

With inverter and majority gate, we can design a complete set of QCA logic gates, from which more complex and general-purpose systems can be obtained. Figure 4c shows AND gate from majority logic in which a cell C of majority gate has fixed polarization of 0 to get AND logic. Figure 4d shows OR gate from majority logic in which a cell C of majority gate has a fixed polarization of 1 to get OR logic.

3 QCA Basic Gates

Logic units are the basic building blocks of many computational operations. In the case of computing with QCA, we also require universal logic design. One such logic design is a QCA majority gate. Basic gate inverter, AND, and OR from majority logic described in the previous section can be implemented using QCA cells (Figs. 4 and 5). By combining AND and OR logic with inverter, NAND, and NOR gates which are universal gates can be implemented [7]. Figure 5a, b shows NOR, NAND. Similarly, XOR and XNOR logic gates also can be designed. Figure 5c, d represents XOR and XNOR logic, here cells with different clocks are used to reduce the number of cells required for the design [8–12]. The Table 2 explains the number of cells required in all basic logic gates.

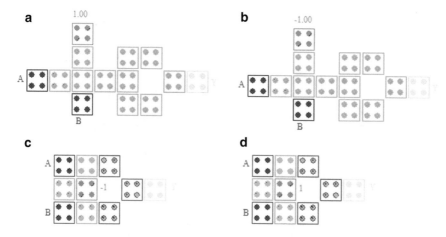

Fig. 5 **a** NOR gate. **b** NAND gate. **c** XOR gate. **d** XNOR gate

4 Proposed Design

Traditional Boolean logic has two inputs and one output. Here the proposed logic gate has two inputs and two outputs logic AND and OR operation both. We make use of rotating the nearby cell 90° to get the invertor design shown in Fig. 6a and the concept of 8-neighbourhood is shown in Fig. 6b.

8-neighbourhood: The finite neighborhood of a QCA cell $N = \{n_1, n_2, n_3...n_r\}$ has an effect on the current state of the cell. For example, consider a symmetric QCA with a neighborhood of size 3 then if three consecutive cells are in the states q_1, q_2, and q_3 at time t, then at time $t + 1$ the middle cell will contain the effect of all three neighbor cells. The 8-neighbors of QCA cell E, shown in Fig. 6b include the four 4-neighbors (B, D, F, and H) and four cells (A, C, E, and G) along the diagonal direction. The current state of cell E depends on the previous state of all these eight neighbors. This is also known as Moore's neighborhood.

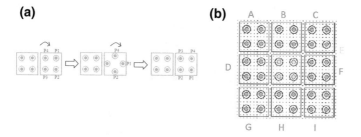

Fig. 6 **a** Invertor design with 90° rotation of cell. **b** Representation of all neighborhood cells

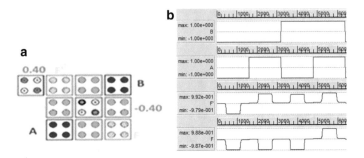

Fig. 7 **a** AND-OR logic design. **b** Simulation result of AND-OR

Table 3 Truth table of AOR logic

A	B	F = AB	F = A + B
0	0	0	0
0	1	0	1
1	0	0	1
1	1	1	1

4.1   AOR Gate Design

Proposed combined AOR(AND-OR) logic design with QCA has two inputs (A, B) and two outputs (AB, A + B) (Fig. 7a). Center cell is provided with fixed polarization of −0.4 which is diagonal neighbor to both output cell F and F′ and 0.4 fixed polarization is given to one immediate neighbor cell of F′ as shown in Fig. 7a. When input is provided we get output OR at F and AND at F′. So both immediate and diagonal neighbor cells affect the output. Table 3 shows the truth table of OR-AND combined logic and Fig. 7b shows simulation results of OR-AND logic design.

Comparison with Existing System

In comparison with the existing AOR implementation [13], this design has less number of cells, it consumes less area, and switching speed is high. Table 4 illustrates a comparison of proposed design and existing design.

Table 4 Comparison table

Design	No. of cells	Consumed area (nm^2)
Existing [13]	13	5184
Proposed	10	3240

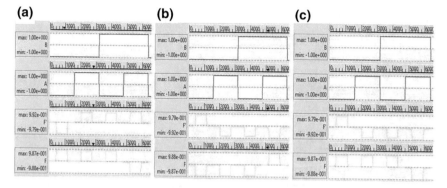

Fig. 8 Simulation result of **a** OR-NAND logic. **b** AND-NOR logic. **c** NAND-NOR logic

From the same logic (Fig. 7a) by rotating output cells F and F′ cells by 90° individually we can obtain NAND-OR and AND-NOR, respectively. If we rotate both F and F′ cell by 90° together we get NAND-NOR logic. The simulation results are obtained in Fig. 8a–c.

4.2 OR-NOR and AND-NAND Logic Design

Here combined OR-NOR logic and AND-NAND logic with QCA are proposed using the concept of neighborhood cell effect (Fig. 6b). In Fig. 9a fixed polarization −1 is given to the center cell, when inputs A and B are given, we will get output OR at O cell and NOR at O′ cell. In Fig. 9b output cell F′ is rotated by 90° so we get output OR from F cell and NOR from F′ cell.

If we provide fixed polarization of 1 to center cell as shown in Fig. 9c then we get output AND at O cell and NAND at O′ cell. In Fig. 9d output cell F′ is rotated by 90° then we get output AND from F cell and NAND from F′ cell. Figure 10a, b shows the simulation results of OR-NOR and AND-NAND logic.

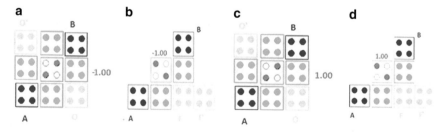

Fig. 9 **a** OR-NOR logic design. **b** Reduced OR-NOR logic design. **c** AND-NAND logic design. **d** Reduced AND-NAND logic design

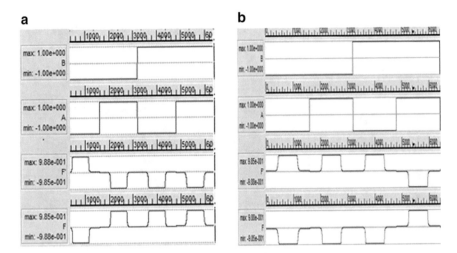

Fig. 10 a Simulation result of OR-NOR logic. **b** Simulation result of AND-NAND logic

5   Conclusion

In this paper, a conception of multi-logic gates using QCA is proposed and described. Architectures and simulation results are also proposed. An important step in designing QCA circuits is reducing the number of required cells. In the proposed design, the reduction of the number of cells in the QCA structure is based on the interaction forces between the neighborhood cells, and also the 90° rotation of cell is used to attain multi-logic design with less number of cells. In this work, we have optimized the number of QCA cells and area required for multi-logic design.

References

1. Aoun B, Tarifi M Quantum cellular automata, quant-ph/0401123
2. Porod W, Lent CS, Bernstein GH Quantum cellular automata. In: Annual report for the period 1 July 1993–30 June 1994, Department of Electrical Engineering University of Notre Dame
3. Bilal B, Ahmed S, Kakkar V (2018) An insight into beyond CMOS next generation computing using quantum-dot cellular automata nanotechnology. Int J Eng Manuf (IJEM) **8** (1):25–37. https://doi.org/10.5815/ijem.2018.01.03
4. Michael Thaddeus Niemier BS (2004) Designing digital systems in quantum cellular automata. In: A thesis submitted to the Graduate School of the University of Notre Dame
5. Laajimi R (2018) Nanoarchitecture of quantum-dot cellular automata (QCA) using small area for digital circuits, intechopen
6. Lee Ai Lim A., Ghazali SC, Yan T, Fat CC (2012) Sequential circuit design using quantum-dot cellular automata (QCA). In 2012 IEEE international conference on circuits and systems (ICCAS), Kuala Lumpur, pp 162–167

7. Raj M, Gopalakrishnan L (2019) High speed memory cell with data integrity in QCA. In: 2019 3rd international conference on electronics, communication and aerospace technology (ICECA), Coimbatore, India, pp 926–929
8. Walus K, Schulhof G, Jullien GA (2004) High level exploration of quantum-dot cellular automata (QCA). In: Conference record of the thirty-eighth Asilomar conference on signals, systems and computers, Pacific Grove, CA, 2004, vol 1, pp 30–33
9. Anumula SK, Xiong X (2019) Design and simulation of 4-bit QCA BCD full-adder. In: 2019 IEEE long island systems, applications and technology conference (LISAT), pp 1–6. Farmingdale, NY
10. Dhare V, Mehta U (2019) A simple synthesis process for combinational QCA circuits: QSynthesizer. In: 2019 32nd international conference on VLSI design and 2019 18th international conference on embedded systems (VLSID), pp 498–499. Delhi, NCR
11. Blair E (2019) Electric-field inputs for molecular quantum-dot cellular automata circuits. IEEE Trans Nanotechnol 18:453–460
12. Isaksen B, Lent CS (2003) Molecular quantum-dot cellular automata. In: 2003 third IEEE conference on nanotechnology, 2003. IEEE-NANO 2003, vol 2, pp 5–8. San Francisco, CA
13. Chattopadhyay T, Sarkar T (2013) Design of AND-OR gate using quantum cellular automata and its cell-cell response using computer simulation. In: 2013 proceedings of 3rd national conference on electronics, communication and signal processing—NCECS

Hyper-parameter Optimization on Viola Jones Algorithm for Gesture Recognition

Aditya Pande, B. K. Rout$^{(\boxtimes)}$, and Sangram K. Das

Department of Mechanical Engineering, Center for Robotics and Intelligent Systems, Birla Institute of Technology and Science, Pilani Campus, Pilani, Rajasthan, India
f2013573p@alumni.bits-pilani.ac.in,
rout@pilani.bits-pilani.ac.in,
sagram.das@pilani.bits-pilani.ac.in

Abstract. The problem of features, objects, gestures, and face detection has been tackled using a numerous vision-based algorithms available in literature. Each of these algorithms requires a set of hyper-parameters, which need to be set on the basis of trial and error such that the results provide best performance to a situation. Mostly, researchers use trial and error approach to satisfactory result and solve the above problems. In this work, an approach has been suggested to determine an optimum set of hyper-parameters, which will provide a starting point for anyone using Viola Jones algorithm for hand gesture recognition or similar endeavors. This will reduce the time spent in searching for the best combination of hyper-parameters.

Keywords: Viola jones · Hyper-parameters · Computer vision

1 Introduction

Computer vision is one of the emerging subdomains of machine learning and image processing. Object and feature detection is a generalized problem, with applications in autonomous vehicles, face detection, smile shutters in cameras, 3D reconstruction, and security. A variety of methods are used for the same, and this paper focuses specifically on Viola Jones algorithm [1]. Given a set of images which satisfy or do not satisfy a criterion (labeled as positive and negative images respectively) and four input hyper-parameters, the algorithm trains and produces an output file (xml format). Depending on hyper-parameters supplied, performance of the classifier changes, i.e., its training speed (computational power), detection speed, and accuracy. Values of these hyper-parameters are taken by trial and error and depend on the application, nature of image dataset supplied (uniqueness, image backgrounds), etc. There is a conflict of interest involved, where we want the detection to be fast and accurate, both of which are difficult to satisfy simultaneously. The aim here is to provide a starting point of these values for researchers, especially for tasks like human hand gesture recognition. The optimum values may differ for more intricate objects like human faces or road signs.

© Springer Nature Singapore Pte Ltd. 2020
N. Goel et al. (eds.), *Modelling, Simulation and Intelligent Computing,*
Lecture Notes in Electrical Engineering 659,
https://doi.org/10.1007/978-981-15-4775-1_68

2 Background of Research for Gesture Recognition

Over time, a lot of methods have been used to solve the problem of recognizing hand gestures. The general approach is as follows: Voila Jones algorithm [1] is quite commonly used for identifying gestures, and since then, newer techniques have been developed. Brethes and Ludovic [2] used a combination of color-based segmentation, contours, and cascade classifier to build their gesture recognition system. Another method called Histogram of oriented gradients (HOG) was used by Misra et al. [3]. Contour detection was explored and improved by Argyos and Lourakis [4], in which they used a pair of stereo cameras to reconstruct the contours of hand in three dimensions. Voila Jones algorithm is relatively old and fails to capture difficult objects, that is those which can have abstract shapes or more intricacy. For this reason, deep learning, one of the latest, state-of-the-art machine learning algorithms, has been explored. Lawrence [5] and his team have attempted to identify human faces using neural networks. A convolutional neural network and color segmentation-based approach were used by Nagi [6] and his team to identify hand gestures using a colored glove for mobile robots. Ren et al. [7] used Finger Earth Movers distance method to measure dissimilarities between hand shapes (different from template matching). They used a Kinect sensor to obtain image and depth data. Quite recently, leap motion controller has been explored by Bachmann and Weichert [8]. All of these algorithms require some hyper-parameters, for a neural network it might be its weights, number of neurons, architecture, and activation functions. For an image processing-based algorithm, it might be size used for blurring, image color ranges, and so on. If the algorithm is made too "rigid," it cannot capture the complexity of problem, and if made too "flexible," it can lead to overfitting. As mentioned earlier, Viola Jones being a common algorithm was chosen for finding optimum hyper-parameter.

3 Proposed Methodology

Training and Testing: Initial work involves "training" the classifier files, i.e., generating xml files from the given dataset and hyper-parameters. Second step involves testing the classifiers on a separate "testing only" dataset, a collection of images that were not included during training. Third step involves analysis of performance of the classifiers. Finally, we draw meaningful conclusions from the same. MATLAB has an inbuilt function called train CascadeObjectDetector() which takes six arguments as hyper-parameters, out of which four were selected to be optimized. Local binary patterns (LBP) feature was selected, as it trains much faster than "Haar" based features. The work has been implemented in Python and partly in MATLAB (training classifiers) environment [9–11].

3.1 Identified Hyper-parameters to Be Optimized

The hyper-parameters to be optimized are identified and presented in Table 1.

Once trained, the classifiers are tested for their performance on a separate dataset on the basis of parameters on the right. The importance of the is as follows:

Table 1 Parameters to be optimized and evaluated

Parameters	Testing metrics
False alarm rate	Training time
Number of cascade stage	False positives
Object size	True positives
True positive rate	False negatives

False Alarm Rate: Acceptable false alarm rate at each stage has a value in the range (0 1]. It is the fraction of negative training samples incorrectly classified as positive samples.

Number of Stages: Number of cascade stages to train. Increasing the number of stages may result in a more accurate detection but also increases training time. More stages can require more training images, because at each stage, some number of positive and negative samples are eliminated.

Object Size: Before training, the function resizes the positive and negative samples to this size in pixels. For optimal detection accuracy, specify an object training size close to the expected size of the object in the image. However, for faster training and detection, set the object training size to be smaller than the expected size of the object in the image.

True Positive Rate: Minimum true positive rate required at each stage. It has a value in the range (0 1]. The true positive rate is the fraction of correctly classified positive training samples.

For Testing Parameters:

Time Taken: Classifier should not take too long to train.

False Positives: Number of detections that do not contain the desired object and are classified as positive.

True Positives: Number of detections that are in fact the desired object and get classified as positives.

False Negatives: Number of detections that do contain the actual object but are classified as negatives.

3.2   Dataset Preparation

Four gestures were selected, namely "UP," "DOWN," "WAIT," and "FIST." A large dataset of images was created by clicking pictures of the work area and hand gestures and cropping, labeling manually. Figure 1 shows the sample images used for the investigation. Table 2 shows the numbers of such attempts and Table 3, provides the discrete values for hyper-parameters to be used for calibration.

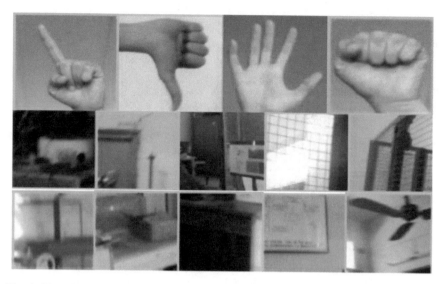

Fig. 1 **Top**: Sample Images from UP, DOWN, WAIT and FIST dataset, **Bottom**: negative

Table 2 Number of images in each category used as dataset

Name of gesture	Positive training	Negative training	Negatives testing	Positives testing
UP	495	1946	157	45
DOWN	332	1946	152	50
WAIT	520	1946	155	47
FIST	465	1946	142	60

Table 3 Discrete values for hyper-parameters for Optimization

Hyper-parameter	Values parameter
False alarm rate	0.3, 0.5, 0.7
No. of stages	5, 10, 15, 20
Object size	10, 20, 30, 40, 50
True positive rate	0.7, 0.8, 0.9, 0.99

3.3 Step Wise Approach for Training and Testing

The four parameters to be optimized can take any discrete integer values. It is not practically possible to train a huge number of classifiers resulting from combination of all discrete values of hyper-parameters. Hence, small number of values for each parameter was chosen and the resulting Cartesian product was used for training the

Table 4 Belief description of steps for training and testing of classifiers

Step	Action
1	• Generate combination of parameters to be trained • Store combination in the spreadsheet
2	• Supply the combination to MATLAB • Train over inputs using these combinations if possible • Store .xml files and training log
3	• Verify if all combinations have been trained successfully • Get time taken for training in the sheet from logs
4	• Test over "negatives" folder • Generate report in a spreadsheet
5	• Copy step four testing results to main results file
6	• Evaluate over positives folder using intersection threshold method • Fill up "true positives" and "false positives" columns in spreadsheet
7	• Fill up "False negatives" column

classifiers. However, not all such combinations can be trained, and some do not complete the training process.

The work was split into seven distinct scripts, such that if any step fails, troubleshooting is easy and work can be restarted from previous data log. Table 4 gives a brief idea of the same.

At the end, we trained 240 classifiers each for fours gestures, i.e., a total of 960 classifiers. Figure 2 shows the current status. It can be observed that some of the attempts have failed, and some could not reach the specified number of stages. Corrections for the same have been done appropriately.

3.4   Analysis of Test Scores

We have the combinations and their scores for the dataset ready in the spreadsheet. The next procedure was to analyze the scores, and ranking the xmls based on performance. This is a multi-objective problem, requiring some results are to be minimized, some are to be maximized, and some to be near a particular value. The score generates a single scalar for performance, which takes into account all the multiple objectives. This is achieved by using Euclidean distance between the desired scores and actual scores. The "scores" can be imagined as four-dimensional "vectors." The distance between two vectors $[a_1; a_2; a_3; a_4]$ and $[b_1; b_2; b_3; b_4]$ is given by:

$$d = \sqrt{(a_1 - b_1)^2 + (a_2 - b_2)^2 + (a_3 - b_3)^2 + (a_4 - b_4)^2} \qquad (1)$$

Sr. NO	False Alram Rate	Num Stages	Object Size	True Positive Rate	Status	Time (s)	NS*	FP-N	FP-P	TP-P	FN-P
1	0.3	5	10	0.7	Success	53.8	5	3103	831	72	10
2	0.3	5	10	0.8	Success	23.4	5	3956	1196	72	10
3	0.3	5	10	0.9	Success	22.1	5	5983	1754	195	0
4	0.3	5	10	0.99	Success	20.9	5	11216	3481	171	0
5	0.3	5	20	0.7	Success	20.7	5	6650	2074	152	0
6	0.3	5	20	0.8	Success	30.3	5	7536	2189	111	4
7	0.3	5	20	0.9	Success	28.6	5	13828	4238	134	0
8	0.3	5	20	0.99	Success	22.0	5	18265	6057	147	3
9	0.3	5	30	0.7	Success	22.4	5	8750	2740	93	0
10	0.3	5	30	0.8	Success	40.0	5	9270	2935	95	2
11	0.3	5	30	0.9	Success	23.1	5	18480	5597	177	2
12	0.3	5	30	0.99	Success	27.9	5	19662	6731	117	9
13	0.3	5	40	0.7	Success	27.5	5	16521	5402	191	0
14	0.3	5	40	0.8	Success	29.6	5	13432	4481	170	1
15	0.3	5	40	0.9	Success	43.0	5	13533	4522	144	3
16	0.3	5	40	0.99	Success	59.8	5	20222	6082	211	1
17	0.3	5	50	0.7	Error	NA	NA	NA	NA	NA	NA
18	0.3	5	50	0.8	Error	NA	NA	NA	NA	NA	NA
19	0.3	5	50	0.9	Error	NA	NA	NA	NA	NA	NA
20	0.3	5	50	0.99	Error	NA	NA	NA	NA	NA	NA
21	0.3	10	10	0.7	Success	28.4	10	0	0	0	50
22	0.3	10	10	0.8	Success	23.3	10	0	0	0	50

Fig. 2 Data obtained after training and testing. Green columns shot the hyper-parameters used to train, orange ones give the scores to evaluate upon

A. Priority Order

The desired score is obviously, zero false positives, zero false negatives, and true positives to be equal to the exact number of objects in the dataset, and time to be as low as possible. However, not all of these are equally important. False positive can be disastrous in an industrial environment. An unintended action should not trigger a procedure or stop the work. Second comes true positives. False negatives can be tolerated to a certain extent, as it would require some more attempts to set the trigger, but it should not cause major problems. This of course is specific to the action to be triggered by the gesture. If it is supposed to be an interrupt based gesture, then the order may be reversed, or given equal priority. Time for training is the least important.

B. Steps for Analysis

As it has been shown in previous section, analysis was also done step wise for easy troubleshooting.

Step 1: Step one copies the useful data from the main sheet to a separate file, excluding the combinations that could not be trained successfully.

Step 2: Output scores are normalized between 0 and 1. That is, the values of "false positives," "true positives," "training time," and "false negatives" now all lie between 0 and 1. Then these values are scaled by a factor directly related to their priority. That is, "false positives" is multiplied by 2, "true positives" by 1.5, false negatives by 1, and time by 0.5. Let us consider P as a score type, say "false positives." Normalization formula for P for ith classifier is given by Eq. 2.

$$nP_i = P_i / \sum_{i=1}^{n} P_i \qquad (2)$$

After normalization, score vector is computed by:

$$V_i = [2 * nFP, 1.5 * nTP, 1 * nFN, 0.5 * nT] \qquad (3)$$

where, nFP—False positives normalized, nTP—True positives normalized, nFN—False negatives normalized, nT—Time taken to train normalized.

Step 3: Euclidean distance is computed with respect to ideal performance and the top 10 xmls are ranked. Lesser the distance means better the performance.

4 Experimental Results and Discussion

Optimum parameters for each gesture vary a little, but a general pattern can be observed. **UP Gesture**: It is seen from Fig. 3 that low object size (10, 20) and high true positive rates (0.99) are performing better than the rest. This may be due to the fact that the positive dataset supplied is easy to differentiate from the negatives and that the effect of intricacies is not large. The algorithm is finding it easy to devise LBP classifiers that correctly classify images and hence larger images are not relevant. **WAIT Gesture**: It is observed in Fig. 4, that high object size (40) and low true positive rates (0.7) are performing better than the rest. **FIST Gesture**: It has been observed from Fig. 5 that medium object size (20, 30) and medium true positive rates (0.8, 0.9) are performing better than the rest. It can be seen that the shape of closeness bar chart has changed for 81–96 combination numbers. This says that the features found are quite

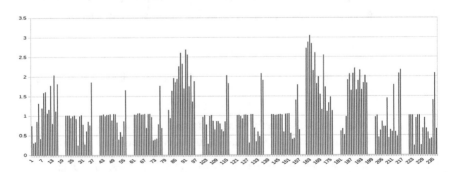

Fig. 3 UP gesture: deviation of score vector from ideal performance for all trained cascades from 1 to 240 (excluding those which could not be trained)

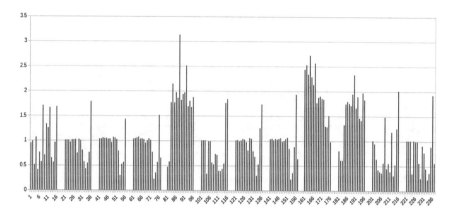

Fig. 4 WAIT gesture: deviation of score vector from ideal performance for all trained cascades from 1 to 240 (excluding those which could not be trained)

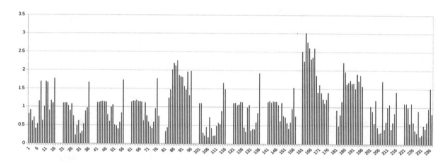

Fig. 5 FIST gesture: deviation of score vector from ideal performance for all trained cascades from 1 to 240 (excluding those which could not be trained)

different from that of wait or up. Higher combinations like 161–176 are performing similar to up or wait ones. **DOWN Gesture**: It shows the deviation plot. It can be seen from Fig. 6 that large object size (40) and medium true positive rates (0.8, 0.9) are performing better than the rest. The shape of closeness plot is similar to first plot, meaning that the features detected are similar. Here, intricacies are playing an important role due to the nature of dataset. For this, we have analyzed the top ten performing xmls for all four gestures. The set of central tendencies of each parameter would serve as the optimum combination. Number of occurrences of each hyperparameter in all 40 top-performing classifiers with 0.5 false alarm rate, shows up 40% of time. The corresponding distribution is plotted in Fig. 7.

The central tendencies for top-performing (40 among 960) classifiers are shown in Table 5. Looking at these central tendencies, the optimal combination should lie around these values, depending on the dataset quality and quantity. The optimal combination should lie around these values, depending on the dataset quality and quantity. These are the findings based on the dataset developed and the results may vary depending on the application. Preprocessing of the images may further affect these values.

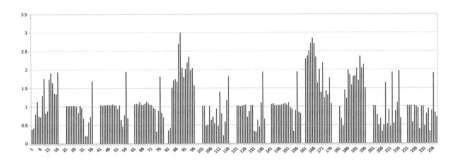

Fig. 6 DOWN gesture: deviation of score vector from ideal performance for all trained cascades from 1 to 240 (excluding those which could not be trained)

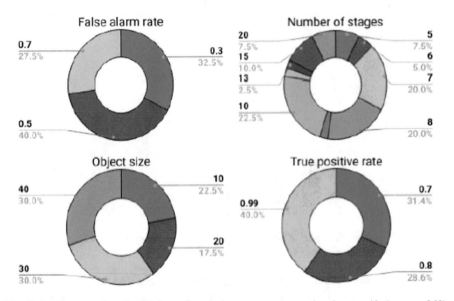

Fig. 7 Percentage-wise distribution of each hyper-parameter value in top 40 (among 960) performing classifiers

Table 5 Central tendencies of hyperparameters of top-performing classifiers

Central tendency	False alarm rate	No. of stages	Object size	True positive rate
Mode	0.5	10	30	0.99
Mean	0.49	9.78	26.75	0.85
Median	0.5	8	30	0.8

The entire training, testing, and validation were performed using Python 3.2, running on Windows 10 on HP Pavilion 15 N012TX. The entire procedure, from declaring labels for training to displaying charts for results (for just one gesture), took eight hours to train on the machine, with no extra major processes running. Optimal values of hyper-parameters from above investigation are: False alarm = 0.5, number of stages = 10, object size (square) = 30 and true positive = 0.9.

5 Conclusion

The objective of the current work was to obtain optimum parameters for an overall gesture recognition problem. Depending on hyper-parameters supplied, performance of the classifier changes, i.e., its training speed (computational power), detection speed, and accuracy. Optimal values of these hyper-parameters are obtained for gesture recognition. These results are dependent on the application, nature of image dataset supplied (uniqueness, image backgrounds), etc. Present work proposes a weight-based multiobjective optimization approach to handle conflict of interest. The results obtained are best starting point for tasks like human hand gesture recognition.

References

1. Viola P, Jones M (2001) Rapid object detection using a boosted cascade of simple features. In: Proceedings of the 2001 IEEE computer society conference on computer vision and pattern recognition, 2001. CVPR 2001, vol 1
2. Brethes L, Menezes P, Lerasle F, Hayet J Face tracking and hand gesture recognition for human-robot interaction. In: Proceedings. ICRA '04, vol 2, pp 1901–1906
3. Misra A, Abe T, Deguchi K Hand gesture recognition using histogram of oriented gradients and partial least squares regression. In: MVA, pp 479–482
4. Argyros AA, Lourakis MI (2006) Vision-based interpretation of hand gestures for remote control of a computer mouse. In: European conference on computer vision, pp 40–5120. Springer, Berlin Heidelberg
5. Lawrence S (1997) Face recognition: a convolutional neural network approach. IEEE Trans Neural Networks 8:98–113
6. Nagi J, Ducatelle F, Di Caro GA, Cirean D, Meier U, Giusti A, Gambardella LM (2011) Max-pooling convolutional neural networks for vision-based hand gesture recognition. In: Signal and Image Processing Applications (ICSIPA), pp 342–347
7. Ren Z, Meng J, Yuan J, Zhang Z (2011) Robust hand gesture recognition with kinect sensor. In: Proceedings of the 19th ACM international conference on multimedia, pp 759–760
8. Bachmann D, Weichert F, Rinkenauer G (2014) Evaluation of the leap motion controller as a new contact-free pointing device. Sensors 15(1):214–233
9. Jones E, Oliphant E, Peterson P et al (2001) SciPy: open source scientific tools for Python. http://www.scipy.org/. Accessed 19 Mar 2019
10. Stéfan van der Walt S, Colbert C, Varoquaux G (2011) The numpy array: a structure for efficient numerical computation, computing in science & engineering, vol 13, pp 22–30
11. Hunter JD (2007) Matplotli, A 2-D graphics environment. Comput Sci Eng 9:90–95

Printed in the United States
by Baker & Taylor Publisher Services